APPROXIMATE SOLUTION OF OPERATOR EQUATIONS WITH APPLICATIONS

APPROXIMATE SOLUTION OF OPERATOR EQUATIONS WITH APPLICATIONS

Ioannis K Argyros
Cameron University, USA

World Scientific

NEW JERSEY · LONDON · SINGAPORE · BEIJING · SHANGHAI · HONG KONG · TAIPEI · CHENNAI

Published by

World Scientific Publishing Co. Pte. Ltd.
5 Toh Tuck Link, Singapore 596224
USA office: 27 Warren Street, Suite 401-402, Hackensack, NJ 07601
UK office: 57 Shelton Street, Covent Garden, London WC2H 9HE

British Library Cataloguing-in-Publication Data
A catalogue record for this book is available from the British Library.

APPROXIMATE SOLUTION OF OPERATOR EQUATIONS WITH APPLICATIONS

Copyright © 2005 by World Scientific Publishing Co. Pte. Ltd.

All rights reserved. This book, or parts thereof, may not be reproduced in any form or by any means, electronic or mechanical, including photocopying, recording or any information storage and retrieval system now known or to be invented, without written permission from the Publisher.

For photocopying of material in this volume, please pay a copying fee through the Copyright Clearance Center, Inc., 222 Rosewood Drive, Danvers, MA 01923, USA. In this case permission to photocopy is not required from the publisher.

ISBN-13 978-981-256-365-1
ISBN-10 981-256-365-2

Printed in Singapore

To Diana Mina Argyros

Preface

Researchers in computational sciences are faced with the problem of solving a variety of equations. A large number of problems are solved by finding the solutions of certain equations. For example, dynamic systems are mathematically modelled by difference or differential equations, and their solutions represent usually the states of the systems. For the sake of simplicity, assume that a time-invariant system is driven by the equation $x' = f(x)$, where x is the state, then the equilibrium states are determined by solving the equations $f(x) = 0$. Similar equations are used in the case of discrete systems. The unknowns of engineering equations can be functions (difference, differential, integral equations), vectors (systems of linear or nonlinear algebraic equations), or real or complex numbers (single algebraic equations with single unknowns). Except special cases, the most commonly used solutions methods are iterative, when starting from one or several initial approximations a sequence is constructed, which converges to a solution of the equation. Iteration methods are applied also for solving optimization problems. In such cases the iteration sequences converge to an optimal solution of the problem in hand. Since all of these methods have the same recursive structure, they can be introduced and discussed in a general framework.

To complicate the matter further, many of these equations are nonlinear. However, all may be formulated in terms of operators mapping a linear space into another, the solutions being sought as points in the corresponding space. Consequently, computational methods that work in this general setting for the solution of equations apply to a large number of problems, and lead directly to the development of suitable computer programs to obtain accurate approximate solutions to equations in the appropriate space.

This book is intended for researchers, practitioners and students in computational sciences. The goal is to introduce these powerful concepts and techniques at the earliest possible stage. The reader is assumed to have had basic courses in numerical analysis, computer programming, computational linear algebra, and an introduction to real, complex, and functional analysis.

We have divided the material into several chapters. Each section of a chapter is as independent from another as possible, so the reader interested in a particular method/result can access directly the information without studying previous or following sections. Each chapter contains several new theoretical results and important applications in engineering, in dynamic economic systems, in input-output systems, in the solution of nonlinear and linear differential equations, and optimization problems. Sections have been written as independent of each other as possible. Hence the interested reader can go directly to a certain section and understand the material without having to go back and forth in the whole textbook to find related material.

There are three basic problems connected with iterative methods.

Problem 1 *Show that the iterates are well defined. For example, if the algorithm requires the evaluation of F at each x_n, it has to be guaranteed that the iterates remain in the domain of F. It is, in general, impossible to find the exact set of all initial data for which a given process is well defined, and we restrict ourselves to giving conditions which guarantee that an iteration sequence is well defined for certain specific initial guesses.*

Problem 2 *Concerns the convergence of the sequences generated by a process and the question of whether their limit points are, in fact, solutions of the equation. There are several types of such convergence results. The first, which we call a local convergence theorem, begins with the assumption that a particular solution x^* exists, and then asserts that there is a neighborhood U of x^* such that for all initial vectors in U the iterates generated by the process are well defined and converge to x^*. The second type of convergence theorem, which we call semilocal, does not require knowledge of the existence of a solution, but states that, starting from initial vectors for which certain-usually stringent-conditions are satisfied, convergence to some (generally nearby) solutions x^* is guaranteed. Moreover, theorems of this type usually include computable (at least in principle) estimates for the error $x_n - x^*$, a possibility not afforded by the local convergence theorems. Finally, the third and most elegant type of convergence result, the global theorem, asserts that starting anywhere in a linear space, or at least in a large part of it, convergence to a solution is assured.*

Problem 3 *Concerns the economy of the entire operations, and, in particular, the question of how fast a given sequence will converge. Here, there are two approaches, which correspond to the local and semilocal convergence theorems. As mentioned above, the analysis which leads to the semilocal type of theorem frequently produces error estimates, and these, in turn, may sometimes be reinterpreted as estimates of the rate of convergence of the sequence. Unfortunately, however, these are usually overly pessimistic. The second approach deals with the behavior of the sequence $\{x_n\}$ when n is large, and hence when x_n is near the solutions x^*. This behavior may then be determined, to a first approximation, by the properties of the iteration function near x^* and leads to so-called asymptotic rates of convergence.*

We have included a variety of new results dealing with problems 1–3.

This textbook is an outgrowth of research work undertaken by us and complements/updates earlier works of ours focusing on in depth treatment of convergence theory for iterative methods [68]-[99], and the references there. Such a comprehensive study of optimal iterative procedures appears to be needed and should benefit not only those working in the field but also those interest in, or in need of, information about specific results or techniques. We have endeavored to make the main text as self contained as possible, to prove all results in full detail and to include a number of exercises throughout the textbook. In order to make the study useful as a reference source, we have complemented each section with a set of "Remarks" in which literature citations are given, other related results are discussed, and various possible extensions of the results of the text are indicated. For completion, the book ends with a comprehensive list of references. Because we believe our readers come from diverse backgrounds and have varied interests, we provide "recommended reading" throughout the textbook. Often a long textbook summarized knowledge in a field. This textbook, however, may be viewed as a report on work in progress. We provide a foundation for a scientific field that is rapidly changing. Therefore we list numerous conjectures and open problems as well as alternative models which need to be explored.

I. K. Argyros

Contents

Preface vii

1. Linear Spaces 1
   - 1.1 Linear Operators 1
   - 1.2 Continuous Linear Operators 2
   - 1.3 Equations 4
   - 1.4 Computing the Inverse of a Linear Operator 5
   - 1.5 Fréchet Derivatives 6
   - 1.6 Integration 10
   - 1.7 Exercises 11

2. Divided Differences 17
   - 2.1 Partially Ordered Topological Spaces 17
   - 2.2 Divided Differences in a Linear Space 20
   - 2.3 Divided Difference in a Banach Space 21
   - 2.4 Divided Difference and Monotone Convergence 27
   - 2.5 Divided Differences and Fréchet-derivatives 31
   - 2.6 Enclosing the Root of a Nonlinear Equation 37
   - 2.7 Exercises 44

3. Fundamental Fixed Point Theory 51
   - 3.1 Fixed Points of Operators 51
   - 3.2 Examples 53
   - 3.3 Integral Equations Arising in Newton Transport 55
   - 3.4 An Efficient Contractive Method 65

4. Solving Equations — 83

- 4.1 Linearization of Equations — 83
- 4.2 The Convergence of Newton's Method — 84
- 4.3 Local Convergence — 89
- 4.4 Approximating Distinct Solutions — 91
- 4.5 Approximation Using Finite Rank Operators — 106
- 4.6 Projection Methods for Approximating Fixed Points — 114
- 4.7 Solving Nonlinear Equations with a Nondifferentiale Term — 121
- 4.8 Iteration Converging Faster than Newton's Method — 127
- 4.9 Exercises — 135

5. Two-Step Newton Methods and Their Applications — 157

- 5.1 Two-Step Newton Methods — 157
- 5.2 Monotone Convergence — 175
- 5.3 Exercises — 179

6. The Secant Method — 193

- 6.1 The Modified Secant Method — 193
- 6.2 Error Bounds for the Secant Method — 203
- 6.3 Exercises — 215

7. Newton-Like Methods — 225

- 7.1 Stirling's Method — 225
- 7.2 Convergence for a Certain Class of Newton-Like Methods — 242
- 7.3 Newton-Like Methods Under Mild Differentiability Conditions — 261
- 7.4 Perturbed Newton-Like Methods — 274
- 7.5 Projection Methods and Inexact Newton-like Iterations — 285
- 7.6 Exercises — 291

8. Two-Point Newton-Like Methods — 307

- 8.1 Two-Point Newton-Like Methods in Banach Space — 307
- 8.2 A Fast Convergent Method — 333
- 8.3 Exercises — 349

9. Variational Inequalities — 363

- 9.1 Generalized Equations Using Newton's Method — 363

	9.2 Exercises	377
10.	Special Topics	387
	10.1 Methods Involving Outer or Generalized Inverses	387
	10.2 Exercises	397
11.	Operator Equations and Their Discretizations	405
	11.1 The Mesh Independence Principle Under Hölder continuity	405
	11.2 Exercises	420
12.	Convergence on Generalized Spaces	431
	12.1 Iterative Methods on Banach Spaces with a Convergence Structure	431
	12.2 Exercises	440
13.	Dynamic Processes	451
	13.1 On Time Dependent Multistep Dynamic Processes	451
	13.2 The Monotone Convergence of General Newton-Like Methods	460
	13.3 Convergence Methods and Point to Point Mappings	473
	13.4 Exercises	489
Appendix A Glossary of Symbols		493
Bibliography		495
Index		511

Chapter 1

Linear Spaces

The basic background for solving equations is introduced here.

1.1 Linear Operators

Some mathematical operations have certain properties in common. These properties are given in the following definition.

Definition 1.1 An operator T which maps a linear space X into a linear space Y over the same scalar field S is said to be additive if

$$T(x+y) = T(x) + T(y), \qquad \text{for all } x, y \in X,$$

and homogeneous if

$$T(sx) = sT(x), \qquad \text{for all } x \in X, s \in S.$$

An operator that is additive and homogeneous is called a linear operator.

Many examples of linear operators exist.

Example 1.1 Define an operator T from a linear space X into it self by $T(x) = sx$, $s \in S$. Then T is a linear operator.

Example 1.2 The operator $D = \frac{d}{dt}$ mapping $X = C^1[0,1]$ into $Y = C[0,1]$ given by

$$D(x) = \frac{dx}{dt} = y(t), \ 0 \le t \le 1,$$

is linear.

If X and Y are linear spaces over the same scalar field S, then the set $L(X,Y)$ containing all linear operators from X into Y is a linear space over S if addition is defined by

$$(T_1 + T_2)(x) = T_1(x) + T_2(x), \qquad \text{for all } x \in X,$$

and scalar multiplication by

$$(sT)(x) = s(T(x)), \qquad \text{for all } x \in X,\ s \in S.$$

We may also consider linear operators B mapping X into $L(X,Y)$. For an $x \in X$ we have

$$B(x) = T,$$

a linear operator from X into Y. Hence, we have

$$B(x_1, x_2) = (B(x_1))(x_2) = y \in Y.$$

B is called a bilinear operator from X into Y. The linear operators B from X into $L(X,Y)$ form a linear space $L(X, L(X,Y))$. This process can be repeated to generate j-linear operators ($j > 1$ an integer).

Definition 1.2 A linear operator mapping a linear space X into its scalar S is called a linear functional in X.

Definition 1.3 An operator Q mapping a linear space X into a linear space Y is said to be nonlinear if it is not a linear operator from X into Y.

1.2 Continuous Linear Operators

Some metric concepts of importance are introduced here.

Definition 1.4 An operator F from a Banach space X into a Banach space Y is continuous at $x = x^*$ if

$$\lim_{n\to\infty} \|x_n - x^*\|_X = 0 \implies \lim_{n\to\infty} \|F(x_n) - F(x^*)\|_Y = 0$$

Theorem 1.1 *If a linear operator T from a Banach space X into a Banach space Y is continuous at $x^* = 0$, then it is continuous at every point x of space X.*

Proof. We have $T(0) = 0$, and from $\lim_{n\to\infty} \|x_n\| = 0$ we get $\lim_{n\to\infty} \|T(x_n)\| = 0$. If sequence $\{x_n\}\,(n \geq 0)$ converges to x^* in X, by setting $y_n = x_n - x^*$ we obtain $\lim_{n\to\infty} \|y_n\| = 0$. By hypothesis this implies that

$$\lim_{n\to\infty} \|T(x_n)\| = \lim_{n\to\infty} \|T(x_n - x^*)\| = \lim_{n\to\infty} \|T(x_n) - T(x^*)\| = 0.$$
□

Definition 1.5 An operator F from a Banach space X into a Banach space Y is bounded on the set A in X if there exists a constant $c < \infty$ such that

$$\|F(x)\| \le c \|x\|, \qquad \text{for all } x \in A.$$

The greatest lower bound (infimum) of numbers c satisfying the above inequality is called the bound of F on A. An operator which is bounded on a ball (open) $U(z,r) = \{x \in X \mid \|x - z\| < r\}$ is continuous at z. It turns out that for linear operators the converse is also true.

Theorem 1.2 *A continuous linear operator T from a Banach space X into a Banach space Y is bounded on X.*

Proof. By the continuity of T there exists $\varepsilon > 0$ such that $\|T(z)\| < 1$, if $\|z\| < \varepsilon$. For $0 \ne z \in X$

$$\|T(z)\| \le \tfrac{1}{\varepsilon} \|z\|, \qquad (1.1)$$

since $\|cz\| < \varepsilon$ for $|c| < \frac{\varepsilon}{\|z\|}$, and $\|T(cz)\| = |c| \cdot \|T(z)\| < 1$. Letting $c = \varepsilon^{-1}$ in (1.1), we conclude that operator T is bounded on X. □

The bound on X of a linear operator T denoted by $\|T\|_X$ or simply $\|T\|$ is called the norm of T. As in Theorem 1.2 we get

$$\|T\| = \sup_{\|x\|=1} \|T(x)\|. \qquad (1.2)$$

Hence, for any bounded linear operator T

$$\|T(x)\| \le \|T\| \cdot \|x\|, \qquad \text{for all } x \in X. \qquad (1.3)$$

From now on, $L(X,Y)$ denotes the set of all bounded linear operators from a Banach space X into another Banach space Y. It also follows immediately that $L(X,Y)$ is a linear space if equipped with the rules of addition and scalar multiplication introduced in Section 1.1.

The proof of the following result is left as an exercise (see also [101], [124]).

Theorem 1.3 *The set $L(X,Y)$ is a Banach space for the norm (1.2).*

1.3 Equations

In a Banach space X solving a linear equation can be stated as follows: given a bounded linear operator T mapping X into itself and some $y \in X$, find an $x \in X$ such that

$$T(x) = y. \tag{1.4}$$

The point x (if it exists) is called a solution of Equation (1.4).

Definition 1.6 If T is a bounded linear operator in X and a bounded linear operator T_1 exists such that

$$T_1 T = T T_1 = I, \tag{1.5}$$

where I is the identity operator in X (i.e., $I(x) = x$ for all $x \in X$), then T_1 is called the inverse of T and we write $T_1 = T^{-1}$. That is,

$$T^{-1} T = T T^{-1} = I. \tag{1.6}$$

If T^{-1} exists, then Equation (1.4) has the unique solution

$$x = T^{-1}(y). \tag{1.7}$$

The proof of the following result is left as an exercise (see also [140], [185], [188]).

Theorem 1.4 *(Banach Lemma on Invertible Operators). If T is a bounded linear operator in X, T^{-1} exists if and only if there is a bounded linear operator P in X such that P^{-1} exists and*

$$\|I - PT\| < 1. \tag{1.8}$$

If T^{-1} exists, then

$$T^{-1} = \sum_{n=0}^{\infty} (I - PT)^n P \qquad \text{(Neumann Series)} \tag{1.9}$$

and

$$\|T^{-1}\| \le \frac{\|P\|}{1 - \|I - PT\|}. \tag{1.10}$$

Based on Theorem 1.4 we can immediately introduce a computational theory for Equation (1.4) composed by three factors:

(A) *Existence and Uniqueness.* Under the hypotheses of Theorem 1.4 Equation (1.4) has a unique solution x^*.

(B) *Approximation.* The iteration

$$x_{n+1} = P(y) + (I - PT)(x_n) \quad (n \geq 0) \tag{1.11}$$

gives a sequence $\{x_n\}$ $(n \geq 0)$ of successive approximations, which converges to x^* for any initial guess $x_0 \in X$.

(C) *Error Bounds.* Clearly the speed of convergence of iteration $\{x_n\}$ $(n \geq 0)$ to x^* is governed by the estimate:

$$\|x_n - x^*\| \leq \frac{\|I - PT\|^n}{1 - \|I - PT\|}\|P(y)\| + \|I - PT\|^n \|x_0\|. \tag{1.12}$$

1.4 Computing the Inverse of a Linear Operator

Let T be a bounded linear operator in X. One way to obtain an approximate inverse is to make use of an operator sufficiently close to T.

Theorem 1.5 *If T is a bounded linear operator in X, T^{-1} exists if and only if there is a bounded linear operator P_1 in X such that P_1^{-1} exists, and*

$$\|P_1 - T\| \leq \|P_1^{-1}\|^{-1}. \tag{1.13}$$

If T^{-1} exists, then

$$T^{-1} = \sum_{n=0}^{\infty} \left(I - P_1^{-1}T\right)^n P_1^{-1} \tag{1.14}$$

and

$$\|T^{-1}\| \leq \frac{\|P^{-1}\|}{1 - \|I - P_1^{-1}T\|} \leq \frac{\|P_1^{-1}\|}{1 - \|P_1^{-1}\| \|P_1 - T\|}. \tag{1.15}$$

Proof. Let $P = P_1^{-1}$ in Theorem 1.4 and note that by (1.13)

$$\|I - P_1^{-1}T\| = \|P_1^{-1}(P_1 - T)\| \leq \|P_1^{-1}\| \cdot \|P_1 - T\| < 1. \tag{1.16}$$

That is, (1.8) is satisfied. The bounds (1.15) follow from (1.10) and (1.16). That proves the sufficiency. The necessity is proved by setting $P_1 = T$, if T^{-1} exists. □

The following result is equivalent to Theorem 1.4.

Theorem 1.6 *A bounded linear operator T in a Banach space X has an inverse T^{-1} if and only if linear operators P, P^{-1} exist such that the series*

$$\sum_{n=0}^{\infty} (I - PT)^n P \qquad (1.17)$$

converges. In this case we have

$$T^{-1} = \sum_{n=0}^{\infty} (I - PT)^n P.$$

Proof. If series (1.17) converges, then it converges to T^{-1} (see Theorem 1.4). The existence of P, P^{-1} and the convergence of series (1.17) is again established as in Theorem 1.4, by taking $P = T^{-1}$, when it exists. □

Definition 1.7 A linear operator N in a Banach space X is said to be nilpotent if

$$N^m = 0, \qquad (1.18)$$

for some positive integer m.

Theorem 1.7 *A bounded linear operator T in a Banach space X has an inverse T^{-1} and only if there exist linear operators P, P^{-1} such that $I - PT$ is nilpotent.*

Proof. If P, P^{-1} exists and $I - PT$ is nilpotent, then series

$$\sum_{n=0}^{\infty} (I - PT)^n P = \sum_{n=0}^{m-1} (I - PT)^n P$$

converges to T^{-1} by Theorem 1.6. Moreover, if T^{-1} exists, then $P = T^{-1}$, $P^{-1} = T$ exists, and $I - PT = I - T^{-1}T = 0$ is nilpotent. □

1.5 Fréchet Derivatives

The computational techniques to be considered later make use of the derivative in the sense of Fréchet [185], [186], [229].

Definition 1.8 Let F be an operator mapping a Banach space X into a Banach space Y. If there exists a bounded linear operator L from X into

Y such that

$$\lim_{\|\Delta x\| \to 0} \frac{\|F(x_0 + \Delta x) - F(x_0) - L(\Delta x)\|}{\|\Delta x\|} = 0, \quad (1.19)$$

then P is said to be Fréchet differentiable at x_0, and the bounded linear operator

$$P'(x_0) = L \quad (1.20)$$

is called the first Fréchet-derivative of F at x_0. The limit in (1.19) is supposed to hold independently of the way that Δx approaches 0. Moreover, the Fréchet differential

$$\delta F(x_0, \Delta x) = F'(x_0) \Delta x \quad (1.21)$$

is an arbitrary close approximation to the difference $F(x_0 + \Delta x) - F(x_0)$ relative to $\|\Delta x\|$, for $\|\Delta x\|$ small.

If F_1 and F_2 are differentiable at x_0, then

$$(F_1 + F_2)'(x_0) = F_1'(x_0) + F_2'(x_0). \quad (1.22)$$

Moreover, if F_2 is an operator from a Banach space X into a Banach space Z, and F_1 is an operator from Z into a Banach space Y, their composition $F_1 \circ F_2$ is defined by

$$(F_1 \circ F_2)(x) = F_1(F_2(x)), \quad \text{for all } x \in X. \quad (1.23)$$

It follows from Definition 1.8 that $F_1 \circ F_2$ is differentiable at x_0 if F_2 is differentiable at x_0 and F_1 is differentiable at $F_2(x_0)$ of Z, with (chain rule):

$$(F_1 \circ F_2)'(x_0) = F_1'(F_2(x_0))F_2'(x_0). \quad (1.24)$$

In order to differentiate an operator F we write:

$$F(x_0 + \Delta x) - F(x_0) = L(x_0, \Delta x)\Delta x + \eta(x_0, \Delta x), \quad (1.25)$$

where $L(x_0, \Delta x)$ is a bounded linear operator for given $x_0, \Delta x$ with

$$\lim_{\|\Delta x\| \to 0} L(x_0, \Delta x) = L, \quad (1.26)$$

and

$$\lim_{\|\Delta x\| \to 0} \frac{\|\eta(x_0, \Delta x)\|}{\|\Delta x\|} = 0. \quad (1.27)$$

Estimates (1.26) and (1.27) give

$$\lim_{\|\Delta x\|\to 0} L(x_0, \Delta x) = F'(x_0). \tag{1.28}$$

If $L(x_0, \Delta x)$ is a continuous function of Δx in some ball $U(0, R)$ $(R > 0)$, then

$$L(x_0, 0) = F'(x_0). \tag{1.29}$$

We need the definition of a mosaic:

Higher-order derivatives can be defined by induction:

Definition 1.9 If F is $(m-1)$-times Fréchet-differentiable ($m \geq 2$ an integer), and an m-linear operator A from X into Y exists such that

$$\lim_{\|\Delta x\|\to 0} \frac{\left\|F^{(m-1)}(x_0 + \Delta x) - F^{(m-1)}(x_0) - A(\Delta x)\right\|}{\|\Delta x\|} = 0, \tag{1.30}$$

then A is called the m-Fréchet-derivative of F at x_0, and

$$A = F^{(m)}(x_0) \tag{1.31}$$

Higher partial derivatives in product spaces can be defined as follows: Define

$$X_{ij} = L(X_j, X_i), \tag{1.32}$$

where $X_1, X_2, \ldots$ are Banach spaces and $L(X_j, X_i)$ is the space of bounded linear operators from X_j into X_i. The elements of X_{ij} are denoted by L_{ij}, etc. Similarly,

$$X_{ijm} = L(X_m, X_{ij}) = L(X_m, L(X_j, X_i)) \tag{1.33}$$

denotes the space of bounded bilinear operators from X_k into X_{ij}. Finally, we write

$$X_{ij_1 j_2 \cdots j_m} = L\left(X_{jk}, X_{ij_1 j_2 \cdots j_{m-1}}\right), \tag{1.34}$$

which denotes the space of bounded linear operators from X_{jm} into $X_{ij_1 j_2 \cdots j_{m-1}}$. The elements $A = A_{ij_1 j_2 \cdots j_m}$ of $X_{ij_1 j_2 \cdots j_m}$ are a generalization of m-linear operators [10], [54].

Consider an operator F_i from space

$$X = \prod_{p=1}^{n} X_{j_p} \tag{1.35}$$

into X_i, and that F_i has partial derivatives of orders $1, 2, \ldots, m-1$ in some ball $U(x_0, R)$, where $R > 0$ and

$$x_0 = \left(x_{j_1}^{(0)}, x_{j_2}^{(0)}, \ldots, x_{j_n}^{(0)}\right) \in X. \tag{1.36}$$

For simplicity and without loss of generality we renumber the original spaces so that

$$j_1 = 1, j_2 = 2, \ldots, j_n = n. \tag{1.37}$$

Hence, we write

$$x_0 = (x_1^{(0)}, x_2^{(0)}, \ldots, x_n^{(0)}). \tag{1.38}$$

A partial derivative of order $(m-1)$ of F_i at x_0 is an operator

$$A_{iq_1q_2\cdots q_{m-1}} = \frac{\partial^{(m-1)} F_i(x_0)}{\partial x_{q_1} \partial x_{q_2} \cdots \partial x_{q_{m-1}}} \tag{1.39}$$

(in $X_{iq_1q_2\cdots q_{m-1}}$) where

$$1 \leq q_1, q_2, \ldots, q_{m-1} \leq n. \tag{1.40}$$

Let $P(X_{q_m})$ denote the operator from X_{q_m} into $X_{iq_1q_2\cdots q_{m-1}}$ obtained from (1.39) by letting

$$x_j = x_j^{(0)}, \quad j \neq q_m, \tag{1.41}$$

for some q_m, $1 \leq q_m \leq n$. Moreover, if

$$P'(x_{q_m}^{(0)}) = \frac{\partial}{\partial x_{q_m}} \cdot \frac{\partial^{m-1} F_i(x_0)}{\partial x_{q_1} \partial x_{q_2} \cdots \partial x_{q_{m-1}}} = \frac{\partial^m F_i(x_0)}{\partial x_{q_1} \cdots \partial x_{q_m}}, \tag{1.42}$$

exists, it will be called the partial Fréchet-derivative of order m of F_i with respect to $x_{q_1}, \ldots, x_{q_m}$ at x_0.

Furthermore, if F_i is Fréchet-differentiable m times at x_0, then

$$\frac{\partial^m F_i(x_0)}{\partial x_{q_1} \cdots \partial x_{q_m}} x_{q_1} \cdots x_{q_m} = \frac{\partial^m F_i(x_0)}{\partial x_{s_1} \partial x_{s_2} \cdots \partial x_{s_m}} x_{s_1} \cdots x_{s_m} \tag{1.43}$$

for any permutation $s_1, s_2, \ldots, s_m$ of integers $q_1, q_2, \ldots, q_m$ and any choice of points $x_{q_1}, \ldots, x_{q_m}$, from $X_{q_1}, \ldots, X_{q_m}$ respectively. Hence, if $F = (F_1, \ldots, F_t)$ is an operator from $X = X_1 \times X_2 \times \cdots \times X_n$ into $Y = Y_1 \times Y_2 \times \cdots \times Y_t$, then

$$F^{(m)}(x_0) = \left(\frac{\partial^m F_i}{\partial x_{j_1} \cdots \partial x_{j_m}}\right)_{x=x_0} \tag{1.44}$$

$i = 1, 2, \ldots, t$, $j_1, j_2, \ldots, j_m = 1, 2, \ldots, n$, is called the m-Fréchet derivative of F at $x_0 = (x_1^{(0)}, x_2^{(0)}, \ldots, x_n^{(0)})$.

1.6 Integration

In this section we state results concerning the mean value theorem, Taylor's theorem, and Riemannian integration. The proofs are left out as exercises.

The mean value theorem for differentiable real functions f:

$$f(b) - f(a) = f'(c)(b - a), \qquad (1.45)$$

where $c \in (a, b)$, does not hold in a Banach space setting. However, if F is a differentiable operator between two Banach spaces X and Y, then

$$\|F(x) - F(y)\| \leq \sup_{\bar{x} \in L(x,y)} \|F'(\bar{x})\| \cdot \|x - y\|, \qquad (1.46)$$

where

$$L(x, y) = \{z : z = \lambda y + (1 - \lambda)x, \ 0 \leq \lambda \leq 1\}. \qquad (1.47)$$

Set

$$z(\lambda) = \lambda y + (1 - \lambda)x, \quad 0 \leq \lambda \leq 1, \qquad (1.48)$$

and

$$F(\lambda) = F(z(\lambda)) = F(\lambda y + (1 - \lambda)x). \qquad (1.49)$$

Divide the interval $0 \leq \lambda \leq 1$ into n subintervals of lengths $\Delta \lambda_i$, $i = 1, 2, \ldots, n$, choose points λ_i inside corresponding subintervals and as in the real Riemann integral consider sums

$$\sum_\sigma F(\lambda_i) \Delta \lambda_i = \sum_{i=1}^n F(\lambda_i) \Delta \lambda_i, \qquad (1.50)$$

where σ is the partition of the interval, and set

$$|\sigma| = \max_{(i)} \Delta \lambda_i. \qquad (1.51)$$

Definition 1.10 If

$$S = \lim_{|\sigma| \to 0} \sum_\sigma F(\lambda_i) \Delta \lambda_i \qquad (1.52)$$

exists, then it is called the Riemann integral from $F(\lambda)$ from 0 and 1, denoted by

$$S = \int_0^1 F(\lambda)\, d\lambda = \int_x^y F(\lambda)\, d\lambda. \tag{1.53}$$

Definition 1.11 A bounded operator $P(\lambda)$ on $[0,1]$ such that the set of points of discontinuity is of measure zero is said to be integrable on $[0,1]$.

We now state the famous Taylor theorem [161].

Theorem 1.8 *If F is m-times Fréchet-differentiable in $U(x_0, R)$, $R > 0$, and $F^{(m)}(x)$ is integrable from x to any $y \in U(x_0, R)$, then*

$$F(y) = F(x) + \sum_{n=1}^{m-1} \tfrac{1}{n!} F^{(n)}(x)(y-x)^n + R_m(x,y), \tag{1.54}$$

$$\left\| F(y) - \sum_{n=0}^{m-1} \tfrac{1}{n!} F^{(n)}(x)(y-x)^n \right\| \leq \sup_{\bar{x} \in L(x,y)} \left\| F^{(m)}(\bar{x}) \right\| \frac{\|y-x\|^m}{m!}, \tag{1.55}$$

where

$$R_m(x,y) = \int_0^1 F^{(m)}(\lambda y + (1-\lambda)x)(y-x)^m \tfrac{(1-\lambda)^{m-1}}{(m-1)!} d\lambda. \tag{1.56}$$

1.7 Exercises

1.1 Show that the operators introduced in Examples 1.1 and 1.2 are indeed linear.

1.2. Show that the Laplace transform

$$\Delta = \frac{\partial^2}{\partial x_1^2} + \frac{\partial^2}{\partial x_2^2} + \frac{\partial^2}{\partial x_3^2}$$

is a linear operator mapping the space of real functions $x = x(x_1, x_2, x_3)$ with continuous second derivatives on some subset D of R^3 into the space of continuous real functions on D.

1.3. Define $T : C''[0,1] \times C'[0,1] \to C[0,1]$ by

$$T(x,y) = \left(\alpha \frac{d^2}{dt^2} \; \beta \frac{d}{dt} \right) \begin{pmatrix} x \\ y \end{pmatrix} = \alpha \frac{d^2 x}{dt^2} + \beta \frac{dy}{dt}, \quad 0 \leq t \leq 1.$$

Show that T is a linear operator.

1.4. In an inner product $\langle \cdot, \cdot \rangle$ space show that for any fixed z in the space
$$T(x) = \langle x, z \rangle$$
is a linear functional.

1.5. Show that an additive operator T from a real Banach space X into a real Banach space Y is homogeneous if it is continuous.

1.6. Show that matrix $A = \{a_{ij}\}$, $i, j = 1, 2, \ldots, n$ has an inverse if
$$|a_{ii}| > \frac{1}{2} \sum_{j=1}^{n} |a_{ij}| > 0, \quad i = 1, 2, \ldots, n.$$

1.7. Show that the linear integral equation of second Fredholm kind in $C[0, 1]$
$$x(s) - \lambda \int_0^1 K(s,t)x(t)dt = y(s), \quad 0 \leq \lambda \leq 1,$$
where $K(s,t)$ is continuous on $0 \leq s, t \leq 1$, has a unique solution $x(s)$ for $y(s) \in C[0,1]$ if
$$|\lambda| < \left[\max_{[0,1]} \int_0^1 |K(s,t)|dt \right]^{-1}.$$

1.8. Prove Theorem 1.3.

1.9. Prove Theorem 1.4.

1.10. Show that the operators defined below are all linear.

(a) Identity operator. The identity operator $I_X : X \to X$ given by $I_X(x) = x$, for all $x \in X$.

(b) Zero operator. The zero operator $O : X \to Y$ given by $O(x) = 0$, for all $x \in X$.

(c) Integration. $T : C[a,b] \to C[a,b]$ given by $T(x(t)) = \int_0^1 x(s)ds$, $t \in [a,b]$.

(d) Differentiation. Let X be the vector space of all polynomials on $[a,b]$. Define T on X by $T(x(t)) = x'(t)$.

(e) Vector algebra. The cross product with one factor kept fixed. Define $T_1 : \mathbb{R}^3 \to \mathbb{R}^5$. Similarly, the dot product with one fixed factor. Define $T_2 : \mathbb{R}^3 \to \mathbb{R}$.

(f) Matrices. A real matrix $A = \{a_{ij}\}$ with m rows and n columns. Define $T : \mathbb{R}^n \to \mathbb{R}^m$ given by $y = Ax$.

1.11. Let T be a linear operator. Show:

(a) the $R(T)$ (range of T) is a vector space;
(b) if $\dim(T) = n < \infty$, then $\dim R(T) \le n$;
(c) the null/space $N(T)$ is a vector space.

1.12. Let X, Y be vector spaces, both real or both complex. Let $T : D(T) \to Y$ (domain of T) be a linear operator with $D(T) \subseteq X$ and $R(T) \subseteq Y$. Then, show:

(a) the inverse $T^{-1} : R(T) \to D(T)$ exists if and only if
$$T(x) = 0 \Rightarrow x = 0;$$

(b) if T^{-1} exists, it is a linear operator;
(c) if $\dim D(T) = n < \infty$ and T^{-1} exists, then $\dim R(T) = \dim D(T)$.

1.13. Let $T : X \to Y$, $P : Y \to Z$ be bijective linear operators, where X, Y, Z are vector spaces. Then, show: the inverse $(ST)^{-1} : Z \to X$ of the product ST exists, and
$$(ST)^{-1} = T^{-1} S^{-1}.$$

1.14. If the product (composite) of two linear operators exists, show that it is linear.

1.15. Let X be the vector space of all complex 2×2 matrices and define $T : X \to X$ by $T(x) = cx$, where $c \in X$ is fixed and cx denotes the usual product of matrices. Show that T is linear. Under what conditions does T^{-1} exist?

1.16. Let $T : X \to Y$ be a linear operator and $\dim X = \dim Y = n < \infty$. Show that $R(T) = Y$ if and only if T^{-1} exists.

1.17. Define the integral operator $T : C[0,1] \to C[0,1]$ by $y = T(x)$, where $y(t) = \int_0^1 k(x,s) x(s) ds$ and k is continuous on $[0,1] \times [0,1]$. Show that T is linear and bounded.

1.18. Show that the operator T defined in 10(f) is bounded.

1.19. If a normed space X is finite dimensional then show that every linear functional on X is bounded.

1.20. Let $T : D(T) \to Y$ be a linear operator, where $D(T) \subseteq X$ and X, Y are normed spaces. Show:

(a) T is continuous if and only if it is bounded;
(b) if T is continuous at a single point, it is continuous.

1.21. Let T be a bounded linear operator. Show:

(a) $x_n \to x$ (where $x_n, x \in D(T)$) $\Rightarrow T(x_n) \to T(x)$;
(b) the null space $N(T)$ is closed.

1.22. If $T \neq 0$ is a bounded linear operator, show that for any $x \in D(T)$ such that $\|x\| < 1$, we have $\|T(x)\| < \|T\|$.

1.23. Show that the operator $T : \ell^\infty \to \ell^\infty$ defined by $y = (y_i) = T(x)$, $y_i = \frac{x_i}{i}$, $x = (x_i)$, is linear and bounded.

1.24. Let $T : C[0,1] \to C[0,1]$ be defined by

$$y(t) = \int_0^t x(s)ds.$$

Find $R(T)$ and $T^{-1} : R(T) \to C[0,1]$. Is T^{-1} linear and bounded?

1.25. Show that the functionals defined on $C[a,b]$ by

$$f_1(x) = \int_a^b x(t)y_0(t)dt \quad (y_0 \in C[a,b])$$
$$f_2(x) = c_1 x(a) + c_2 x(b) \quad (c_1, c_2 \text{ fixed})$$

are linear and bounded.

1.26. Find the norm of the linear functional f defined on $C[-1,1]$ by

$$f(x) = \int_{-1}^0 x(t)dt - \int_0^1 x(t)dt.$$

1.27. Show that

$$f_1(x) = \max_{t \in J} x(t), \quad f_2(x) = \min_{t \in J} x(t), \quad J = [a,b]$$

define functionals on $C[a,b]$. Are they linear? Bounded?

1.28. Show that a function can be additive and not homogeneous. For example, let $z = x + iy$ denote a complex number, and let $T : \mathbb{C} \to \mathbb{C}$ be given by

$$T(z) = \bar{z} = x - iy.$$

1.29. Show that a function can be homogeneous and not additive. For example, consider the operator $T : \mathbb{R}^2 \to \mathbb{R}$ given by

$$T((x_1, x_2)) = \frac{x_1^2}{x_2}.$$

1.30. Let F be an operator in $C[0,1]$ defined by
$$F(x)(s) = x(s)\int_0^1 \frac{s}{s+t}x(t)dt, \quad 0 \le \lambda \le 1.$$
Show that for $x_0, z \in C[0,1]$
$$F'(x_0)z = x_0(s)\int_0^1 \frac{s}{s+t}z(t)dt + z(s)\int_0^1 \frac{s}{s+t}x_0(t)dt.$$

1.31. Find the Fréchet-derivative of the operator F in $\mathbb{R}_\infty^2$ given by
$$F\begin{pmatrix}x\\y\end{pmatrix} = \begin{pmatrix}x^2 + 7x + 2xy - 3\\ x + y^3\end{pmatrix}.$$

1.32. Find the first and second Fréchet-derivatives of the Uryson operator
$$U(x) = \int_0^1 k(s,t,x(t))dt$$
in $C[0,1]$ at $x_0 = x_0(s)$.

1.33. Find the Fréchet-derivative of the Riccati differential operator
$$R(z) = \frac{dz}{dt} + p(t)z^2 + q(t)z + r(t),$$
from $C'[0,s]$ into $C[0,s]$ at $z_0 = z_0(t)$ in $C'[0,s]$.

1.34. Find the first two Fréchet-derivatives of the operator
$$F\begin{pmatrix}x\\y\end{pmatrix} = \begin{pmatrix}x^2 + y^2 - 3\\ x\sin y\end{pmatrix} \quad \text{in } \mathbb{R}^2.$$

1.35. Consider the partial differential operator
$$F(x) = \Delta x - x^2$$
from $C^2(I)$ into $C(I)$, the space of all continuous function on the square $0 \le \alpha, \beta \le 1$. Show that
$$F'(x_0)z = \Delta z(\alpha,\beta) - 2x_0(\alpha,\beta)z(\alpha,\beta),$$
where Δ is the usual Laplace operator.

1.36. Let $F(L) = L^3$, in $L(x)$. Show:
$$F'(L_0) = L_0[\ ]L_0 + L_0^2[\ ] + [\ ]L_0.$$

1.37. Let $F(L) = L^{-1}$, in $L(x)$. Show:
$$F'(L_0) = -L_0^{-1}[\ ]L_0^{-1},$$
provided that L_0^{-1} exists.

1.38. Show estimates (1.45) and (1.46).

1.39. Show Taylor's Theorem 1.8.

1.40. Integrate the operator
$$F(L) = L^{-1} \quad \text{in } L(X)$$
from $L_0 = I$ to $L_1 = A$, where $\|I - A\| < 1$.

Chapter 2

Divided Differences

This chapter introduces the fundamentals of the theory of divided differences of a nonlinear operator. Several results are also provided differences as well as Fréchet derivatives satisfying Lipschitz or monotone-type conditions that will be used later.

2.1 Partially Ordered Topological Spaces

Let X be a linear space. We introduce the following definition:

Definition 2.1   a partially ordered topological linear space (POTL-space) is a locally convex topological linear space X which has a closed proper convex cone.

A proper convex cone is a subset K such that $K + K \subset K$, $\alpha K \subset K$ for $\alpha > 0$, and $K \cap (-K) = \{0\}$. Thus the order relation $\leq$, defined by $x \leq y$ if and only if $y - x \in K$, gives a partial ordering which is compatible with the linear structure of the space. The cone K which defines the ordering is called the positive cone since $K = \{x \in X \mid x \geq 0\}$. The fact that K is closed implies also that intervals, $[a, b] = \{z \in X \mid a \leq z \leq b\}$, are closed sets.

Example 2.1   Some simple examples of POTL-spaces are:

(1) $X = E^n$, n-dimensional Euclidean space, with

$$K = \{(x_1, x_2, ..., x_n) \in E^n \mid x_i \geq 0,\ i = 1, 2, ..., n\};$$

(2) $X = E^n$ with

$$K = \{(x_1, x_2, ..., x_n) \in E^n \mid x_i \geq 0,\ i = 1, 2, ..., n-1, x_n = 0\};$$

(3) $X = C^n[0,1]$, continuous functions, maximum norm topology, pointwise ordering;
(4) $X = C^n[0,1]$, n-times continuously differentiable functions with

$$\|f\| = \sum_{k=0}^{n} \max \left| f^{(K)}(t) \right|, \text{ and point wise ordering;}$$

(5) $C = L^p[0,1]$, $0 \leq p \leq \infty$ usual topology, $K = \{f \in L^p[0,1] \mid f(t) \leq 0 \text{ a.e.}\}$.

Remark 2.1 *Using the above examples, it is easy to see that the closedness of the positive cone is not, in general, a strong enough connection between the ordering and the topology. Consider, for example, the following properties of sequences of real numbers:*

(1) $x_1 \leq x_2 \leq \cdots \leq x^*$, *and* $\sup\{x_n\}$ x^* *implies* $\lim_{n \to \infty} x_n = x^*$;
(2) $\lim_{n \to \infty} x_n = 0$ *implies that there exists a sequence* $\{y_n\}$ *with* $y_1 \geq y_2 \geq \cdots \geq 0$, $\inf\{y_n\} = 0$ *and* $-y_n \leq x_n \leq y_n$;
(3) $0 \leq x_n \leq y_n$, *and* $\lim_{n \to \infty} y_n = 0$ *imply* $\lim_{n \to \infty} x_n = 0$.

Unfortunately, these statements are not trues for all POTL-spaces:

(a) In $X = C[0,1]$ let $x_n(t) = -t^n$. Then $x_1 \leq x_2 \leq \cdots \leq 0$, and $\sup\{x_n\} = 0$, but $\|x_n\| = 1$ for all n, so $\lim_{n \to \infty} x_n$ does not exist. Hence (1) does not hold.
(b) In $X = L^1[0,1]$ let $x_n(t) = n$ for $\frac{1}{n+1} \leq t \leq \frac{1}{n}$ and zero elsewhere. Then $\lim_{n \to \infty} \|x_n\| = 0$ but clearly property (2) does not hold.
(c) In $X = C^1[0,1]$ let $x_n(t) = \frac{t^n}{n}$, $y_n(t) = \frac{1}{n}$. then $0 \check{S} x_n \leq y_n$, and $\lim_{n \to \infty} y_n = 0$, but $\|x_n\| = \max\left|\frac{t^n}{n}\right| + \max\left|t^{n-1}\right| = \frac{1}{n} + 1 > 1$; hence x_n does not converge to zero.

We will now devote a brief discussion of certain types of POTL spaces in which some of the above statements are true.

Definition 2.2 *A POTL-space is called regular if every order-bounded increasing sequence has a limit.*

Remark 2.2 *Examples of regular POTL-spaces are E^n and L^p, $0 \leq p \leq \infty$, where as $C[0,1]$, $C^n[0,1]$ and $L^\infty[0,1]$ are not regular, as was shown in (a) of the above remark. If $\{x_n\}$ $n \geq 0$ is a monotone increasing sequence and $\lim_{n \to \infty} x_n = x^*$ exists, then for any k_0, $n \geq k_0$ implies $x_n \geq x_{k_0}$. Hence*

$x^* = \lim_{n\to\infty} x_n \geq x_{k_0}$, i.e., x^* is an upper bound on $\{x_n\}$ $n = 0$. Moreover, if y is any other upper bound, then $x_n \leq y$, and hence $x^* = \lim_{n\to\infty} x_n \leq y$, i.e., $x^* = \sup\{x_n\}$. This shows that in any POTL-space, the closedness of the positive cone guarantees that, if a monotone increasing sequence has a limit, then it is also a supremum. In a regular space, the converse of this is true; i.e., if a monotone increasing sequence has a supremum, then it also has a limit. It is important to note that the definition of regularity involves both an order concept (monotone boundedness) and a topological concept (limit).

Definition 2.3 A POTL-space is called normal if, given a local base U for the topology, there exists a positive number η so that if $0 \leq x \in V \in U$ then $[0, x] \subseteq \eta^U$.

Remark 2.3 *If the topology of a POTL-space is given by a norm then this space is called a partially ordered normed space (PON)-space. If a PON-space is complete with respect to its topology then it is called a partially ordered Banach space (POB)-space. According to Definition 2.3. A PON-space is normal if and only if there exists a positive number α such that*

$$\|x\| \leq \alpha \|y\| \quad \text{for all} \quad x, y \in X \quad \text{with} \quad 0 \leq x \leq y.$$

Let us note that any regular POB-space is normal. The converse is not true. For example, the space $C[0,1]$, ordered by the cone of nonnegative functions, is normal but is not regular. All finite dimensional POTL-spaces are both normal and regular.

Remark 2.4 *Let us now define some special types of operators acting between two POTL-spaces. First we introduce some notation if X and Y are two linear spaces then we denote by (X, Y) the set of all operators from X into Y and by $L(X, Y)$ the set of all linear operators from X into Y. If X and Y are topological linear spaces then we denote by $LB(X, Y)$ the set of all continuous linear operators from X into Y. for simplicity the spaces $L(X, X)$ and $LB(X, X)$ will be denoted by $L(X)$ and $LB(X)$. Now let X and Y be two POTL-spaces and consider an operator $G \in (X, Y)$. G is called isotone (resp. antitone) if $x \geq y$ implies $G(x) \leq G(y)$ (resp. $G(x) \leq G(y)$). G is called nonnegative if $x \geq 0$ implies $G(x) \geq 0$. For linear operators the nonnegativity is clearly equivalent with the isotony. Also, a linear operator is inverse nonnegative if and only if it is invertible and its inverse is nonnegative. If G is a nonnegative operator then we write $G \geq 0$. If G and H are two operators from X into Y such that $H - G$ is nonnegative* then we write $G \leq H$. If Z is a linear space then we denote by $I = I_z$ the identity operator in Z (i.e., $I(x) = x$ for all $x \in Z$). If Z is a POTL-space then we have obviously $I \geq 0$. Suppose that X and Y are two POTL-spaces and consider the operators $T \in L(X,Y)$ and $S \in L(Y,X)$. If $ST \leq I_x$ (resp. $ST \geq I_x$) then S is called a left subinverse (resp. superinverse) of T and T is called a right sub-inverse (resp. superinverse) of S. We say that S is a subinverse of T is S is a left-as a right subinverse of T.

We finally end this section by noting that for the theory of partially ordered linear spaces the reader may consult M.A. Krasnosel'skii [184]-[187], Vandergraft [272] or Argyros and Szidarovszky [99].

2.2 Divided Differences in a Linear Space

The concept of a divided difference of a nonlinear operator generalizes the usual notion of a divided difference of a scalar function in the same way in which the Fréchet-derivative generalizes the notion of a derivative of a function.

Definition 2.4 Let F be a nonlinear operator defined on a subset D of a linear space X with values in a linear space Y, i.e., $F \in (D,Y)$ and let x, y be two points of D. A linear operator from X into Y, denoted $[x, y]$, which satisfies the condition

$$[x,y](x-y) = F(x) - F(y) \tag{2.1}$$

is called a divided difference of F at the points x and y.

Remark 2.5 *If X and Y are topological linear spaces then we shall always assume the continuity of the linear operator $[x,y]$. (Generally, $[x,y] \in L(X,Y)$ if X,Y are POTL-spaces then $[x,y] \in LB(X,Y)$).*

Obviously, condition (2.1) does not uniquely determine the divided difference, with the exception of the case when X is one-dimensional. An operator $[\cdot,\cdot] : D \times D \to L(X,Y)$ satisfying (2.1) is called a divided difference of F on D. If we fix the first variable, we get an operator

$$[x^0, \cdot] : D \to L(X,Y). \tag{2.2}$$

Let x^1, x^2 be two points of D. A divided difference of the operator (2.2) at the points x^1, x^2 will be called a divided difference of the second order of F at the points x^0, x^1, x^2 and will be denoted by $[x^0, x^1, x^2]$. We have by definition

$$[x^0, x^1, x^2]\,(x^1 - x^2) = [x^0, x^1] - [x^0, x^2]. \tag{2.3}$$

Obviously, $[x^0, x^1, x^2] \in L(X, L(X, Y))$.

Let us now state a well known result due to Kantorovich concerning the location of fixed points which will be used extensively later.

Theorem 2.1 *Let X be a regular POTL-space and let x, y be two points of X such that $x \leq y$. If $H : [x, y] \to X$ is a continuous isotone operator having the property that $x \leq H(x)$ and $y \geq H(y)$, then there exists a point $z \in [x, y]$ such that $H(z) = z$.*

2.3 Divided Difference in a Banach Space

In this section we will assume that X and Y are banach spaces. Accordingly we shall have $[x, y] \in LB(X, Y)$, $[x, y, z] \in LB(X, LB(X, Y))$. As we will see in later chapters, most convergence theorems in a Banach space require that the divided differences of F satisfy Lipschitz conditions of the form:

$$\|[x, y] - [x, z]\| \leq c_0 \|y - z\| \tag{2.4}$$

$$\|[y, x] - [z, x]\| \leq c_1 \|y - z\| \tag{2.5}$$

$$\|[x, y, z] - [u, y, z]\| \leq c_2 \|x - y\| \quad \text{for all} \quad x, y, z, u \in D. \tag{2.6}$$

It is a simple exercise to show that if $[\cdot, \cdot]$ is a divided difference of F satisfying (2.4) or (2.5) then F is Fréchet differentiable on D and we have

$$F'(x) = [x, x] \quad \text{for all} \quad x \in D. \tag{2.7}$$

Moreover, if (2.4) and (2.5) are both satisfied then the Fréchet derivative F' is Lipschitz continuous on D with Lipschitz constant $l = c_0 + c_1$.

At the end of this section we shall give an example of divided differences of the first and of the second order in the finite dimensional case. We shall consider the space $|R^q$ equipped with the Chebysheff norm which is given by

$$\|x\| = \max\{|x_i| \in |R| \, 1 \leq I \leq q\} \quad \text{for} \quad x = (x_1, x_2, ..., x_q) \in |R^q. \tag{2.8}$$

It follows that the norm of a linear operator $L \in LB(|R^q)$ represented by the matrix with entries I_{ij} is given by

$$\|L\| = \max\left\{\sum_{j=1}^{q} |I_{ij}| \mid 1 \leq i \leq q\right\}. \tag{2.9}$$

We cannot give a formula for the norm of a bilinear operator. However, if B is a bilinear operator with entries b_{ijk} then we have the estimate

$$\|B\| \leq \max\left\{\sum_{j=1}^{q}\sum_{k=1}^{q} |b_{ijk}| \mid 1 \leq i \leq q\right\}. \tag{2.10}$$

Let U be an open ball of $|R^q$ and let F be an operator defined on U with values in $|R^q$. We denote by $f_1, ..., f_q$ the components of F. For each $x \in U$ we have

$$F(x) = (f_1(x), ..., f_q(x))^T. \tag{2.11}$$

Moreover, we introduce the notation

$$D_j f_i(x) = \frac{\partial f(x)}{\partial x_j}, \quad D_{kj} f_i(x) = \frac{\partial^2 f_i(x)}{\partial x_j \partial x_k}. \tag{2.12}$$

Let x, y be two points of U and let us denote by $[x, y]$ the matrix with entries

$$[x,y]_{ij} = \frac{1}{x_j - y_j}\left(f_i(x_1, ..., x_j, y_{j+1}, ..., y_q) - f_i(x_1, ..., x_{j-1}, y_j, ..., y_q)\right). \tag{2.13}$$

The linear operator $[x,y] \in LB(|R^q)$ defined in this way obviously satisfies condition (2.1). If the partial derivatives $D_j f_i$ satisfy some Lipschitz conditions of the form

$$|D_j f_i(x_1, ..., x_k + t, ..., x_q) - D_j f_i(x_1, ..., x_k, ..., x_q)| \leq p_{jk}^i |t| \tag{2.14}$$

then condition (2.4) and (2.5) will be satisfied with

$$c_0 = \max\left\{\tfrac{1}{2}\sum_{j=1}^{q}\left(p_{jj}^i + \sum_{k=j+1}^{q} p_{jk}^i\right) \mid 1 \leq i \leq q\right\} \tag{2.15}$$

and

$$c_1 = \max\left\{\tfrac{1}{2}\sum_{j=1}^{q}\left(p_{jj}^i + \sum_{k=1}^{j-1} p_{jk}^i\right) \mid 1 \leq i \leq q\right\}. \tag{2.16}$$

We shall prove (2.4) only since (2.5) can be proved similarly.

Let x, y, z be three points of U. We shall have in turn

$$[x,y]_{ij} - [x,z] = \sum_{k=1}^{q} \left\{ [x,(y_1,...,y_k,z_{k+1},...,z_q)]_{ij} \right.$$
$$\left. - [x(y_1,...,y_{k-1},z_k,...,z_q)]_{ij} \right\}. \quad (2.17)$$

by (2.13).

If $k \leq j$ then we have

$$[x,(y_1,...,y_k,z_{k+1},...,z_q)]_{ij} - [x,(y_1,...,y_{k-1},z_k,...,z_q)]_{ij}$$
$$= \frac{1}{x_j-z_j} \{f_i(x_1,...,x_j,z_{j+1},...,z_q) - f_i(x_1,...,x_{j-1},z_j,...,z_q)\}$$
$$- \frac{1}{x_j-z_j} \{f_i(x_1,...,x_j,z_{j+1},...,z_q) - f_i(x_1,...,x_{j-1},z_j,...,z_q)\} = 0.$$

For $k = j$ we have

$$\left| [x,(y_1,...,y_j,z_{j+1},...,z_q)]_{ij} - \left[x,(y_1,...,y_{j-1},z_j,...,z_q)_{ij}\right] \right|$$
$$= \left| \frac{1}{x_j-y_j} \{f_i(x_1,...,x_j,z_{j+1},...,z_q) - f_i(x_1,...,x_{j-1},y_i,z_{j+1},...,z_q)\} \right.$$
$$\left. - \frac{1}{x_j-y_j} \{f_i(x_1,...,x_j,z_{j+1},...,z_q) - f_i(x_1,...,x_{j-1},z_j,...,z_q)\} \right|$$
$$= \left| \int_0^1 \{D_j f_i(x_1,...,x_j,y_j+t(x_j-y_j),z_{j+1},...,z_q) \right.$$
$$\left. - D_j f_i(x_1,...,x_j,z_j+t(x_j-z_j),z_{j+1},...,z_q)\} dt \right|$$
$$\leq |y_j - z_j| p_{jj}^i \int_0^1 t\,dt = \tfrac{1}{2} |x_j - z_j| p_{jj}^i$$

(by (2.14)).

Finally for $k > j$ we have using (2.13) and (2.17) again

$$\left| [x,(y_1,...,y_k,z_{k+1},...,z_q)]_{ij} - [x,(y_1,...,y_{k-1},z_k,...,z_q)]_{ij} \right|$$
$$= \left| \frac{1}{x_j-y_j} \{f_i(x_1,...,x_j,y_{j+1},...,y_k,z_{k+1},...,z_q) \right.$$
$$-f_i(x_1,...,x_{j-1},y_j,...,y_k,z_{k+1},...,z_q)$$
$$-f_i(x_1,...,x_j,y_{j+1},...,y_{k-1},z_k,...,z_q)$$
$$\left. +f_i(x_1,...,x_{j-1},y_j,...,y_{k-1},z_k,...,z_q)\} \right|$$
$$= \left| \int_0^1 \{f_i(x_1,...,x_{j-1},y_j+t(x_j-y_j),y_{j+1},...,y_k,z_{k+1},...,z_q) \right.$$
$$\left. -f_i(x_1,...,x_{j-1},y_j+t(x_j-y_j),y_{j+1},...,y_{k-1},z_k,...,z_q)\} dt \right|$$
$$\leq |y_k - z_k| p_{jk}^i.$$

By adding all the above we get

$$\left| [x,y]_{ij} - [x,z]_{ij} \right| \le \frac{1}{2} |y_j - z_j| p_{jj}^i + \sum_{k=j+1}^{q} |y_k - z_k| p_{jk}^i$$

$$\le \|y - z\| \left\{ \frac{1}{2} \sum_{j=1}^{q} \left(p_{jj}^i + \sum_{k=j+1}^{q} p_{jk}^i \right) \right\}.$$

Consequently condition (2.4) is satisfied with c_0 given by (2.15). If each f_j has continuous second order partial derivatives which are bounded on U we have

$$p_{jk}^i = \sup \left\{ |D_{jk} f_i(x)| \ |x \in U \right\}.$$

In this case $p_{jk}^i = p_{kj}^i$ so that $c_0 = c_1$.

Moreover, consider again three points x, y, z of U. Similarly with (2.17) the second divided difference of F at x, y, z is the bilinear operators defined by

$$[x,y,z]_{ijk} = \frac{1}{y_k - z_k} \left\{ [x, (y_1, ..., y_k, z_{k+1}, ..., z_q)]_{ij} \right.$$
$$\left. - [x, (y_1, ..., y_{k-1}, z_k, ..., z_q)]_{ij} \right\}. \qquad (2.18)$$

It is easy to see as before that $[x,y,z]_{ijk} = 0$ for $k < j$. For $k = j$ we have

$$[x,y,z]_{ijj} = [x_j, y_j, z_j]_t f_i(x_1, ..., x_{i-1}, t, z_{j+1}, ..., z_q) \qquad (2.19)$$

where the right hand side of (2.19) represents the divided difference of $f_i(x_1, ..., x_{j-1}, t, z_{j+1}, ..., z_q)$ as a function of t, at the points x_j, y_j, z_j. Using Genocchi's integral representation of divided differences of scalar functions we get

$$[x,y,z]_{ijj} = \int_0^1 \int_0^1 t D_{jj} f_i(x_1, ..., x_{j-1}, x_j + t(y_j - x_j)$$
$$+ ts(z_j - y_j), z_{j+1}, ..., z_q) \, ds \, dt. \qquad (2.20)$$

Hence, for $k > j$ we obtain

$$[x, y, z]_{ijk}$$
$$= \frac{1}{(y_k - z_k)(x_j - y_j)} \{ f_i(x_1, ..., x_j, y_{j+1}, ..., y_k, z_{k+1}, ..., z_q)$$
$$- f_i(x_1, ..., x_j, x_{j+1}, ..., y_{k-1}, z_k, ..., z_q)$$
$$- f_i(x_1, ..., x_{j-1}, y_j, ..., y_k, z_{k+1}, ..., z_q)$$
$$+ f_i(x_1, ..., x_{j-1}, y_j, ..., y_{k-1}, z_k, ..., z_q) \}$$
$$\frac{1}{x_j - y_j} \int_0^1 \{ D_k f_i(x_1, ..., x_j, y_{j+1}, ..., y_{k-1}, z_k + t(y_k - z_k), z_{k+1}, ..., z_q)$$
$$- D_k f_i(x_1, ..., x_{j-1}, y_j, ..., y_{k-1}, z_k + t(y_k - z_k), z_{k+1}, ..., z_q) \} \, dt$$
$$= \int_0^1 \int_0^1 D_{kj} f_i(x_1, ..., x_{j-1}, y_j + s(x_j - y_i), y_{j+1}, ..., y_{k-1}, z_k$$
$$+ t(y_k - z_k), z_{k+1}, ..., z_q) \, ds \, dt. \qquad (2.21)$$

We now want to show that if

$$|D_{kj} f_i(v_1, ..., v_m + t, ..., v_q) - D_{kj} f_i(v_1, ..., v_m, ... v_q)| \leq q_{km}^{ij} |t|$$
$$\text{for all} \quad v = (v_1, ..., v_q) \in U, \quad 1 \leq i, j, k, m \leq q, \qquad (2.22)$$

then the divided difference of F of the second order defined by (2.18) satisfies condition (2.6) with the constant

$$c_2 = \max_{1 \leq i \leq q} \sum_{j=1}^q \left\{ \frac{1}{6} q_{jj}^{ij} + \frac{1}{2} \sum_{m=1}^{j-1} q_{jm}^{ij} + \frac{1}{2} \sum_{k=j+1}^q q_{kj}^{ij} + \sum_{k=j+1}^q \sum_{m=1}^{j-1} q_{km}^{ij} \right\}. \qquad (2.23)$$

Let u, x, y, z be four points of U. Then using (2.18) we can easily have

$$[x, y, z]_{ijk} - [u, y, z]_{ijk} = \sum_{m=1}^q \left\{ [(x_1, ..., x_m, u_{m+1}, ..., u_q), y, z]_{ijk} \right.$$
$$\left. [(x_1, ..., x_{m-1}, u_m, ..., u_q), y, z]_{ijk} \right\}. \qquad (2.24)$$

If $m = j$ the terms in (2.24) vanish so that using (2.21) and (2.22) we deduce that for $k > j$

$$\left|[x,y,z]_{ijk} - [u,y,z]_{ijk}\right|$$
$$= \left|\sum_{m=1}^{j-1} \int_0^1 \int_0^1 \{D_{kj}f_i(x_1,...,x_m,u_{m+1},...,u_{j-1},y_j\right.$$
$$+s(x_j-y_j),y_{j+1},...,y_{k-1},z_k+t(y_k-z_k),z_{k+1},...,z_q)$$
$$-D_{kj}f_i(x_1,...,x_{m-1},u_m,...,u_{j-1},y_j$$
$$+s(x_j-y_j),y_{j+1},...,y_{k-1}z_k+t(y_k-z_k),z_{k+1},...,z_q)\} \, dsdt$$
$$+ \int_0^1 \int_0^1 \{D_{kj}f_i(x_1,...,x_{j-1},y_j+s(x_j-y_j),y_{j+1},...,y_{k-1},z_k$$
$$+t(y_k-z_k),z_{k+1},...,z_q)-D_{kj}f_i(x_1,...,x_{j-1},y_j$$
$$\left.+s(u_j-y_j),y_{j+1},...,y_{k-1},z_k+t(y_k-z_k),z_{k+1},...,z_q)\}\, dsdt\right|$$
$$\leq \frac{1}{2}|x_j-u_j|q_{kj}^{ij} + \sum_{m=1}^{j-1} |x_m-u_m|q_{km}^{ij}.$$

Similarly for $k = j$ we obtain in turn

$$\left|[x,y,z]_{ijj} - [u,y,z]_{ijj}\right|$$
$$= \left|\int_0^1 \int_0^1 t\{D_{jj}f_i(x_1,...,x_{j-1},x_j+t(y_j-x_j)+ts(z_j-y_j),z_{j+1},...,z_q)\right.$$
$$-D_{jj}f_i(x_1,...,x_{j-1},u_j+t(y_j-u_j)+ts(z_j-y_j),z_{j+1},...,z_q)\} \, dsdt$$
$$+\sum_{m=1}^{j-1}\int_0^1\int_0^1 t\{D_{jj}f_i(x_1,...,x_m,u_{m+1},...,u_{j-1},x_j+t(x_j-y_j)$$
$$+ts(z_j-y)j,z_{j+1},...,z_q)$$
$$-D_{jj}f_i(x_1,...,x_{m-1},u_m,...,u_{j-1},x_j+t(y_j-x_j)$$
$$\left.+ts(z_j-y_j),z_{j+1},...,z_q)\}\, dsdt\right|$$
$$\leq \tfrac{1}{6}|x_j-u_j|q_{jj}^{ij} + \tfrac{1}{2}\sum_{m=1}^{j-1}|x_m-u_m|q_{jm}^{ij}.$$

Finally using the estimate (2.10) of the norm of a bilinear operator, we deduce that condition (2.6) holds with c_2 given by (2.23).

2.4 Divided Difference and Monotone Convergence

In this section we make an introduction to the problem of approximating a locally unique solution x^* of the nonlinear operator equation $F(x) = 0$, in a POTL-space X. In particular, consider an operator $F : D \subseteq X \to Y$ where X is a POTL-space with values in a POTL-space Y. Let x_0, y_0, y_{-1} be three points of D such that

$$x_0 \leq y_0 \leq y_{-1}, \quad [x_0, y_{-1}],$$

and denote

$$D_1 = \{(x,y) \in X^2 \mid x_0 \leq x \leq y \leq y_0\},$$
$$D_2 = \{(y, y_{-1}) \in X^2 \mid x_0 \leq u \leq y_0\},$$
$$D_3 = D_1 \cup D_2. \qquad (2.25)$$

Assume there exist operators $A_0 : D_3 \to LB(X,Y), A : D_1 \to L(X,Y)$ such that:

(a)
$$F(y) - F(x) \leq A_0(w,z)(y-x) \; forall \; (x,y), (y,w) \in D_1, (w,z) \in D_3; \qquad (2.26)$$

(b) the linear operator $A_0(u,v)$ has a continuous nonsingular nonnegative left subinverse;

(c)
$$F(y) - F(x) \geq A(x,y)(y-x) \quad \text{for all} \quad (x,y) \in D_1; \qquad (2.27)$$

(d) the linear operator $A(x,y)$ has a nonegative left superinverse for each $(x,y) \in D_1$

$$F(y) - F(x) \leq A_0(y,z)(y-x) \quad \text{for all} \quad x,y \in D_1, (y,z) \in D_3. \qquad (2.28)$$

Moreover, let us define approximations

$$F(y_n) + A_0(y_n, y_{n-1})(y_{n+1} - y_n) = 0 \qquad (2.29)$$
$$F(x_n) + A_0(y_n, y_{n-1})(x_{n+1} - x_n) = 0 \qquad (2.30)$$
$$y_{n+1} = y_n - B_n F(y_n) \; n \geq 0 \qquad (2.31)$$
$$x_{n+1} = x_n - B_n^1 F(x_n) \quad n \geq 0, \qquad (2.32)$$

where B_n and B_n^1 are nonnegative subinverses of $A_0(y_n, y_{n-1})\; n \geq 0$.

Under very natural conditions, hypotheses the form (2.26) or (2.27) or (2.28) have been used extensively to show that the approximations (2.26) and (2.30) or (2.31) and (2.32) generate two sequences $\{x_n\}$ $n \geq 1$, $\{y_n\}$ $n \geq 1$ such that

$$x_0 \leq x_1 \leq \cdots \leq x_n \leq x_{n+1} \leq y_{n+1} \leq y_n \leq \cdots \leq y_1 \leq y_0 \quad (2.33)$$

$$\lim_{n \to \infty} x_n = x^* = y^* = \lim_{n \to \infty} y_n \quad \text{and} \quad F(x^*) = 0. \quad (2.34)$$

For a complete survey on these results we refer to the works of [237], Argyros and Szidarovszky [97]-[99].

Here we will use similar conditions (i.e. like (2.26), (2.27), (2.28)) for two-point approximations of the form (2.29) and (2.30) or (2.31) and (2.32).

Consequently a discussion must follow on the possible choices of the linear operators A_0 and A.

Remark 2.6 *Let us now consider an operator* $F : D \subseteq X \to Y$, *where* X, Y *are both POTL-spaces. The operator* F *is called order-complex on an interval* $[x_0, y_0] \subseteq D$ *if*

$$F(\lambda x + (1-\lambda) y) \leq \lambda F(x) + (1-\lambda) F(y) \quad (2.35)$$

for all comparable $x, y \in [x_0, y_0]$ *and* $\lambda \in [0, 1]$. *If* F *has a linear* G-*derivative* $F'(x)$ *at each point* $[x_0, y_0]$ *then* (2.35) *holds if and only if*

$$F'(x)(y-x) \leq F(y) - F(x) \leq F'(y)(y-x) \quad \text{for} \quad x_0 \leq x \leq y \leq y_0. \quad (2.36)$$

(See e.g. Ortega and Rheinboldt for the properties of the Gateux-derivative).

Hence, for order-convex G-differentiable operators conditions (2.27) and (2.28) are satisfied with $A_0(y, v) = A(y, v) = F'(u)$. In the unidimensional case (2.36) is equivalent with the isotony of the operator $x \to F'(x)$ but in general the latter property is stronger. Assuming the isotony of the operator $x \to F'(x)$ it follows that

$$F(y) - F(x) \leq F'(w)(y-x) \quad \text{for} \quad x_0 \leq x \leq y \leq w \leq y_0$$

Hence, in this case condition (2.26) is satisfied for $A_0(w, z) = F'(w)$.

The above observations show to choose A and A_0 for single or two-step Newton-methods. We note that the iterative algorithm (2.29)-(2.30) with

$$A_0(u, v) = F'(u)$$

is the algorithm proposed by Fourier in 1818 in the unidimensional case and extended by Baluev in 1952 in the general case. The idea of using an algorithm of the form (2.31)-(2.32) goes back to Slugin [256]. In Ortega and Rheinboldt [227] it is shown that with B_n properly chosen (2.31) reduces to a general Newton-SOR algorithm. In particular, suppose (in the finite dimensional case) that $F'(y_n)$ is an M-matrix and let $F'(y) = D_n - L_n - U_n$ be the partition of $F'(y_n)$ into diagonal, strictly lower - and strictly upper-triangular parts respectively for all $n \geq 0$. Consider an integer $m_n \geq 1$, a real parameter $w_n \in (0, 1]$ and denote

$$P_n = w_n^{-1}(D_n - w_n L_n), \quad Q_n = w_n^{-1}[(1 - w_n)D_n + w_n U_n], \quad (2.37)$$
$$H_n = P_n^{-1} Q_n, \quad \text{and} \quad B_n = (I + H_n + \cdots + H_n^{m_n - 1}) P_n^{-1}. \quad (2.38)$$

It can easily be seen that B_n $n \geq 0$ is a nonnegative subinverse of $F'(y_n)$ (see, also [99]). If $f : [a, b] \to |R$ is a real function of a real variable, then f is (order) convex if and only if

$$\frac{f(x) - f(y)}{x - y} \leq \frac{f(u) - f(v)}{u - v}$$

for all x, y, u, v from $[a, b]$ such that $x \leq u$ and $y \leq v$.

This fact motivates the notion of convexity with respect to a divided difference considered by J.W. Schmidt and H. Leonhardt [253]. Let $F : D \subseteq X \to Y$ be a nonlinear operator between the POTL-spaces X and Y. Assume that the nonlinear operator F has a divided difference $[\cdot, \cdot]$ on D. F is called convex with respect to the divided difference $[\cdot, \cdot]$ on D if

$$[x, y] \leq [u, v] \quad \text{for all} \quad x, y, u, v \in D \quad \text{with} \quad x \leq u \quad \text{and} \quad y \leq v. \quad (2.39)$$

In the above quoted study, Schmidt and Leonhardt studied (2.29)-(2.30) with $A_0(u, v) = [u, v]$ in case the nonlinear operator F is convex with respect to $[\cdot, \cdot]$. Their result was extended by N. Schneider [254] who assumed instead of (1.2.1) the milder condition

$$[u, v](u - v) \geq F(u) - F(v) \quad \text{for all comparable} \quad u, v \in D. \quad (2.40)$$

An operator $[\cdot, \cdot] : D \times D \to L(X, Y)$ satisfying (2.40) is called a generalized divided difference of F on D. If both (2.39) and (2.40) are satisfied, then we say that F is convex with respect to the generalizeddivided difference of $[\cdot, \cdot]$. It is easily seen that if (2.39) and (2.40) are satisfied on $D = [x_0, y_{-1}]$ then conditions (2.26) and (2.27) are satisfied with $A = A_0 = [\cdot, \cdot]$. Indeed for $x_0 \leq x \leq y \leq w \leq z \leq y_{-1}$ we have

$$[x, y](y - x) \leq F(y) - F(x) \leq [y, x](y - x) \leq [w, z](y - x).$$

Moreover, concerning general secant-SOR methods, in case the generalized difference $[y_n, y_{n-1}]$ is an M-matrix and if B_n $n \geq 0$ is computed according to (2.37) and (2.38) where $[y_n, y_{n-1}] = D_n - L_n - U_n$ $n \geq 0$ is the partition of $[y_n, y_{n-1}]$ into its diagonal, strictly lower- and strictly upper-triangular parts.

We remark that an operator which is convex with respect to a generalized divided difference is also order-convex. To see that, consider $x, y \in D$, $x \leq y$, $\lambda \in [0,1]$ and set $z = \lambda x + (I - \lambda) y$. Observing that $y - x = (1 - \lambda)^{-1} (z - x) = \lambda^{-1} (y - z)$ and applying (2.40) we have in turn:

$$(1 - \lambda)^{-1} (F(z) - F(x)) \leq (1 - \lambda)^{-1} [z, x] (z - x)$$
$$= [z, x] (y - x) \leq [z, y] (y - x)$$
$$= \lambda^{-1} [z, y] (y - z) \leq^1 (F(y) - F(z)).$$

By the first and last term we deduce that $F(z) \leq \lambda F(x) + (1 - \lambda) F(y)$. Thus, Schneider's result can be applied only to order-convex operators and its importance resides in the fact that the use of a generalized divided difference instead of the G-derivative may be more advantageous from a numerical point of view. We note however, that conditions (2.26) and (2.27) do not necessarily imply convexity. For example, if f is a real function of a real variable such that

$$\inf_{x,y \in [x_0, y_0]} \frac{f(x) - f(y)}{x - y} = m > 0, \qquad \sup_{x,y \in [x_0, y_0]} \frac{f(x) - f(y)}{x - y} = M < \infty$$

then (2.26) and (2.27) are satisfied for $A_0(u,v) = M$ and $A(u,v) = m$. It is not difficult to find examples of nonconvex operators in the finite (or even in the infinite) dimensional case satisfying a condition of the form

$$A(y - x) \leq F(y) - F(x) \leq A_0(y, x), \quad x_0 \leq x \leq y \leq y_0$$

where A_0 and A are fixed linear operators. If A_0 has a continuous nonsingular nonnegative left subinverse and A has a nonnegative right superinverse, then convergence of the algorithm (2.29)-(2.30) can be discussed. This algorithm becomes extremely simple in this case. The monotone convergence of such an iterative procedure seems to have been first investigated by S. Slugin [256].

In the end of this section we shall consider a class of nonconvex operators which satisfy condition (2.26) but do not necessarily satisfy condition (2.27). Consequently from convergence theorems involving (2.29) and (2.30) it will follow that Jacobi-Newton and the Jacobi-secant methods have monotonous convergence for operators belonging to this class (see, also the elegenat papers by F. Potra [235]-[237].

Let $F = (f_1, ..., f_q)^{T0}$ be an operator acting in the finite dimensional space $|R^q$, endowed with the natural (componentwise) partial ordering. Let us denote by e_i the i–th coordinate vector of $|R^q$. We say that F is off-diagonally antitone if the functions

$$g_{ij} : |R \to |R, \quad g_{ij}(t) = f_i(x + te_j), \quad i \neq j, \quad i,j = 1, ..., q$$

are antitone. Suppose that at each point x belonging to an interval $U = [x_0, y_{-1}]$ the partial derivatives $\partial_i F_i(x)$, $i = 1, 2, ..., q$, exist and are positive. For any two points $x, y \in U$ we consider the quotients

$$[x,y]_i = \begin{cases} \frac{f_i(y) - f_i\left(y - e_i^T(y-x)e_i\right)}{e_i^T(y-x)}, & \text{if } e_i^T(y-x) \neq 0 \\ \partial_i F_i(x) & \text{if } e_i^T(y-x) = 0. \end{cases} \quad (2.41)$$

Let us denote $\Delta[x,y]$ the diagonal matrix having as elements the number $[x,y]_i$, $i = 1, 2, ..., q$. For the diagonal matrix $\Delta[x,y]$ formed by the partial derivatives $\partial_i f_i(x)$, $i = 1, 2, ..., q$, we shall also use the notation $DF(x)$.

Suppose now that F is off-diagonally antitone and that the operator $DF : U \to LB(|R^q)$ is isotone (i.e. all functions $\partial_i F_i : |R \to |R)$ are isotone). In this case for all $x_0 \leq x \leq y \leq w \leq z \leq y_{-1}$ and all $i \in \{1, 2, ..., q\}$ there exist $\lambda, \mu \in [0, 1]$ such that

$$f_i(y) - f_i(x) \leq f_i(y) - f_i\left(y - e_i^T(y-x)e_i\right) = \partial_i f_i\left(y - \lambda e_i^T(y-x)e_i\right),$$
$$e_i^T(y-x) \leq \partial_i f_i(y) e_i^T(y-x) \leq \partial_i f_i(w) e_i^T(y-x)$$
$$\leq \partial_i f_i\left(z - \mu e_i^T(z-w)\right) e_i^T(y-x) = [w,z]_i e_i^T(y-x).$$

It follows that condition (2.26) is satisfied for $A_0(w,z) = DF(w)$ as well as for $A_0(w,z) = [\Delta w, z]$. With the choice $A_0(w,z) = \Delta[w,z]$ the iterative procedure (2.29) is a Jacobi-secant method while with the choice $A_0(w,z) = DF(w)$ it reduces to the Jacobi-Newton method. For some applications of the latter method, see W. Torning [263].

2.5 Divided Differences and Fréchet-derivatives

Let F be a nonlinear operator defined on an open subset D of a Banach space X with values in a Banach space Y. Choose also $x_0 \in D$ to be fixed.

Definition 2.5 The operator $F : D \subseteq X \to Y$ is Fréchet - differentiable at the point $x_0 \in D$ if there exists a linear operator $L \in LB(X,Y)$ such that

$$\lim_{\|h\| \to 0} \frac{1}{\|h\|} \|F(x_0 + h) - F(x_0) - A(h)\| = 0.$$

It is easy to see that such an operator is unique if it exists. This operator L is called the first Fréchet-derivative of F at the point x_0 and is denoted by $F'(x_0)$.

The following facts are left as exercises:

1. If F is Fréchet differentiable at the point x_0 then F is continuous at this point.
2. If $A \in LB(X,Y)$ and $F(x)$ for all $x \in X$, then F is Fréchet-differentiable at every point $x \in X$ and $F'(x) = A$.
3. Let F and G be two operators defined on D with values in Y. If F and G are Fréchet-differentiable at a point $x_0 \in D$, then the operator $F + G$ is also Fréchet-differentiable at x_0 and $(F+G)'(x_0) = F'(x_0) + G'(x_0)$.
4. Let X, y and Z be three Banach spaces and consider the operators $F : D_1 \subseteq X \to Y$, $G : D_2 \subseteq Y \to Z$. Let $x_0 \in D_1^0$ such that $y_0 = F(x_0)$ is a point in D_2^0. If F is Fréchet-differentiable at x_0 and G is Fréchet differentiable at y_0 then the mapping $H = G_0 F$ is Fréchet-differentiable at x_0 and $H'(x_0) = G'(y_0) F'(x_0)$.

If the operator F is Fréchet-differentiable at all $x \in D$, then we shall say that F is Fréchet-differentiable on D. In this case we may consider an operator $F' : D \to LB(X,Y)$ which associates to each point $x \in D$ the Fréchet-derivative of F at x. This operator will be called the first Fréchet-derivative of F.

Let x, y be two points in D and suppose that the segment

$$S = \{x + t(y-x) \,|\, t \in [0,1]\} \subseteq D.$$

Let y' be a continuous linear functional, set $h = y - x$ and define

$$\varphi(t) = (F(x+th), y').$$

If F is fréchet-differentiable at each point of the segment S, then φ is differentiable on $[0,1]$ and

$$\varphi(t) = (F'(x+th), y').$$

Let us now suppose that

$$\alpha = \sup_{t\in[0,1]} \|F'(x+t(y-x))\| < \infty;$$

then we have

$$\|(F(y)-F(x),y')\| = \|\varphi(1)-\varphi(0)\|$$
$$\leq \sup_{t\in[0,1]} \|\varphi'(t)\| \leq \alpha \|y'\| \|y-x\|.$$

But, we also have

$$\|F(y)-F(x)\| = \sup_{\|y'\|\leq 1} \|(F(y)-F(x),y')\|,$$

we deduce that

$$\|F(y)-F(x)\| \leq \sup_{t\in[0,1]} \|F'(x+t(y-x))\| \cdot \|y-x\| \tag{2.42}$$

so, we proved:

Theorem 2.2 *Let D be a convex subset of a Banach space X and $F: D \subseteq X \to Y$. If F is Fréchet-differentiable on D and if there exists a constant c such that*

$$\|F'(x)\| \leq M \quad \text{for all} \quad x \in D \Rightarrow \|F(x)-F(y)\| \leq c\|x-y\|$$
$$\text{for all} \quad x \in D. \tag{2.43}$$

The estimate (2.42) is the analogue of the famous mean value formula from real analysis. If the operator F' is Riemann integrable on the segment S we can give the following integral representation of the mean value formula

$$F(x)-F(y) = \int_0^1 F'(x+t(y-x))\,dt\,(x-y). \tag{2.44}$$

Let now D be a convex open subset of X and let us suppose that we have associated to each pair (x,y) of distinct points from D a divided difference $[x,y]$ of F at these points. In applications one often has to require that the operator $(x,y) \to [x,y]$ satisfy a Lipschitz condition (see also Section 2.3). We suppose that there exists a nonnegative $c > 0$ such that

$$\|[x,y]-[x_1,y_1]\| \leq c(\|x-x_1\|+\|y-y_1\|) \tag{2.45}$$

for all $x, y, x_1, y_1 \in D$ with $x \neq y$ and $x_1 = y_1$.

We say in this case that F has a Lipschitz continuous difference on D. This condition allows us to extend by continuity W the operator $(x,y) \to [x,y]$ to the whole Cartesian product $D \times D$. From (1.2.1) and (2.45) it follows that F is Fréchet-differentiable on D and that $[x,x] = F'(x)$. It also follows that

$$\|F'(x) - F'(y)\| \le c_1 \|x-y\| \quad \text{with} \quad c_1 = 2c \tag{2.46}$$

and

$$\|[x,y] - F'(z)\| \le c(\|x-z\| + \|y-z\|) \tag{2.47}$$

for all $x, y \in D$. conversely if we assume that F is Fréchet-differentiable on D and that its Frechet derivative satisfies (2.46) then it follows that F has a Lipschitz continuous divided difference on D. We can certainly take

$$[x,y] = \int_0^1 F'(x + t(y-x))\,dt. \tag{2.48}$$

We now want to give the definition of the second Frechet derivative of F. We must first introduce the definition of bounded multilinear operators (which will also be used later).

Definition 2.6 Let X and Y be two Banach spaces. An operator $A: X^n \to y$ $(n \in N)$ will be called n-linear operator from X to Y if the following conditions are satisfied:

(a) The operator $(x_1, ..., x_n) \to A(x_1, ..., x_n)$ is linear in each variable x_k $k = 1, 2, ..., n$.
(b) There exists a constant c such that

$$\|A(x_1, x_2, ..., x_n)\| \le c \|x_1\| \dots \|x_n\| \tag{2.49}$$

The norm of a bounded n-linear operator can be defined by the formula

$$\|A\| = \sup\{\|A(x_1, ..., x_n)\| \mid \|x_n\| = 1\}. \tag{2.50}$$

Set $LB^{(1)}(X,Y) = LB(X,Y)$ and define recursively

$$LB^{(k+1)}(X,Y) = LB\left(X, LB^{(k)}(X,Y)\right), \quad k \ge 0. \tag{2.51}$$

In this way we obtain a sequence of Banach spaces $LB^{(n)}(X,Y)$ $(n \geq 0)$. Every $A \in LB^{(n)}(X,Y)$ can be viewed as a bounded n-linear operator if one takes

$$A(x_1, ..., x_n) = (...(A(x_1)(x_2)(x_3))...)(x_n). \qquad (2.52)$$

In the right hand side of (2.52) we have

$$A(x_1) \in LB^{(n-1)}(X,Y), \quad (A(x_1))(x_2) \in LB^{(n-2)}(X,Y), \quad \text{etc.}$$

Conversely, any bounded n-linear operator A drom X to Y can be interpreted as an element of $B^{(n)}(X,Y)$. Moreover the norm of A as a bounded n-linear operator coincides with the norm as an element of the space $LB^{(n)}(X,Y)$. Thus we may identify this space with the space of all bounded $n-$linear operators from X to Y. In the sequel we will identify $A(x, x, ..., x) = Ax^n$, and

$$A(x_1)(x_2)...(x_n) = A(x_1, x_2, ..., x_n) = Ax_1x_2...x_n.$$

Let us now consider a nonlinear operator $F : D \subseteq X \to Y$ where D is open. Suppose that F is Fréchet differentiable on D. Then we may consider the operator $F' : D \to LB(X, Y)$ which associated to each point x the Fréchet derivative of F at x. If the operator F' is Fréchet differentiable at a point $x_0 \in D$ then we say that F is twice Fréchet differentiable at x_0. The Fréchet derivative of F' at x_0 will be denoted by $F''(x_0)$ and will be called the second Fréchet derivative of F at x_0. Note that $F''(x_0) \in LB^{(2)}(X,Y)$. Similarly we can define Fréchet derivatives of higher order. Finally by analogy with (2.48)

$$[x_0, ..., x_k] =$$
$$= \int_0^1 \cdots \int_0^1 t_1^{k-1} t_2^{k-2} \cdots t_{k-1} F(x_0 + t_1(x_1 - x_0) + t_1 t_2(x_2 - x_1)$$
$$+ \cdots + t_1 t_2, ..., t_k(x_k - x_{k-1})) dt_1 dt_2 ... dt_k. \qquad (2.53)$$

It is easy to see that the multilinear operators defined above verify

$$[x_0, ..., x_{k-1}, x_k, x_{k+1}](x_k - x_{k+1}) = [x_0, ..., x_{k-1}, x_{k+1}]. \qquad (2.54)$$

We note that throughout this sequel a 2−linear operator will also be called bilinear.

Finally, we will also need the definition of a n−linear symmetric operator. Given a n-linear operator $A : X^n \to Y$ and a permutation $i = (i_1, i_2, ..., i_n)$ of the integers $1, 2, ..., n$, the notation $A(i)$ (or $A_n(i)$ if we want to emphasize the n−linearity of A) can be used for the n−linear operator $A(i) = A_n(i)$ such that

$$A(i)(x_1, x_2, ..., x_n) = A_n(i)(x_1, x_2, ..., x_n)$$
$$= A_n(x_{i_1}, x_{i_2}, ..., x_{i_n}) A_n x_{i_1} x_{i_2} ... x_{i_n} \quad (2.55)$$

for all $x_1, x_2, ..., x_n \in X$. Thus, there are $n!$ n−linear operators $A(i) = A_n(i)$ associated with a given n−linear operator $A = A_n$.

Definition 2.7 A n−linear operator $A = A_n : X^n \to Y$ is said to be symmetric if

$$A = A_n = A_n(i) \quad (2.56)$$

for all i belonging in R_n which denotes the set of all permutations of the integers $1, 2, ..., n$. The symmetric n−linear operator

$$\overline{A} = \overline{A}_n = \frac{1}{n!} \sum_{i \in R_n} A_n(i) \quad (2.57)$$

is called the mean of $A = A_n$.

Definition 2.8 A n−linear operator $A = A_n : X^n \to Y$ is said to be symmetric if

$$A = A_n = A_n(i)$$

for all i belonging in R_n which denotes the set of all permutations of the integers $1, 2, ..., n$. The symmetric n−linear operator

$$\overline{A} = \overline{A}_n = \frac{1}{n!} \sum_{i \in R_n} A_n(i) \quad (2.58)$$

is called the mean of $A = A_n$.

Notation 2.1 The notation

$$A_n X^P = A_n x x ... x \text{ (}p\text{-times)} \quad (2.59)$$

$p \leq n$, $A = A_n : X^n \to Y$, for the result of applying A_n to $x \in X$ p−times will be used. If $p < n$, then (2.58) will represent a $(n - p)$−linear operator. For $p = n$, note that

$$A_k x^k = \overline{A}_k x^k = A_k(i) x^k \quad (2.60)$$

for all $i \in R_k$, $x \in X$. It follows from (2.59) that whenever we are dealing with an equation involving n-linear operators A_n, we may assume that they are symmetric without loss of generality, since each A_n may be replaced by $\overline{A}_n$ without changing the value of the expression at hand.

2.6 Enclosing the Root of a Nonlinear Equation

We consider the equation

$$L(x) = T(x) \qquad (2.61)$$

where L is a linear operator and T is a nonlinear operator defined on some convex subset D of a linear space E with values in a linear space Y.

We study the convergence of the iterations

$$L(y_{n+1}) = T(y_n) + A_n(y,x)(y_{n+1} - y_n) \qquad (2.62)$$

and

$$L(x_{n+1}) = T(x_n) + A_n(y,x)(x_{n+1} - x_n) \qquad (2.63)$$

to a solution x^* of equation (2.61), where $A_n(y,x)$, $n \geq 0$ is a linear operator.

If P is a linear projection operator $(P^2 = P)$, that projects the space Y into $Y_P \subseteq Y$, then the operator PT will be assumed to be Fréchet differentiable on D and its derivative $PT'_x(x)$ corresponds to the operator $PB(y,x)$, $y,x \in D \times D$, $PT'_X(x) = PB(x,x)$ for all $x \in D$.

We will assume that

$$A_n(y,x) = APB(y_n, x_n), \quad n \geq 0.$$

Iterations (2.62) and (2.63) have been studied extensively under several assumptions [68], [99], [101], [227], [236], [237], when $P = L = I$, the identity operator on D. However, the iterates $\{x_n\}$ and $\{y_n\}$ can rarely be computed in infinite dimensional spaces in this case. But if the space Y_P is finite dimensional with $\dim(Y_P) = N$, then iterations (2.62) and (2.63) reduce to systems of linear algebraic equations of order at most N. This case has been studied in [101], [182] and in particular in [194]. The assumptions in [199] involve the positivity of the operators $PB(y,x) - APB(y,x)$, $QT'_x(x)$ with $Q = I - P$ and $L(y) - A_n(y,x)y$ on some interval $[y_0, x_0]$, which is difficult to verify.

In this section we simplify the above assumptions and provide some further conditions for the convergence of iterations (2.62) and (2.63) to a solution x^* of equation (2.61).

We finally illustrate our results with an example.

We can now formulate our main result.

Theorem 2.3 *Let $F = D \subset X \to Y$, where X is a regular POTL-space and Y is a POTL-space. Assume*

(a) *there exist points $x_0, y_0, y_{-1} \in D$ with*

$$x_0 \leq y_0 \leq y_{-1}, \quad [x_0, y_{-1}] \subset D, \quad L(x_0) - T(x_0) \leq 0 \leq L(y_0) - T(y_0).$$

Set

$$S_1 = \{(x, y) \in X^2; x_0 \leq x \leq y \leq y_0\},$$
$$S_2 = \{(u, y_{-1}) \in X^2; x_0 \leq u \leq y_0\}$$

and

$$S_3 = S_1 \cup S_2.$$

(b) *Assume that there exists an operator $A: S_3 \to B(X, Y)$ such that*

$$(L(y) - T(y)) - (L(x) - T(x)) \leq A(w, z)(y - x) \qquad (2.64)$$

for all $(x, y), (y, w) \in S_2, (w, z) \in S_3$.

(c) *Suppose that for any $(u, v) \in S_3$ the linear operator $A(u, v)$ has a continuous nonsingular nonnegative left subinverse.*

Then there exist two sequences $\{x_n\}, \{y_n\}, n \geq 1$ satisfying (2.62), (2.63),

$$L(x_n) - T(x_n) \leq 0 \leq L(y_n) - T(y_n), \qquad (2.65)$$
$$x_0 \leq x_1 \leq \cdots \leq x_n \leq x_{n+1} \leq y_{n+1} \leq y_n \leq \cdots \leq y_1 \leq y_0 \qquad (2.66)$$

and

$$\lim_{n \to \infty} x_n = x^*, \quad \lim_{n \to \infty} y_n = y^*. \qquad (2.67)$$

Moreover, if the operators $A_n = A(y_n, y_{n-1})$ are inverse nonnegative then any solution of the equation (2.61) from the interval $[x_0, y_0]$ belongs to the interval $[x^*, y^*]$.

Proof. Let us define the operator $M : [0, y_0 - x_0] \to X$ by
$$M(x) = x - L_0(L(x_0) - T(x_0) + A_0(x))$$
where L_0 is a continuous nonsingular nonnegative left subinverse of A_0. It can easily be seen that M is isotone, continuous with
$$M(0) = -L_0(L(x_0) - T(x_0)) \geq 0$$
and
$$\begin{aligned} M(y_0 - x_0) &= y_0 - x_0 - L_0(L(y_0) - T(y_0)) \\ &\quad + L_0[(L(y_0) - T(y_0)) - (L(x_0) - T(x_0)) - A_0(y_0 - x_0)] \\ &\leq y_0 - x_0 - L_0(L(y_0) - T(y_0)) \\ &\leq y_0 - x_0. \end{aligned}$$

It now follows from the theorem of L.V. Kantorovich on fixed points that the operator M has a fixed point $w \in [0, y_0 - x_0]$. Set $x_1 = x_0 + w$ to get
$$L(x_0) - T(x_0) + A_0(x_1 - x_0) = 0, \quad x_0 \leq x_1 \leq y_0.$$

By (2.64), we get
$$L(x_1) - T(x_1) = (L(x_1) - T(x_1)) - (L(x_0) - T(x_0)) + A_0(x_0 - x_1) \leq 0.$$

Let us now define the operator $M_1 : [0, y_0 - x_1] \to X$ by
$$M_1(x) = x + L_0(L(y_0) - T(y_0) - A_0(x)).$$

It can easily be seen that M_1 is continuous, isotone with
$$M_1(0) = L_0(L(y_0) - T(y_0)) \geq 0$$
and
$$\begin{aligned} M_1(y_0 - x_1) &= y_0 - x_1 + L_0(L(x_1) - T(x_1)) \\ &\quad + L_0[(L(y_0) - T(y_0)) - (L(x_1) - T(x_1)) - A_0(y_0 - x_1)] \\ &\leq y_0 - x_1 + L_0(L(x_1) - T(x_1)) \\ &\leq y_0 - x_1. \end{aligned}$$

As before, there exists $z \in [0, y_0 - x_1]$ such that $M_1(z) = z$. Set $y_1 = y_0 - z$ to get
$$L(y_0) - T(y_0) + A_0(y_1 - y_0) = 0, \quad x_1 \leq y_1 \leq y_0.$$

But from (2.64) and the above

$$L(y_1) - T(y_1) = (L(y_1) - T(y_1)) - (L(y_0) - T(y_0)) - A_0(y_1 - y_0) \geq 0.$$

Using induction on n we can now show, following the above technique, that there exist sequences $\{x_n\}$, $\{y_n\}$, $n \geq 1$ satisfying (2.61), (2.62), (2.65) and (2.66). since the space X is regular (2.66) we get that there exist $x^*, y^* \in E$ satisfying (2.67), with $x^* \leq y^*$. Let $x_0 \leq z \leq y_0$ and $L(z) - T(z) = 0$ then we get

$$\begin{aligned} A_0(y_1 - z) &= A_0(y_0) - (L(y_0) - T(y_0)) - A_0(z) \\ &= A_0(y_0 - z) - [(L(y_0) - T(y_0)) - (L(z) - T(z))] \geq 0 \end{aligned}$$

and

$$\begin{aligned} A_0(x_1 - z) &= A_0(x_0) - (L(x_0) - T(x_0)) - A_0(z) \\ &= A_0(x_0 - z) - [(L(x_0) - T(x_0)) - (L(z) - T(z))] \leq 0. \end{aligned}$$

If A_0 is inverse isotone, then $x_1 \leq z \leq y_1$ and by induction $x_n \leq z \leq y_n$. Hence $x^* \leq z \leq y^*$.

That completes the proof of the theorem. □

Using (2.61), (2.62), (2.65), (2.66) and (2.67) we can easily prove the following theorem which gives us natural conditions under which the points x^* and y^* are solutions of the equation (2.61).

Theorem 2.4 *Let $L - T$ be continuous at x^* and y^* and the hypotheses of Theorem 2.3 be true. Assume that one of the conditions is satisfied:*

(a) $x^* = y^*$;
(b) X *is normal and there exists an operator $H : X \to Y$ ($H(0) = 0$) which has an isotone inverse continuous at the origin and $A_n \leq H$ for sufficiently large n;*
(c) Y *is normal and there exists an operator $G : X \to Y$ ($G(0) = 0$) continuous at the origin and such that $A_n \leq G$ for sufficiently large n.*
(d) *The operators L_n, $n \geq 0$ are equicontinuous.*

Then $L(x^*) - T(x^*) = L(y^*) - T(y^*) = 0$.

Moreover, assume that there exists an operator $G_1 : S_1 \to L(X, Y)$ such that $G_1(x, y)$ has a nonnegative left superinverse for each $(x, y) \in S_1$ and

$$L(y) - T(y) - (L(x) - T(x)) \geq G_1(x, y)(y - x) \quad \text{for all} \quad (x, y) \in S_1.$$

Then if $(x^*, y^*) \in S_1$ and $L(x^*) - T(x^*) = L(y^*) - T(y^*) = 0$ then $x^* = y^*$.

We now complete this paper with an application.

APPLICATIONS

Let $X = Y = \mathbb{R}^k$ with $k = 2N$. We define a projection operator P_N by

$$P_N(v) = \begin{cases} v_i, & i = 1, 2, \ldots, N \\ 0, & i = N+1, \ldots, k, \end{cases} \quad v = (v_1, v_2, \ldots, v_k) \in X.$$

We consider the system of equations

$$v_i = f_i(v_1, \ldots, v_k), \quad i = 1, 2, \ldots, k. \tag{2.68}$$

Set $T(v) = \{f_i(v_1, \ldots, v_k)\}, \quad i = 1, 2, \ldots, k,$ then

$$P_N T(v) = \begin{cases} f_i(v_1, \ldots, v_k) & i = 1, \ldots, N, \\ 0 & i = N+1, \ldots, k, \end{cases}$$

$$P_N T'(v) u = \begin{cases} \sum_{j=1}^{k} f'_{ij}(v_1, \ldots, v_k) u_j & i = 1, 2, \ldots, N, \\ 0 & i = N+1, \ldots, k, \end{cases} \quad f'_{ij} = \frac{\partial f_i}{\partial v_j},$$

$$P_N B(w, z) u = \begin{cases} \sum_{j=1}^{k} F_{ij}(w_1, \ldots, w_k, z_1, \ldots, z_k) u_j, & i = 1, \ldots, N \\ 0, & i = N+1, \ldots, k \end{cases}$$

$$= C_N^i(w, z) u,$$

where $F_{ij}(v_1, \ldots, v_k, v_1, \ldots, v_k) = \partial f_i(v_1, \ldots, v_k) / \partial v_j$. Choose

$$A_n(y, x) = C_N^i(y, x),$$

then iterations (2.62) and (2.63) become

$$y_{i,n+1} = f_i(y_{1,n}, \ldots, y_{k,n}) + C_N^i(y_{i,n}, x_{i,n})(y_{i,n+1} - y_{i,n}) \tag{2.69}$$

$$x_{i,n+1} = f_i(x_{1,n}, \ldots, x_{k,n}) + C_N^i(y_{i,n}, x_{i,n})(x_{i,n+1} - x_{i,n}). \tag{2.70}$$

Let us assume that the determinant D_n of the above N-th order linear systems is nonzero, then (2.69) and (2.70) become

$$y_{i,n+1} = \frac{\sum_{m=1}^{N} D_{im} F_m^1(y_n, x_n)}{D_n}, \quad i = 1, \ldots, N \tag{2.71}$$

$$y_{i,n+1} = f_i(y_{1n}, \ldots, y_{kn}), \quad i = N+1, \ldots, k \tag{2.72}$$

and

$$x_{i,n+1} = \frac{\sum_{m=1}^{N} D_{im} F_m^2(y_n, x_n)}{D_n}, \quad i = 1, \ldots, N, \qquad (2.73)$$

$$x_{i,n+1} = f_i(x_{1,n}, \ldots, x_{k,n}), \quad i = N+1, \ldots, k \qquad (2.74)$$

respectively. Here D_{im} is the cofactor of the element in the i-th column and m-th row of D_n and $F_m^1(y_n, x_n)$, $i = 1, 2$ are given by

$$F_m^1(y_n, x_n) = f_m(y_{1,n}, \ldots, y_{k,n}) + \sum_{i=N+1}^{k} \alpha_{mj}^n f_j(y_{1,n}, \ldots, y_{k,n}) - \sum_{j=1}^{k} \alpha_{mj}^n y_{j,n},$$

and

$$F_m^2(y_n, x_n) = f_m(x_{1,n}, \ldots, x_{k,n}) + \sum_{i=N+1}^{k} \alpha_{mj}^n f_j(x_{1,n}, \ldots, x_{k,n}) - \sum_{j=1}^{k} \alpha_{mj}^n x_{j,n},$$

where $\alpha_{mj}^n = F_{mj}(y_n, x_n)$.

If the hypotheses of Theorem 2.3 and 2.4 are now satisfied for the equation (2.68) then the results apply to obtain a solution x^* of equation (2.68) $[y_0, x_0]$.

In particular consider the system of differential equations

$$q_i'' = f_i(t, q_1, q_2), \quad i = 1, 2, \quad 0 \le t \le 1 \qquad (2.75)$$

subject to the boundary conditions

$$q_i(0) = d_i, \quad q_i(1) = e_i, \quad i = 1, 2, . \qquad (2.76)$$

The functions f_1 and f_2 are assumed to be sufficiently smooth. For the discretization a uniform mesh

$$t_j = jh, \quad j = 0, 1, \ldots, N+1, \quad h = \frac{1}{N+1}$$

and the corresponding central-difference approximation of the second derivatives are used. Then the discretized given by

$$x = T(x) \qquad (2.77)$$

with

$$T(x) = (B+I)(x) = h^2 \varphi(x) - b, \quad x \in \mathbb{R}^{2N}$$

where

$$B = \begin{bmatrix} A+I & 0 \\ 0 & A+I \end{bmatrix}, \quad A + \begin{bmatrix} 2 & -1 & 0 & \\ -1 & 2 & \ddots & \\ & \ddots & \ddots & -1 \\ 0 & & -1 & 2 \end{bmatrix},$$

$$\varphi(x) = \begin{bmatrix} \varphi_1(x) \\ \varphi_2(x) \end{bmatrix}, \quad \varphi_i(x) = (f_i(t, x_j, x_{n+j})), \quad j = 1, 2, ..., N, \quad i = 1, 2,$$

$x \in \mathbb{R}^{2N}$, and $b \in \mathbb{R}^{2N}$ is the vector of boundary values that has zero components except for $b_1 = d_1$, $b_n = e_1$, $b_{n+1} = d_2$, $b_{2n} = e_2$. That is (2.77) plays the role of (2.68) (in vector form).

As a numerical example, consider the problem (2.75)-(2.76) with

$$f_1(t, q_1, q_2) = q_1^2 + q_1 + .1q_2^2 - 1.2$$
$$f_2(t, q_1, q_2) = .2q_1^2 + q_2^2 + 2q_2 - .6$$
$$d_1 = d_2 = e_1 = e_2 = 0.$$

Choose $N = 49$ and starting points

$$x_{i,0} = 0, \quad y_{i,0} = (t_j(1 - t_j)), \quad j = 1, ..., N), \text{ with } t = .1(.1).5.$$

It is trivial to check that the hypotheses of Theorem 2.3 are satisfied with the above values. Furthermore the components of the first two iterates corresponding to the above values using the procedure described above and (2.71)-(2.72), (2.73)-(2.74) we get the following values.

	t	$p=1$	$p=2$
	.1	.0478993317265	.0490944353538
	.2	.0843667040291	.0866974354188
$x_{1,p}$	.3	.1099518629493	.1132355483832
	.4	.1251063839064	.1290273001442
	.5	.1301240123325	.1342691068706
	.1	.0219768501208	.0227479238400
	.2	.0384462112803	.0399528292723
$x_{2,p}$	.3	.0498537074028	.0519796383151
	.4	.0565496306877	.0590905187490
	.5	.0587562590344	.0614432572165
	.1	.0494803602542	.0490951403091
	.2	.0874507511044	.0866988216544
$y_{1,p}$	.3	.1142981809478	.1132375255317
	.4	.1302974325097	.1290296859551
	.5	.1356123753407	.1342716394060
	.1	.0235492475283	.0227486289905
	.2	.0415200498433	.0399542200344
$y_{2,p}$	.3	.0541939935471	.0519816281202
	.4	.0617399319012	.0590929252230
	.5	.0642461600398	.0614458137439

The computations were performed in double precision on a PRIME-850

2.7 Exercises

2.1. Show that the spaces defined in Examples 2.1 are POTL.

2.2. Show that any regular POB-space is normal but the converse is not necessarily true.

2.3. Prove Theorem 2.1.

2.4. Show that if (2.4) or (2.5) are satisfied then $F'(x) = [x,x]$ for all $x \in D$. Moreover show that if both (2.4) and (2.5) are satisfied, then F' is Lipschitz continuous with $l = c_0 + c_1$.

2.5. Find sufficient conditions so that estimates (2.33) and (2.34) are both satisfied.

2.6. Show that B_n ($n \geq 0$) in (2.38) is a nonnegative subinverse of $F'(y_n)$ ($n \geq 0$).

2.7. Let $x_0, x_1, ..., x_n$ be distinct real numbers, and let f be a given real-valued function. Show that:

$$[x_0, x_1, ..., x_n] = \sum_{j=0}^{n} \frac{f(x_j)}{g'_n(x_j)}$$

and

$$[x_0, x_1, ..., x_n](x_n - x_0) = [x_1, ..., x_n] - [x_0, ..., x_{n-1}]$$

where

$$g_n(x) = (x - x_0) ... (x - x_n).$$

2.8. Let $x_0, x_1, ..., x_n$ be distinct real numbers, and let f be n times continuously differentiable function on the interval $I\{x_0, x_1, ..., x_n\}$. Then show that

$$[x_0, x_1, ..., x_n] = \int_{T_n} \cdots \int f^{(n)}(t_0 x_0 + \cdots + t_n x_n) dt_1 \cdots dt_n$$

in which

$$T_n = \{(t_1, ..., t_n) \mid t_1 \geq 0, ..., t_n \geq 0, \sum_{i=1}^{n} t_i \leq 1\}$$
$$t_0 = 1 - \sum_{i=1}^{n} t_i.$$

2.9. If f is a real polynomial of degree m, then show:

$$[x_0, x_1, ..., x_n, x] = \begin{cases} \text{polynomial of degree } m - n - 1, & n \leq m - 1 \\ a_m & n = m - 1 \\ 0 & n > m - 1 \end{cases}$$

where $f(x) = a_m x^n +$ lower-degree terms.

2.10. The tensor product of two matrices $M, N \in L(\mathbf{R}^n)$ is defined as the $n^2 \times n^2$ matrix $M \times N = (m_{ij} N \mid i, j = 1, ..., n)$, where $M = (m_{ij})$. Consider two F-differentiable operators $H, K : L(\mathbf{R}^n) \to L(\mathbf{R}^n)$ and set $F(X) = H(X) K(X)$ for all $X \in L(\mathbf{R}^n)$. Show that $F'(X) = [H(X) \times I] K'(X) + \left[I \times K(X)^T \right] H'(X)$ for all $X \in L(\mathbf{R}^n)$.

2.11. Let $F : \mathbf{R}^2 \to \mathbf{R}^2$ be defined by $f_1(x) = x_1^3$, $f_2(x) = x_2^2$. Set $x = 0$ and $y = (1, 1)^T$. Show that there is no $z \in [x, y]$ such that $F(y) - F(x) = F'(z)(y - x)$.

2.12. Let $F : D \subset \mathbf{R}^n \to \mathbf{R}^m$ and assume that F is continuously differentiable on a convex set $D_0 \subset D$. For and $x, y \in D_0$, show that

$$\|F(y) - F(x) - F'(x)(y - x)\| \leq \|y - x\| w(\|y - x\|),$$

where w is the modulus of continuity of F' on $[x, y]$. That is

$$w(t) = \sup\{\|F'(x) - F'(y)\| \mid x, y \in D_0, \|x - y\| \leq t\}.$$

2.13. Let $F : D \subset \mathbf{R}^n \to \mathbf{R}^m$. Show that F'' is continuous at $z \in D$ if and only if all second partial derivatives of the components $f_1, ..., f_m$ of F are continuous at z.

2.14. Let $F : D \subset \mathbf{R}^n \to \mathbf{R}^m$. Show that $F''(z)$ is symmetric if and only if each Hessian matrix $H_1(z), ..., H_m(z)$ is symmetric.

2.15. Let $M \in L(\mathbf{R}^n)$ be symmetric, and define $f : \mathbf{R}^n \to \mathbf{R}$ by $f(x) = x^T M x$. Show, directly from the definition that f is convex if and only if M is positive semidefinite.

2.16. Show that $f : D \subset \mathbf{R}^n \to \mathbf{R}$ is convex on the set D if and only if, for any $x, y \in D$, the function $g : [0,1] \to \mathbf{R}$, $g(t) = g(tx + (1-t)y)$, is convex on $[0,1]$.

2.17. Show that if $g_i : \mathbf{R}^n \to \mathbf{R}$ is convex and $c_i \geq 0$, $i = 1, 2, ..., m$, then $g = \sum_{t=1}^{m} c_i g_i$ is convex.

2.18. Suppose that $g : D \subset \mathbf{R}^n \to \mathbf{R}$ is continuous on a convex set $D_0 \subset D$ and satisfies

$$\frac{1}{2} g(x) + \frac{1}{2} g(y) - g\left(\frac{1}{2}(x+y)\right) \geq \gamma \|x - y\|^2$$

for all $x, y \in D_0$. Show that g is convex on D_0 if $\gamma = 0$.

2.19. Let $M \in L(\mathbf{R}^n)$. Show that M is a nonnegative matrix if and only if it is an isotone operator.

2.20. Let $M \in L(\mathbf{R}^n)$ be diagonal, nonsingular, and nonnegative. Show that $\|x\| = \|D(x)\|$ is a monotonic norm on $\mathbf{R}^n$.

2.21. Let $M \in L(\mathbf{R}^n)$. Show that M is invertible and $M^{-1} \geq 0$ if and only if there exist nonsingular, nonnegative matrices $M_1, M_2 \in L(\mathbf{R}^n)$ such that $M_1 M M_2 = 1$.

2.22. Let $[\cdot, \cdot] : D \times D$ be an operator satisfying conditions (2.1) and (2.44). The following two assertions are equivalent:

(a) Equality (2.48) holds for all $x, y \in D$.
(b) For all points $u, v \in D$ such that $2v - u \in D$ we have

$$[u, v] = 2[u, 2v - u] - [v, 2v - u].$$

2.23. If $\delta\, F$ is a consistent approximation F' on D show that each of the following four expressions in an estimate for

$$\|F(x) - F(y) - \delta F(u,v)(x-y)\|$$
$$c_1 = h\left(\|x-u\| + \|y-u\| + \|u-v\|\right)\|x-y\|,$$
$$c_2 = h\left(\|x-v\| + \|y-v\| + \|u-v\|\right)\|x-y\|,$$
$$c_3 = h\left(\|x-y\| + \|y-u\| + \|y-v\|\right)\|x-y\|,$$

and

$$c_4 = h\left(\|x-y\| + \|x-u\| + \|x-v\|\right)\|x-y\|.$$

2.24. Show that the integral representation of $[x_0, ..., x_k]$ is indeed a divided difference of k-th order of F. Let us assume that all divided differences have such an integral representation. In this case for $x_0 = x_1 = ... = x_k = x$ we shall have

$$\underbrace{[x, x, ..., x]}_{k+1 \text{ times}} = \frac{1}{k} f^{(k)}(x).$$

Suppose now that the n-th Fréchet-derivative of F is Lipschitz continuous on D, i.e. there exists a constant c_{n+1} such that

$$\left\|F^{(n)}(u) - F^{(n)}(v)\right\| \le c_{n+1} \|u-v\|$$

for all $u, v \in D$. In this case, set

$$R_n(y) = \left([x_0, ..., x_{n-1}, y] - [x_0, ..., x_{n-1}, x_n]\right)(y - x_{n-1}), ..., (y - x_0)$$

and show that

$$\|R_n(y)\| \le \frac{c_{n+1}}{(n+1)!} \|y - x_n\| \cdot \|y - x_{n-1}\| \cdots \|y - x_0\|$$

and

$$\left\|F(x+h) - \left(F(x) + F'(x)h + \frac{1}{2}F''(x)h^2 + \cdots + \frac{1}{n!}F^{(n)}(x)h^n\right)\right\|$$
$$\le \frac{c_{n+1}}{(n+1)!} \|h\|^{n+1}.$$

2.25. We recall the definitions:

(a) An operator $F: D \subset \mathbf{R}^n \to \mathbf{R}^m$ is Gateaux - (or G -) differentiable at an interior point x of D if there exists a linear operator $L \in$

$L(\mathbf{R}^n, \mathbf{R}^m)$ such that, for any $h \in \mathbf{R}^n$

$$\lim_{t \to 0} \frac{1}{t} \|F(x+th) - F(x) - tL(h)\| = 0.$$

L is denoted by $F'(x)$ and called the G-derivative of F at x.
(b) An operator $F : D \subset \mathbf{R}^n \to \mathbf{R}^m$ is hemicontinuous at $x \in D$ if, for any $h \in \mathbf{R}^n$ and $\varepsilon > 0$, there is a $\delta = \delta(\varepsilon, h)$ so that whenever $|t| < \delta$ and $x + th \in D$, then $\|F(x+th) - F(x)\| < \varepsilon$.
(c) If $F : D \subset \mathbf{R}^n \to \mathbf{R}^m$ and if for some interior point x of D, and $h \in \mathbf{R}^n$, the limit

$$\lim_{t \to 0} \frac{1}{t}[F(x+th) - F(x)] = A(x,h)$$

exists, then F is said to have a Gateaux-differential at x in the direction h.
(d) If the G-differential exists at x for all h and if, in addition

$$\lim_{h \to 0} \frac{1}{\|h\|} \|f(x+H) - F(x) - A(x,h)\| = 0,$$

then F has a Fréchet differential at x.
Show:

(i) The linear operator L is unique;
(ii) If $F : D \subset \mathbf{R}^n \to \mathbf{R}^m$ is G-differentiable at $x \in D$, then F is hemicontinuous at x.
(iii) G-differential and "uniform in h" implies F-differential;
(iv) F-differential and "linear in h" implies F-derivative;
(v) G-differential and "linear in h" implies G-derivative;
(vi) G-derivative and "uniform in h" implies F-derivative;
Here "uniform in h" indicated the validity of (d).
Linear in h means that $A(x,h)$ exists for all $h \in \mathbf{R}^n$ and

$$A(x,h) = M(x)h, \quad \text{where} \quad M(x) \in L(\mathbf{R}^n, \mathbf{R}^m).$$

Define $F : \mathbf{R}^2 \to \mathbf{R}$ by $F(x) = \text{sgn}(x_2) \min(|x_1|, |x_2|)$.
Show that, for any $h \in \mathbf{R}^2$ $A(0,h) = F(h)$, bu F does not have a G-derivative at 0.
(viii) Define $F : \mathbf{R}^2 \to \mathbf{R}$ by $F(0) = 0$ if $x = 0$ and

$$F(x) = x_2 (x_1^2 + x_2^2)^{\frac{3}{2}} \Big/ \left[(x_1^2 + x_2^2)^2 + x_2^2\right], \quad \text{if} \quad x \neq 0.$$

Show that F has a G-derivative at 0, but not an F-derivative. Show, moreover, that the G-derivative is hemicontinuous at 0.

(ix) If the g-differential $A(x, h)$ exists for all x in an open neighborhood of an interior point x_0 of D and for all $h \in \mathbf{R}^n$, then F has an F-derivative at x_0 provided that for each fixed h, $A(x, h)$ is continuous in x at x_0.

(e) Assume that $F : D \subset \mathbf{R}^n \to \mathbf{R}^m$ has a G-derivative at each point of an open set $D_0 \subset D$. If the operator $F' : D_0 \subset \mathbf{R}^n \to L(\mathbf{R}^n, \mathbf{R}^m)$ has a G-derivative at $x \in D_0$, then $(F')'(x)$ is denoted by $F''(x)$ and called the second G-derivative of F at x.
Show:

(i) If $F : \mathbf{R}^n \to \mathbf{R}^m$ has a G-derivative at each point of an open neighborhood of x, then F' is continuous at x if and only if all partial derivatives $\partial_i F_i$ are continuous at x.

(ii) F'' is continuous at $x_0 \in D$ if and only if all second partial derivatives of the components $f_1, ..., f_m$ of F are continuous at x_0. $F''(x_0)$ is symmetric if and only if each Hessian matrix $H_1(x_0), ..., H_m(x_0)$ is symmetric.

Chapter 3

Fundamental Fixed Point Theory

The ideas of a contraction operator and its fixed points are fundamental to many questions in applied mathematics. In this chapter we outline the essential ideas.

3.1 Fixed Points of Operators

Definition 3.1 Let F be an operator mapping a set X into itself. A point $x \in X$ is called a fixed point of F if

$$x = F(x). \tag{3.1}$$

Equation (3.1) leads naturally to the construction of the method of successive approximations or substitutions

$$x_{n+1} = F(x_n) \ (n \geq 0) \ x_0 \in X. \tag{3.2}$$

If sequence $\{x^n\}$ $(n \geq 0)$ converges to some point $x^* \in X$ for some initial guess $x_0 \in X$, and F is a continuous operator in a Banach space X, we can have

$$x^* = \lim_{n \to \infty} x_{n+1} = \lim_{n \to \infty} F(x_n) = F\left(\lim_{n \to \infty} x_n\right) = F(x^*).$$

That is, x^* is a fixed point of operator F. Hence, we showed:

Theorem 3.1 *If F is a continuous operator in a Banach space X, and the sequence $\{x_n\}$ $(n \geq 0)$ generated by (3.2) converges to some point $x^* \in X$ for some initial guess $x_0 \in X$, then x^* is a fixed point of operator F.*

We need information on the uniqueness of x^* and the distances $\|x_n - x^*\|$, $\|x_{n+1} - x_n\|$ $(n \geq 0)$. That is why we introduce the concept:

Definition 3.2 Let $(X, \|\ \|)$ be a metric space and F a mapping of X into itself. The operator F is said to be a contraction or a contraction mapping if there exists a real number c, $0 \leq c < 1$, such that

$$\|F(x) - F(y)\| \leq c \|x - y\|, \quad \text{for all } x, y \in X. \tag{3.3}$$

It follows immediately from (3.3) that every contraction mapping F is uniformly continuous. Indeed, F is Lipschitz continuous with a Lipschitz constant c. The point c is called the contraction constant for F.

We now arrive at the Banach contraction mapping principle.

Theorem 3.2 *Let $(X, \|\cdot\|)$ be a Banach space, and $F : X \to X$ be a contraction mapping. Then F has a unique fixed point.*

Proof. *Uniqueness*: Suppose there are two fixed points x, y of F. Since F is a contraction mapping,

$$\|x - y\| = \|F(x) - F(y)\| \leq c \|x - y\| < \|x - y\|,$$

which is impossible. That shows uniqueness of the fixed point of F.

Existence: Using (3.2) we obtain

$$\|x_2 - x_1\| \leq c \|x_1 - x_0\|$$
$$\|x_3 - x_2\| \leq x \|x_2 - x_1\| \leq c^2 \|x_1 - x_0\| \tag{3.4}$$
$$\vdots$$
$$\|x_{n+1} - x_n\| \leq c^n \|x_1 - x_0\|$$

and, so

$$\|x_{n+m} - x_n\| \leq \|x_{n+m} - x_{n+m-1}\| + \cdots + \|x_{n+1} - x_n\|$$
$$\leq \left(c^{m+1} + \cdots + c + 1\right) c^n \|x_1 - x_0\|$$
$$\leq \frac{c^n}{1-c} \|x_1 - x_0\|.$$

Hence, sequence $\{x_n\}(n \geq 0)$ is Cauchy in a Banach space X and such it converges to some x^*. The rest of the theorem follows from Theorem 3.1. $\square$

Remark 3.1 *It follows from (3.1), (3.3) and $x^* = F(x^*)$ that*

$$\|x_n - x^*\| = \|F(x_{n-1}) - F(x^*)\| \leq c \|x_{n-1} - x^*\| \leq c^n \|x_0 - x^*\| \quad (n \geq 1). \tag{3.5}$$

Inequality (3.5) describes the convergence rate. This is convenient only when an a priori estimate for $x_0 - x^*$ is available. Such an estimate can be derived from the inequality

$$\|x_0 - x^*\| \leq \|x_0 - F(x_0)\| + \|F(x_0) - F(x^*)\| \leq \|x_0 - F(x_0)\| + c\|x_0 - x^*\|,$$

which leads to

$$\|x_0 - x^*\| \leq \frac{1}{1-c} \|x_0 - F(x_0)\|. \tag{3.6}$$

By (3.5) and (3.6), we obtain

$$\|x_n - x^*\| \leq \frac{c^n}{1-c} \|x_0 - F(x_0)\| \quad (n \geq 1). \tag{3.7}$$

Estimates (3.5) and (3.7) can be used to determine the number of steps needed to solve Equation (3.1). For example if the error tolerance is $\varepsilon > 0$, that is, we use $\|x_n - x^*\| < \varepsilon$, then this will certainly hold if

$$n > \frac{1}{\ln c} \ln \frac{\varepsilon(1-c)}{\|x_0 - F(x_0)\|}. \tag{3.8}$$

3.2 Examples

These are examples of operators with fixed points that are not contraction mappings:

Example 3.1 Let $F : \mathbb{R} \to \mathbb{R}$, $q > 1$, $F(x) = qx + 1$. Operator F is not a contraction, but it has a unique fixed point $x = (1-q)^{-1}$.

Example 3.2 Let $F : X \to X$, $x = (0, \frac{1}{\sqrt{6}}]$, $F(x) = x^3$. We have

$$|F(x) - F(y)| = |x^3 - y^3| \leq \left(|x|^2 + |x| \cdot |y| + |y|^2\right) |x - y| \leq \tfrac{1}{2} |x - y|$$

That is, F is a contraction with $c = \frac{1}{2}$, with no fixed points in X. This is not violating Theorem 3.2 since X is not a Banach space.

Example 3.3 Let $F : [a, b] \to [a, b]$, F differentiable at every $x \in (a, b)$ and $|F'(x)| \leq c < 1$. By the mean value theorem, if $x, y \in [a, b]$ there exists a point z between x and y such that

$$F(x) - F(y) = F'(z)(z - y),$$

from which it follows that F is a contraction with constant c.

Example 3.4 Let $F : [a,b] \to \mathbb{R}$. Assume there exist constants p_1, p_2 such that $p_1 p_2 < 0$ and $0 < p_1 < F'(x) \leq p_2^{-1}$ and assume that $F(a) < 0 < F(b)$. How can we find the zero of $F(x)$ guaranteed to exist by the intermediate value theorem? Define a function P by

$$P(x) = x - p_2 F(x).$$

Using the hypotheses, we obtain $P(a) = a - p_2 F(a) > a$, $P(b) = b - F(b) < b$, $P'(x) = 1 - p_2 F'(x) \geq 0$, and $P'(x) \leq 1 - p_1 p_2 < 1$. Hence P maps $[a,b]$ into itself and $|p'(x)| \leq 1$ for all $x \in [a,b]$. By Example 3.3, $P(x)$ is a contraction mapping. Hence, P has a fixed point which is clearly a zero of F.

Example 3.5 (Fredholm integral equation). Let $K(x,y)$ be a continuous function on $[a,b] \times [a,b]$, $f_0(x)$ be continuous on $[a,b]$ and consider the equation

$$f(x) = f_0(x) + \lambda \int_a^b K(x,y) f(y) \, dy.$$

Define the operator $P : C[a,b] \to C[a,b]$ by $p(f) = g$ given by

$$g(x) = f_0(x) + \lambda \int_a^b K(x,y) f(y) \, dy.$$

Note that a fixed point of P is a solution of the integral equation. We get

$$\|P(q_1) - P(q_2)\| = \sup_{x \in [a,b]} |p(q_1(x)) - P(q_2(x))|$$

$$= |\lambda| \sup_{x \in [a,b]} \left| \int_a^b k(x,y) (q_1(y) - q_2(y)) \, dy \right|$$

$$\leq |\lambda| \delta \left| \int_a^b (q_1(y) - q_2(y)) \, dy \right| \quad \text{(by Wierstrass theorem)}$$

$$\leq |\lambda| \delta (b-a) \sup_{x \in [a,b]} |q_1(y) - q_2(y)|$$

$$\leq |\lambda| \delta (b-a) \|q_1 - q_2\|.$$

Hence, P is a contraction mapping if there exists a $c > 1$ such that

$$|\lambda| \delta (b-a) \leq c$$

3.3 Integral Equations Arising in Newton Transport

We apply the method of continuation to study the structure of the solutions of quadratic integral equations of the form

$$x(s) = y(s) + \frac{1}{2}\lambda x(s) \int_s^1 k(s,t) x(t) dt \qquad (3.9)$$

in the space $C[s,1]$ of all functions continuous on the interval $0 \leq s \leq 1$, with norm

$$\|x\| = \max_{0 \leq s \leq 1} |x(s)|.$$

Here we assume that λ is a real number called the "albedo" for scattering and the kernel $k(s,t)$ is a continuous function of two variables s, t with $0 \leq s, t \leq 1$ and satisfying

$$\begin{array}{ll} (i) & 0 < k(s,t) < 1, \quad 0 \leq s, \quad t \leq 1 \\ (ii) & k(s,t) + k(t,s) = 1, \quad 0 \leq s, \quad t \leq 1 \end{array} \qquad (3.10)$$

The function $y(s)$ is a continuous given function on $[s,1]$ and finally $x(s)$ is the unknown function sought in $C[s,1]$.

Equations of this type arise in the theories of radiative transfer, neutron transport and in the kinetic theory of gases.

There exists an extensive literature on equations like (3.9) under various assumptions on the kernel $k(s,t)$ if the lower limit of integration in (3.9) is zero and λ a real or complex number. One can refer to the recent work in [7], [54] and the references there.

Here, we use the method of continuation to find solutions of (3.9) in order, on the one hand, to suggest a new method for solving equations like (3.9) and, on the other hand, to improve the results in [4], [113] and [242].

For $\lambda = 0$, $x(s) \equiv y(s)$ is the unique solution of (3.9). Using this simple observation, we show we can extend this solution for $\lambda \in (\lambda_1, 1)$ where $\lambda_1 < 0$.

The result in [242], for $k(s,t) = \frac{s}{s+t}$, $0 \leq s, t \leq 1$ requires

$$0 < \lambda < .72134 \cdots \qquad (3.11)$$

whereas, the result in [7], requires

$$0 < \lambda < .848108 \cdots \qquad (3.12)$$

for the existence of a solution of (3.9).

We first need the following abstract results.
Consider the quadratic equations

$$x = y + \lambda B(x,x) \qquad (3.13)$$

in a Banach space X, where $y \in X$ is fixed, λ is a real number and B is a bounded symmetric bilinear operator on X, [10], [11].

A number of existence theorem has already been proved in [7], [113], for (3.13).

Here we prove a number of new results for (3.13). We first introduce the definitions.

Definition 3.3 Let B be a bilinear operator on X and $x_0 \in X$ be fixed. Denote by $B(x_0)$ the linear operator on X defined by

$$B(x_0)(x) = B(x_0, x) \quad \text{for all} \quad x \in x.$$

Note that if B is symmetric, then

$$B(x_1)(x_2) = B(x_1, x_2) = B(x_2, x_1) = B(x_2),(x_1) \quad \text{for all} \quad x_1, x_2 \in x.$$

Definition 3.4 Let B be a symmetric bilinear operator on X. Let x_0 be a solution of (3.13) corresponding to $\lambda = \lambda_0$, where λ_0^{-1} is not a characteristic value of the linear operator $2B(x_0)$ on x. Fix $\lambda_1 \in \mathbf{R}$. Define the numbers $k_1, k_2, \bar{r}$ and the real functions $f_1(r)$, $f_2(r)$, $f_3(k)$ or $\mathbf{R}$ by

$$k_1 = \left\| (I - 2\lambda_0 B(x_0))^{-1} \right\|,$$

$$k_2 = \frac{1}{2\left(|\lambda_1|\,\|B(x_0)\| + \sqrt{|\lambda_1|\,\|B\| \cdot \|B(x_0, x_0)\|\,|\lambda_0 + \lambda_1|}\right)},$$

$$\bar{r} = \frac{1 - 2|\lambda_1| \cdot k_1\,\|B(x_0)\|}{2|\lambda_0 + \lambda_1|\,\|B\|},$$

$$f_1(r) = a_2 r^2 + b_1 r + c_1,$$

$$f_2(r) = a_2 r + b_2, \quad \text{and}$$

$$f_3(k) = a_3 k^2 + b_3 k + c_3$$

where

$$a_1 = |\lambda_0 + \lambda_1| \, \|B\| \, k_1,$$
$$b_1 = 2|\lambda_1| \cdot \|B(x_0)\| \, k_1 - 1,$$
$$c_1 = |\lambda_1| \, \|B(x_0, x_0)\| \, k_1,$$
$$a_2 = 2|\lambda_0 + \lambda_1| \, \|B\| \, k_1,$$
$$b_2 = -1 + 2|\lambda_1| \, \|B(x_0)\| \, k_1,$$
$$a_3 = 4|\lambda_1| \left(|\lambda_1| \, \|B(x_0)\|^2 - \|B(x_0, x_0)\| \cdot \|B\| \, |\lambda_0 + \lambda_1| \right),$$
$$b_3 = -4|\lambda_1| \, \|B(x_0)\|, \quad \text{and}$$
$$c_3 = 1.$$

It is easy to check that $f_1(r) \leq 0$ and $f_2(r) < 0$ if $r \in [r_1, \overline{r})$ and $k_1 \in (0, k_2)$, where r_1 is the small solution of the equation $f_1(r) = 0$.

Obviously, for $\lambda = 0$, $x(s) \equiv y(s)$ is the unique solution of equation (3.13). The above solution motivates us to introduce the following theorem on the extension of solutions. A similar theorem which makes extensive use of the topological degree of an operator has been proved in [185]. In our proof, we do not use the notion of the topological degree of an operator and we introduce certain items to be used later.

Theorem 3.3 *Let x_0 be a solution of equation (3.13) corresponding to $\lambda = \lambda_0$, where λ_0^{-1} is not a characteristic value of $2B(x_0)$. Fix $\lambda_1 \in \mathbf{R}$. Assume that b_1, k_1 are such that*

$$b_1 < 0 \quad \text{and} \quad k_1 \in (0, k_2)$$

and choose $r \in [r_1, \overline{r})$, where $r_1, \overline{r}$ and k_2 are as in Definition 3.4.
Then,

(i) *there exists a unique $w \in \overline{U}(r) = \{v \in X | \, \|v\| \leq r\}$ such that*

$$x = x_0 + w$$

is a solution of (3.13) corresponding to

$$\lambda = \lambda_0 + \lambda_1.$$

(ii) *The element $w \in X$ is continuous in norm as a function of λ_1 and $\|w\| \to 0$ as $\lambda_1 \to 0$.*

Proof. The element $x_0 + w \in X$ is a solution of (3.13) if

$$x_0 + w = y + (\lambda_1 + \lambda_0) B(x_0 + w, x_0 + w),$$

or if

$$w = (I - 2\lambda_0 B(x_0))^{-1} [\lambda_1 B(x_0, x_0) + 2\lambda_1 B(x_0, w) + (\lambda_0 + \lambda_1) B(w, w)]$$
$$= T(w),$$

since x_0 is a solution of (3.13) and λ_0^{-1} is not a characteristic value of $2B(x_0)$. Let $w_1, w_2 \in \overline{U}(r)$. □

Claim 1. T is a contraction operator on $\overline{U}(r)$.
We have,

$$\|T(w_1) - T(w_2)\| =$$
$$= \left\|(I - 2\lambda_0 B(x_0))^{-1} [2\lambda_1 B(x_0, w_1 - w_2) + (\lambda_0 + \lambda_1) b(w_1 - w_2, w_1 + w_2)]\right\|$$
$$\leq k_1 (2|\lambda_1| \|B(x_0)\| + |\lambda_0 + \lambda_1| \|B\| \cdot 2r) \|w_1 - w_2\|$$

so, T is a contraction on $\overline{U}(r)$ if

$$f_2(r) < 0$$

which is true, since $r \in [r_1, \overline{r}]$ and $k_1 \in (0, k_2)$.

Claim 2. T maps $\overline{U}(r)$ into $\overline{U}(r)$.
Let $w \in \overline{U}(r)$, then $\|T(w)\| \leq r$ if

$$\|T(w)\| \leq k_1 (|\lambda_1| \|B(x_0, x_0)\|) + 2|\lambda_1| \|B(x_0)\| r + |\lambda_0 + \lambda_1| \|B\| \cdot r^2 \leq r,$$

or, $f_1(r) \leq 0$, which is true since, $r \in [r_1, \overline{r}]$ and $k \in (0, k_2)$.

The result (i) now follows from the contraction mapping principle.
By the assumption on b_1 and k_1 there exists $\epsilon > 0$ such that

$$|\lambda_1| \leq \epsilon.$$

Let $\Delta\lambda$ be such that

$$|\lambda_1 + \Delta\lambda| < \epsilon,$$

then

$$\|w_{\lambda_1 + \Delta\lambda} - w_{\lambda_1}\| =$$
$$= \|T_{\lambda_1 + \Delta\lambda}(w_{\lambda_1 + \Delta\lambda}) - T_{\lambda_1}(w_{\lambda_1})\|$$
$$\leq \|T_{\lambda_1 + \Delta\lambda}(w_{\lambda_1 + \Delta\lambda}) - T_{\lambda_1 + \Delta\lambda}(w_{\lambda_1})\| + \|T_{\lambda_1 + \Delta\lambda}(w_{\lambda_1}) - T_{\lambda_1}(w_{\lambda_1})\|$$
$$\leq (f_2(r) - 1) \|w_{\lambda_1 + \Delta\lambda} - w_{\lambda_1}\| + \|T_{\lambda_1 + \Delta\lambda}(w_{\lambda_1}) - T_{\lambda_1}(w_{\lambda_1})\|,$$

so that

$$\|w_{\lambda_1+\Delta\lambda} - w_{\lambda_1}\| \leq (2 - f_2(r))^{-1} \|T_{\lambda_1+\Delta\lambda}(w_{\lambda_1}) - T_{\lambda_1}(w_{\lambda_1})\| \to 0$$

as $\Delta\lambda \to 0$. Moreover, $\|w\| \to 0$. This proves (ii) and the proof of the theorem is completed.

Remark 3.2 *We observe from Thorem 3.3 that w, λ_2 are such that*

$$|\lambda_1| \leq \epsilon \quad \text{and} \quad \|w\| \leq r_1.$$

Assuming as in Theorem 3.3 that $[I - 2\lambda_0 B(x_0)]^{-1}$ exists, we can choose $\epsilon > 0$ and $r_1 > 0$ small enough so that $[r - 2(\lambda_0 + \lambda_1) B(x_0 + w)]^{-1}$ exists, provided $|\lambda_1| \leq \epsilon$, $\|w\| \leq r_1$. If we combine this requirement on ϵ, r_1 with the hypotheses of Theorem 1 [193, p. 92], then the solution x of equation (3.9) given by Theorem 3.3 is such that $(I - 2\lambda B(x))^{-1}$ exists for any

$$\lambda\epsilon[\lambda_0 - \delta, \lambda_0 + \delta].$$

We can now apply Theorem 3.3 for $\lambda = \lambda_0 \pm \delta$. This will wxtend the solution to two adjacent intervals overlapping the original one until we can find a pair (λ_C, x_C) such that $x_C \in x$ and λ_C is a characteristic value of $2B(x_C)$.

we now return to equation (3.9). We work in the Banach space $C[s, 1]$. We define a bounded linear operator L and a bounded bilinear operator $\overline{B}$ on $C[s, 1]$ by

$$(Lv)(s) = \frac{1}{2}\int_s^1 k(s, t) v(t) dt,$$
$$\overline{B}(v, w)(s) = v(s)(Lw)(s) + w(s)(Lv)(s)$$

where $v, w \in C[s, 1]$, $0 \leq s \leq 1$. It is standard to show that $\overline{B}(v, w) \in C[s, 1]$, [266].

Let λ be a fixed real number, define an operator F on $C[s, 1]$ by

$$F(v) = v - y - \frac{1}{2}\lambda\overline{B}(v, v).$$

The, equation (3.9) can be written as

$$F(x) = 0.$$

Finally, define the Fréchet derivative of F with respect to v by

$$F'(v) w = w - \lambda\overline{B}(v, w). \tag{3.14}$$

Note that $F'(v)$ is a bounded linear operator on $C[s,1]$.

We now prove the following lemma.

Lemma 3.1 *Let x be a solution of (3.9) for some fixed λ. Set*

$$A_s = \int_s^1 x(s)\,ds, \quad 0 \le s < 1$$

and

$$D_s = \int_s^1 y(s)\,ds, \quad 0 \le s < 1.$$

Then,

$$\frac{1}{2}\lambda A_s^2 - 2A_s + 2D_s = 0. \tag{3.15}$$

Proof. Integrating (3.9) and using the identity

$$k(s,t) = \frac{1}{2}\{1 + k(s,t) - k(t,s)\}$$

we have

$$A_s = D_s + \frac{1}{4}\lambda A_s^2 + \frac{1}{2}\int_s^1 \int_s^1 \{k(s,t) - k(t,s)\} x(s) x(t)\,ds dt,$$

and the last integral is obviously zero.

From now on we take $y(s) = 1$, i.e., $D_s = 1 - s$ for simplicity.

Since $\lambda A_s^2 - 4A_s + 4D_s = 0$ has no real solutions when $\lambda > \frac{1}{1-s}$, a fundamental interruption in the continuation of the solution $s_\lambda(s)$ must occur at some boundary point $\lambda_C \le \frac{1}{1-s} \le 1$. equation (3.9) can have no real solutions for $\lambda > \frac{1}{1-s}$. □

Moreover, we can show the following:

Theorem 3.4 *Let $x_{\lambda_0}(s)$ be any solution of (3.9) with $\lambda = \lambda_0$. Then the linear operator $F'(x_{\lambda_0})$ given by (3.14) is an invertible operator on $C[s,1]$ with bounded inverse if $\lambda_0 \ne 1$.*

Proof. Let w be such that

$$F'(v)(w) = 0, \quad v, w \in C[s,1].$$

If the lower limit of integration in (3.9) is different than zero, we set
$$x_{\lambda_0}(t) = 0$$
$$v(t) = 0 \quad \text{for} \quad 0 \le t < s.$$
$$w(t) = 0$$

Then by integrating (3.14), we obtain
$$\int_0^1 w(s)\,ds = \frac{1}{2}\lambda_0 \int_0^1 B(x_{\lambda_0}, w)(s)\,ds$$
$$= \frac{1}{2}\lambda_0 \left[\int_0^1 x_{\lambda_0}(s)\,ds\right]\left[\int_0^1 w(s)\,ds\right].$$

If $\lambda_0 \ne 1$, using Lemma 3.1 in the above equation, we get
$$\int_0^1 w(s)\,ds = 0. \tag{3.16}$$

Let $p(s) = \frac{w(s)}{x_{\lambda_0}(s)}$, to obtain
$$p = \frac{1}{2}\lambda_0 x_{\lambda_0} L(x_{\lambda_0} p)$$

and
$$\int_0^1 p(s) x_{\lambda_0}(s)\,ds = 0,$$
$$p = -\frac{1}{2}\lambda x_{\lambda_0} Q(x_{\lambda_0} p) \tag{3.17}$$

where,
$$(Qw)(s) = \int_0^1 \frac{t}{s+t} w(t)\,dt.$$

Define a subspace s of $C[0,1]$ by
$$S = \left\{ p \in C[0,1] \quad \int_0^1 p(s) x_{\lambda_0}(s)\,ds = 0 \right\}.$$

we will show that (3.17) has only the trivial solution in S. Let $q \in C[0,1]$. Define $E(q)$ by
$$E(q) = -\frac{1}{2}\lambda \left(1 - \frac{1}{2}\lambda r\right)^{-1} \int_0^1 x_{\lambda_0}(s) s q(s)\,ds,$$

where r is a root of (3.15) for the lower limit of integration in (3.9) being zero. The functional E is well defined since $1 - \frac{1}{2}\lambda r \neq 0$.

define the operator R on $C[0,1]$ by

$$R(q(s)) = s(sq(s) - E(q)).$$

Since,

$$\begin{aligned} x_{\lambda_0} L(x_{\lambda_0} R(p)) &= x_{\lambda_0} L[s(sx_{\lambda_0}p(s) - x_{\lambda_0}E(p))] \\ &= x_{\lambda_0} Q[sx_{\lambda_0}(sp(s) - E(p))]; \end{aligned}$$

(3.16) gives,

$$\begin{aligned} sp(s) &= -\tfrac{1}{2}\lambda s(x_{\lambda_0} Q(x_{\lambda_0} p(s))) \\ &= -\tfrac{1}{2}\lambda s x_{\lambda_0}(Q(x_{\lambda_0} p(s))) \\ &= -\tfrac{1}{2}\lambda x_{\lambda_0} L(sx_{\lambda_0} p(s)) \\ &= \left(1 - \tfrac{1}{2}\lambda r\right) E(p(s)) x_{\lambda_0} + \tfrac{1}{2}\lambda x_{\lambda_0} Q(sx_{\lambda_0}). \end{aligned} \qquad (3.18)$$

Hence, as x_{λ_0} is a solution of (3.9),

$$\tfrac{1}{2}\lambda x_{\lambda_0} Q(x_{\lambda_0}) = \tfrac{1}{2}\lambda r x_{\lambda_0} - \tfrac{1}{2}\lambda x_{\lambda_0} L x_{\lambda_0} = \tfrac{1}{2}\lambda r x_{\lambda_0} + 1 - x_{\lambda_0} = 1 - \left(1 - \tfrac{1}{2}\lambda r\right) x_{\lambda_0}$$

we have, by (3.17)

$$\begin{aligned} &\tfrac{1}{2}\lambda x_{\lambda_0} Q[x_{\lambda_0}(sp(s) - E(p))] \\ &= sp(s) - \left(1 - \tfrac{1}{2}\lambda r\right) E(p) x_{\lambda_0} - E(p) \tfrac{1}{2}\lambda x_{\lambda_0} Q(x_{\lambda_0}) \\ &= sp(s) - E(p). \end{aligned}$$

Therefore, if $\lambda_0 \neq 1$ and p is a solution of (3.16), then $p \in S$, $R(g) \in S$ and $R(g)$ is a solution of (3.16). Now, as L is compact, (3.16) has a finite dimensional solution space. hence, there exist an integer $n \geq 1$ and numbers $\{c_i\}_{i=1}^n$ with $c_n \neq 0$, so that

$$\sum_{i=1}^n c_i R^i(g) = 0.$$

The above equation implies that there exist polynomials D_1 and D_2 so that

$$D_2(s^2) p(s) = s D_1(s^2).$$

The linearity of R implies that

$$\sum_{i=1}^{n} c_i R^{i+1}(p) = D.$$

Hence $R(p) = p$, so

$$p(s) = \frac{sE(p)}{s^2 - 1},$$

that is, $p = 0$ as $p \in C[0,1]$. The proof of the theorem is now completed □

Proposition 3.1 *There exists a sheet of sokutions x_λ for λ such that $0 \leq \lambda \leq \lambda_C$ and $x_0 \equiv 1$. The seet x_λ is continuous as a function of λ uniformly over $0 \leq s \leq 1$.*

Proof. Obviously, for $\lambda_0 = 0$, $x_0 \equiv 1$ is the unique solution of (3.9). since $I - 2\lambda_0 B(x_0) = I$ is nonsingular, we can apply Theorem 3.3 to generate x_λ. The element w is a continuous function of λ in each extension. Therefore x_λ is continuous as a function of λ uniformly over $0 \leq s \leq 1$. □

A similar proposition can be stated for some negative λ values.

Proposition 3.2 *The following are true:*

(a) *The solutions $x_\lambda, \lambda > 0$ are such that*

$$x_\lambda(s) > 0, \quad 0 \leq s \leq 1;$$

(b) *For fixed $s, 0 \leq s \leq 1$, $x_\lambda(s)$ is a monotone increasing function of λ.*

Proof. (a) For $\lambda = 0$ and for small positive λ since $x_0(s) \equiv 1$, the result is true by continuity. Let $p_\lambda = \inf x_\lambda(s)$. Then

$$|p_{\lambda+\Delta\lambda} - p_\lambda| \leq |\sup(p_{\lambda+\Delta\lambda} - p_\lambda)| \leq \sup|p_{\lambda+\Delta\lambda} - p_\lambda| \to 0 \quad \text{as} \quad \Delta\lambda \to 0$$

that is, p_λ is a continuous function of λ. If x_λ becomes non-positive, then there exists $\lambda_0 > 0$ such that

$$p_{\lambda_0} \equiv 0.$$

Therefore,

$$x_{\lambda_0}(s) \geq 0,$$

with zero actually attained. Moreover, at $\lambda = \lambda_0$

$$x_{\lambda_0}(s) = 1 + \frac{1}{2}\lambda_0 x_{\lambda_0}(s) \int_s^1 x_{\lambda_0}(t) k(s,t) dt \geq 1,$$

which provides a contradiction.

(b) For $0 \leq \lambda \leq \lambda_C < 1$, $[I - 2\lambda B(x_\lambda)]^{-1}$ does exists and

$$[I - 2\lambda B(x_\lambda)]^{-1} = I + 2\lambda B(x_\lambda) + 4\lambda^2 (B(x_\lambda))^2 + \cdots.$$

By (a) the above series, i.e., the above inverse, is a positive operator. Moreover, w is the limit of the iterations

$$w_{n+1} = T(w_n), \quad n = 0, 1, 2, ...,$$

where w_0 might be chosen positive. Hence, $w > 0$ at each extension step of Proposition 3.1. This proves (b). $\square$

Proposition 3.3 *The solution x_λ of equation (3.9) exists and is positive at least for $\lambda \in (-1, 0]$.*

Proof. By Proposition 3.2, $x_\lambda > 0$ for small $|\lambda|$ by continuity. Suppose either that continuation ceases or that p_λ vanishes at $\lambda_1 < 0$. Therefore, x_λ exists and is positive on $(\lambda_1, 0)$. We have

$$x_\lambda(s) = 1 - \frac{1}{2}|\lambda| x_\lambda(s) \int_s^1 k(s,t) x(t) dt, \quad \lambda_1 < \lambda \leq 0 \qquad (3.19)$$

so, $\|x_\lambda(s)\|$ is finite if λ_1 is a leftward limit of continuation and $\lambda \to \lambda_1$. $\square$

Claim. $\lambda_1 \leq -1$.

If $\lambda_1 > -1$, or $|\lambda_1| < 1$. Since $p_{\lambda_1} = 0$, $x_{\lambda_1}(s)$ has a zero s_1. At $s = s_1$, we have

$$0 = x_\lambda(s_1) = 1 - \frac{1}{2}|\lambda_1| x_\lambda(s_1) \int_{s_1}^1 k(s_1, t) x_\lambda(t) dt >$$

$$> 1 - \frac{1}{2} x_\lambda(s_1) \int_{s_1}^1 k(s_1, t) x_\lambda(t) dt \geq 0,$$

since

$$\frac{1}{2} x_\lambda(s) \int_s^1 k(s,t) x_\lambda(t) dt \leq 1,$$

by (3.19). This contradiction justifies the claim and the proposition is proved.

Remark 3.3 *In references [242] and [7], the range is*

$$0 \leq \lambda < .72134 \cdots$$
$$0 \leq \lambda < .848108 \cdots$$

respectively, for

$$k(s,t) = \frac{s}{s+t}, \quad 0 \leq s, \quad t \leq 1$$

the lower limit of integration in (3.9) being zero and $y(s) = 1$.

Here, it follows from Theorems 3.3, 3.4 and Proposition 3.1 that the range for λ is at least such that

$$\lambda_1 \leq \lambda < 1,$$

where λ_1 by Proposition 3.3 is such that

$$\lambda_1 \leq -1.$$

That is, there exists solutions x_λ by continuation such that

$$.848108 \cdots \leq \lambda < 1, \text{ for example.}$$

This justifies the claim made at the introduction.

3.4 An Efficient Contractive Method

Consider the equation

$$F(x) = 0 \qquad (3.20)$$

where F is twice Fréchet-differentiable, nonlinear operator mapping a subset U of a Banach space X into a Banach space Y. We shall find it convenient to assume that U is a ball. Suppose that the approximation x_n has been found. To determine the next approximation x_{n+1} we replace (3.20) by the equation

$$F(x_n) + F'(x_n)(x - x_n) + \frac{1}{2}F''(x_n)(x - x_n, x - x_n) = 0 \qquad (3.21)$$

if the linear operators $\left[F'(z) - \frac{1}{2}F''(z)(x_n)\right]^{-1}$, $\left[F'(z) - \frac{1}{2}F''(z)(z)\right]^{-1}$ exists, then (3.21) suggests the iteration

$$x_{n+1} = x_n - \left[F'(z) - \frac{1}{2}F''(z)(x_n)\right]^{-1} F(x_n), \quad n = 0, 1, 2, \cdots \quad (3.22)$$

or the modified version of (3.22)

$$x_{n+1} = x_n - \left[F'(z) - \frac{1}{2}F''(z)(z)\right]^{-1} F(x_n), \quad n = 0, 1, 2, \cdots. \quad (3.23)$$

The above iterations convergence to a solution x of (3.20) if the operators

$$T(x) = x - [F'(z) + B(x)]^{-1} F(x) \quad (3.24)$$

or the modified version of (3.24)

$$P(x) = x - [F'(z) + L]^{-1} F(x) \quad (3.25)$$

have a fixed point in X, where B is a bounded symmetric bilinear operator on $X \times X$ and L is a bounded linear operator on X (usually, but not necessarily $B = -\frac{1}{2}F''(z)$, $L = -\frac{1}{2}F''(z)(z)$).

In this section we give sufficient conditions for T and P to have unique fixed points in a closed ball centered at a specific $z \in X$ and then we compare (3.22) and (3.23) with the modified Newton's method

$$x_{n+1} = x_n - (F'(z))^{-1} F(x_n), \quad n = 0, 1, 2, \ldots \quad (3.26)$$

using a simple scalar equation to show that (3.22) or (3.23) can converge to a solution of (3.20) faster than (3.26).

Lemma 3.2 *If the linear operator $(F'(z) + B(z))^{-1}$ exists for some $z \in X$, then the linear operator*

$$\left[I + (F'(z))^{-1} B(x - z)\right]^{-1} \quad \text{exists for every } x \in U(z, r),$$

where r is such that

$$0 < r < \frac{1}{\left\|(F'(z) + B(z))^{-1}\right\| \|B\|}.$$

Proof. By the Banach Lemma on invertible operators it is enough to show

$$\left\|(F'(z) + B(z))^{-1}\right\| \cdot \|B(x-z)\| < 1,$$

or

$$\left\|(F'(z) + B(z))^{-1}\right\| \cdot \|B\| \cdot r < 1$$

which is true by hypothesis since $x \in U(z,r)$. $\square$

Definition 3.5 Let z be fixed in X. Assume that the linear operator $C = [F'(z) + B(z)]^{-1}$ exists and set $d = \|C\|$, $e = \|B\|$. define the linear operator A on $U(z,r) = \{x \in X | \|x - z\| < r < \frac{1}{de}\}$ by

$$A(x) = \left[I + (F'(z) + B(z))^{-1} B(x-z)\right]^{-1}.$$

Assume now that

$$\left.\begin{array}{r}\|F'(x) - F'(y)\| \leq \ell_1 \|x - y\|, \\ \|CF(x)\| \leq n,\end{array}\right\} \quad (3.27)$$

where $x, y \in \overline{U}(z,r)$ and ℓ_1, n are nonnegative numbers. Define the numbers

$$h = \|CB\|$$
$$p = \|CB(z)\|$$
$$m = \|CF(z)\|.$$

Note that $\|A(z)\| \leq \frac{1}{1-der}$.

$$\|CB(x)(x-z)\| \leq \|CB(x-z, x-z)\| + \|CB(z, x-z)\|$$
$$\leq hr^2 + pr.$$

Define the real polynomials on $\mathbb{R}$ by

$$f(r) = ar^2 + br + c,$$
$$g(r) = a'r^2 + b'r + c'$$

where
$$a = de(de + d\ell_2 + h), \quad \ell_2 = 3\ell_1$$
$$b = dep - h - d\ell_2 - 2de, \quad \ell_3 = \frac{\ell_1}{2}$$
$$c = 1 - p - hn$$
$$a' = (\ell_3 + e)d + h$$
$$b' = p - 1$$
$$c' = n.$$

Finally, note that for any $x, y \in \overline{U}(z, r)$
$$\begin{aligned}&T(x) - T(y)\\ &= [F'(z) + B(x)]^{-1}\left[F'(z)(x-y) - (F(x) - F(y))\right.\\ &\quad + B(x-z)(x-y) + B(z)(x-y)\\ &\quad \left. + B(x-y)(F'(z) + B(y))^{-1}F(y)\right],\end{aligned}$$

and
$$F'(z)(x-y) - (F(x) - F(y)) = \int_0^1 (F'(z) - F'(x + t(y-x)))(x-y)\, dt.$$

Theorem 3.5 *Assume:*

(i) *The conditions (3.27) are satisfied for some $z \in X$.*
(ii) *There exists r such that $f(r) > 0$ and $g(r) \leq 0$.*

Then (3.24) has a unique fixed point in $\overline{U}(z, r)$.

Proof. T is well defined in $\overline{U}(z, r)$ by Lemma 3.2. □

Claim 1. T is a contraction operator on $\overline{U}(z, r)$
If $x, y \in \overline{U}(z, r)$, then we get
$$\|T(x) - T(y)\| \leq \frac{1}{1 - der}\left[d\ell_2 r + hr + p + \frac{hdn}{1 - der}\right]\|x - y\|.$$

Now T is a contraction if
$$\frac{1}{1 - der}\left[d\ell_2 r + hr + p + \frac{hdr}{1 - der}\right] < 1$$
or $f(r) > 0$, which is true by (ii).
Claim 2. T maps $\overline{U}(z, r)$ into $\overline{U}(z, r)$.

If $x \in \overline{U}(z,r)$,

$$T(x) - z = A(x)\left[C\int_0^1 (F'(z) - F'(z + t(x-z)))(x-z)\,dt \right.$$
$$\left. + CB(x)(x-z) - CF(z)\right],$$

then

$$\|T(x) - z\| \leq r$$

if

$$\frac{1}{1 - der}\left(d\ell_3 r^2 + hr^2 + pr + n\right) \leq r$$

or

$$g(r) \leq 0,$$

which is true by (ii).

The result now follows from the contraction mapping principle.

We now state a theorem for the modified equation (3.25). The proof as similar to the proof of Theorem 3.5 is omitted.

Theorem 3.6 *Let z be as in Definition 3.5 and assume that there exists $r_0 \in [s,t]$ where*

$$S = \frac{1 - \|(F'(z)+L)^{-1}L\| - \left[\left(1 - \|(F'(z)+L)^{-1}L\|\right)^2 - 2\|(F'(z)+L)^{-1}\|\|(F'(z)+L)^{-1}F(z)\|\ell_1\right]^{1/2}}{\|(F'(z)+L)^{-1}\cdot\ell_1\|}$$

and

$$t = \frac{1 - \|(F'(z)+L)^{-1}L\|}{\|(F'(z)+L)^{-1}\|\ell_1},$$

provided that the quantity under the radical is positive and

$$\|(F'(z)+L)^{-1}\| < 1.$$

Then (3.20) has a unique solution x in $\overline{U}(z,r_0)$. Moreover, the rate of convergence $q(r_0) \in [u,1)$ where

$$u = 1 - \left[\left(1 - \|(F'(z)+L)^{-1}L\|\right)^2 - 2\|(F'(z)+L)^{-1}\|\|(F'(z)+L)^{-1}F(z)\|\ell_1\right]^{1/2}.$$

Assume that z is sufficient for the application of Newton's method and Theorem 3.6, then if q_N is the rate of convergence in Newton's method the iteration

$$x_{n+1} = x_n - [F'(z) + L]^{-1} F(x_n), \quad n = 0, 1, 2, \ldots \quad (3.28)$$

converges faster to a solution x of (3.20) if

$$q(r_0) < q_N = 1 - \left[1 - 2\ell_1 \left\|F'(z)^{-1}\right\| \left\|F'(z)^{-1} F(z)\right\|\right]^{1/2}. \quad (3.29)$$

Denote by D, D_1 the quantities under the radicals in Theorem 3.6 and (3.29) respectively then, we have the following theorem.

Theorem 3.7 *If the hypothesis in Newton's method [4] are satisfied then Theorem 3.6 can be applied also in (3.20) if*

$$\|L\| < \frac{1 - [1 - D_1]^{1/2}}{2 \left\|F'(z)^{-1}\right\|}. \quad (3.30)$$

Moreover, if $q(r_0) < q_N$ then the iteration in (3.28) converges faster to a solution x of (3.20) than the iteration in Newton's method (3.26).

Proof. By Lemma 3.2

$$\left\|(F'(z) + L)^{-1} L\right\| = \left\|\left(I + F'(z)^{-1} L\right)^{-1} (F'(z))^{-1} L\right\|$$

$$\leq \frac{\left\|F'(z)^{-1}\right\| \cdot \|L\|}{1 - \left\|F'(z)^{-1}\right\| \cdot \|L\|} < 1 \quad \text{by (3.30).} \quad (3.31)$$

Now, using (3.30) and (3.31)

$$D \geq \left(\frac{1 - 2\left\|F'(z)^{-1}\right\| \cdot \|L\|}{1 - \left\|F'(z)^{-1}\right\| \cdot \|L\|}\right)^2 - \frac{2\ell_1 \left\|F'(z)^{-1}\right\| \left\|F'(z)^{-1} F(z)\right\|}{\left(1 - \left\|F'(z)^{-1}\right\| \cdot \|L\|\right)^2}$$

$$= \frac{D_1 - 4\left\|F'(z)^{-1}\right\| \cdot \|L\| \left(1 - \left\|F'(z)^{-1}\right\| \cdot \|L\|\right)}{\left(1 - \left\|F'(z)^{-1}\right\| \cdot \|L\|\right)^2}$$

so, $D > 0$ if

$$D_1 > 4\left\|F'(z)^{-1}\right\| \cdot \|L\| \left(1 - \left\|F'(z)^{-1}\right\| \cdot \|L\|\right)$$

Fundamental Fixed Point Theory

which is true by (3.30). Therefore, Theorem 3.6 can be applied. The rest follows from the discussion after Theorem 3.6. □

The following simple example justifies Theorems 3.23 and 3.7.

Example 3.6 Let $x = \mathbb{R} \times \mathbb{R}$, be equipped with the max-norm. Define a bilinear operator B on x by the following calculation scheme:

$$B(x,y) = \left\{ (x_1, x_2) \begin{bmatrix} b_1^{11} & b_1^{12} \\ b_1^{21} & b_1^{22} \\ b_2^{11} & b_2^{12} \\ b_2^{21} & b_2^{22} \end{bmatrix} \begin{bmatrix} y_1 \\ y_2 \end{bmatrix} \right\}$$

$$= \begin{bmatrix} b_1^{11} x_1 + b_1^{21} x_2 & b_1^{12} x_1 + b_1^{22} x_2 \\ b_2^{11} x_1 + b_2^{21} x_2 & b_2^{12} x_1 + b_2^{22} x_2 \end{bmatrix} \begin{bmatrix} y_1 \\ y_2 \end{bmatrix}$$

$$= \begin{bmatrix} b_1^{11} x_1 y_1 + b_1^{21} x_2 y_1 + b_1^{12} x_1 y_2 + b_1^{22} x_2 y_2 \\ b_2^{11} x_1 y_1 + b_2^{21} x_2 y_1 + b_2^{12} x_1 y_2 + b_2^{22} x_2 y_2 \end{bmatrix},$$

$$x = \begin{bmatrix} x_1 \\ x_2 \end{bmatrix}, \quad y = \begin{bmatrix} y_1 \\ y_2 \end{bmatrix} \in x. \tag{3.32}$$

It can easily be checked that B is a bilinear operator on x and we can define the norm of B on x by

$$\|B\| = \sup_{\|x\|=1} \max_{(i)} \sum_{j=1}^{2} \left| \sum_{k=1}^{2} b_i^{jk} \xi_k \right|. \tag{3.33}$$

Define the linear operator $B(x)$ on X by

$$B(x)(y) = B(x,y)$$

where,

$$B(x) = \begin{bmatrix} b_1^{11} x_1 + b_1^{21} x_2 & b_1^{12} x_1 + b_1^{22} x_2 \\ b_2^{11} x_1 + b_2^{21} x_2 & b_2^{12} x_1 + b_2^{22} x_2 \end{bmatrix}.$$

Let us now consider the quadratic system on X given by

$$F(x) = B(x,x) + L_1(x) + L_1(x) + y = 0 \tag{3.34}$$

where,

$$B \sim \begin{bmatrix} \frac{1}{2} & 1 \\ 1 & -\frac{3}{2} \\ \frac{3}{2} & -1 \\ -1 & \frac{1}{2} \end{bmatrix},$$

L_1 is a linear operator on X given by

$$L_1 = \frac{5}{2}I \quad (I, \text{ the identity operator on } x)$$

and

$$y = \begin{bmatrix} \frac{1}{16} \\ -\frac{1}{16} \end{bmatrix}.$$

Equation (3.34) can also be written using (3.32) as

$$\frac{1}{2}x_1^2 + 2x_1x_2 - \frac{3}{2}x_2^2 + \frac{5}{2}x_1 + \frac{1}{16} = 0 \qquad (3.35)$$
$$\frac{3}{4}x_1^2 - 2x_1x_2 + \frac{1}{2}x_2^2 + \frac{5}{2}x_2 - \frac{1}{16} = 0.$$

Let us choose

$$L = \frac{1}{10}I$$

and

$$z = \begin{bmatrix} 0 \\ 0 \end{bmatrix}.$$

Then, obviously

$$\|L_1\| = \frac{5}{2},$$
$$\|y\| = \frac{1}{16},$$
$$\|L\| = \frac{1}{10},$$

and using (3.33)

$$\|B\| = 4$$

and

$$\ell_1 = 2\|B\|.$$

Note that

$$F'(z) = 2B(z) + L_1 = L_1.$$

Then we can easily compute the quantities

$$D_1 = .84$$
$$q_N = .0834849$$
$$D = .9230769$$
$$u = .0392311$$
$$s = .0025008$$
$$t = 3.125.$$

Note that the hypotheses of Theorem 3.7 are satisfied for $r_0 \in [s, t]$ and by choosing $q(r_0) = u$ we observe that

$$q(r_0) < q_N$$

therefore, iteration (3.28) converges faster to a unique solution x of (3.34) in $U(0, r_0)$ than Newton's iteration.

Indeed iteration (3.28) and (3.26) for solving (3.35) can now be written

$$x_{n+1} = x_n - \frac{10}{26} F(x_n) \tag{3.36}$$

and

$$\overline{x}_{n+1} = \overline{x}_n - \frac{10}{25} F(\overline{x}_n) \tag{3.37}$$

respectively, where

$$x_n = \begin{bmatrix} x_{1,n} \\ x_{2,n} \end{bmatrix}, \quad \overline{x}_n = \begin{bmatrix} \overline{x}_{1,n} \\ \overline{x}_{2,n} \end{bmatrix}, \quad n = 0, 1, 2, ...$$

and

$$x_0 = \overline{x}_0 = \begin{bmatrix} 0 \\ 0 \end{bmatrix}.$$

Let $\epsilon = (.5) 10^{-2}$ be the desired error tolerance that is

$$\|x - x_n\| \leq \epsilon \quad \text{for} \quad n \geq N,$$

and

$$\|x - \overline{x}_n\| \leq \epsilon \quad \text{for} \quad n \geq \overline{N}.$$

Then the true solution $x = \begin{bmatrix} x_1 \\ x_2 \end{bmatrix}$ is given by

$$x_1 = -(24302916852540)\,10^{-2},$$
$$x_2 = (24062003442371)\,10^{-2}.$$

Moreover, we have by (3.36) and (3.37)

$$x_{1,1} = -(24038461538462)\,10^{-2},$$
$$x_{2,1} = (23705087903960)\,10^{-2},$$
$$\overline{x}_{1,1} = -(25)\,10^{-2}$$
$$\overline{x}_{2,1} = (24625)\,10^{-2},$$
$$\overline{x}_{1,2} = -(24268665625)\,10^{-2},$$

and

$$\overline{x}_{2,2} = (24047248283457)\,10^{-2}.$$

We now observe that the number of steps N in (3.36) required to achieve the desired accuracy ϵ is

$$N = 1,$$

whereas the number of steps $\overline{N}$ in (3.37) required to achieve the same accuracy ϵ is

$$\overline{N} = 2.$$

Exercises

3.1. Consider the problem of approximating a solution $y \in C'[0, t_0]$ of the nonlinear ordinary differential equation

$$\frac{dy}{dt} = K(t, y(t)), \quad 0 \le t \le t_0, \; y(0) = y_0.$$

The above equation may be turned into a fixed point problem of the form

$$y(t) = y_0 + \int_0^t K(s, y(s))\,ds, \quad 0 \le t \le t_0.$$

Assume $K(x,y)$ is continuous on $[0,t_0] \times [0,t_0]$ and satisfies the Lipschitz condition

$$\max_{[0,t_0]} |K(s,q_1(s)) - K(s,q_2(s))| \leq M\|q_1 - q_2\|, \text{ for all } q_1, q_2 \in C[0,t_0].$$

Note that the integral equation above defines an operator P from $C[0,t_0]$ into itself. As in Example 3.2.5 find a sufficient condition for P to be a contraction mapping.

3.2. Let F be a contraction mapping on the ball $\bar{U}(x_0, r)$ in a Banach space X, and let

$$\|F(x_0) - x_0\| \leq (1-c)r.$$

Show F has a unique fixed point in $\bar{U}(x_0, r)$.

3.3. Under the assumptions of Theorem 3.1, show that the sequence generalized by (3.2) minimizes the functional

$$f(x) = \|x - F(x)\|$$

for any x_0 belonging to a closed set A such that $F(A) \subseteq A$.

3.4. Let the equation $F(x) = x$ have a unique solution in a closed subset A of a Banach space X. Assume that there exists an operator F_1 that $F_1(A) \subseteq A$ and F_1 commutes with F on A. Show the equation $x = F_1(x)$ has at least one solution in A.

3.5. Assume that operator F maps a closed set A into a compact subset of itself and satisfies

$$\|F(x) - F(y)\| < \|x - y\| \quad (x \neq y), \text{ for all } x,y \in A.$$

Show F has a unique fixed point in A. Apply these results to the mapping $F(x) = x - \frac{x^2}{2}$ of the interval $[0,1]$ into itself.

3.6. Show that operator F defined by

$$F(x) = x + \frac{1}{x}$$

maps the half line $[1, \infty)$ into itself and satisfies

$$\|F(x) - F(y)\| < \|x - y\|$$

but has no fixed point in this set.

3.7. Consider condition

$$\|F(x) - F(y)\| \leq \|x - y\| \quad (x, y \in A).$$

Let A be either an interval $[a, b]$ or a disk $x^2 + y^2 \leq r^2$. Find conditions in both cases under which F has a fixed point.

3.8. Consider the set c_0 of null sequences $x = \{x_1, x_2, \ldots\}$ ($x_n \to 0$) equipped with the norm $\|x\| = \max_n |x_n|$. Define the operator F by

$$F(x) = \left\{\tfrac{1}{2}(1 + \|x\|), \tfrac{3}{4}x_1, \tfrac{7}{8}x_2, \ldots, \left(1 - \tfrac{1}{2^{n+1}}\right) x_n, \ldots\right\}.$$

Show that $F : \bar{U}(0, 1) \to \bar{U}(0, 1)$ satisfies

$$\|F(x) - F(y)\| < \|x - y\|,$$

but has no fixed points.

3.9. Repeat Exercise 3.8 for the operator F defined in c_0 by $F(x) = \{y_1, \ldots, y_n, \ldots\}$, where $y_n = \tfrac{n-1}{n}x_n + \tfrac{1}{n}\sin(n)$ ($n \geq 1$).

3.10. Repeat Exercise 2.8 for the operator F defined in $C[0, 1]$ by

$$Ax(t) = (1 - t)x(t) + t\sin\left(\tfrac{1}{t}\right).$$

3.11. Let F be a nonlinear operator on a Banach space X which satisfies (3.3) on $\bar{U}(0, r)$. Let $F(0) = 0$. Define the resolvent $R(x)$ of F by

$$F(x)f = xFR(x)f + f.$$

Show:

(a) $R(x)$ is defined on the ball $\|f\| \leq (1 - |x|c)r$ if $|x| < c^{-1}$;
(b) $\tfrac{1}{1+|x|c}\|f - g\| \leq \|R(x)f - R(x)g\| \leq \tfrac{1}{1-|x|c}\|f - g\|$;
(c) $\|R(x)f - R(y)f\| \leq \tfrac{c\|f\| |x-y|}{(1-|x|c)(1-|y|c)}$.

3.12. Let A be an operator mapping a closed set A of a Banach space X into itself. Assume that there exists a positive integer m such that A^m is a contraction operator. Prove that sequence (3.2) converges to a unique fixed point of F in A.

3.13. Let F be an operator mapping a compact set $A \subseteq X$ into itself with $\|F(x) - F(y)\| < \|x - y\|$ ($x \neq y$, all $x, y \in A$). Show that sequence (3.2) converges to a fixed point of (3.1).

3.14. Let F be a continuous function on $[0, 1]$ with $0 \leq f(x) \leq 1$ for all $x \in [0, 1]$. Define the sequence

$$x_{n+1} = x_n + \tfrac{1}{n+1}(F(x_n) - x_n).$$

Show that for any $x_0 \in [0, 1]$ sequence $\{x_n\}$ ($n \geq 0$) converges to a fixed point of F.

3.15. Show:

(a) A system $x = Ax + b$ of n linear equations in n unknowns $x_1, x_2, \ldots, x_n$ (the components of x) with $A = \{a_{jk}\}$, $j, k = 1, 2, \ldots, n$, b given, has a unique solution x^* if
$$\sum_{k=1}^{n} |a_{jk}| < 1, \qquad j = 1, 2, \ldots, n.$$

(b) The solution x^* can be obtained as the limit of the iteration $(x^{(0)}, x^{(1)}, x^{(2)}, \ldots)$, where $x^{(0)}$ is arbitrary and
$$x^{(m+1)} = Ax^{(m)} + b \quad (m \geq 0).$$

(c) The following error bounds hold:
$$\|x^{(m)} - x^*\| \leq \frac{c}{1-c} \|x^{(m-1)} - x^{(m)}\| \leq \frac{c^m}{1-c} \|x^{(0)} - x^{(1)}\|,$$
where
$$c = \max_{j} \sum_{k=1}^{n} |a_{jk}| \quad \text{and} \quad \|x - z\| = \max_{j} |x_i - z_i|, \; j = 1, 2, \ldots, n.$$

3.16. (Gershgorin's theorem: If λ is an eigenvalue of a square matrix $A = \{a_{jk}\}$, then for some j, where $1 \leq j \leq n$,
$$|a_{jj} - \lambda| \leq \sum_{\substack{k=1 \\ k \neq j}}^{n} |a_{jk}|.$$

Show that $x = Ax + b$ can be written $Bx = b$, where $B = I - A$, and $\sum_{k=1}^{n} |a_{jk}| < 1$ together with the theorem imply that 0 is not an eigenvalue of B and A has spectral radius less than 1.

3.17. Let (X, d), (X, d_1), (X, d_2) be metric spaces with $d(x, z) = \max_j |x_j - z_j|$, $j = 1, 2, \ldots, n$,
$$d_1(x, z) = \sum_{j=1}^{n} |x_j - z_j| \quad \text{and} \quad d_2(x, z) = \left[\sum_{j=1}^{n} (x_j - z_j)^2 \right]^{1/2},$$
respectively. Show that instead of $\sum_{k=1}^{n} |a_{jk}| < 1$, $j = 1, 2, \ldots, n$, we obtain the conditions
$$\sum_{j=1}^{n} |a_{jk}| < 1, \; k = 1, 2, \ldots, n \quad \text{and} \quad \sum_{j=1}^{n} \sum_{k=1}^{n} a_{jk}^2 < 1.$$

3.18. Let us consider the ordinary differential equation of the first order (ODE)

$$x' = f(t, x), \quad x(t_0) = x_0,$$

where t_0 and x_0 are given real numbers. Assume:

$$|f(t, x)| \leq c_0$$

on $R = \{(t, x) \mid |t - t_0| \leq a, |x - x_0| \leq b\}$,

$$|f(t, x) - f(t, v)| \leq c_1 |x - v|, \quad \text{for all } (t, x), (t, v) \in R.$$

Then show: the (ODE) has a unique solution on $[t_0 - c_2, t_0 + c_2]$, where

$$c_2 < \min\left\{a, \frac{b}{c_0}, \frac{1}{c_1}\right\}.$$

3.19. Show that f defined by $f(x, y) = |\sin y| + x$ satisfies a Lipschitz condition with respect to the second variable (on the whole xy-plane).

3.20. Does f defined by $f(t, x) = |x|^{1/2}$ satisfy a Lipschitz condition?

3.21. Apply Picard's iteration $x_{n+1}(t) = \int_{t_0}^{t} f(s, x_n(s)) ds$ used for the (ODE) $x' = f(t, x)$, $x(t_0) = x_0 + 0$, $x' = 1 + x^2$, $x(0) = 0$. Verify that for x_3, the terms involving $t, t^2, \ldots, t^5$ are the same as those of the exact solution.

3.22. Show that $x' = 3x^{2/3}$, $x(0) = 0$ has infinitely many solutions x.

3.23. Assume that the hypotheses of the contraction mapping principle hold, then show that x^* is accessible from any point $\bar{U}(x_0, r_0)$.

3.24. Define the sequence $\{\bar{x}_n\}$ $(n \geq 0)$ by $\bar{x}_0 = x_0$, $\bar{x}_{n+1} = F(\bar{x}_n) + \varepsilon_n$ $(n \geq 0)$. Assume:

$$\|\varepsilon_n\| \leq \lambda^n \varepsilon \quad (n \geq 0) \quad (0 \leq \lambda < 1);$$

F is a c-contraction operator on $U(x_0, r)$. Then show sequence $\{\bar{x}_n\}$ $(n \geq 0)$ converges to the unique fixed point x^* of F in $\bar{U}(x_0, r)$ provided that

$$r \geq r_0 + \frac{\varepsilon}{1-c}.$$

3.25.
(a) Let $F: D \subseteq X \to X$ be an analytic operator. Assume:

- there exists $\alpha \in [0,1)$ such that
$$\|F'(x)\| \leq \alpha \quad (x \in D); \tag{1}$$

-
$$\gamma = \sup_{\substack{k>1 \\ x \in D}} \left\| \tfrac{1}{k!} F^{(k)}(x) \right\|^{\frac{1}{k-1}} \quad \text{is finite};$$

- there exists $x_0 \in D$ such that
$$\|x_0 - F(x_0)\| \leq \eta \leq \tfrac{3-\alpha-2\sqrt{2-\alpha}}{\gamma}, \quad \gamma \neq 0;$$

- $\bar{U}(x_0, r_1) \subseteq D$, where, r_1, r_2 with $0 \leq r_1 \leq r_2$ are the two zeros of function f, given by
$$f(r) = \gamma(2-\alpha)r^2 - (1+\eta\gamma-\alpha)r + \eta.$$

Show: method of successive substitutions (3.2) is well defined, remains in $\bar{U}(x_0, r_1)$ for all $n \geq 0$ and converges to a fixed point $x^* \in \bar{U}(x_0, r_1)$ of operator F.

Moreover, x^* is the unique fixed point of F in $\bar{U}(x_0, r_2)$. Furthermore, the following error bounds hold for all $n \geq 0$:
$$\|x_{n+2} - x_{n+1}\| \leq \beta \|x_{n+1} - x_n\|$$

and
$$\|x_n - x^*\| \leq \frac{\beta^n}{1-\beta} \eta,$$

where
$$\beta = \frac{\gamma\eta}{1-\gamma\eta} + \alpha.$$

The above result is based on the assumption that the sequence
$$\gamma_k = \left\| \tfrac{1}{k!} F^{(k)}(x) \right\|^{\frac{1}{k-1}} \quad (x \in D), \quad (k > 1)$$

is bounded above by γ. This kind of assumption does not always hold. Let us then not assume sequence $\{\gamma_k\}$ ($k > 1$) is bounded and define "function" f_1 by
$$f_1(r) = \eta - (1-\alpha)r + \sum_{k=2}^{\infty} \gamma_k^{k-1} r^k.$$

(b) Let $F: D \subseteq X \to X$ be an analytic operator. Assume for $x_0 \in D$ function f_1 has a minimum positive zero r_3 such that

$$\bar{U}(x_0, r_3) \subseteq D.$$

Show: method of successive substitutions is well defined, remains in $\bar{U}(x_0, r_3)$ for all $n \geq 0$ and converges to a unique fixed point $x^* \in \bar{U}(x_0, r_3)$ of operator F. Moreover the following error bounds hold for all $n \geq 0$

$$\|x_{n+2} - x_{n+1}\| \leq \beta_1 \|x_{n+1} - x_n\|$$

and

$$\|x_n - x^*\| \leq \frac{\beta_1^n}{1 - \beta_1} \eta,$$

where,

$$\beta_1 = \sum_{k=2}^{\infty} \gamma_k^{k-1} \eta^{k-1} + \alpha.$$

3.26.
(a) It is convenient to define:

$$\gamma = \sup_{k>1} \left\| \tfrac{1}{k!} F^{(k)}(x^*) \right\|^{\frac{1}{k-1}}$$

with $\gamma = \infty$, if the supremum does not exist. Let $F: D \subseteq X \to X$ be an analytic operator and $x^* \in D$ be a fixed point of F. Moreover, assume that there exists α such that

$$\|F'(x^*)\| \leq \alpha, \tag{2}$$

and

$$\bar{U}(x^*, r^*) \subseteq D,$$

where,

$$r^* = \begin{cases} \infty, & \text{if } \gamma = 0 \\ \frac{1}{\gamma} \cdot \frac{1-\alpha}{2-\alpha}, & \text{if } \gamma \neq 0. \end{cases}$$

Then, if

$$\beta = \alpha + \frac{\gamma r^*}{1 - \gamma r^*} < 1,$$

show: the method of successive substitutions remains in $\bar{U}(x^*, r^*)$ for all $n \geq 0$ and converges to x^* for any $x_0 \in U(x^*, r^*)$. Moreover, the following error bounds hold for all $n \geq 0$:

$$\|x_{n+1} - x^*\| \leq \beta_n \|x_n - x^*\| \leq \beta \|x_n - x^*\|,$$

where,

$$\beta_0 = 1, \quad \beta_{n+1} = \alpha + \frac{\gamma r^* \beta_n}{1 - \gamma r^* \beta_n} \quad (n \geq 0).$$

The above result was based on the assumption that the sequence

$$\gamma_k = \left\| \tfrac{1}{k!} F^{(k)}(x^*) \right\|^{\frac{1}{k-1}} \quad (k \geq 2)$$

is bounded γ. In the case where the assumption of boundedness does not necessarily hold, we have the following local alternative.

(b) Let $F \colon D \subseteq X \to X$ be an analytic operator and $x^* \in D$ be a fixed point of F. Moreover, assume: $\max_{r>0} \sum_{k=2}^{\infty} (\gamma_k r)^{k-1}$ exists and is attained at some $r_0 > 0$. Set

$$p = \sum_{k=2}^{\infty} (\gamma_k r_0)^{k-1};$$

there exist α, δ with $\alpha \in [0, 1)$, $\delta \in (\alpha, 1)$ such that

$$p + \alpha - \delta \leq 0$$

and

$$\bar{U}(x^*, r_0) \subseteq D.$$

Show: the method of successive substitutions $\{x_n\}$ $(n \geq 0)$ generated by (2) remains in $\bar{U}(x^*, r_0)$ for all $n \geq 0$ and converges to x^* for any $x_0 \in \bar{U}(x^*, r_0)$. Moveover the following error bounds hold for all $n \geq 0$:

$$\|x_{n+1} - x^*\| \leq \alpha \|x_n - x^*\| + \sum_{k=2}^{\infty} \gamma_k^{k-1} \|x_n - x^*\|^k \leq \delta \|x_n - x^*\|.$$

Chapter 4

Solving Equations

In this chapter we are concerned with the problem of approximating locally unique solution of an equation in a Banach space. The Newton–Kantorovich method is undoubtedly the most popular method for solving such equations.

4.1 Linearization of Equations

Let F be a Fréchet-differentiable operator mapping a subset of a Banach space X into a Banach space Y. Consider the equation

$$F(x) = 0. \tag{4.1}$$

The principal method for constructing successive approximations x_n to a solution x^* (if it exists) of Equations (4.1) is based on successive linearization of the equation.

Assuming that an approximation x_n has been found, we compute the next x_{n+1} by replacing Equation (4.1) by

$$F(x_n) + F'(x_n)(x_{n+1} - x_n) = 0. \tag{4.2}$$

If $F'(x_n)^{-1} \in L(Y, X)$, then approximation x_{n+1} is given by

$$x_{n+1} = x_n - F'(x_n)^{-1} F(x_n) \quad (n \geq 0). \tag{4.3}$$

The iterative procedure generated by (4.3) is the famous Newton–Kantorovich method [183].

We are concerned about the following aspects:

(a) Finding effectively verifiable conditions for its applicability;
(b) computing convergence rates and a priori error estimates are useful;

(c) choosing an initial approximation x_0 for which the method converges, and
(d) the degree of "stability" of the method.

4.2 The Convergence of Newton's Method

Define the operator P by

$$P(x) = x - F'(x)^{-1} F(x) \qquad (4.4)$$

Then the Newton–Kantorovich method may be regarded as the usual iterative method

$$x_{n+1} = P(x_n) \quad (n \geq 0), \qquad (4.5)$$

for approximating solution x^* of the equation

$$x = P(x) \qquad (4.6)$$

Consequently, all the results of the previous chapters involving Equation (4.6) are applicable.

Suppose that

$$\lim_{n \to \infty} x_n = x^*. \qquad (4.7)$$

We would like to know under what conditions on F and F' the point x^* is a solution of Equation (4.1).

Proposition 4.1 *If F' is continuous at $x = x^*$, then we have*

$$F(x^*) = 0. \qquad (4.8)$$

Proof. The approximations x_n satisfy the equation

$$F'(x_n)(x_{n+1} - x_n) = -F(x_n). \qquad (4.9)$$

Since the continuity of F at x^* follows from the continuity of F', taking the limit as $n \to \infty$ in (4.9) we obtain (4.8). □

Proposition 4.2 *If*

$$\|F'(x)\| \leq b \qquad (4.10)$$

in some closed ball which contains $\{x_n\}$, then x^* is a solution of $F(x) = 0$.

Proof. By (4.10) we get

$$\lim_{n \to \infty} F(x_n) = F(x^*), \tag{4.11}$$

and since

$$\|F(x_n)\| \leq b \|x_{n+1} - x_n\|, \tag{4.12}$$

(4.8) is obtained by taking the limit as $n \to \infty$ in (4.7). $\square$

Proposition 4.3 *If*

$$\|F''(x)\| \leq K \tag{4.13}$$

in some closed ball $\bar{U}(x_0, r), 0 < r < \infty$, which contains $\{x_n\}$, then x^* is a solution of equation $F(x) = 0$.

Proof. By (4.13)

$$\|F'(x) - F'(x_0)\| \leq K \|x - x_0\| \leq Kr, \tag{4.14}$$

for all $x \in \bar{U}(x_0, r)$. Moreover, we can write

$$\|F'(x)\| \leq \|F'(x_0)\| + \|F'(x) - F'(x_0)\|, \tag{4.15}$$

so the conditions of Proposition 4.2 hold with

$$b = \|F'(x_0)\| + Kr. \tag{4.16}$$

As in Kantorovich (see, e.g., [149], [152], [197] consider constants B_0, η_0 such that

$$\left\|[F'(x_0)]^{-1}\right\| \leq B_0, \tag{4.17}$$

$$\|x_1 - x_0\| \leq \eta_0, \tag{4.18}$$

respectively. $\square$

Theorem 4.1 *If*

$$\|F''(x)\| \leq K \tag{4.19}$$

in some closed ball $\bar{U}(x_0, r)$ and

$$h_0 = B_0 \eta_0 K \leq \tfrac{1}{2}, \tag{4.20}$$

then the Newton sequence (4.3), starting from x_0, converges to a solution x^* of Equation (4.1) which exists in $\bar{U}(x_0, r)$, provided that

$$r \geq r_0 = \frac{1 - \sqrt{1 - 2h_0}}{h_0} \eta_0. \tag{4.21}$$

Proof. We first show that (4.3) is well defined. By (4.19) we get

$$\|F'(x_1) - F'(x_0)\| \leq K \|x_1 - x_0\| = K\eta_0. \tag{4.22}$$

Using (4.20), we obtain

$$\|F'(x_1) - F'(x_0)\| \leq K\eta_0 \leq \frac{1}{2B_0} < \frac{1}{\|[F'(x_0)]^{-1}\|}. \tag{4.23}$$

By Theorem 1.4, $[F'(x_1)]^{-1}$ exists, and

$$\|[F'(x_1)]^{-1}\| \leq \frac{\|[F'(x_0)]^{-1}\|}{1 - \|[F'(x_0)]^{-1}\| \cdot \|F'(x_1) - F'(x_0)\|} \tag{4.24}$$

or

$$\|[F'(x_1)]^{-1}\| \leq \frac{B_0}{1 - B_0\eta_0 K} = \frac{B_0}{1 - h_0} = B_1. \tag{4.25}$$

Hence, x_2 exists. To estimate $\|x_2 - x_1\|$, note that

$$\|x_2 - x_1\| = \|[F'(x_1)]^{-1} F(x_1)\| \tag{4.26}$$

and

$$[F'(x_1)]^{-1} = \left(\sum_{n=0}^{\infty} \left\{ I - [F'(x_0)]^{-1} F'(x_1) \right\}^n \right) [F'(x_0)]^{-1}$$

$$= \left(\sum_{n=0}^{\infty} \left\{ [F'(x_0)]^{-1} [F'(x_0) - F'(x_1)] \right\}^n \right) [F'(x_0)]^{-1}.$$

$$\tag{4.27}$$

Consequently

$$\|[F'(x_1)]^{-1} F(x_1)\| \leq \frac{1}{1 - B_0\eta_0 K} \|[F'(x_0)]^{-1} F(x_1)\| \tag{4.28}$$

or

$$\|x_2 - x_1\| \leq \frac{1}{1 - h_0} \|[F'(x_0)]^{-1} F(x_1)\|. \tag{4.29}$$

To estimate $\left\|[F'(x_0)]^{-1} F(x_1)\right\|$ consider the operator

$$F_1(x) = x - [F'(x_0)]^{-1} F(x). \tag{4.30}$$

Note

$$F_1'(x_0) = I - [F'(x_0)]^{-1} F(x_0) = 0. \tag{4.31}$$

By (4.30)

$$[F'(x_0)]^{-1} F(x_1) = x_1 - F_1(x_1), \tag{4.32}$$

and since

$$x_1 = F_1(x_0) \tag{4.33}$$

(4.31) may be used to write (4.32) as

$$[F'(x_0)]^{-1} F(x_1) = -[F_1(x_1) - F_1(x_0) - F_1'(x_0)(x_1 - x_0)], \tag{4.34}$$

and

$$\left\|[F'(x_0)]^{-1} F(x_1)\right\| \leq \sup_{\bar{x} \in L(x_0, x_1)} \|F_1''(\bar{x})\| \frac{\|x_1 - x_0\|^2}{2}. \tag{4.35}$$

Using (4.30)

$$F_1''(x) = -[F'(x_0)]^{-1} F''(x). \tag{4.36}$$

Since $x_1 \in \bar{U}(x_0, r)$, if (4.31) holds, then

$$\|F_1''(x)\| \leq B_0 K \tag{4.37}$$

on $L(x_0, x_1)$. From (4.35)

$$\left\|[F'(x_0)]^{-1} F(x_1)\right\| \leq \frac{B_0 K \eta_0^2}{2} = \frac{h_0 \eta_0}{2}, \tag{4.38}$$

and from (4.29)

$$\|x_2 - x_1\| \leq \frac{1}{2} \frac{h_0}{1 - h_0} \eta_0 = \eta_1 \tag{4.39}$$

and obviously $\eta_1 \leq \frac{1}{2}\eta_0$. The constant

$$h_1 = B_1 \eta_1 K = \frac{B_0}{1 - h_0} \frac{1}{2} \frac{h_0}{1 - h_0} \eta_0 K = \frac{1}{2} \frac{h_0^2}{(1 - h_0)^2} \leq 2h_0^2 \leq \frac{1}{2}, \tag{4.40}$$

so that (4.20) is satisfied if x_0 is replaced by x_1. Now if for

$$r_1 = \frac{1 - \sqrt{1 - 2h}}{h_1} \eta_1 \qquad (4.41)$$

it can be shown that

$$\bar{U}(x_1, r_1) \subset \bar{U}(x_0, r_0), \qquad (4.42)$$

then condition (4.19) will hold with x_0 replaced by x_1. By direct substitution

$$\sqrt{1 - 2h_1} = \left(1 - \frac{h_0^2}{(1 - h_0)^2}\right)^{1/2} = \frac{1}{1 - h_0}\sqrt{1 - 2h_0} \qquad (4.43)$$

and

$$r_1 = \frac{1 - (1 - h_0)^{-1}\sqrt{1 - 2h_0}}{\frac{1}{2}\left[h_0^2/(1 - h_0)^2\right]} \frac{1}{2} \frac{h_0}{1 - h_0} \eta_0$$

$$= \left(\frac{1 - \sqrt{1 - 2h_0}}{h_0} - 1\right)\eta_0 = r_0 - \eta_0. \qquad (4.44)$$

Consequently, if $x \in \bar{U}(x_1, r_1)$, then $\|x - x_1\| \leq r_1 = r_0 - \eta_0$ and

$$\|x - x_0\| \leq \|x - x_1\| + \|x_1 - x_0\| \leq (r_0 - \eta_0) + \eta_0 = r_0, \qquad (4.45)$$

so that $x \in \bar{U}(x_0, r_0)$, which establishes (4.42).

It follows by mathematical induction that the Newton process (4.3) generates an infinite sequence $\{x_n\}$, starting from an x_0 at which the hypotheses of the theorem are satisfied. It remains to be shown that this sequence converges to a solution x^* of (4.1). Along with $\{x_n\}$ the sequences of numbers $\{B_n\}, \{\eta_n\}$, and $\{h_n\}$ defined by

$$B_n = \frac{B_{n-1}}{1 - h_{n-1}}, \qquad (4.46)$$

$$\eta_n = \frac{1}{2}\frac{h_{n-1}\eta_{n-1}}{1 - h_{n-1}}, \qquad (4.47)$$

$$h_n = \frac{1}{2}\frac{h_{n-1}^2}{(1 - h_{n-1})^2}, \qquad (4.48)$$

respectively, are obtained for $n = 1, 2, \ldots$. We have

$$h_n \leq 2h_{n-1}^2 \leq \tfrac{1}{2}(2h_{n-2})^4 \leq \cdots \leq \tfrac{1}{2}(2h_0)^{2^n} \tag{4.49}$$

from (4.40) and

$$\eta_n = \tfrac{1}{2}\tfrac{h_{n-1}\eta_{n-1}}{1-h_{n-1}} \leq h_{n-1}\eta_{n-1} \leq h_{n-1}h_{n-2}\eta_{n-2} \leq \cdots$$
$$\leq \tfrac{1}{2^n}(2h_0)^{2^{n-1}}(2h_0)^{2^{n-2}}\cdots(2h_0)\eta_0 = \tfrac{1}{2^n}(2h_0)^{2^n-1}\eta_0. \tag{4.50}$$

For any positive integer p, $x_{n+p} \in \bar{U}(x_n, r_n)$, so that

$$\|x_{n+p} - x_n\| \leq r_n = \tfrac{1-\sqrt{1-2h_n}}{h_n}\eta_n \leq 2\eta_n, \tag{4.51}$$

and thus from (4.50)

$$\|x_{n+p} - x_n\| \leq \tfrac{1}{2^{n-1}}(2h_0)^{2^n-1}\eta_0. \tag{4.52}$$

Therefore $\{x_n\}$ is a Cauchy sequence which has a limit $x^* \in \bar{U}(x_0, r_0)$, and x^* is a solution of (4.1) by Proposition 4.3. $\square$

4.3 Local Convergence

We can now show a local result for (4.3).

Theorem 4.2 *If x^* is a simple zero of F, $B^* \geq \|[F'(x^*)]^{-1}\|$, and*

$$A = \left\{x : \|x - x^*\| < 1/(B^*K)\right\} \subset \bar{U}(x_0, r) \tag{4.53}$$

then the hypotheses (i) with $h < \tfrac{1}{2}$ and (ii) of the Kantorovich theorem are satisfied at each $x_0 \in B$, where

$$B = \left\{x : \|x - x^*\| < (2-\sqrt{2})/(2B^*K)\right\}. \tag{4.54}$$

Proof. For $x_0 \in B$,

$$\|F'(x_0) - F'(x^*)\| \leq K\|x_0 - x^*\| < (2-\sqrt{2})/(2B^*) < \|[F'(x^*)]^{-1}\|^{-1}, \tag{4.55}$$

so that $[F'(x_0)]^{-1}$ exists, and

$$B = \frac{B^*}{1 - B^*K\|x_0 - x^*\|} \geq \|[F'(x_0]^{-1}\| \tag{4.56}$$

by Theorem 1.4. By the fundamental theorem of calculus we obtain

$$F(x^*) - F(x_0) = \int_0^1 F'(x_0 + \theta(x^* - x_0))(x^* - x_0) d\theta$$
$$= F'(x_0)(x^* - x_0) + \qquad (4.57)$$
$$+ \int_0^1 [F'(x_0 + \theta(x^* - x_0)) - F'(x_0)](x^* - x_0) d\theta.$$

Since $F(x^*) = 0$,

$$-[F'(x_0)]^{-1} F(x_0) = x^* - x_0 + [F'(x_0)]^{-1} \cdot \qquad (4.58)$$
$$\cdot \int_0^1 [F'(x_0 + \theta(x^* - x_0)) - F'(x_0)](x^* - x_0) d\theta,$$

and hence

$$\left\| [F'(x_0)]^{-1} F(x_0) \right\| \leq \left\{ 1 + BK \|x^* - x_0\| \int_0^1 \theta d\theta \right\} \|x^* - x_0\|. \qquad (4.59)$$

By (4.56) we get

$$\eta = \frac{1 - \frac{1}{2} B^* K \|x^* - x_0\|}{1 - B^* K \|x^* - x_0\|} \|x^* - x_0\| \geq \left\| [F'(x_0)]^{-1} F(x_0) \right\|. \qquad (4.60)$$

It follows that for $x_0 \in A$,

$$h = BK\eta = \frac{1 - \frac{1}{2} B^* K \|x^* - x_0\|}{(1 - B^* K \|x^* - x_0\|)^2} B^* K \|x^* - x_0\| < \tfrac{1}{2}, \qquad (4.61)$$

hence (4.20) holds with $h < \tfrac{1}{2}$. Using the values of h and η given above,

$$\left(1 - \sqrt{1 - 2h}\right) \frac{\eta}{h} < \frac{\eta}{h} = \frac{1}{B^* K} - \|x^* - x_0\|, \qquad (4.62)$$

and thus $U(x_0, r_0) \subseteq A \subseteq U(x_0, r) \subset \Omega_* \subset \Omega$ satisfying (4.2). $\square$

Remark 4.1 *The value $(2 - \sqrt{2})(2B^*K)$ for the radius of Ω^* is best possible, as the following scalar example shows. Take $x^* > 0$ real, and consider the quadratic*

$$F(x) = \tfrac{1}{2} K \left(x^2 - x^{*2} \right),$$

where $K > 0$. One has $(2 - \sqrt{2})/(2B^*K) = \left(1 - \tfrac{1}{2}\sqrt{2}\right) x^*$, and for

$$x_0 = x^* - \left(1 - \tfrac{1}{2}\sqrt{2}\right) x^* = \tfrac{1}{2} \sqrt{2} x^*,$$

it follows that

$$B = \left(\tfrac{1}{2}\sqrt{2}Kx^*\right)^{-1}, \quad \eta = \tfrac{1}{4}\sqrt{2}x^*,$$

from which $h = BK\eta = \tfrac{1}{2}$.

4.4 Approximating Distinct Solutions

We provide sufficient convergence conditions for the convergence of Newton's method to distinct solutions of the quadratic equation in Banach space.

Consider the equation

$$x = y + B(x,x) \tag{4.63}$$

in a Banach space X over the field $\mathbb{R}$ of real numbers, where $B: E \times E \to Y$ is a bounded bilinear operator with values in a Banach space Y and $y \in X$ is fixed. We introduce the iteration

$$x_{n+1} = B(x_n)^{-1}(x_n - y), \quad n = 0, 1, 2, \ldots \tag{4.64}$$

for approximating solutions x^* of equation (4.63). For each fixed $x \in X$, $B(x)$ denotes a linear operator from X to Y such that $B(x)(y) = B(x,y)$, for all $y \in X$.

Special cases of (4.63) appear in many interesting problems arising in astrophysics, in the kinetic theory of gases as well as the theory of ordinary and partial differential equations [4], [54], [113], [129]. Equation (4.63) has been studied extensively. The continued fraction approach, the contraction mapping theorem technique and the famous Newton-Kantorovich method have been used to find a solution x^* of equation [7], [9], [54].

A common hypothesis for the above techniques is the estimate $4\|B\|\|y\| < 1$.

It turns out that under this hypothesis the previous mentioned techniques approximate a small solution v^* of equation (4.63) for any starting point x_0 close enough to the solution. The obtained solution v^* is such that $v^* = v^*(y) \to 0$. We make use of the "theory of majorants" and under assumptions similar to the ones introduced in the above mentioned techniques, iteration (4.64) can be used to approximate a second solution x^* of (4.63) with $x^* \neq v^*$ and $x^* = x^*(y) \to 0$ as $y \to 0$. Moreover, under the same assumptions we show that the Newton-Kantorovich method can be used to obtain a solution $z_N^* = x^*$ also. This result is not known not even for quadratic systems in $\mathbb{R}^n$, $n > 1$. Some sufficient conditions are also given for the existence of more than one distinct solutions of (4.63). Our results are illustrated with the solution of a quadratic system in $X = \mathbb{R}^2$ as well as the solution of a Riccati differential equation.

Definition 4.1 An operator $B : X \times Y \to Z$ is called **bilinear** if it is linear in each variable separately and **symmetric** if $X = Y$ and $B(x, y) = B(y, x)$ for all $(x, y) \in X \times Y$.

Definition 4.2 The **mean** $\overline{B}$ of B on $X \times Y$ is defined by

$$\overline{B}(x, y) = \frac{1}{2}(B(x, y) + B(y, x)) \quad \text{for all} \quad (x, y) \in X \times Y.$$

Definition 4.3 A bilinear operator $B : X \times Y \to Z$ is said to be **bounded** if there exists $q > 0$ such that

$$\|B(x, y)\| \leq q \|x\| \cdot \|y\| \quad \text{for all} \quad (x, y) \in X \times Y.$$

The quantity $\|B\| = \sup_{\|x\| \leq 1, \|y\| \leq 1} \|B(x, y)\|$ is called the **norm** of B. Note that, for B symmetric,

$$\overline{B}(x, x) = B(x, x) \quad \text{for all} \quad x \in X. \tag{4.65}$$

Without loss of generality due to (4.65) we may assume that the operator B in (4.63) is symmetric.

From now on $X = Y = X$ and $Z = Y$. We can now prove a theorem for the existence of a solution x^* of equation (4.63).

Theorem 4.3 *Let B be a bounded symmetric bilinear operator on $X \times X$ and suppose that $x_0, y \in E$ with $x_0 \neq 0$ and $x_0 \neq y$. Assume:*

(i) *The inverse of the linear operator $B(x_0) : X \to X$ with $B(x_0)(x) = B(x_0, x)$ for all $x \in E$ exists and is bounded.*
(ii) *The estimates:*

$$0 \leq c < 1 \tag{4.66}$$

and

$$0 \leq d < \frac{(1-c)^2}{4ab} \tag{4.67}$$

are true, where we have denoted

$$a \geq \left\|B(x_0)^{-1}\right\|, \tag{4.68}$$

$$b \geq \|B\|, \tag{4.69}$$

$$c \geq \left\|B(x_0)^{-1}(I - B(x_0))\right\| \tag{4.70}$$

and

$$d \geq \left\|B(x_0)^{-1}(B(x_0, x_0) + y - x_0)\right\|. \tag{4.71}$$

Then:
(a) The real sequence $\{t_n\}$, $n = 0, 1, 2, \ldots$ given by

$$t_{n+2} = t_{n+1} - \tfrac{1+c-2abt_{n+1}}{1+c+2abt_{n+1}}(t_n - t_{n+1}), \quad n = 0, 1, 2, \ldots,$$

$$t_0 = \frac{1-c}{2ab}, \quad t_1 = \left[\tfrac{1+c}{2}\right] t_0, \tag{4.72}$$

is positive and decreasing converges to zero.
(b) The sequence $\{x_n\}$, $n = 0, 1, 2, \ldots$ generated by (4.64) is well defined, remains in $U(x_0, r_0)$ with $r_0 = \tfrac{1-c}{2ab}$ and converges to a unique solution $x^* \in \overline{U}(x_0, r_0)$ of equation (4.63).

Moreover, the following estimates are true for all $n = 0, 1, 2, \ldots$,

$$\|x_{n+1} - x_n\| \leq t_n - t_{n+1}$$

and

$$\|x_n - x^*\| \leq t_n \leq \left[\tfrac{1+c}{2}\right]^n t_0.$$

Proof. (a) It can easily be seen by (4.72) that the sequence $\{t_n\}$, $n = 0, 1, 2, \ldots$ is certainly nonnegative if

$$(1+c)t_{k+1} + 2abt_k t_{k+1} - (1+c)t_k \geq 0 \quad \text{for all} \quad k = 0, 1, 2, \ldots. \tag{4.73}$$

Inequality (4.73) is true as equality for $k = 0$. Let us assume that it is true for $k = 0, 1, 2, \ldots, n$. We shall show that it is true for $k = n + 1$. Using (4.72), the left hand side of inequality (4.73) for $k = n + 1$ becomes

$$\frac{2ab(1+c+2abt_k)t_{k+1}^2 + (1+c)^2 t_{k+1} - (1+c)^2 t_k}{1+c+2abt_{k+1}}$$

which is nonnegative if $t_{k+1} \geq \tfrac{(1+c)t_k}{1+c+2abt_k}$ and that is true by our assumption. By the choice of t_0, t_1 and (4.71), $t_0 - t_1 > 0$.

Let us assume that

$$t_k - t_{k+1} > 0, \quad k = 0, 1, 2, ..., n. \tag{4.74}$$

Using (4.72), we see that (4.74) is true for $k = n + 1$ if

$$1 + c - 2abt_{k+1} > 0 \quad \text{for} \quad k = 0, 1, 2, ..., n. \tag{4.75}$$

Inequality (4.75) is true for $k = 0$ by the choice of t_1. Let us assume that (4.75) is true for $k = 0, 1, 2, ..., n$. To show (4.75) for $k = n + 1$ it suffices to show $t_{k+2} < \frac{1+c}{2ab}$ or by (4.72)

$$2ab\left[2(1+c)t_{k+1} + 2abt_k t_{k+1} - (1+c)t_k\right] \le (1+c)^2 + 2ab(1+c)t_{k+1}$$

or $t_{k+1} \le \frac{1+c}{2ab}$ which is true by hypothesis.

we have now showed that the real sequence $\{t_n\}$, $n = 0, 1, 2, ...$ is positive and decreasing and as such it converges to some $t^* \ge 0$. But using simple induction and (4.72) we can easily show that $t_{n+1} \le \left[\frac{1+c}{2}\right]t_n \le \left[\frac{1+c}{2}\right]^{n+1} t_0$. That is $t^* = 0$.

(b) Let us observe that the linear operator $B(x)$ is invertible for all $x \in U(x_0, r_0)$. Indeed we have

$$\left\|B(x_0)^{-1} B(x - x_0)\right\| \le \left\|B(x_0)^{-1}\right\| \cdot \|B\| \cdot \|x - x_0\| \le ab\|x - x_0\| < 1$$

so that according to Banach's Lemma on invertible operators

$$\left\|B(x)^{-1}\right\| = \left\|\left[I + B(x_0)^{-1} B(x - x_0)\right]^{-1} B(x_0)^{-1}\right\| \le \frac{a}{1 - ab\|x - x_0\|}. \tag{4.76}$$

We shall prove that

$$\|x_n - x_{n+1}\| \le t_n - t_{n+1} \quad \text{for} \quad n = 0, 1, 2, ... \tag{4.77}$$

By (a) it follows that if (4.64) is well defined for $n = 0, 1, 2, ..., k$ and if (4.77) holds for $n \ge k$ then $\|x_0 - x_n\| \le t_0 - t_n < t_0 - t^*$ for $n \le k$. This shows that (4.76) is satisfied for $x = x_i$, $i \le k$. Thus (4.64) will be defined for $n = k + 1$, too. By (4.64) and (4.71) $\|x_1 - x_0\| \le d \le t_0 - t_1$. That is, (4.77) is true for $n = 0$. Suppose (4.77) holds for $n = 0, 1, 2, ..., k$. Observing that

$$B(x_{k+1})(x_{k+1} - x_{k+2})$$
$$= B(x_{k+1}, x_{k+1}) + y - x_{k+1} - B(x_k, x_k) - y + x_k + B(x_k)(x_k - x_{k+1})$$
$$= B(x_{k+1} - x_k, x_{k+1} + x_k) + x_k - x_{k+1} + B(x_k)(x_k - x_{k+1})$$
$$= B(x_{k+1} - x_k, x_{k+1} - x_k) + B(x_k - x_{k+1}, x_{k+1} - x_k)$$
$$+ B(x_{k+1}, x_{k+1} - x_k) - (x_{k+1} - x_k)$$
$$= (B(x_{k+1}) - I)(x_{k+1} - x_k),$$

we get

$$x_{k+1} - x_{k+2} = B(x_{k+1})^{-1}[B(x_{k+1} - x_0) + B(x_0) - I](x_{k+1} - x_k). \tag{4.78}$$

By taking norms in (4.78) and using (4.76) we obtain

$$\|x_{k+1} - x_{k+2}\| \leq \frac{[c + ab(t_0 - t_{k+1})](t_k - t_{k+1})}{1 - ab(t_0 - t_{k+1})} = t_{k+1} - t_{k+2}$$

by choice of t_0. Inequality (4.77) shows that $\{x_n\}$, $n = 0, 1, 2, ...$ is a Cauchy sequence in a banach space X and as such it converges to some $x^* \in X$. By taking yhe limit as $n \to \infty$ in (4.64) we get $x^* = y + B(x^*, x^*)$. That is x^* is a solution of equation (4.63). Fix n and let $p = 0, 1, 2, ...$. Then

$$\|x_n - x^*\| \leq \|x_n - x_{n+p}\| + \|x_{n+p} - x^*\| \leq t_n - t_{n+p} + \|x_{n+p} - x^*\|. \tag{4.79}$$

By letting $p \to \infty$ we obtain

$$\|x_n - x^*\| \leq t_n - t^*, \quad n = 0, 1, 2, ... \tag{4.80}$$

By (4.80) for $n = 0$ we get

$$\|x_0 - x^*\| \leq t_0 - t^* = \frac{1-c}{2ab} - t^* = \frac{1-c}{2ab}.$$

That is $x^* \in \overline{U}(x_0, r_0)$.

Finally, let us assume that there exists a second solution $z^* \in U(x_0, r_0)$ of equation (4.63).

By (4.64) we have

$$x_{n+1} - z^* = x_n - B(x_n)^{-1}(y + B(x_n, x_n) - x_n) - z^*$$
$$= B(x_n)^{-1}[B(x_n)(x_n - z^*) + x_n - y - B(x_n, x_n)$$
$$+ y + B(z^*, z^*) - z^*]$$
$$= -B(x_n)^{-1}[B(z^* - z_0) + B(x_0) - I](x_n - z^*).$$

By taking the norms in the above identity and using (4.76) we obtain

$$\|x_{n+1} - z^*\| \leq 2 \left[\frac{c + ab\|z^* - z_0\|}{1 + c + 2abt_{n+1}} \right] \|x_n - z^*\|.$$

By the choice of r_0 the factor of $\|x_n - z^*\|$ is less than 1 so that $\|x_n - z^*\|$ goes to zero as $n \to \infty$; hence $z^* = \lim_{n \to \infty} x_n = x^*$.

That completes the proof of the theorem. □

Moreover, we can show the following theorem:

Theorem 4.4 *Let B a bounded symmetric bilinear operator in $X \times X$ and suppose that $x_0, y \in X$ with $x_0 \neq 0$, $x_0 \neq y$. Assume:*

(i) *The following estimate is true*

$$4be < 1 \tag{4.81}$$

where

$$e \geq \|y\|. \tag{4.82}$$

(ii) *The hypotheses of Theorem 4.3 are satisfied for some $x_0 \in E$ such that*

$$\|x_0\| > p \text{ with a certain } p \in (p_1, p_2), \tag{4.83}$$

where p_1 and p_2 are the two positive solutions of the scalar quadratic equation

$$bz^2 - z + e = 0. \tag{4.84}$$

Then:

(a) *The iteration*

$$v_{n+1} = yB(v_n, v_n) \tag{4.85}$$

remains in $U(0, p_1)$ and converges to a unique solution v^* of equation (4.63) in $U\left(0, \frac{1}{2\|B\|}\right)$ for any $v_0 \in \overline{U}(0, p_1)$. Moreover, for all $n = 0, 1, 2, ...$

$$\|v_n - v^*\| \leq p_1 - e \sum_{j=0}^{n} \frac{(2j)!}{j!(j+1)!} (eb)^j.$$

(b) *The solution x^* of equation (4.63) obtained via iteration (4.64) is such that $x^* \neq v^*$.*

Proof. (a) The first part of the result in (a) follows from Corollary1 in [7], whereas the second part follows can be found in [54].

(b) We shall show that $\|x_n\| > p$ for a certain $p \in (p_1, p_2)$. By (4.64) we obtain $\|x_n - y\| = \|B(x_n, x_{n+1})\| \leq \|B\| \cdot \|x_n\| \|x_{n+1}\|$ or

$$\|x_{n+1}\| \geq \frac{\|x_n - y\|}{\|B\| \cdot \|x_n\|}.$$

Assume that $\|x_k\| > p$ for all $k = 0, 1, 2, ..., n$. since

$$\|x_n\| > p > e \tag{4.86}$$

it is enough to show

$$\frac{\|x_n\| - e}{b \|x_n\|} > p \tag{4.87}$$

or

$$\|x_n\| > \frac{e}{1 - pb}.$$

By (4.86) it finally suffices to show $p > \frac{e}{1-pb}$ which is true for $p \in (p_1, p_2)$. By taking the limit as $n \to \infty$ in (4.86) we get $\|x^*\| \geq p$. Therefore, we obtain $x^* \neq v^*$.

That completes the proof of the theorem. □

Furthermore, we can prove the following theorem concerning the number of solutions of equation (4.63).

Theorem 4.5 *Let B be a bounded symmetric bilinear operator on $X \times X$ and suppose that $x_0, y \in X$ with $x_0 \neq y$ and $y \neq 0$.*
Assume:

(i) *the point $x_0 \in E$ is such that*

$$B(x_0) = I; \tag{4.88}$$

(ii) *the inequality (4.81) is true.*

Then the elements v^*, x^*, $x_0 - x^*$ and $x_0 - v^*$ are solutions of equation (4.63) with

$$x^* \neq x_0 - x^* \tag{4.89}$$

and

$$v^* \neq x_0 - v^*. \tag{4.90}$$

Proof. It follows by (i) that the hypotheses of Theorem 4.3 are satisfied. That is x^* is a solution of equation (4.63). By (ii) v^* is a solution of equation (4.63). For $z = x_0 - x^*$ we have

$$y + B(z, z) = y + B(x_0 - x^*, x_0 - x^*)$$
$$= y + B(x_0, x_0) - 2B(x_0, x^*) + B(x^*, x^*)$$
$$= x^* + x_0 - 2x^* = x_0 - x^*.$$

Similarly we show that $x_0 - v^*$ is a solution of equation (4.63).

Let us assume now that

$$x_0 - x^* - x^*. \tag{4.91}$$

Then by (4.91) and (4.63) we have

$$x_0 2 (y + B(x^*, x^*)) = 2\left(y + \frac{1}{4}B(x_0, x_0)\right)$$

which implies

$$x_0 = 4y.$$

That is

$$x^* = 2y. \tag{4.92}$$

But then by (4.92) and (4.63) $2y = y + B(2y, 2y)$ or $4\|B\|\|y\| \geq 1$ since $y \neq 0$ contradicting (ii). This shows (4.89). Similarly we show (4.90) and that completes the proof of the theorem. $\square$

We can show the following.

Proposition 4.4 *Let B be a bounded symmetric bilinear operator on $X \times X$ and suppose that $x_0, y \in X$ with $x_0 \neq 0$, $x_0 \neq y$. Assume:*

(i) *The hypotheses of Theorem 4.3 are stisfied.*
(ii) *The inequality (4.81) is true.*
(iii) *The inequality*

$$\|x_0 - y\| > \frac{1 - 2b(e - r_0) - \sqrt{1 - 4eb}}{2b} = R \tag{4.93}$$

is true.

Then the solutions x^* and v^* obtained via Theorems 4.3 and 4.4 are distinct.

Proof. Assume that $x^* = v^*$. The solution x^* is such that

$$\|B(x^*, x^*)\| = \|x^* - y\| \tag{4.94}$$

and since $R.r_0$, (4.94) gives

$$\frac{1 - \sqrt{1 - 4eb}}{2b} \geq \|x^*\| \geq \sqrt{\frac{\|x_0 - y\| - r_0}{b}}. \tag{4.95}$$

But then from (4.95) we deduce

$$\|x_0 - y\| \leq R$$

contradicting (4.93).

That completes the proof of the proposition. □

Note that under the hypotheses of Theorem 4.5 and the above proposition it follows immediately that

$$x_0 - x^* \neq x_0 - v^*.$$

Remark 4.2

(a) *It can easily be seen that (4.68) and (4.69) can be replaced by the weaker condition*

$$\left\| B(x_0)^{-1} B \right\| \leq q. \tag{4.96}$$

(b) *If we know the constants a, b, c, d then we may compute the sequence $\{t_n\}$, $n = 0, 1, 2, \ldots$ before obtaining the sequence $\{x_n\}$, $n = 0, 1, 2, \ldots$ via the iterative algorithm (4.64). Therefore the estimates on the distances $\|x_n - x^*\|$ and $\|x_{n+1} - x_n\|$ obtained in Theorem 4.3 may be called apriori error estimates. Moreover the convergence of iteration (4.64) to a solution x^* of equation (4.63) is only linear. Let us assume that the linear operator*

$$\Gamma_0 = (I - 2B(x_0))^{-1} \tag{4.97}$$

exists for some $x_0 \in E$ and

$$\|\Gamma_0\| \leq b_0, \quad \|\Gamma_0(x_0 - y - B(x_0, x_0))\| \leq \eta_0, \quad h_0 = 2b_0 \|B\| \eta_0 \leq \frac{1}{2}. \tag{4.98}$$

Then the Newton-Kantorovich iteration

$$z_{n+1} = z_n - (I - 2B(z_n))^{-1}(z_n - y - B(z_n, z_n)),$$
$$n = 0, 1, 2, \ldots, \quad z_0 = x_0 \tag{4.99}$$

for solving (4.63) converges to a unique solution z_N^* of equation (4.63) in $U(x_0, r_N)$ with

$$r_N = \frac{1 - \sqrt{1 - 4bb_0\eta_0}}{2bb_0}. \qquad (4.100)$$

Moreover the order of convergence is quadratic. However we do not know if $\|z_n\| > p$ for a certain $p \in (p_1, p_2)$ whenever $\|z_0\| > p$. That is, we do not know if $\|z_N^*\| \not\geq p$ or if $z_N^* \neq v^*$.

It will be show later that whenever the hypotheses of Theorem 4.3 are satisfied then the Newton-Kantorovich hypotheses (4.98) are satisfied also and $x^* = z_N^*$.

That is, if we choose $x_0 = z_0$ with $\|x_0\| \geq p$, then

$$\|z_N^*\| \not\geq p \quad \text{and} \quad z_N^* \neq v^* \qquad (4.101)$$

even if $z_n \geq p$ for some n, $n = 0, 1, 2, \ldots$. Therefore in practice we will prefer to use iteration (4.99) instead of (4.64) to find bounded away from zero solution x^* of equation (4.63), since (4.99) converges faster than (4.64). However our main concern, that is, the property (4.100) could only be proved through iteration (4.64) as the following theorem indicates.

Theorem 4.6 *Under the hypotheses of Theorem 4.4 the Newton-Kantorovich iteration (4.99) for $z_0 = x_0$ converges to a unique solution z_N^* of equation (4.63) in $U(x_0, R_N)$ and $z_N^* = x^*$. Moreover, if $\|z_0\| > p$ for a certain $p \in (p_1, p_2)$ then*

$$\|z_N^*\| \geq p, \qquad (4.102)$$

and

$$\|z_n - z_N^*\| \leq \frac{1}{2^n}(2h_0)^{2^n - 1}\eta_0, \quad n = 0, 1, 2, \ldots .$$

Furthermore, the solution z_N^* can be written as $z_N^* = x_0 + h$ where h is a solution of the quadratic equation

$$h = y_1 + B_1(h, h) \qquad (4.103)$$

with

$$y_1 = (I - 2B(x_0))^{-1}(B(x_0, x_0) + y - x_0) \quad \text{and} \quad B_1 = (I - 2B(x_0))^{-1}B.$$

Proof. By the Banach lemma, the linear operator

$$B(x_0)^{-1} - 2I = \left(B(x_0)^{-1} - I\right) - I$$

is invertible since $\|I\| \cdot \left\|I - B(x_0)^{-1}\right\| \leq c < 1$ and

$$\left\|\left(B(x_0)^{-1} - 2I\right)^{-1}\right\| \leq \frac{1}{1-c}.$$

The equation (4.102) has a solution h if

$$4 \|y_1\| \|B_1\| \leq 4 \left[\frac{1}{1-c} \cdot d\right] \left[\frac{1}{1-c} \cdot ab\right] < 1 \qquad (4.104)$$

which is true by (4.67). It can easily be seen now that $w^* = x_0 + h$ is a solution of equation (4.63) if and only if h is a solution of equation (4.103). The linear operator $(I - 2B(x_0))^{-1}$ exists since

$$(I - 2B(x_0))^{-1} = \left(B(x_0)^{-1} - 2I\right)^{-1} B(x_0)^{-1}. \qquad (4.105)$$

The Newton-Kantorovich hypotheses (4.98) are now satisfied and by the definition of r_N, (4.81) and (4.82) we deduce that $z_N^* = w^*$. By the uniqueness of the solutions x^* and z_N^* in the balls $U(x_0, r_N)$ it follows that $z_N^* = x^*$ (the balls have the same center).

The rest of the theorem follows from part (ii) of Theorem 4.4 and Theorem 11.3 in [7, pp. 142].

all the results obtained in Theorems 4.4-4.5 and in the proposition can apply to iteration (4.99). Note that the result (4.102) is not known not even for quadratic systems in $\mathbb{R}^n$, $n > 1$.

To cover the cases when B is not symmetric we can state the following theorem whose proof as identical to that of Theorem 4.3 is omitted. $\square$

Theorem 4.7 *Let B be a bounded bilinear operator on $X \times X$ and suppose that $x_0, y \in X$ with $x_0 \neq 0$, $x_0 \neq y$. Further, let*

$$\overline{a} \geq \left\|B(x_0)^{-1} B\right\|, \quad \overline{b} \geq \left\|B(x_0)^{-1} (2\overline{B} - B)\right\|,$$
$$\overline{c} \geq \left\|B(x_0)^{-1} (2\overline{B} - B)(x_0) - I\right\|,$$

and let A, B be defined as

$$A = \frac{-[\overline{b}(\overline{a}+\overline{b})\overline{t}_0 + 2(\overline{a}\cdot\overline{c}+\overline{b})] + \left\{[\overline{b}(\overline{a}+\overline{b})\overline{t}_0 + 2(\overline{a}\overline{c}+\overline{b})]^2 + 4(\overline{a}\overline{c}+\overline{b})\overline{t}_0\left(\overline{a}^2 - \overline{b}^2\right)\right\}^{1/2}}{2\left(\overline{a}^2 - \overline{b}^2\right)},$$

$$\overline{t}_0 = \frac{1-\overline{c}}{\overline{a}+\overline{b}},$$

and
$$B = \frac{(1+\bar{c})(1-\bar{c})}{2\bar{a}(\bar{c}+3)}.$$

Assume:

(i) *The inverse of the linear operator* $B(x_0) : X \to X$ *with* $B(x_0)(x) = B(x_0, x)$ *for all* $x \in X$ *exists and is bounded;*

(ii) *The following estimates are true:*

$$\bar{a} \geq \bar{b},$$
$$0 \leq c < 1,$$
$$0 \leq d < \bar{t}_0 - A \quad \text{if} \quad \bar{a} > \bar{b}$$

and

$$0 \leq d < \bar{t}_0 - B \quad \text{if} \quad \bar{a} = \bar{b}.$$

Then

(a) *the real sequence* $\{\bar{t}_n\}$, $n = 0, 1, 2, \ldots$ *given by*

$$\bar{t}_{n+2} = \bar{t}_{n+1} - \frac{\bar{c} + \bar{b}\bar{t}_0 - \bar{b}\bar{t}_{n+1}}{1 - a\bar{t}_0 + \bar{a}\bar{t}_{n+1}} (\bar{t}_n - \bar{t}_{n+1}), \quad n = 0, 1, 2, \ldots$$
$$\bar{t}_1 = B \quad \text{if} \quad \bar{a} > \bar{b}$$

and

$$\bar{t}_1 = B \quad \text{if} \quad \bar{a} = \bar{b}$$

is positive and decreasingly converges to zero.

(b) *The sequence* $\{x_n\}$, $n = 0, 1, 2, \ldots$ *generated by* (4.64) *is well defined, remains in* $U(x_0, \bar{r}_0)$ *and converges to a unique solution* $x^* \in \overline{U}(x_0, \bar{r}_0)$ *of equation* (4.63) *with* $\bar{r}_0 = \bar{a}^{-1}$.

Moreover, the following estimates are true for all $n = 0, 1, 2, \ldots$

$$\|x_{n+1} - x_n\| \leq \bar{t}_n - \bar{t}_{n+1} \quad \text{and} \quad \|x_n - x^*\| \leq \bar{t}_n.$$

Remarks similar to the ones made after the proposition can now easily follow for Theorem 4.7.

The results obtained in the next three examples can also be obtained through the use of iteration (4.99). However we will only use iteration (4.64) for demonstrational purposes.

Example 4.1 Let $X = R^2$ and define a bilinear operator on X by

$$B(w, v) = \left\{ (w_1, w_2) \begin{bmatrix} \begin{bmatrix} b_{111} & b_{112} \\ b_{121} & b_{122} \\ b_{211} & b_{212} \\ b_{221} & b_{222} \end{bmatrix} \begin{bmatrix} v_1 \\ v_2 \end{bmatrix} \end{bmatrix} \right\}$$

$$(B(w))(v) = \begin{bmatrix} b_{111}w_1 + b_{121}w_2 & b_{112}w_1 + b_{122}w_2 \\ b_{211}w_1 + b_{221}w_2 & b_{212}w_1 + b_{222}w_2 \end{bmatrix} \begin{bmatrix} v_1 \\ v_2 \end{bmatrix}$$

$$= \begin{bmatrix} b_{111}w_1v_1 + b_{121}w_2v_1 + b_{112}w_1v_2 \\ b_{211}w_1v_1 + b_{221}w_2v_1 + b_{212}w_1v_2 + b_{222}w_2v_2 \end{bmatrix}.$$

Consider the quadratic equation on X given by

$$w = y + B(w, w)$$

or equivalently

$$w_1 = \frac{1}{48} - 3w_1^2 + 2w_1w_2 - w_2^2 \qquad (4.106)$$

$$w_2 = -\frac{1}{48} + w_1^2 - 2w_1w_2 - w_2^2$$

where

$$\begin{aligned} b_{111} &= -3 & b_{221} &= -1 \\ b_{112} &= 1, & b_{222} &= -1 \\ b_{121} &= 1, & y &= \begin{bmatrix} y_1 \\ y_2 \end{bmatrix}, \quad w = \begin{bmatrix} w_1 \\ w_2 \end{bmatrix} \\ b_{122} &= -1, \\ b_{211} &= 1, & y_1 &= \tfrac{1}{48} \\ b_{212} &= -1, & \text{and } y_2 &= -\tfrac{1}{48}. \end{aligned}$$

For $x \in X$, let $\|x\| = \max_{(i)} |x_i|$, $i = 1, 2, .$ Using the norm on $L(X, X)$ one can define the norm of B on X [10] by

$$\|B\| = \sup_{\|x\|=1} \max_{(i)} \sum_{j=1}^{2} \sum_{k=1}^{2} \left| \sum_{k=1}^{2} b_{ijk}x_k \right|,$$

from which it follows at once that $\|B\| \le \max_{(i)} \sum_{i=1}^{2} \sum_{k=1}^{2} |b_{ijk}|$. Let $x_0 = \begin{bmatrix} -.5 \\ -.5 \end{bmatrix}$. With the above values it can easily be seen that B is a bounded, symmetric operator on X and

$$B(x_0) = I, \quad e = d = \|y\| = \frac{1}{48}, \quad b = 6, \quad a = 1, \quad c = 0,$$

$$r_0 = \frac{1}{12}, \quad R = .08690776$$

and

$$\|x_0 - y\| = .520833333.$$

According to Theorem 4.4 (i), equation (4.106) has a small solution $v^* \in \overline{U}(0, p_1)$ which can be found to be $v^* = \begin{bmatrix} .0200308 \\ .0200308 \end{bmatrix}$ using the iteration (4.85) for v_0 for $v_0 = y$. We took $v_8 = v^*$. According to Theorem 4.3, equation (4.106) has a solution in $U(x_0, r_0)$ which can be found to be $x^* = \begin{bmatrix} -.5200308 \\ -.5200308 \end{bmatrix}$ using the iteration (4.64) for $x_0 = \begin{bmatrix} -.5 \\ -.5 \end{bmatrix}$. We took $x_9 = x^*$. Since $\|x_0 - y_0\| > R$, it was known before actually computing v^* and x^* that $x^* \neq v^*$. Note however that $x^* - x_0 - v^*$.

It can easily be seen that $v_1^* = \begin{bmatrix} -.25 \\ .1318813 \end{bmatrix}$, is the third solution of (4.106).

Finally, the forth solution x_1^* of equation (4.106) is given by $x_1^* = x_0 - v_1^*$. We have now found all four solutions of equation (4.106).

A more interesting example is given by the following.

Example 4.2 Cnsider the Riccati differential equation

$$x^2(t) + 2z(t)x(t) + y_1(t) - \frac{dx}{dt} = 0, \quad 0 \leq t < T < 1, \quad x(0) = 0. \quad (4.107)$$

As X takes $C_0'[0, T]$, the space of all continuously differentiable function $x = x(t)$, such that $x(0) = 0$, and as Y take the space $C[0, T]$ of all continuous real functions. Let us equip the above spaces with the usual sup-norm. That is

$$\|x\| = \sup_{0 \leq t \leq T} |x(t)| \quad \text{for} \quad x \in X \quad (\text{or } Y).$$

Equation (4.107) is a quadratic equation of the form (4.63) with $B(x_1, x_2) = B(x_1)(x_2)$ where $B(x_1)$ is a linear operator for fixed x_1 given by

$$B(x_1)(w)(t)$$
$$= \left[\left[\frac{d}{dt}2z\right]^{-1} x_1 w\right](t)$$
$$= \left[\exp\left[\int_0^t 2z(q)\,dq\right]\right] \int_0^t \exp\left[-\int_0^s 2z(q)\,dq\right] x_1(s)\,w(s)\,ds,$$

for all $w \in X$ and $0 \le t \le T$, and

$$y = \left[\frac{d}{dt} - 2z\right]^{-1} Y_1.$$

The linear operator $\frac{d}{dt} - 2z$ is indeed invertible for all $x \in X$, in fact, the inverse transformation $u = \left[\frac{d}{dt} - 2z\right]^{-1} v$ has the explicit representation

$$u(t) = \left[ext\left[\int_0^t 2z(q)\,dq\right]\right] \int_0^t \exp\left[-\int_0^s 2z(q)\,dq\right] v(s)\,ds, \quad 0 \le t \le T,$$

where $u \in X$ for $v \in Y$. It can easily be seen that the bilinear operator B defined above is bounded and symmetric. Using Definition 4.3 we deduce for $T = \frac{1}{2}$,

$$\|B\| = \frac{1}{2} \sup_{0 \le t \le T} \left|(1-t^2)\ln(1-t^2)\right| \le .375 \quad \text{for} \quad z(5) = -\frac{1}{1-t^2}.$$

Take $y_1(t) = -.14\frac{1+t^2}{1-t^2}$ then easily, $y(t) = -.14t$ for all $0 \le t \le T$ and $\|y\| = .07$.

The condition (i) in Theorem 4.4 is now satisfied. Moreover if the condition (ii) in Theorem 4.4 is satisfied for some x_0, then using iterations (4.64) and (4.85) we can obtain the solutions x^* and v^*, respectively, with $x^* \ne v^*$.

Example 4.3 There are examples of interesting linear operators satisfying condition (4.88). Indeed, with the notation of the previous example, let us define a linear operator $B(\cdot)$ by $B(v) = \left[\frac{d}{dt} - 2z\right]^{-1}(v)$. Choose z as before and $v(t) = x_0(t) = \frac{1+t^2}{1-t^2}$. It can then easily be seen that $B(x_0)(t) = I(t) = t$ for all $0 \le t \le T$, that is $B(x_0) = I$. Therefore the differential equation $\frac{du}{dt} - 2z(t)u(t) = v(t)$, $u(0) = 0$, has the unique solution u given by $u(t) = t$, $0 \le t \le T$.

Example 4.4 Consider the scalar equation $x = \delta + \beta x^2$ with $\delta, b > 0$ and $1 - 4\delta\beta > 0$.

Let us choose $\frac{1}{2\beta} < x_0 < \frac{1+\sqrt{1-4\delta\beta}}{2\beta}$. The conditions (4.66), (4.67) and (4.83) become, respectively,

$$x_0 \geq \frac{1}{2\beta}$$

$$\frac{2\beta + \sqrt{2(1-4\delta\beta)}}{2\beta} < x_0,$$

$$x_0 > p \quad \text{for} \quad p \in (p_1, p_2),$$

$$p_1 = \frac{1 - \sqrt{1-4\delta\beta}}{2\beta}, \quad p_2 = \frac{1 + \sqrt{1-4\delta\beta}}{2\beta}.$$

That is, x_0 must be chosen such that

$$\frac{2\beta + \sqrt{2(1-4\delta\beta)}}{4\beta} < x_0 < \frac{1+\sqrt{1-4\delta\beta}}{2\beta}.$$

The large solution of the scalar quadratic equation can now be obtained using iteration (4.64) for the above choice of x_0.

4.5 Approximation Using Finite Rank Operators

Consider the quadratic equation

$$x = y + B(x,x) \tag{4.108}$$

in a Banach space x, where $y \in x$ is fixed and B is a bounded symmetric bilinear operator on x [4]. We choose $z \in x$ and F to be a bounded symmetric bilinear operator on x in such a way that the following auxiliary quadratic equation is satisfied

$$z = y + F(z,z). \tag{4.109}$$

We then use the solutions of (4.109) to approximate the fixed points of (4.108).

We make use of the following version of Theorem 4.1.

Theorem 4.8 *Let P be a nonlinear operator defined on $D \subset X$ such that P is twice Fréchet differentiable on D. Let $z \in D$ be such that:*

(i) $\Gamma_0 = (P'(z))^{-1}$ *exists and is bounded;*
(ii) $\|P(z)\| \|\| \leq v$;

(iii) $\|P''(x)\|\| \leq b$ if $\|x-z\| < r$, $U(z,r) \subset D$;
(iv) $h = \|\Gamma_0\|^2 vb \leq \frac{1}{2}$;
(v) $r_0 = \left(1 - \sqrt{1-2h}\right) v \|\Gamma_0\|/h < r$.

Then there exists $x \in U(z, r_0)$ such that $P(x) = 0$. Furthermore, x is the only solution of P contained in $U(z,r) \cap U(z, r_1)$, where

$$r_1 = \left(1 + \sqrt{1-2h}\right) \|\Gamma_0\| v/h.$$

Definition 4.4 Let $z \in X$ be such that

$$z = y + F(z,z) \tag{4.110}$$

for some auxiliary bounded symmetric bilinear operator F defined on D. define the operator P on D by

$$P(x) = x - z + F(z,z) - B(x,x). \tag{4.111}$$

Then every solution x of (4.111) is a solution of (4.108).

Note that

$$P'(x) = I - 2B(x) \quad \text{and} \quad P''(x) = -2B.$$

The following theorem now follows easily from Theorem 4.8 and the above observations.

Theorem 4.9 *Let P, z be as in definition and such that:*

(i) $(I - 2B(z))^{-1}$ exists and is bounded;
(ii) $\|P(z)\| = \|(F-B)(z,z)\| \leq \|F - B\| \cdot \|z\|^2 = v$;
(iii) $\|P''(x)\| \leq 2\|B\| = b$ if $\|x - z\| < r$, $U(z,r) \subset D$;
(iv) $\overline{h} = \left\|(I - 2B(z))^{-1}\right\|^2 v \cdot b \leq \frac{1}{2}$;
(v) $r_0 = \left(1 - \sqrt{1 - 2\overline{h}}\right) v \cdot \left\|(I - 2B(z))^{-1}\right\|/\overline{h} < r$.

Then there exists $x \in U(z, r_0)$ such that $x = y + B(x,x)$ and x is unique in $U(z,r) \cap U(z, r_1)$, where

$$r_1 = \left(1 + \sqrt{1 - 2\overline{h}}\right) \left\|v(I - 2B(z))^{-1}\right\|/\overline{h}.$$

Note that if z is such that

$$\|z\| < \frac{1}{\|2B\|},$$

then the linear operator $(I - 2B(z))^{-1}$ exists and

$$\left\| (I - 2B(z))^{-1} \right\| \leq \frac{1}{1 - 2\|B\| \cdot \|z\|}.$$

In the above case, (iv) can be replaced by

$$\left(\frac{1}{1 - 2\|B\| \cdot \|z\|} \right)^2 \|F - B\| \cdot \|z\|^2 \, 2\|B\| \leq \frac{1}{2},$$

or

$$\|z\| \leq \left[2\sqrt{\|B\|} \left(\sqrt{\|B\|} + \sqrt{\|B - F\|} \right) \right]^{-1}. \tag{4.112}$$

We can prove the theorem.

Theorem 4.10 *Let B be defined on $D \subset X$ such that $B(x)$ is compact for each $x \in D$. Let $F(z)$ be a linear operator on D for some $z \in X$ such that*

$$z = y + F(z, z).$$

Assume:

(i) $(I - 2F(z))^{-1}$ *exists and is bounded above by some $K > 0$;*
(ii) $4\|F(z)B(z) - B(z)B(z)\| \leq \frac{1}{\|(I-2F(z))^{-1}\|}$;
(iii) $\|P(z)\| \leq v$;
(iv) $2\|B\| \leq b$ *if* $\|x - z\| < r$, $U(z, r) \subset D$;
(v) $h = K^2 v \cdot b$, $K = \frac{1 + 2\|(I-2F(z))^{-1}\| \cdot \|B(z)\|}{1 - 4\|(I-2F(z))^{-1}\| \|F(z)B(z) - B(z)B(z)\|}$;
(vi) $r_0 - (1 - \sqrt{1 - 2h}) K \cdot v/h < r$.

Then there exists $x \in U(z, r_0)$ such that $x = y + B(x, x)$ and x is unique in $U(z, r) \cap U(z, r_1)$, where

$$r_1 = \left(1 + \sqrt{1 - 2h} \right) K \cdot v/h.$$

Proof. We obviously have that $(I - 2B(z))^{-1}$ exists and is bounded above by K according to the lemma, (i), (ii) abd the compactness of $B(z)$. The rest follows by applying Theorem 4.8 to

$$P(x) = x - z + F(z, z) - B(x, x).$$

The natural question arises now, what are the best choices for F and z?

(a) Fpr $F = 0$, (4.109) gives $z = y$ and (4.112) requires $4\|B\| \cdot \|y\| \leq 1$.

(b) For $F = B$, (4.112) requires $\|z\| \leq \frac{1}{2\|B\|}$.

The best choice however for F and z must be such that

$$z = y + F(z, z).$$

The difficulties in finding solutions of the above auxiliary equation may be equivalent to those of finding solutions x of (4.108). However, if Q is the unique symmetric quadratic operator associated with F such that

$$Q(x) = F(x, x) \quad \text{for all} \quad x \in X$$

then (4.109) can be written as

$$z = y + Q(z). \tag{4.113}$$

Now assume that Q is of finite rank $v = \dim(span(Rang(Q)))$ and set $x = z - y$ to obtain

$$x = Q(x + y).$$

The above equation implies that the problem of solving the auxiliary equation can be translated to a finite dimensional one since x must lie in rang(Q). □

Definition 4.5 Let A denote the set of all bounded quadratic operators Q in X such that Q has finite rank. Denote by E, the set of all bounded quadratic functionals f on x.

Let $f \in E$, $d \in X$; the operator $f \otimes d : X \to X$ sending $x \in X$ to $f(x)d \in X$ is bounded quadratic operator of rank one. Thus

$$Q = \sum_{i=1}^{n} f_i \otimes d_i \in A$$

for any $f_i \in E$, $i = 1, 2, ..., n$, $d_i \in x$, $i = 1, 2, ..., n$.

Note that if $Q = X \to Y$ is a bounded quadratic operator and $L : Y \to Z$ is a bounded linear operator, then $L \circ Q : X \to Z$ is a bounded quadratic operator. (Q and L need not be of finite rank.)

Definition 4.6 Denote by $E \otimes X$ the vector subspace generated in the space of all bounded quadratic operators by the set $\{Q \in A \,/\, Q = f \otimes d, f \in E, d \in X\}$ so $Q \in E \otimes X$ if and only if

$$Q = \sum_{i=1}^{n} f_i \otimes d_i.$$

Theorem 4.11 $A = E \otimes X$.

Proof. Let $\{d_1, ..., d_n\}$ be a basis for $rang(Q)$ and choose g_i such that $g_i(d_i) = \delta_{ij}$, $i.j = 1, 2, ..., n$. since $rang(Q)$ is finite dimensional, the $\{g_i\}$, $i = 1, 2, ..., n$ functionals are bounded and by the Hahn-Banach theorem they can be extended linear functionals on x without increasing their norms. Let

$$f_i = g_i \circ Q, \quad i = 1, 2, ..., n.$$

Then the f_i, $i = 1, 2, ..., n$ are bounded quadratic functionals and

$$Q = \sum_{i=1}^{n} f_i \otimes d_i.$$

$\square$

Definition 4.7 Let f_i^*, $i = 1, 2, ..., n$ denote the symmetric bilinear functionals associated with the f_i, $i = 1, 2, ..., n$, given by

$$f_i^*(x, y) = \frac{1}{4}(f_i(x+y) - f_i(x-y)).$$

Denote by C' the matrix of the linear transformation $2B(y)(\circ)$ restricted to $rang(Q)$ relative to the basis $d_1, ..., d_n$. define the $n \times n$ matrix C, by

$$C = I - C',$$

$$\underline{\ell} = \begin{bmatrix} \ell_1 \\ \vdots \\ \ell_n \end{bmatrix}, \quad \text{by} \quad \ell_i = f_i(y), \quad i = 1, 2, ..., n,$$

the block of matrices $\underline{\underline{C}}$, $\underline{\underline{C}} = \begin{bmatrix} C_1 \\ \vdots \\ C_n \end{bmatrix}$ by $C_i = \{c_i^{jk}\}$ where $c_i^{jk} = f_i^*(d_j, d_k)$, $i, j, k = 1, 2, ..., n$.

define $\underline{v}$ by $\underline{v} = C^{-1}\underline{\ell}$ if $|C| \neq 0$ and the block of matrices $\underline{\underline{M}} = \begin{bmatrix} M_1 \\ \vdots \\ M_n \end{bmatrix}$

with $M_k = |C|^{-1} M'_k$ where each M'_k, $k = 1, 2, ..., n$ is the $n \times n$ matrix which results from the determinant of the matrix C if we replace the kth column by $\begin{bmatrix} C_1 \\ \vdots \\ V_n \end{bmatrix}$. Define $C\underline{M}$ by $\begin{bmatrix} CM_1 \\ \vdots \\ CM_n \end{bmatrix}$.

Note that M'_k, $k = 1, 2, ..., n$ is indeed a $n \times n$ matrix. for the case $n = 2$,

$$M'_1 = \begin{vmatrix} C_1 & c_{12} \\ C_2 & c_{22} \end{vmatrix} = c_{22}C_1 - c_{12}C_2.$$

$$M'_2 = \begin{vmatrix} c_{11} & C_1 \\ c_{21} & C_2 \end{vmatrix} = c_{11}C_2 - c_{21}C_1.$$

Theorem 4.12 *The point $w \in X$ is a solution of the auxiliary equation (4.113) if and only if*

$$w = y + \sum_{i=1}^{n} \xi_i d_i$$

where the vector $\underline{\xi} = \begin{bmatrix} \xi_1 \\ \vdots \\ \xi_n \end{bmatrix} \in R^n$ *(or $\mathbb{C}^n$) is a solution of*

$$\underline{x} = \underline{\ell} + C'\underline{x} + \underline{x}^{+r}\underline{C}\underline{x} \quad \text{in} \quad R^n \quad (\text{or } \mathbb{C}^n). \tag{4.114}$$

Moreover, if $|C| = |I - C'| \neq 0$ the Cramer's rule transforms the above to

$$\underline{x} = \underline{v} + \underline{x}^{+r}\underline{M}\underline{x} \quad \text{in} \quad R^n \quad (\text{or } \mathbb{C}^n). \tag{4.115}$$

Proof. Assume that (4.113) has a solution $w \in X$. Then

$$w - y + Q(w)$$
$$= y + \sum_{i=1}^{n} f_i(w) d_i.$$

Apply $f_1, f_2, ..., f_n$ in turn to this vector identity to obtain for $p =$

$1, 2, ..., n$

$$f_p(w) = f_p\left(y + \sum_{k=1}^{n} f_i(w) d_i\right)$$
$$= f_p(y) + \sum_{k=1}^{n} f_k^2(w) f_p(d_k) + 2\sum_{k=1}^{n} f_k(w) f_p^*(y, d_k)$$
$$+ 2\sum_{i \neq j}^{n} f_i(w) f_j(w) f_p^*(d_i, d_j).$$

Letting

$$f_i(w) = x_i, \quad i = 1, 2, ..., n$$

and writing these equations in vector form, we obtain

$$\underline{x} = \underline{\ell} + C'\underline{x} + \underline{x}^{+r}\underline{\underline{C}}\underline{x}$$

or

$$C\underline{x} = \underline{\ell} + \underline{x}^{+r}\underline{\underline{C}}\underline{x}.$$

since $|C| \neq 0$, we obtain (4.115) by composing both sides of the above equation by C^{-1}. □

Conversely, given (4.115), assume (4.114) has a solution vector $\underline{\xi} = \begin{bmatrix} \xi_1 \\ \vdots \\ \xi_n \end{bmatrix}$. Let $w \in X$ be defined as

$$w = y + \sum_{i=1}^{n} \xi_i d_i.$$

Apply $f_1, f_2, ..., f_n$ in turn to this vector identity to obtain for $p = 1, 2, ..., n$,

$$f_p(w) = f_p(y) + \sum_{k=1}^{n} \xi_k^2 f_p(d_k) + 2\sum_{k=1}^{n} \xi_k f_p^*(y, d_k) + 2\sum_{k=1}^{n} \xi_k f_p^*(y, d_k)$$
$$+ 2\sum_{i \neq j}^{n} \xi_i \xi_j f_p^*(d_i, d_i),$$

or in matrix notation,

$$\underline{f(w)} = \underline{\ell} + C'\underline{\xi} + \underline{\xi}^{+r}\underline{\underline{C}}\underline{\xi}.$$

Now since $\underline{\xi}$ satisfies (4.114) we have

$$\underline{\xi} = \underline{\ell} + C'\underline{\xi} + \underline{\xi}^{+r}\underline{\underline{C}}\underline{\xi}.$$

Comparing the last two equations, we get

$$\xi_i = f_i(w), \quad i = 1, 2, ..., n,$$

so

$$w = y + \sum_{i=1}^{n} f_i(w) d_i,$$

or

$$w = y + Q(w).$$

Therefore, w is a solution of (4.113) and the theorem is proved.

Example 4.5 Let $X = C[0, 1]$ and consider the equation

$$x(s) = s + s \int_0^1 x^2(t) \, dt$$

where $s \in [0, 1]$. This equation is of the form (4.113), with $rank(Q) = 1$,

$$y(s) = s$$
$$d = s, \quad \text{and}$$
$$d(s) = \int_0^1 x^2(t) \, dt.$$

Using the formula,

$$f^*(v, w) = \frac{1}{4}(f(v+w) - f(v-w)),$$

we have

$$C = 1 - 2f^*(y, d) = 1 - 2\frac{1}{4} \int_0^1 4s^2 ds = \frac{1}{3}$$

$$\underline{\ell} = f(y) = f(s) = \int_0^1 s^2 ds = \frac{1}{3}$$

$$\underline{C} = f(d) = f(s) = \int_0^1 s^2 ds = \frac{1}{3}$$

$$\underline{v} = 3 \cdot \frac{1}{3} = 1$$

$$\underline{M} = 3 \cdot \frac{1}{3} = 1.$$

Therefore, (4.114) becomes

$$\xi = 1 + \xi^2 \quad \text{in} \quad \mathbb{C} \quad \text{with solutions} \quad \frac{1 \pm i\sqrt{3}}{2};$$

since $x = y + \xi d$, we finally have

$$x(s) = \left(\frac{3 \pm i\sqrt{3}}{2}\right) S.$$

Now note that if the linear operator $F(z)$ is of finite rank n then the linear operator $I - 2F(z)$ is invertible if and only if for every fixed $v \in X$ there exists $w \in X$ such that

$$w - 2F(z, w) = v.$$

Since $F(z)$ is of finite rank n, the above equation can be translated exactly as in Rheorem 4.12, for the quadratic case to a linear system in R^n, or $\mathbb{C}^n$, similar to system (4.115).

4.6 Projection Methods for Approximating Fixed Points

Consider the problem of approximating a fixed point x^* of the operator equation

$$x = T(x) \tag{4.116}$$

where $T(x)$ is a nonlinear operator defined on a subset D of a Banach space X with values in a Banach space Y.

we study the convergence of the Newton methods

$$x_{n+1} = T(x_n) - PT'(x_n)(x_n - x_{n+1}), \quad n \geq 0 \tag{4.117}$$

and

$$y_{n+1} = T(y_n) - PT'(x_0)(y_n - y_{n+1}), \quad x_0 = y_0, \quad n \geq 0 \tag{4.118}$$

to x^*, where $T'(x_n)$ is the Fréchet derivative of T evaluated at x_n and P is a linear projection operator projecting X on its subspace X_P. If X_P is a finite dimensional space with $\dim(X_P) = N$, then the iterates (4.117) and (4.118) can be computed at each step by solving a system of linear algebraic equations of order at most N. The case when $P = I$, the identity operator on X, has been examined by many authors, under different assumptions, [68], [99]. The iterates, however, can rarely be computed in infinite dimensional spaces, since it may be very difficult or impossible to find the inverses of the linear operators $I - T'(x_n)$, $n \geq 0$.

In this section, we provide sufficient conditions for the convergence of iterations (4.117) and (4.118) to a locally unique fixed point x^* of equation (4.116).

Finally, we illustrate our results with an example.

We can now formulate our main theorem concerning iteration (4.117).

Theorem 4.13 *Let $T : D \subset X \to Y$ and assume*

(a) *the inverse of the linear operator $I - PT'(x_0)$ exists and*

$$\left\| (I - PT'(x_0))^{-1} (x_0 - T(x_0)) \right\| \leq \eta; \qquad (4.119)$$

(b) *the following inequalities are true for all*

$$x, y \in U(x_0, r) = \{ x \in E / \|x - x_0\| < r \} :$$

$$\left\| (I - PT'(x_0))^{-1} (PT'(x) - PT'(y)) \right\| \leq M \|x - y\|^\lambda \qquad (4.120)$$

and

$$\left\| (I - PT'(x_0))^{-1} (QT(x) - QT(y)) \right\| \leq q \|x - y\|^\lambda,$$
$$Q = I - P, \quad \lambda \in [0, 1). \qquad (4.121)$$

(c) *The conditions*

$$(\eta d)^\lambda < 1, \qquad (4.122)$$

$$Mr^\lambda < 1, \qquad (4.123)$$

$$\eta + \frac{ed^{-1}}{1-e} \leq r \qquad (4.124)$$

are satisfied, where

$$e = (d\eta)^\lambda, \quad d^{\lambda - 1} = c$$

and

$$c(r) = c = \frac{1}{1 - Mr^\lambda} \left(\frac{2Mr}{1+\lambda} + q \right).$$

(d) The ball $\overline{U}(x_0, r) \subset D$.

Then, equation (4.116) has a fixed point x^* in $\overline{U}(x_0, r)$ where r is chosen to be the minimum number $r > 0$ satisfying (4.123) – (4.124). Moreover, the following estimates are true

$$\|x_n - x^*\| \leq d^{-1} \frac{e^n}{1 - e}, \quad n \geq 0 \tag{4.125}$$

and

$$\|x_{n+1} - x_n\| \leq c \|x_n - x_{n-1}\|^\lambda, \quad n \geq 1. \tag{4.126}$$

Furthermore, if

$$(dr)^\lambda < 1 \tag{4.127}$$

then x^* is the unique fixed point of equation (4.116) in $\overline{U}(x_0, r)$.

Proof. From (4.117) and (4.118) we get the identity

$$(I - PT'(x_n))(x_{n+1} - x_n) = T(x_n) - T(x_{n-1})$$
$$- PT'(x_{n-1})(x_n - x_{n-1}), \quad n \geq 1. \tag{4.128}$$

By the Banach lemma on invertible operators, (4.120) and (4.123), it follows that $I - PT'(x)$ is invertible for all $x \in U(x_0, r)$ and

$$\left\|(I - PT'(x))^{-1}(I - PT'(x_0))\right\| \leq \frac{1}{1 - M\|x - x_0\|^\lambda} \leq \frac{1}{1 - Mr^\lambda}. \tag{4.129}$$

Let us assume that $x_0, x_1, ..., x_n \in U(x_0, r)$, then from (4.119)-(4.121), (4.128) and (4.129) we get

$$\|x_{n+1} - x_n\| \leq \left\|(I - PT'(x_n))^{-1}(I - PT'(x_0))\right\|$$
$$\cdot \left[\left\|(I - PT'(x_0))^{-1}(PT(x_n) - PT(x_{n-1}) - PT'(x_{n-1})(x_n - x_{n-1}))\right\|\right.$$
$$\left. + \left\|(I - PT'(x_0))^{-1}(QT(x_n) - QT(x_{n-1}))\right\|\right]$$
$$\leq \tfrac{1}{1 - Mr^\lambda} \left[\left\|(I - PT'(x_0))^{-1} \int_0^1 PT'(x_{n-1} + t(x_n - x_{n-1}))\right.\right.$$
$$\left.\left. - PT'(x_{n-1})(x_n - x_{n-1}) \, dt\right\| + q \|x_n - x_{n-1}\|^\lambda\right]$$
$$\leq \tfrac{1}{1 - Mr^\lambda} \left[\tfrac{M}{1 + \lambda} \|x_n - x_{n-1}\| + q\right] \|x_n - x_{n-1}\|^\lambda$$
$$\leq c \|x_n - x_{n-1}\|^\lambda, \quad \text{which shows (4.126).} \tag{4.130}$$

From (4.126), we get

$$\begin{aligned}
\|x_0 - x_{n+1}\| &\leq \|x_1 - x_0\| + \|x_2 - x_1\| + \cdots + \|x_n - x_{n+1}\| \\
&\leq \eta + c\eta^\lambda + c^{1+\lambda}\eta^{\lambda^2} + \cdots + c^{1+\lambda+\cdots+\lambda^{n-1}}\eta^{\lambda^n} \\
&\leq \eta + d^{-1}\left[(d\eta)^\lambda + (d\eta)^{\lambda^2} + \cdots + (d\eta)^{\lambda^n}\right] \\
&\leq \eta + d^{-1}\left[(d\eta)^\lambda + (d\eta)^{2\lambda} + \cdots + (d\eta)^{n\lambda}\right] \\
&\leq \eta + d^{-1}e\left(1 + e + e^2 + \cdots + e^{n-1}\right) \\
&\leq \eta + d^{-1}e\frac{1-e^n}{1-e} \leq \eta + d^{-1}e\frac{1}{1-e} \leq r \quad \text{(by (4.124))}.
\end{aligned}$$

hence, $x_{n+1} \in U(x_0, r)$. For $p \geq 1$,

$$\begin{aligned}
\|x_n - x_{n+p}\| &\leq \|x_n - x_{n+1}\| + \|x_{n+1} - x_{n+2}\| + \cdots + \|x_{n+p-1} - x_{n+p}\| \\
&\leq d^{-1}(d\eta)^{\lambda^n} + d^{-1}(d\eta)^{\lambda^{n+1}} + \cdots + d^{-1}(d\eta)^{\lambda^{n+p}} \\
&\leq d^{-1}e^n\left[1 + e + \cdots + e^{p-1}\right] = d^{-1}e^n\frac{1-e^p}{1-e}. \quad (4.131)
\end{aligned}$$

It now follows from that the sequence $\{x_n\}$ is a Cauchy sequence in a Banach space and as such it converges to some $x^* \in \overline{U}(x_0, r)$. By letting $p \to \infty$ in (4.131) we obtain (4.125), whereas by letting $n \to \infty$ in (4.117) we get $x^* = T(x^*)$. To show uniqueness let us assume that z^* is any fixed point of T in $U(x_0, r)$ and use the identity

$$(I - PT'(x_n))(x_{n+1} - z^*) = T(x_n) - T(z^*) - PT'(x_n)(x_n - z^*)$$

to get

$$\begin{aligned}
\|x_{n+1} - z^*\| &\leq c\|x_n - z^*\|^\lambda \leq \cdots \leq d^{-1}(dr)^{\lambda^n} \\
&\leq d^{-1}(dr)^{\lambda^n} \leq d^{-1}(dr)^{\lambda^n} \to 0
\end{aligned}$$

as $n \to \infty$ from (4.127). Hence $x^* = \lim_{n \to \infty} x_n = z^*$.

That completes the proof of the theorem. $\square$

Note that for $\lambda = 1$ the proof of the previous theorem can be repeated, but (4.122) becomes $c < 1$, (4.124) becomes $\frac{\eta}{1-c} \leq r$, $e = c$, (4.125) becomes $\|x_n - x^*\| \leq \frac{e^n}{1-e}\eta$ and (4.127) becomes $c < 1$.

The proof of the following theorem concerning iteration (4.118) is omitted as similar to the proof of Theorem 4.13.

Theorem 4.14 *Let $T : D \subset X \to Y$ and assume*

(a) *the following inequalities are true:*

$$\left\|(I - PT'(x_0))^{-1}(x_0 - T(x_0))\right\| \leq \eta,$$

$$\left\|(I - PT'(x_0))^{-1}(PT'(x) - PT'(y))\right\| \leq M \|x-y\|^\lambda$$

and

$$\left\|(I - PT'(x_0))^{-1}(QT(x) - QT(y))\right\| \leq q \|x-y\|^\lambda,$$
$$Q = I - P, \lambda \in [0,1),$$

for all $x, y \in U(x_0, R)$.

(b) *The conditions*

$$(\eta d_1)^\lambda < 1,$$

$$\eta + \frac{e_1 d_1^{-1}}{1 - e_1} \leq R$$

are satisfied, where

$$e_1 = (d_1 \eta)^\lambda, \quad d_1^{\lambda-1} = c_1$$

and

$$c_1(r) = c_1 = 2^{1-\lambda} MR + q.$$

(c) *The ball* $\overline{U}(x_0, R) \subset D$.

The equation (4.116) has a fixed point x^* in $\overline{U}(x_0, R)$ where R is chosen to be the minimum number $R > 0$ satisfying conditions (b). Moreover, the following estimates are true

$$\|y_n - x^*\| \leq d_1^{-1} \frac{e_1^n}{1 - e_1}, \quad n \geq 0$$

and

$$\|y_{n+1} - y_n\| \leq c_1 \|y_n - y_{n-1}\|^\lambda, \quad n \geq 1.$$

Furthermore if

$$(d_1 R)^\lambda \leq 1$$

then x^* is the unique fixed point of equation (4.116) in $\overline{U}(x_0, R)$.

Note that a remark similar to the one made after Theorem 4.13 for the case $\lambda = 1$ can now easily follow for Theorem 4.14.

We now complete this paper with an application.

Example 4.6 Let us consider the following system in $X = Y = R^k$

$$v_{io} = f_i(v_1, ..., v_k), \quad i = 1, 2, ..., k. \tag{4.132}$$

Set

$$T(v) = \{f_i(v_1, ..., v_k)\}, \quad i = 1, 2, ..., k;$$

$$T'(w)v = \left\{ \sum_{j=1}^{k} f'_{i,j}(w_1, ..., w_k) v_j \right\}, \quad i = 1, 2, ..., k;$$

$$PT'(w)v = \begin{cases} \sum_{j=1}^{k} f'_{i,j}(w_1, ..., w_k) v_j, & i = 1, 2, ..., N \\ 0, & i = N+1, ..., k, \end{cases}$$

where the symbol f'_{ij} denotes $\partial f_i / \partial v_j$.

Iterations (4.117) and (4.118) can be written as

$$v_{i,n+1} = f_i(v_{1,n}, \cdots, v_{k,n})$$
$$+ \sum_{j=1}^{k} f'_{ij}(v_{1,n}, \cdots, v_{k,n})(v_{j,n+1} - v_{j,n}), \quad i = 1, \cdots, N$$

$$v_{i,n+1} = f_i(v_{1,n}, \cdots, v_{k,n}), \quad i = N+1, \cdots, k \tag{4.133}$$

and

$$\overline{v}_{i,n+1} = f_i(\overline{v}_{1,n}, \cdots, \overline{v}_{k,n})$$
$$+ \sum_{j=1}^{k} f'_{ij}(\overline{v}_{1,0}, \cdots, \overline{v}_{n,0})(\overline{v}_{j,n+1}, \cdots, \overline{v}_{j,n}), \quad i = 1, ..., N$$

$$\overline{v}_{i,n+1} = f_i(\overline{v}_{1,n}, \cdots, \overline{v}_{k,n}), \quad i = N+1, ..., k, \tag{4.134}$$

respectively.

If the determinants $D(x_n)$ and D_0 of (4.133) and (4.134) respectively, are nonzero, then we have

$$v_{i,n+1} = \frac{\sum_{m=1}^{N} D_{im}(v_n) \overline{f}_m(v_n)}{D(v_n)}, \quad i = 1, 2, ..., N,$$

$$v_{i,n+1} = f_i(v_n), \quad i = N+1, \cdots, k$$

for system (4.133) and

$$\bar{v}_{i,n+1} = \frac{\sum_{m=1}^{N} D_{im}(v_0) \bar{f}_m(v_0)}{D_0}, \quad i = 1, ..., N,$$

$$\bar{v}_{i,n+1} = f_i(\bar{v}_n), \quad i = N+1, ..., k$$

for system (4.134).

Here

$$\bar{f}_m(v_n) = f_m(v_n) - \sum_{i=1}^{k} f'_{mj}(v_n) v_{j,n} + \sum_{i=N+1}^{k} f'_{mj}(v_n) f_j(v_n),$$

$$\bar{f}_m(v_0) = f_m(v_0) - \sum_{j=1}^{k} f'_{mj}(v_0) v_{j,n} + \sum_{i=N+1}^{k} f'_{mj}(v_0) f_j(v_0),$$

$m = 1, 2, ..., k$, where $D_{im}(v_n)$, $D_{im}(v_0)$ are the cofactors of the elements at the intersection of the m-th row and i-th column of the determinants $D(x_n)$ and D_0, respectively.

We assume that the following conditions are satisfied on some region under consideration.

$$|f_i(v_1, ..., v_k) - f_i(w_1, ..., w_k)|$$
$$\leq \sum_{j=1}^{k} t_{ij} |v_j - w_j|^\lambda, \quad i = N+1, ..., k, \quad \lambda \in [0, 1]$$
$$|f'_{ij}(v_1, ..., v_k) - f'_{ij}(w_1, ..., w_k)|$$
$$\leq \sum_{s=1}^{k} b_{ijs} |v_s - w_s|^\lambda, \quad i = 1, ..., N, \quad j = 1, ..., k,$$
$$|D_{im}(v)| \leq a_{im'} |D(v)| \leq a,$$
$$|f'_{ij}(v)| \leq h_{ij}, \quad i = 1, ..., N, \quad j = 1, 2, ..., k.$$

For any $v \in X$, set $\|v\| = \sup_{1 \leq i \leq k} |v_i|$, then the constants q and M appearing in the Theorems 4.13-4.14 can be computed by

$$q \leq \sup_{i=N+1,...,k} \sum_{j=1}^{k} t_{ij} \quad \text{and} \quad M \leq \sup_{i=1,2,...,N} \sum_{j,j=1}^{k} c_{ijs}.$$

4.7 Solving Nonlinear Equations with a Nondifferentiale Term

Consider the fixed point problem

$$T(x) = x \text{ with } T(x) = F(x) + G(x) \qquad (4.135)$$

where F, G are nonlinear operators defined on some convex subset D of a Banach space X with values in a Banach space X. We assume that F is Fréchet-differentiable on D, whereas G is not. Zabrejko-Nguen in [294] and others [298], [68] have proposed the modified Newton-Kantorovich iteration

$$z_{n+1} = z_n - (F'(z_n) - I)^{-1}(F(z_n) + G(z_n) - z_n), \quad z_0 = x_0, \quad n \geq 0, \qquad (4.136)$$

for approximating a fixed point x^* of equation (4.135).

The above authors showed that under certain conditions, iteration (4.136) generates a sequence which converges to x^* for $G = 0$, iteration (4.136) reduces to the classical Newton-Kantorovich method which has been studied in 3.2.

However the iterates $\{z_n\}$, $n \geq 0$ can rarely be computed in infinite-dimensional spaces, since it may be difficult or even impossible to compute the inverses of the linear operators $F'(z_n) - I$, $n \geq 0$.

In this section we will make practical use of iteration (4.136), by considering the iteration

$$x_{n+1} = x_n - (I - PF'(x_n))^{-1}(x_n - T(x_n)), \quad n = 0, 1, 2, \ldots \qquad (4.137)$$

where P is a projection operator $(P^2 = P)$ on D.

Let us assume that the inverse of the operator $I - PF'(x_0)$ exists and

$$\left\|(I - PF'(x_0))^{-1}[PF'(x_1) - PF'(x_2)]\right\| \leq K_1(r)\|x_1 - x_2\|, \qquad (4.138)$$

$$\left\|(I - PF'(x_0))^{-1}[(QF(x_1) + G(x_1)) - (QF(x_2) + G(x_2))]\right\|$$
$$\leq K_2(r)\|x_1 - x_2\| \qquad (4.139)$$

for all $x_1, x_2 \in \overline{U}(x_0, r) \subset \overline{U}(x_0, R)$ with $Q = I - P$, where $K_1(r)$ and $K_2(r)$ are nonnegative, nondecreasing functions on $[0, R]$. We note that for $P = I$ the conditions (4.138)-(4.139) reduce to the Zabrejko-Nguen conditions given in [294], [298].

It is easy to see that the solution of iteration (4.137) reduces to solving certain operator equations in the space X_P. If moreover X_P is a finite dimensional space of dimension N, we obtain a system of linear algebraic equations of order at most N.

We will provide sufficient conditions for the convergence of iteration (4.137) to x^* as well as error bounds on the distances $\|x_{n+1} - x_n\|$ and $\|x_n - x^*\|$, $n \geq 0$.

Finally we illustrate our results by considering a nondifferentiable nonlinear integral equation.

We will need to introduce the constant

$$a = \left\|(I - PF'(x_0))^{-1}(x_0 - T(x_0))\right\|$$

and the functions

$$\omega(r) = \int_0^r K_1(t)\,dt,$$

$$\varphi(r) = a + \int_0^r w(t)\,dt - r,$$

$$\psi(r) = \int_0^r K_2(t)\,dt,$$

$$\varkappa(r) = \varphi(r) + \psi(r).$$

We can now prove the main theorem.

Theorem 4.15 *Suppose that the function $\varkappa(r)$ has a unique zero s^* in the interval $[0, R]$ and $\varkappa(R) \leq 0$.*

Then

(a) *the equation (4.135) has a fixed point $x^* \in \overline{U}(x_0, s^*)$, which is unique in $\overline{U}(x_0, R)$;*

(b) *the iterates generated by (4.137) are well defined, remain in $\overline{U}(x_0, s^*)$ for all $n \geq 0$ and satisfy*

$$\|x_{n+1} - x_n\| \leq s_{n+1} - s_n, \quad n \geq 0 \tag{4.140}$$

and

$$\|x_n - x^*\| \leq s^* - s_n, \quad n \geq 0 \tag{4.141}$$

where the sequence $\{s_n\}$, $n \geq 0$ given by

$$s_{n+1} = s_n - \frac{\varkappa(s_n)}{\varphi'(s_n)} = s_n + u(s_n),$$

with $\quad u(r) = -\dfrac{\varkappa(r)}{\varphi'(r)}, \quad s_0 = 0, \quad n \geq 0$

is monotonically increasing and converges to s^*.

Proof. The function $\varkappa(r)$ is positive on $[0, s^*]$, since s^* is the unique zero of $\varkappa(r)$. Exactly as in Proposition 3 in [294, p. 677] we can show that the function $\varphi'(r)$ is negative on $[0, s^*]$ and that the sequence $\{s_n\}$, $n \geq 0$ is monotonically increasing and converges to s^*.

We will only show (4.140), since (4.141) will follow then immediately. We must show that the iterates $\{x_n\}$, $n \geq 0$ belong to $\overline{U}(x_0, s_n) \subset \overline{U}(x_0, s^*)$ and that the inverses $I - PF'(x_n)$, $n \geq 0$ exist. For $n = 0$, (4.140) becomes

$$a = \|x_1 - x_0\| = s_1 - s_0 = a.$$

Hence, (4.140) is true for $n = 0$. Suppose the (4.140) is true for $N < k$; then

$$\|x_k - x_0\| \leq \sum_{j=1}^{k} \|x_j - x_{j-1}\| \leq \sum_{j=1}^{k} (s_j - s_{j-1}) = s_k.$$

but by (4.138) and the result on [294, p. 676]

$$\left\|(I - PF'(x_0))^{-1}(PF'(x_k) - PF'(x_0))\right\| \leq w(s_k) < w(s^*) = \varphi'(s^*) + 1 \leq 1.$$

By the Banach lemma on invertible operators $(I - PF'(x_k))^{-1}$ exists and

$$\left\|(I - PF'(x_k))^{-1}(I - PF'(x_0))\right\| \leq -\frac{1}{\varphi'(s_k)}. \quad (4.142)$$

Using the identity

$$x_{k+1} - x_k = \left[(I - PF'(x_k))^{-1}(I - PF'(x_0))\right]\{(I - PF'(x_0))^{-1}$$
$$[(PF(x_k) - PF(x_{k-1}) - PF'(x_{k-1})(x_k - x_{k-1}))$$
$$+ ((G(x_k) + QF(x_k))) - (G(x_{k-1}) + QF(x_{k-1}))]\}, \quad (4.143)$$

(4.138), (4.139) and (4.142), we get

$$\left\|(I - PF'(x_0))^{-1}[PF(x_k) - PF(x_{k-1}) - PF'(x_{k-1})(x_k - x_{k-1})]\right\|$$
$$\leq \int_0^1 \left\|(I - PF'(x_0))^{-1}(PF'((1-t)x_{k-1} + tx_k) - PF'(x_{k-1}))\right\| \|x_k - x_{k-1}\| dt$$
$$\leq \int_0^1 (w((1-t)s_{k-1} + ts_k) - w(s_{k-1}))(s_k - s_{k-1}) dt$$
$$= \int_{s_{k-1}}^{s_k} w(t) dt - w(s_{k-1})(s_k - s_{k-1})$$

and
$$\left\|(I - PF'(x_0))^{-1}[(G(x_k) + QF(x_k)) - (G(x_{k-1}) + QF(x_{k-1}))]\right\|$$
$$\leq \int_{s_{k-1}}^{s_k} K_2(t)\,dt = \psi(s_k) - \psi(s_{k-1}).$$

With these majorizations, (4.143) becomes

$$\|x_{k+1} - x_k\| \leq -\frac{\varphi(s_k) - \varphi(s_{k-1}) - \varphi'(s_{k-1})(s_k - s_{k-1}) + \psi(s_k) - \psi(s_{k-1})}{\varphi'(s_k)} = s_{k+1} - s_k. \tag{4.144}$$

Hence (4.140) is true for $n = k$. Since, the sequence $\{s_n\}$, $n \geq 0$ majorizes the sequence $\{x_n\}$, $n \geq 0$, there exists $x^* \in \overline{U}(x_0, s^*)$ such that $x^* = \lim_{n \to \infty} x_n$. By taking the limit in (4.137) we get $x^* = T(x^*)$. To show uniqueness we consider the sequences

$$y_{n+1} = y_n - (I - PF'(x_0))^{-1}(y_n - T(y_n)), \quad n \geq 0,$$
$$v_{n+1} = v_n - (I - PF'(v_0))^{-1}(v_n - t(v_n)), \quad n \geq 0, \quad v_0 \in \overline{U}(x_0, R),$$
$$q_{n+1} = d(q_n), \quad n \geq 0, \quad q_0 = 0, \quad d(r) = r + \varkappa(r)$$

and

$$p_{n+1} = d(p_n), \quad n \geq 0, \quad p_0 = R. \tag{4.145}$$

Then it is simple calculus to show that the iteration $\{q_n\}$, $n \geq 0$ is monotonically increasing and converges to s^*, whereas the sequence $\{p_n\}$, $n \geq 0$ is monotonically decreasing and converges to s^* also.

Exactly as we derived (4.144), we get

$$\|y_{n+1} - y_n\| \leq q_{n+1} - q_n, \quad n \geq 0 \tag{4.146}$$

and

$$\|y_n - v_n\| \leq p_n - q_n, \quad n \geq 0. \tag{4.147}$$

From (4.146) we get $\lim_{n \to \infty} y_n = x^* \in \overline{U}(x_0, s^*)$. If for v_0 we choose the second solution $x_1^* \in \overline{U}(x_0, s^*)$ of equation (4.135), we get by (4.147) that $\|x^* - x_1^*\| \leq p_n - q_n$ and hence $x^* = x_1^*$.

That completes the proof of the theorem. □

For completion we will now obtain some further error bounds. Let $r_n = \|x_n - x_0\|$, $K_n(r) = K_1(r_n + r)$ and $X_n(r) = K_2(r_n + r)$ for $r \in [0, R - r_n]$ and set $a_n = \|x_{n+1} - x_n\|$ $b_n = (1 - w(r_n))^{-1}$. Without loss of generality, we may assume that $a_n > 0$. Then by following exactly the same steps as the proof of Theorem 2 in [285, p. 989] and Theorem 2 in [294, p. 680] we can easily prove the theorem.

Theorem 4.16 *Suppose that the hypotheses of Theorem 4.15 are true. Then*

(a) *the equation*

$$r = a_n + b_n \int_0^r \{(r-t)K_n(t) + X_n(t)\}\,dt$$

has a unique positive zero s_n^* in the interval $[0, R - r_n]$ and

$$\|x_n - x^*\| \leq s_n^*, \quad n \geq 0, \quad \text{with } s_0^* = s^*.$$

(b) *Moreover, the following estimates are true:*

$$\|x_{n+1} - x_n\| \leq \Delta^{(n)}(a) = s_{n+1} - s_n, \quad n \geq 0$$

and

$$\|x_n - x^*\| \leq w\left(\Delta^{(n)}(a)\right) = s - s_n, \quad n \geq 0$$

where

$$\Delta^{(0)}(r) = r, \quad \Delta^{(n+1)}(r) = \Delta\left(\Delta^{(n)}(r)\right), \quad n \geq 0$$

and

$$w(r) = \sum_{n=0}^{\infty} \Delta^{(n)}(r).$$

(c) *Furthermore, the following estimates are true:*

$$\|x_n - x^*\} \leq s_n^*$$
$$\leq (s^* - s_n)\,a_n\,/\,\Delta s_n, \quad n \geq 0$$
$$\leq (s^* - s_n)\,a_{n-1}\,/\,\Delta s_{n-1}, \quad n \geq 0,$$
$$\leq s^* - s_n, \quad n \geq 0.$$

We complete this paper with an application.

Example 4.7 Consider the integral equation

$$x(t) = \int_0^1 K(t, s, x(s))\,ds,$$

where the kernel $K(t, s, x(s))$ is nondifferentiable on some convex subset $D \subset E = C[0, 1]$. We set $T(x) = \int_0^1 K(t, s, x(s))\, ds$ and $F(x) = \int_0^1 \overline{K}(t, s, x(s))\, ds$, where $\overline{K}(t, s, x(s))$ is differentiable on D. Then

$$PF'(x) = \int_0^1 \overline{K}'_x(t, s, x(s))\, ds,$$

where

$$\overline{K}(t, s, x(s)) = \sum_{i=1}^{\infty} A_i(t) B_i(s, x(s))$$

is a degenerate kernel approximating the functions $\overline{K}(t, s, x)$, e.g., a portion of the Taylor or Fourier series for the function $\overline{K}(t, s, x)$ if we consider it as a function of t. The modified Newton-Kantorovic iteration (4.137) can now be written as

$$x_{n+1}(t) = \int_0^1 K(t, s, x_n(s))\, ds - \int_0^1 \overline{K}'_x(t, s, x_n(s)) x_n(s)\, ds$$
$$+ \int_0^1 \overline{K}'_x(t, s, x_n(s)) x_{n+1}(s)\, ds. \qquad (4.148)$$

Let

$$f_n(t) = \int_0^1 K(t, s, x_n(s))\, ds - \int_0^1 \overline{K}'_x(t, s, x_n(s)) x_n(s)\, ds,$$

then iteration (4.148) can be written as

$$x_{n+1}(t) = f_n(t) + \sum_{i=1}^{m} A_i(t) \int_0^1 B'_i(s, x_n(s)) x_{n+1}(s)\, ds,$$

which can be solved to give a system of linear algebraic equations

$$\int_0^1 B'_i(s, x_n(s)) x_{n+1}(s)\, ds - \sum_{i=1}^{m} \int_0^1 B'_i(s, x_n(s)) A_j(s)\, ds \int_0^1 B'_i(s, x_n(s)) x_{n+1}(s)\, ds$$
$$= \int_0^1 B'_i(s, x_n(s)) f_n(s)\, ds.$$

Denote by $D(x_n)$ the determinant of the above system and assume $D(x_n) \neq 0$, $n \geq 0$. Then,

$$\int_0^1 B'_i(s, x_n(s)) x_{n+1}(s)\, ds = \frac{1}{D(x_n)} \int_0^1 \sum_{k=1}^{n} D_{ki}(x_n) B'_k(s, x_n(s)) f_n(s)\, ds$$

and
$$x_{n+1}(t) = f_n(t) + \int_0^1 \sum_{i=1}^m \sum_{k=1}^m \frac{A_i(t) D_{ki}(x_n) B'_k(s, x_n(s))}{D(x_n)} f_n(s)\, ds$$

where $D_{ki}(x_n)$ is the cofactor of the element in the i-th row and k-th column of the determinant $D(x_n)$.

Suppose now that the operators $\overline{\overline{K}}_x(t, s, x)$, $Q(t, s, x)$, $G(t, s, x)$ and $L(t, s, x)$, where

$$Q(t, s, x) = \overline{K}(t, s, x) - \overline{\overline{K}}(t, s, x), \quad G(t, s, x) = K(t, s, x) - \overline{K}(t, s, x)$$

and

$$L(t, s, x) = \frac{1}{D(x)} \sum_{i=1}^m \sum_{k=1}^m A_i(t) D_{ki}(x) B'_k(s, x),$$

satisfy the conditions

$$\left|\overline{\overline{K}}'_x(t, s, x) - \overline{\overline{K}}'_x(t, s, y)\right| \le c_1(t, s)|x - y|$$

$$|(Q(t, s, x) - Q(t, s, y)) + (G(t, s, x) - G(t, s, y))| \le c_2(t, s)|x - y|,$$

and $|L(t, s; x)| \le r(t, s)$ on D.

For simplicity set $K_1(r) = K_1$ and $K_2(r) = K_2$ for all $r = [0, R]$ in (4.138)-(4.139). Then the constants K_1 and K_2 can be computed as follows:

$$K_1 \le w \sup_{t \in [0,1]} \int_0^1 c_1(y, s)\, ds, \quad K_2 \le w \sup_{t \in [0,1]} \int_0^1 c_2(t, s)\, ds$$

where

$$w = 1 + \sup_{t \in [0,1]} \int_0^1 r(t, s)\, ds.$$

4.8 Iteration Converging Faster than Newton's Method

Consider the equation

$$F(x) = 0 \tag{4.149}$$

where F is nonlinear operator mapping a subset E of a normed space X into a normed space Y. We assume that f is k-times Fréchet-differentiable on

E. Suppose that an approximation x_n to a solution x^* of equation (4.149) has been found. To determine the next approximation x_{n+1}, we replace (4.149) by the equation

$$F(x_n) + F'(x_n)(x - x_n) + \tfrac{1}{2}F''(x_n)(x - x_n)^2 + \cdots$$
$$+ \tfrac{1}{k!}F^{(k)}(x_n)(x - x_n)^k = 0, \qquad (4.150)$$

where $F^{(j)}(x_n)$, $j = 1, 2, ..., k$ are j-linear operators corresponding to the jth Fréchet-derivative of F at x_n, $n = 0, 1, 2, ...$.

for fixed x_n, $z \in E$, $n = 0, 1, 2, ...$, define the linear operators on E by

$$L_{n,k}(z)(x) = F'(x_n)(x) + \frac{1}{2}F''(x_n)(z - x_n)(x) + \cdots$$
$$+ \frac{1}{k!}F^{(k)}(x_n)(z - x_n)^{k-1}(x), \quad n = 0, 1, 2, ... \qquad (4.151)$$

Using (4.151), (4.150) can equivalently be written as

$$F(x_n) + L_{n,k}(x)(x - x_n) = 0, \quad n = 0, 1, 2, ... \qquad (4.152)$$

Moreover, if we assume that the linear operators $L_{n,k}(x)$ are invertible on E, (4.152) becomes

$$x = T_{n,k}(x) \qquad (4.153)$$

where $T_{n,k}$ are nonlinear operators defined on E by

$$T_{n,k}(x) = x_n - L_{n,k}(x)^{-1} F(x_n), \quad n = 0, 1, 2, ... \qquad (4.154)$$

Equation (4.153) suggests that the approximation x_{n+1} can be found implicitly using the iteration

$$x_{n+1} = T_{n,k}(x_{n+1}), n = 0, 1, 2, ... \qquad (4.155)$$

Note that for $k = 1$ the above iteration becomes explicit and reduces to the Newton-Kantorovich iteration for solving (4.149).

Assuming that the linear operator $L_{n,1}(x_0)$ has a bounded inverse on some $D \subset E$, the Newton-Kantorovich Theorem 4.1 ensures that if

$$a = a(x_0) = 2b\ell \left\| L_{n,1}(x_0)^{-1} F(x_0) \right\| \leq 1, \qquad (4.156)$$

$$r = r(x_0) = \frac{1}{b\ell}\left(1 - \sqrt{1-a}\right), \qquad (4.157)$$

where ℓ is the Lipschitz constants of $L_{n,1}$ on $D \subset E$ and

$$b = b(x_0) = \left\| L_{n,1}(x_0)^{-1} \right\|.$$

Then equation (4.149) has a solution

$$x^* \in \overline{B}(x_0, r) \subset D$$

which is a unique solution of (4.149) in the open ball $B(x_0, \overline{r})$ with radius

$$\overline{r} = \overline{r}(x_0) = \frac{1}{b\ell}\left(1 + \sqrt{1-a}\right). \tag{4.158}$$

Moreover, iteration (4.155), for $k = 1$ converges to x^* quadratically. That is,

$$\|x_{n+1} - x^*\| = 0\left(\|x_n - x^*\|^2\right), \quad n = 0, 1, 2, \ldots. \tag{4.159}$$

Suppose that there exists $x^* \in E$ which is obtained as the limit of the iteration (4.155) as $n \to \infty$ and k is fixed. They by (4.155)

$$F(x^*) = F\left(\lim_{n \to \infty} x_{n+1}\right) = 0, \tag{4.160}$$

that is x^*, so obtained is a solution of the equation (4.149).

Here we provide sufficient conditions for the convergence of iteration (4.155) to a solution x^* of equation (4.149). We also show that if F is $(k+1)$-times Fréchet-differentiable, then the following estimate holds

$$\|x_{n+1} - x^*\| = 0\left(\|x_n - x^*\|^{k+1}\right), \quad n = 0, 1, 2, \ldots. \tag{4.161}$$

The above result improves (4.159) for $k > 1$. However, iteration (4.155) becomes implicit. More precisely, (4.153) becomes a polynomial equation of degree k on E. Polynomial equations have already been studied in [54] and the references there (see also 3.3).

Due to the particular properties of polynomials, equation (4.153) is, in general easier to handle than equation (4.149), especially on finite dimensional spaces. There are problems where the desired error tolerance $\epsilon > 0$ is such that the number of iterations required by (4.155) for $k = 1$ due to (4.159) is very large. It is in those cases where the solution of (4.153) will reduce the number of iterations required to achieve the same accuracy ϵ due to (4.161).

The evaluation of the iterate x_{n+1} in (4.155) will itself require an iteration of the form

$$x_{n+1,m+1} = T_{n,k}(x_{n+1,m}), \quad m = 0, 1, 2, \ldots \qquad (4.162)$$

for fixed n and some initial guess $x_{n+1,0}$.

Because of rounding or discretization error in the evaluation of $T_{n,k'}$, an approximate sequence $z_{n+1,m}$ is produced in place of the exact sequence $x_{n+1,m}$. That is

$$z_{n+1,m+1} = \widetilde{T}_{n,k}(z_{n+1,m}), \quad m = 0, 1, 2, \ldots, \qquad (4.163)$$

where the $\widetilde{T}_{n,k}$ are related with the $T_{n,k}$, $n = 0, 1, 2, \ldots$.

In [227, 12.2.1] it was proved that if the operators $T_{n,k}$ are all contractions on some closed set $D_1 \subset E$ and $\{z_{n+1,m}\} \subset D_1$, then for $x_{n+1,0} \in D_0$ with $T_{n,k}(D_0) \subset D_0 \subset D_1$ iteration (4.162) converges to a unique fixed point x_{n+1}^* of $T_{n,k}$ in D_0.

Moreover,

$$\lim_{n \to \infty} z_{n+1,m} = x_{n+1}^* \text{ if and only if } \lim_{m \to \infty} \|T_{n,k}(z_{n+1,m}) - z_{n+1,m+1}\| = 0. \qquad (4.164)$$

To illustrate the procedures described above a simple example is provided when $x = \mathbb{C}$, the set of complex numbers.

From now on we assume that $x = y$ is a Banach space and state the main result.

Theorem 4.17 *Let $F : E \subset X \to X$ be a nonlinear operator which is $(k+1)$-times Fréchet-differentiable on E. Assume that the linear operators $L_{n,k}(x_{n+1})$ are invertible with bounded inverse on some closed ball $B^* \subset E$ such that $\{x_n\} \subset B^*$, $n = 0, 1, 2, \ldots$.*

Set,

$$\left\| L_{n,k}(x_{n+1})^{-1} \right\| \leq c_n \leq c, \qquad (4.165)$$

and

$$\frac{1}{(k+1)!} \max_{\widetilde{x} \in B^*} \left\| F^{(k+1)}(\widetilde{x}) \right\| \leq d_n \leq d, \quad n = 0, 1, 2, \ldots \qquad (4.166)$$

for some c, c_n, d, $d_n > 0$ guaranteed to exists by the hypotheses on $L_{n,k}(x_{n+1})$, F and the standard estimate (given in [227] for example) for (4.166).

Then if

$$0 < cd < 1 \quad (4.167)$$

the following are true:

(i) *the iteration $\{x_n\}$ given by (4.155) converges to a solution $x^* \in B^*$ of equation (4.149);*
(ii) *moreover*

$$\|x_{n+1} - x^*\| = 0\left(\|x_n - x^*\|^{k+1}\right), \quad n = 0, 1, 2, \dots.$$

Proof. We have by (4.155)

$$\|x_{n+1} - x_n\| = \left\|L_{n,k}(x_{n+1})^{-1} F(x_n)\right\|$$
$$\leq c \left\| F(x_n) - F(x_{n-1}) - F'(x_{n-1})(x_n - x_{n-1}) - \cdots \right.$$
$$\left. - \tfrac{1}{k!} F^{(k)}(x_{n-1})(x_n - x_{n-1})^k \right\|$$
$$\leq c \frac{1}{(k+1)!} \max_{\widetilde{x} \in B^*} \left\|F^{(k+1)}(\widetilde{x})\right\| \cdot \|x_n - x_{n-1}\|^{k+1}$$
$$\leq cd \|x_n - x_{n-1}\|^{k+1}$$
$$\leq (cd)(cd)^{k+1} \|x_{n-1} - x_{n-2}\|$$
$$\cdots$$
$$\leq (cd)^{(k+1)n} \|x_1 - x_0\|. \quad (4.168)$$

Also, for $p = 2, 3, \dots$

$$\|x_{n+p} - x_n\| \leq \|x_{n+p} - x_{n+(p-1)}\| + \|x_{n+(p-1)} - x_n\|. \quad (4.169)$$

Now,

$$\|x_{n+p} - x_{n+(p-1)}\| \leq (cd) \|x_{n+(p-1)} - x_{n+(p-2)}\|^{k+1}$$
$$\cdots$$
$$\leq (cd)^{(k+1)(p-1)+1} \|x_{n+1} - x_n\|^{k+1}, \quad (4.170)$$

$$\|x_{n+(p-1)} - x_n\|$$
$$= \left\|(x_{n+(p-1)} - x_{n+(p-2)}) + (x_{n+(p-2)} - x_{n+(p-3)}) + \cdots + x_{n+1} - x_n\right\|$$
$$\leq \left[(cd)^{(k+1)(p-2)+1} + (cd)^{(k+1)(p-3)+1} + \cdots + 1\right] \|x_{n+1} - x_n\|.$$
$$(4.171)$$

The inequality (4.169) because of (4.168), (4.170) and (4.171) becomes

$$\|x_{n+p} - x_n\| \le \left[\frac{1 - (cd)^{(k+1)p+1}}{1 - cd}\right] (cd)^{(k+1)n} \|x_1 - x_0\|. \qquad (4.172)$$

Letting $n, p \to \infty$ in (4.172) and using (4.167) we obtain that the sequence $\{x_n\}$ is a Cauchy sequence in Banach space X and as such it converges to some $x^* \in B^*$ which, by the discussion made in the introduction, is a solution of equation (4.149).

This proves (i). The second part of the theorem is immediate from (4.170) and the inequality

$$\|x_n - x^*\| \le \frac{(cd)^{(k+1)n}}{1 - cd} \|x_1 - x_0\|, \quad n = 0, 1, 2, \ldots. \qquad (4.173)$$

which follows from (4.172) by letting $p \to \infty$. That completes the proof of the theorem. □

Note that we can produce the "modified" version of (4.155) by introducing the iteration

$$\tilde{x}_{n+1} = T_{n,k}(\tilde{x}_0) F(\tilde{x}_n), \quad n = 0, 1, 2, \ldots \text{ for some } \tilde{x}_0 \in E$$

and then derive a theorem similar to the one stated above.

Example 4.8 Let $X = \mathbb{C}$, the set of complex numbers equipped with the usual Euclidean norm $\|\cdot\|$. Then $(X, \|\cdot\|)$ becomes a Banach space. Consider the equation

$$F(x) = x^3 - 5x^2 + 7x - 3 = 0. \qquad (4.174)$$

Let $D = \overline{B}(.7, 1.2)$, $x_0 = .7$. The linear operators $L_{n,2}(z)$ become

$$L_{n,2}(z)(x) = (3x_n^2 - 10x_n + 7) x + (3x_n - 5)(z - x_n) x$$

and the Newton-Kantorovich method for (4.174) gives

$$x_0 = .7$$
$$x_1 = .840816$$
$$x_2 = .917578$$
$$x_3 = .957989$$
$$x_4 = .978781$$
$$x_5 = .989335$$
$$x_6 = .994653$$
$$x_7 = .997323$$
$$x_8 = .998661$$
$$x_9 = .99933$$
$$x_{10} = .99965$$
$$x_{11} = .999832$$
$$x_{12} = .999916$$
$$x_{13} = .999958$$
$$x_{14} = .999979$$
$$x_{15} = .999989$$
$$x_{16} = .999995$$
$$x_{17} = .999998$$
$$x_{18} = .1.$$

The Newton-Kantorovich theorem guarantees the existences of a solution of (4.174) only after the 14th iterate, since it can then easily be checked that

$$a = a(x_{14}) = .9981914 < 1, \quad \ell = 4.000126.$$

It is well known, however, that Newton-Kantorovich can sometimes converge even if $x_0 \notin \overline{B}(x_{14}, r(x_{14}))$.

We can now observe that for $x_0 = .7$, $k = 2$ iteration (4.155) becomes a quadratic equation for every n, $n = 1, 2, 3, \ldots$.

For $n = 0$, (4.155) gives

$$-2.9x_1^2 + 5.53x_1 - 3.266 = 0$$

with solutions

$$s_1 = .9534482 \pm .465863i.$$

To apply the iteration (4.155) for $n = 1, 2, ...$ we choose

$$x_m = z_m = \operatorname{Re} \ell(s_m), \quad m = 1, 2, ...$$

That is, for $m = 0$, $z_1 = \operatorname{Re} \ell(s_1) = .9534482$ and (4.155) becomes

$$-2.1396554 z_2^2 + 4.2728098 z_2 - 2.2282236 = 0$$

with solutions

$$s_2 = .9984808 \pm .2107835i.$$

The process will be terminated when $m = 4$ and the results can be tabulated as follows:

$$z_0 = .7$$
$$z_1 = .9534482$$
$$z_2 = .9984808$$
$$z_3 = .99999982$$
$$z_4 = 1.$$

We now observe that starting from the same initial guess, iteration (4.155) for $k = 2$ requires almost the one fourth of the number of iterations required from the same iteration (4.155) for $k = 1$ to obtain the solution $x^* = 1$ of equation (4.174).

Moreover, one can easily check that (4.164) is satisfied.

Finally, it is interesting to note that condition (4.167) is violated since,

$$c_n \to \infty \quad \text{as} \quad n \to \infty$$

and

$$d_n = 1, \quad n = 0, 1, 2, ...$$

However, the sequence $\{z_n\}$, $n = 0, 1, 2, ...$ converges to the solution $x^* = 1$ of equation (4.174).

4.9 Exercises

4.1. If $h_0 < \frac{1}{2}$ holds in $U(x_0, r^*)$, where $r^* = \frac{1+\sqrt{1-2h_0}}{h_0}\eta_0$, then show: Equation (4.1) has a unique solution x^* in $U(x_0, r^*)$.

4.2. If the hypotheses of Theorem 4.1 are satisfied and $h_0 = \frac{1}{2}$, then show: There exists a unique solution x^* of Equation (4.1) in $\bar{U}(x_0, r_0) = \bar{U}(x_0, 2\eta_0)$

4.3. Let x^* be a solution of Equation (4.1). If the linear operator $F'(x^*)$ has a bounded inverse, and $\lim_{\|x-x^*\|\to 0} \|F'(x) - F'(x^*)\| = 0$, then show Newton's method (4.3) converges to x^* if x_0 is sufficiently close to x^* and

$$\|x_n - x^*\| \le d\varepsilon^n \quad (n \le 0),$$

where ε is any positive number; d is a constant depending on x_0 and ε.

4.4. The above result cannot be strengthened, in the sense that for every sequence of positive numbers c_n such that: $\lim_{n\to\infty} \frac{c_{n+1}}{c_n} = 0$, there is an equation for which (4.3) converges less rapidly than c_n. Define

$$s_n = \begin{cases} c_{n/2}, & \text{if } n \text{ is even} \\ \sqrt{c_{(n-1)/2}c_{(n+1)/2}}, & \text{if } n \text{ is odd}. \end{cases}$$

Show: $s_n \to 0$, $\frac{s_{n+1}}{s_n} \to 0$, and $\lim_{n\to\infty} \frac{c_n}{s_{n+k}} = 0$, $(k \ge 1)$.

4.5. Assume operator $F'(x)$ satisfies a Hölder condition

$$\|F'(x) - F'(y)\| \le a\|x-y\|^b,$$

with $0 < b < 1$ and $U(x_0, R)$. Define $h_0 = b_0 a \eta_0^b \le c_0$, where c_0 is a root of

$$\left(\frac{c}{1+b}\right)^b = (1-c)^{1+b} \quad (0 \le c \le 1)$$

and let $R \ge \frac{\eta_0}{1-d_0} = r_0$, where $d_0 = \frac{h_0}{(1+b)(1-h_0)}$. Show that Newton's method (3.3) converges to a solution x^* of Equation (4.1) in $U(x_0, r_0)$.

4.6. Let K, b_0, η_0 be as in Theorem 4.1. If $h_0 = b_0 \eta_0 K < \frac{1}{2}$, and

$$r_0 = \frac{1-\sqrt{1-2h_0}}{h_0}\eta_0 \le r.$$

Then show: modified Newton's method

$$x_{n+1} = x_n - F'(x_0)^{-1} F(x_n) \quad (n \geq 0)$$

converges to a solution $x^* \in U(x_0, r_0)$ of Equation (4.1). Moreover, if

$$r_0 \leq r < \frac{1 + \sqrt{1 - 2h_0}}{h_0} \eta_0,$$

then show: Equation (4.1) has a unique solution x^* in $U(x_0, r)$. Furthermore show: $\bar{x}_{n+1} = \bar{x}_n - F'(x_0)^{-1} F(\bar{x}_n)$ $(n \geq 0)$ converges to a solution x^* of Equation (4.1) for any initial guess $\bar{x}_0 \in U(x_0, r)$.

4.7. Under the hypotheses of Theorem 4.1, let us introduce $\bar{U} = \bar{U}(x_1, r_0 - \eta)$, sequence $\{t_n\}$ $(n \geq 0)$, $t_0 = 0$, $t_{n+1} = t_n - \frac{f(t_n)}{f'(t_n)}$, $f(t) = \frac{1}{2}kt^2 - t + \eta$, $\Delta = r^* - r_0$, $\theta = \frac{r_0}{r^*}$, $\nabla t_{n+1} = t_{n+1} - t_n$, $d_n = \|x_{n+1} - x_n\|$, $\Delta_n = \|x_n - x_0\|$, $\bar{U}_0 = \bar{U}, \bar{U}_n = \bar{U}(x_n, r_0 - t_n)$ $(n \geq 1)$, $K_0 = L_0 = K$,

$$K_n = \sup_{\substack{x,y \in \bar{U}_n \\ x \neq y}} \frac{\|F'(x_n)^{-1} (F'(x) - F'(y))\|}{\|x - y\|} \quad (n \geq 1),$$

$$L_n = \sup_{\substack{x,y \in \bar{U} \\ x \neq y}} \frac{\|F'(x_n)^{-1} (F'(x) - F'(y))\|}{\|x - y\|} \quad (n \geq 1),$$

$$\lambda_n = \frac{2d_n}{1 + \sqrt{1 + 2L_n d_n}} \quad (n \geq 0),$$

$$\underline{\lambda}_n = \frac{2d_n}{1 + \sqrt{1 - 2L_n d_n}} \quad (n \geq 0), \quad \underline{\kappa}_n = \frac{2d_n}{1 + \sqrt{1 + 2K_n d_n}},$$

$$k_n = \frac{2d_n}{1 + \sqrt{1 - 2K_n d_n}} \quad (n \geq 0),$$

$$s_0 = 1, \quad s_n = \frac{s_{n-1}^2}{2^{n-1}\sqrt{1 - 2h} + s_{n-1}\left(1 - \sqrt{1 - 2h}\right)^{2^n-1}} \quad (n \geq 0).$$

With the notation introduced above show (Yamamoto [279]):

$$\|x^* - x_n\| \leq \mathcal{K}_n \ (n \geq 0) \leq \lambda_n \ (n \geq 0)$$

$$\leq \frac{2d_n}{1 + \sqrt{1 - 2K(1 - K\Delta_n)^{-1} d_n}} \quad (n \geq 0)$$

$$\leq \frac{2d_n}{1+\sqrt{1-2K(1-Kt_n)^{-1}d_n}} \quad (n \geq 0)$$

$$= \frac{2d_n}{1+\sqrt{1-2KB_n d_n}} \quad (n \geq 0)$$

$$= \begin{cases} \dfrac{2d_n}{1+\sqrt{1-\frac{4}{\Delta}\cdot\frac{1-\theta^{2^n}}{1+\theta^{2^n}}d_n}} & (2h<1) \\ \dfrac{2d_n}{1+\sqrt{1-\frac{2^n}{\eta}d_n}} & (2h=1) \end{cases} \quad (n \geq 0)$$

$$\leq \frac{r_0 - t_n}{\nabla t_{n+1}} d_n \quad (n \geq 0)$$

$$= \frac{2d_n}{1+\sqrt{1-2h_n}} \quad (n \geq 0)$$

$$\leq \frac{KB_n d_{n-1}^2}{1+\sqrt{1-2h_n}} \quad (n \geq 0)$$

$$= \frac{r_0 - t_n}{(\nabla t_n)^2} d_{n-1}^2 \quad (n \geq 0)$$

$$= \begin{cases} \dfrac{1-\theta^{2^n}}{\Delta} d_{n-1}^2 & (2h<1) \\ \dfrac{2^{n-1}}{\eta} d_{n-1}^2 & (2h=1) \end{cases} (n \geq 1)$$

$$\leq \frac{Kd_{n-1}^2}{\sqrt{1-2h}+\sqrt{1-2h+(Kd_{n-1})^2}} \quad (n \geq 1)$$

$$\leq \frac{K\eta_{n-1}d_{n-1}}{\sqrt{1-2h}+\sqrt{1-2h+(K\eta_{n-1})^2}} \quad (n \geq 1)$$

$$= e^{-2^{n-1}\varphi} d_{n-1} \quad (n \geq 1)$$

$$= \theta^{2^{n-1}} d_{n-1} \quad (n \geq 1)$$

$$= \frac{r_0 - t_n}{\nabla t_n} d_{n-1} \quad (n \geq 1)$$

$$\leq r_0 - t_n \quad (n \geq 0)$$

$$= \frac{2\eta_n}{1+\sqrt{1-2h_n}} \quad (n \geq 0)$$

$$= \begin{cases} e^{-2^{n-1}\varphi \frac{\sinh\varphi}{\sinh 2^{n-1}\varphi}} \eta & (2h<1) \\ 2^{1-n}\eta & (2h<1) \end{cases}$$

$$= \begin{cases} \dfrac{\Delta\theta^{2^n}}{1-\theta^{2^n}} & (2h<1) \\ 2^{1-n}\eta & (2h=1) \end{cases} (n \geq 0)$$

$$= \frac{s_n}{2^n K}\left(\frac{2h}{1+\sqrt{1-2h}}\right)^{2^n} \quad (n \geq 0)$$

$$\leq \frac{1}{2^n}K\left(\frac{2h}{1+\sqrt{1-2h}}\right)^{2^n} \quad (n \geq 0)$$

$$\leq \frac{1}{2^{n-1}}(2h)^{2n-1}\eta \quad (n > 0),$$

$$\|x^* - x_n\| \leq \lambda_n \quad (n \geq 0)$$

$$\leq \frac{L_n d_{n-1}^2}{1+\sqrt{1-(L_n d_{n-1})^2}} \quad (n \geq 1)$$

$$\leq \frac{L_{n-1} d_{n-1}^2}{1 - L_{n-1}d_{n-1} + \sqrt{1 - 2L_{n-1}d_{n-1}}} \quad (n \geq 1)$$

$$\leq \frac{L_{n-1} d_{n-1}^2}{1 - L_{n-1}d_{n-1}} \quad (n \geq 1),$$

$$\|x^* - x_n\| \leq \lambda_n$$

$$\leq \frac{2d_n}{1 + \sqrt{1 - 2L_0(1 - L_0\Delta_n)^{-1}d_n}}$$

$$\leq \frac{2\|F'(x_0)^{-1}F(x_n)\|}{1 - L_0\Delta_n + \sqrt{(1-L_0\Delta_n)^2 - 2L_0\|F'(x_0)^{-1}F(x_n)\|}}$$

$$\leq \frac{L_0 d_{n-1}^2}{1 - L_0\Delta_n + \sqrt{(1 - L_0\Delta_n)^2 - (L_0 d_{n-1})^2}},$$

$$\|x^* - x_n\| \geq \underline{\kappa}_n \ (n \geq 0) \geq \underline{\lambda}_n \ (n \geq 0)$$

$$\geq \frac{2d_n}{1 + \sqrt{1 + 2K(1 - K\Delta_n)^{-1}d_n}} \quad (n \geq 0)$$

$$\geq \frac{2d_n}{1 + \sqrt{1 + 2K(1 - Kt_n)^{-1}d_n}} \quad (n \geq 0)$$

$$= \frac{2d_n}{1 + \sqrt{1 + 2KB_n d_n}} \quad (n \geq 0)$$

$$= \frac{2d_n}{1 + \sqrt{1 + 4 \cdot \frac{r_0 - t_{n+1}}{(r_0 - t_n)^2} d_n}} \quad (n \geq 0)$$

$$= \frac{2d_n}{1 + \sqrt{1 + 4 \cdot \frac{\nabla t_{n+1}}{(\nabla t_n)^2} d_n}} \quad (n \geq 0)$$

$$= \frac{2d_n}{1 + \sqrt{1 + \frac{2Kd_n}{\sqrt{1 - 2h + (K\eta_{n-1})^2}}}} \quad (n \geq 0)$$

$$\geq \frac{2d_n}{1 + \sqrt{1 + \frac{2Kd_n}{\sqrt{1 - 2h + (Kd_{n-1})^2}}}} \quad (n \geq 0)$$

$$= \frac{2d_n}{1 + \sqrt{1 + \frac{2d_n}{\sqrt{a^2 + d_{n-1}^2}}}} \quad \left(a = \sqrt{1 - 2h}/K, n \geq 1 \right)$$

$$\geq \frac{2d_n}{1 + \sqrt{1 + \frac{2d_n}{d_n + \sqrt{a^2 + d_n^2}}}} \quad (n \geq 0)$$

$$\geq \frac{2d_n}{1 + \sqrt{1 + 2h_n}} \quad (n \geq 0)$$

$$= \frac{2d_n}{1 + \sqrt{1 + \frac{4\theta^{2^n}}{(1 + \theta^{2^n})^2}}} \quad (n \geq 0),$$

$$\|x^* - x_{n+1}\| \leq \kappa_{n+1} \leq \kappa_n - d_n \leq \frac{r_0 - t_{n+1}}{\nabla t_{n+1}} d_n,$$

$$\|x^* - x_{n+1}\| \leq \lambda_{n+1} \leq \lambda_n - d_n \leq \frac{r_0 - t_{n+1}}{\nabla t_{n+1}} d_n,$$

$$\begin{aligned}
d_n &\leq \tfrac{1}{2} K_n d_{n-1}^2 \\
&\leq \tfrac{1}{2} L_n d_{n-1}^2 \\
&\leq \tfrac{1}{2} K \left(1 - K\Delta_n\right)^{-1} d_{n-1}^2 \\
&\leq \tfrac{1}{2} K \left(1 - K\Delta_{n-1} - Kd_{n-1}\right)^{-1} d_{n-1}^2 \\
&\leq \tfrac{1}{2} K \left(1 - Kt_n\right)^{-1} d_{n-1}^2 \\
&= \tfrac{1}{2} K B_n d_{n-1}^2
\end{aligned}$$

$$= \frac{r_0 - t_{n+1}}{(r_0 - t_n)^2} d_{n-1}^2$$

$$= \frac{\nabla t_{n+1}}{(\nabla t_n)^2} d_{n-1}^2$$

$$= \frac{d_{n-1}^2}{2\sqrt{a^2 + \eta_{n-1}^2}}$$

$$\leq \frac{d_{n-1}^2}{2\sqrt{a^2 + d_{n-1}^2}}$$

$$\leq \frac{\nabla t_{n+1}}{\nabla t_n} d_{n-1}$$

$$= \frac{\eta_n}{\eta_{n-1}} d_{n-1}$$

$$= \frac{1}{2\cosh 2^{n-1}\varphi} d_{n-1}$$

$$\leq \tfrac{1}{2} d_{n-1}$$

$$\leq \tfrac{1}{2} \eta_{n-1} = \tfrac{1}{2} \nabla t_n,$$

and

$$d_n \leq \eta_n$$
$$= \nabla t_{n+1}$$
$$= (r_0 - t_{n+1}) \theta^{-2^n}$$
$$= (r_0 - t_{n+1}) e^{2^n \varphi}$$
$$= \begin{cases} \frac{\sinh \varphi}{\sinh 2^n \varphi} \eta & (2h < 1) \\ 2^{-n} \eta & (2h = 1) \end{cases}$$
$$= \frac{\Delta \theta^{2^n}}{1 - \theta^{2^{n+1}}} \quad (2h < 1).$$

4.8. Let $F : D \subseteq X \to Y$ be m-times Fréchet-differentiable in D ($m \geq 2$ an integer). Assume that for some $x_0 \in D$ and parameters $\eta > 0$,

$\alpha_i \geq 0$ $(i = 2, \ldots, m)$, $a > 0$, $F'(x_0)^{-1}$ exists,

$$\|F'(x_0)^{-1} F(x_0)\| \leq \eta,$$
$$\|F'(x_0)^{-1}[F'(x) - F'(x_0)]\| \leq a\|x - x_0\|,$$
$$\|F'(x_0)^{-1} F^{(i)}(x_0)\| \leq \alpha_i, \quad i = 2, \ldots, m,$$
$$\|F'(x_0)[F^{(m)}(x) - F^{(m)}(x_0)]\| \leq \alpha_{m+1}\|x - x_0\|$$

for all $x \in D$, and

$$2b\eta \leq 1,$$

where,

$$b = \max\left\{a, \alpha_2 + \tfrac{2}{3}\alpha_3\eta + \cdots + \tfrac{2^m \alpha_{m+1}}{(m+1)!}\eta^{m-1}\right\},$$

and

$$U\left(x_0, \tfrac{1}{a}\right) \subseteq D.$$

Show:

(a) the real polynomial

$$p(t) = \tfrac{b}{2}t^2 - t + \eta$$

has two positive roots r_1, r_2 with $r_1 \leq r_2$ and iteration $\{t_n\}$ $(n \geq 0)$ generated by

$$t_0 = 0, \quad t_{n+1} = t_n - \frac{p(t_n)}{p'(t_n)} \quad (n \geq 0)$$

is monotonically increasing with $\lim_{n \to \infty} t_n = r_1$.

(b) Sequence $\{x_n\}$ $(n \geq 0)$ generated by Newton's method is well defined, remains in $\bar{U}(x_0, r_1)$ for all $n \geq 0$, and converges to a solution $x^* \in \bar{U}(x_0, r_1)$ of equation $F(x) = 0$, which is unique in $U(x_0, r_2)$ if $r_1 < r_2$. If $r_1 = r_2$ the solution x^* is unique in $\bar{U}(x_0, r_1)$. Moreover, the following estimates hold for all $n \geq 0$

$$\|x_{n+1} - x_n\| \leq t_{n+1} - t_n,$$

and

$$\|x_n - x^*\| \leq r_1 - t_n = \left(\frac{r_1}{r_2}\right)^{2^n}(r_2 - t_n).$$

Let
$$F(x) = \tfrac{1}{6}x^3 - (\tfrac{2}{6}^{3/2} + .23), \quad x \in [\sqrt{2}-1, \sqrt{2}+1]$$
show that the Newton–Kantorovich hypothesis is violated, but the conditions in this result hold.

4.9. Let F be a Fréchet-differentiable operator defined on some closed convex subset D of a Banach space X with values in a Banach space Y. Assume: there exists $x_0 \in D$ such that $F'(x_0) \in L(X,Y)$, and $F'(x_0)^{-1} \in L(Y,X)$.
Show:

(a) for all $\varepsilon_0 > 0$ there exists $\delta_0 > 0$ such that
$$\|F'(x_0)^{-1}(F'(x_0) - F'(x))\| < \varepsilon_0, \qquad \text{for all } x \in U(x_0, \delta_0).$$

(b) for all $\varepsilon_1 > 0$ there exists $\delta_1 > 0$ such that
$$\|(F'(x_0)^{-1} - F'(x)^{-1})F'(x_0)\| < \varepsilon_1, \qquad \text{for all } x \in U(x_0, \delta_1).$$

Set $\delta = \min\{\delta_0, \delta_1\}$ and $\varepsilon = \max\{\varepsilon_0, \varepsilon_1\}$.

(c) for $\varepsilon > 0$ there exists $\delta > 0$ as defined above such that
$$\|F'(x_0)^{-1}(F'(x_0) - F'(x))\| < \varepsilon,$$
$$\|(F'(x_0)^{-1} - F'(x)^{-1})F'(x_0)\| < \varepsilon,$$
for all $x \in U(x_0, \delta)$.
Define parameters b, c, η by
$$b \geq \tfrac{-1+\sqrt{3}}{2}, \quad c \geq 2\varepsilon^2 + 2\varepsilon, \text{ for a fixed } \varepsilon \in [0, b),$$
$$\|F'(x_0)^{-1}F(x_0)\| \leq \eta.$$

(d) Assume further:
$$\tfrac{\eta}{1-c} \leq \delta, \quad \varepsilon \in (0, b), \quad \text{and} \quad \bar{U}(x_0, \delta) \subseteq D.$$

Show: Sequence $\{x_n\}$ ($n \geq 0$) generated by Newton's method is well defined, remains in $U(x_0, \delta)$ for all $n \geq 0$, and converges to a solution $x^* \in \bar{U}(x_0, \delta)$ of equation $F(x) = 0$. Moreover, x^* is the unique solution of equation $F(x) = 0$ in $\bar{U}(x_0, \delta)$. Furthermore, the following error bounds hold for all $n \geq 0$
$$\|x_{n+1} - x_n\| \leq c^n \|x_1 - x_0\| \leq c^n \eta$$
and
$$\|x_n - x^*\| \leq \tfrac{c^n}{1-c}\|x_1 - x_0\|.$$

(e) Assume: hypotheses on F, F' hold, with x_0 being replaced by a simple solution x^* of equation $F(x) = 0$; $\varepsilon \in (0, b)$.

Show: sequence $\{x_n\}$ ($n \geq 0$) generated by Newton's method is well defined, remains in $U(x^*, \delta)$ for all $n \geq 0$ and converges to x^*, provided that $x_0 \in \bar{U}(x^*, \delta) \subseteq D$. Moreover, the following error bounds hold for all $n \geq 0$

$$\|x_{n+1} - x^*\| \leq c\|x_n - x^*\| \leq c^n\|x_0 - x^*\|.$$

4.10. Let $m \geq 2$ be an integer and F an m-times Fréchet-differentiable operator defined on a convex subset D of a Banach space X with values in a Banach space Y. Assume:

(a) there exists $x_0 \in D$ such that $F'(x_0)^{-1} \in L(Y, X)$;
(b) there exist parameters $\eta > 0$, $\alpha_i \geq 0$, $i = 2, \ldots, m$ such that

$$\|F'(x_0)^{-1} F(x_0)\| \leq \eta,$$
$$\|F'(x_0)^{-1} F^{(i)}(x_0)\| \leq \alpha_i, \quad i = 2, \ldots, m$$

Since F is m-times Fréchet-differentiable for all $\varepsilon > 0$ there exists $\delta_0 > 0$ such that

$$\left\|F'(x_0)^{-1}[F^{(m)}(x) - F^{(m)}(x_0)]\right\| < \varepsilon$$

for all $x \in U(x_0, \delta_0)$;
(c) the positive zeros of p' is such that

$$p(s) \leq 0,$$

where

$$p(t) = \eta - t + \tfrac{\alpha_2}{2!} t^2 + \cdots + \tfrac{(\varepsilon + \alpha_m)}{m!} t^m;$$

Then polynomial p has only two positive zeros denoted by δ_1, δ_2 ($\delta_1 \leq \delta_2$).
(d) $\bar{U}(x_0, \delta) \subseteq D$;
(e) $\delta_0 \in [\delta_1, \delta_2]$ or $\delta_0 > \delta_2$, where

$$\delta = \max\{\delta_0, \delta_1, \delta_2\}.$$

Show: sequence $\{x_n\}$ ($n \geq 0$) generated by Newton's method is well defined, remains in $U(x_0, \delta_1)$ for all $n \geq 0$ and converges to a solution $x^* \in \bar{U}(x_0, \delta_1)$ of equation $F(x) = 0$. The solution x^* is unique in $U(x_0, \delta_0)$ if $\delta_0 \in [\delta_1, \delta_2]$ or x^* is unique in $U(x_0, \delta_2)$ if $\delta_0 > \delta_2$. Moreover, the following error bounds hold for all $n \geq 0$

$$\|x_{n+1} - x_n\| \leq t_{n+1} - t_n$$

$$\|x_n - x^*\| \leq t^* - t_n,$$

where $\{t_n\}$ ($n \geq 0$) is a monotonically increasing sequence converging to t^*, generated by

$$t_0 = 0, \quad t_{n+1} = t_n - \frac{p(t_n)}{p'(t_n)} \quad (n \geq 0).$$

4.11. (a) Let F be a twice Fréchet-differentiable operator defined on a convex subset D of a Banach space X with values in a Banach space Y. Assume that the equation $F(x) = 0$ has a simple zero $x^* \in D$, in the sense that $F'(x^*)$ has an inverse $F'(x^*)^{-1} \in L(Y,X)$. Then for all $\ell_1 > 0$ there exists $r^* > 0$ such that

$$\|F'(x^*)^{-1}[F''(x) - F''(y)]\| \leq \ell_1, \quad \text{for all } x, y \in U(x^*, r^*).$$

Moreover, assume there exists $b_1 > 0$ such that

$$\|F'(x^*)^{-1}F''(x^*)\| \leq b_1.$$

Then if

$$r^* < (3c_1)^{-1}, \quad c_1 = \tfrac{\ell_1 + b_1}{2}$$

show: sequence $\{x_n\}$ ($n \geq 0$) generated by Newton's method is well defined, remains in $U(x^*, r^*)$ for all $n \geq 0$ and converges to x^* provided that $x_0 \in U(x^*, r^*)$. Furthermore, the following error bounds hold for all $n \geq 0$:

$$\|x_{n+1} - x^*\| \leq \tfrac{c_1}{1 - 2c_1\|x_n - x^*\|}\|x_n - x^*\|^2.$$

(b) Assume:

(i) there exists $\eta \geq 0$, $x_0 \in D$ such that $F'(x_0)^{-1} \in L(Y, X)$

$$\|F'(x_0)^{-1}F(x_0)\| \leq \eta;$$

Then, for all $\ell \geq 0$ there exists $r > 0$ such that

$$\|F'(x_0)^{-1}[F''(x) - F''(y)]\| < \ell_0, \quad \forall x, y \in \bar{U}(x_0, r) \subseteq D;$$

(ii) there exists $b_0 \geq 0$ such that
$$\|F'(x_0)^{-1}F''(x_0)\| \leq b_0;$$
(iii) $c_0\eta \leq \frac{2-\sqrt{3}}{2}$, $c_0 = \frac{1}{2}(\ell_0 + b_0)$;
(iv) r is the smallest positive zero of equation
$$3c_0 s^2 - (1 + 2c_0\eta)s + \eta = 0.$$

Show: sequence $\{x_n\}$ $(n \geq 0)$ generated by Newton's method is well defined, remains in $U(x_0, r)$ for all $n \geq 0$ and converges to a unique solution $x^* \in \bar{U}(x_0, r)$ of equation $F(x) = 0$. Moreover, the following error bounds hold for all $n \geq 0$
$$\|x_{n+2} - x_{n+1}\| \leq \frac{c_0}{1-2c_0\|x_{n+1}-x_n\|}\|x_{n+1} - x_n\|^2$$
and
$$\|x_{n+1} - x^*\| \leq \frac{c_0^n}{1-c_0}\|x_n - x^*\|.$$

4.12. (a) Let : $D \subseteq X \to Y$ be a continuously Fréchet-differentiable operator defined on an open convex subset D of a Banach space X with values in a Banach space Y.
Assume:
- there exists $x_0 \in D$ such that $F(x_0) \neq 0$;
- $F'(x)^{-1} \in L(Y, X)$ $(x \in D)$, and there exists $b > 0$ such that:
$$\|F'(x)^{-1} F'(x_0)\| \leq b \quad (x \in D;)$$
- for each fixed $p \in D$ with $F(p) \neq 0$, there exists
$$\delta = \delta(p, t) \geq t\|F'(p)^{-1} F(p)\|, \quad t \in [0, 1]$$
such that
$$\|F'(x_0)^{-1}[F'(p) - F'(y_t)]\| < t\|F'(p)^{-1} F(p)\|$$
for all
$$y_t \in U(p, \delta) = \{z \in X \mid \|z - p\| < \delta\} \subseteq D,$$
and y_t collinear to p;

- there exists $\eta > 0$ such that:
$$\|F'(x_0)^{-1} F(x_0)\| \leq \eta,$$
$$c = \tfrac{b\eta}{2} \in (0,1);$$
and $U = U(x_0, \delta^*) \subseteq D$, where
$$\delta^* \geq \tfrac{\eta}{1-c}.$$

Show: sequence $\{x_n\}$ ($n \geq 0$) generated by Newton's method is well defined, remains in U for all $n \geq 0$ and converges to a unique solution $x^* \in \bar{U}$ of equation $F(x) = 0$. Moreover the following error bounds hold:
$$\|x_{n+1} - x_n\| < \tfrac{b}{2}\|x_n - x_{n-1}\|^2 \quad (n \geq 1),$$
and
$$\|x_n - x^*\| \leq a_n \|x_n - x_{n-1}\|^2 \quad (n \geq 1),$$
where
$$a_n = \tfrac{b}{2} \sum_{j=0}^{\infty} \left(c^{2^n}\right)^{2^j - 1} \leq \tfrac{b}{2(1 - c^{2^n})}.$$

(b) Assume:

- there exists a simple zero $x^* \in D$ of F in the sense that $F'(x^*)^{-1} \in L(Y, X)$;
- $F'(x)^{-1} \in L(Y, X)$ ($x \in D$) and there exists $q > 0$ such that:
$$\|F'(x)^{-1} F'(x^*)\| \leq q \quad (x \in D);$$
- for each fixed $p \in D$ with $F(p) \neq 0$ there exists $\delta \geq t\|p - x^*\|$, $t \in (0,1]$ such that:
$$\|F'(x^*)^{-1} [F'(p) - F'(p_t)]\| < t\|p - x^*\|$$
for all $p_t = x + t(x^* - x) \in U(p, \delta) \subseteq D$,
- $U^* = U(x^*, r^*) \subseteq D$, where
$$r^* > \tfrac{2}{q}.$$

Show: sequence $\{x_n\}$ ($n \geq 0$) generated by Newton's method is well defined, remains in U^* for all $n \geq 0$ and converges to x^* provided that $x_0 \in U^*$. Moreover, the following error bounds hold for all $n \geq 0$:

$$\|x_{n+1} - x^*\| \leq \tfrac{q}{2} \|x_n - x^*\|^2.$$

4.13. (a) Suppose that F', F'' are uniformly bounded by non-negative constants $\alpha < 1$, K respectively, on a convex subset D of X and the ball

$$\bar{U}(x_0, r_0 \equiv \tfrac{2\|x_0 - F(x_0)\|}{1 - \alpha}) \subseteq D.$$

Moreover, if

$$h_S = \tfrac{K}{2} \tfrac{1 + 2\alpha}{1 - \alpha} \tfrac{\|x_0 - F(x_0)\|}{1 - \alpha} < 1,$$

holds, then Stirling's method

$$x_{n+1} = x_n - \bigl(I - F'(F(x_n))\bigr)(x_n - F(x_n)) \quad (n \geq 0) \quad (4.175)$$

converges to the unique fixed point x^* of F in $\bar{U}(x_0, r_0)$. Moreover, the following error bounds hold for all $n \geq 0$:

$$\|x_n - x^*\| \leq h_S^{2^n - 1} \frac{\|x_0 - F(x_0)\|}{1 - \alpha}. \tag{4.176}$$

(b) Let $F \colon D \subseteq X \to Y$ be analytic. Assume:

$$\|F'(x)\| \leq \alpha < 1, \quad \text{for all } x \in D,$$
$$x_0 \neq F(x_0), \quad x_0 \in D,$$
$$\tfrac{\gamma(1+2\alpha)\|x_0 - F(x_0)\|}{(1-\alpha)^2} < 1,$$
$$r_0 < r_1,$$
$$U(x_0, r_1) \subseteq D,$$

and

$$0 \neq \gamma \equiv \sup_{\substack{k > 1 \\ x \in D}} \bigl\| \tfrac{1}{k!} F^{(k)}(x_0) \bigr\|^{\frac{1}{k-1}} < \infty,$$

where,

$$r_1 = \tfrac{1}{\gamma}\left[1 - \sqrt[3]{\tfrac{\gamma(1+2\alpha)\|x_0 - F(x_0)\|}{(1-\alpha)^2}}\right].$$

Show: sequence $\{x_n\}$ ($n \geq 0$) generated by Stirling's method (4.175) is well defined, remains in $U(x_0, r_0)$ for all $n \geq 0$ and converges to a unique fixed point x^* of operator F at the rate given by (4.176) with

$$K = \frac{2\gamma}{(1-\gamma r_0)^3}.$$

(c) Let $X = D = \mathbb{R}$ and define function F on D by

$$F(x) = \begin{cases} -\frac{1}{3}x, & x \leq 3 \\ \frac{1}{45}(x^2 - 7x - 33), & 3 \leq x \leq 4 \\ \frac{1}{3}(x-7), & x > 4. \end{cases}$$

Using Stirling's method for $x_0 = 3$ we obtain the fixed point $x^* = 0$ of F in one iteration, since $x_1 = 3 - (1+\frac{1}{3})^{-1}(3+1) = 0$. Show: Newton's method fails to converge.

4.14. It is convenient for us to define certain parameters, sequences and functions. Let $\{t_n\}$ ($n \geq 0$) be a Fibonacci sequence given by

$$t_0 = t_1 = 1, \quad t_{n+1} = t_n + t_{n-1} \quad (n \geq 1).$$

Let also c, ℓ, η be non-negative parameters and define:

- the real function f by

$$f(x) = \frac{1}{1-x}, \quad x \in [0, 1),$$

- sequences $\{s_n\}$ ($n \geq -1$), $\{a_n\}$ ($n \geq -1$), $\{A_n\}$ ($n \geq -1$) by

$$s_{-1} = \frac{\eta}{c+\eta}, \quad s_0 = \ell(c+\eta), \quad s_n = f^2(s_{n-1})a_{n-1}a_{n-2} \quad (n \geq 1),$$

$$a_{-1} = a_{-2} = 0, \quad a_{n-2} = \sum_{j=0}^{n-1} c_j, \quad c_j = t_0 + t_1 + \cdots + t_{j+1},$$

$$A_n = [x_n, x_{n-1}; F],$$

for $x_n \in X$, and
- parameters b, d, r_0 by

$$b = \max\left\{\frac{\ell n}{(1-s_0)^2 s_0}, \frac{s_0}{(1-s_0)^2}\right\}, \quad d = \frac{s_0}{1-s_0}, \quad r_0 = \frac{\eta}{1-d}.$$

Let $F\colon D \subseteq X \to Y$ be a nonlinear operator. Assume there exist $x_{-1}, x_0 \in D$ and non-negative parameters c, ℓ, η such that:

$$A_0^{-1} \text{ exists},$$
$$\|x_0 - x_{-1}\| \leq c,$$
$$\|A_0^{-1} F(x_0)\| \leq \eta,$$
$$\|A_0^{-1}([x,y;F] - [z,w;F])\| \leq \ell(\|x-z\| + \|y-w\|),$$
$$\forall x,y,z \in D, x \neq y, w \neq z,$$
$$s_0 < \tfrac{3-\sqrt{5}}{2},$$
$$\ell\eta < (1-s_0)^2 s_0 = \alpha,$$

and

$$\bar{U}(x_0, r_0) = \{x \in X \mid \|x - x_0\| \leq r_0\} \subseteq D.$$

Show: sequence $\{x_n\}$ $(n \geq -1)$ generated by the Secant method

$$x_{n+1} = x_n - [x_n, x_{n-1}; F]^{-1} F(x_n) \quad (n \geq 0) \; (x_{-1}, x_0 \in D)$$

is well defined, remains in $U(x_0, r_0)$ for all $n \geq 0$ and converges to a unique solution $x^* \in \bar{U}(x_0, r_0)$ of equation $F(x) = 0$. Moreover the following error bounds hold for all $n \geq 1$:

$$\|x_n - x^*\| \leq \tfrac{d^n}{1-d} b^{a_{n-2}} \|x_1 - x_0\|,$$

and

$$\|L_0^{-1} F(x_{n+1})\| \leq b^{c_{n-1}} s_0.$$

Furthermore, let

$$r^* = \tfrac{1}{\ell} - r_0 - \eta.$$

Then $r^* > r_0$ and the solution x^* is unique in $U(x_0, r^*)$.

4.15. Let F be a nonlinear operator defined on an open convex subset D of a Banach space X with values in a Banach space Y and let $A(x) \in L(X, Y)$ $(x \in D)$. Assume:

- there exists $x_0 \in D$ such that $A(x_0)^{-1} \in L(Y, X)$;

- there exist non-decreasing, non-negative functions a, b such that:

$$\|A(x_0)^{-1}[A(x) - A(x_0)]\| \leq a(\|x - x_0\|),$$
$$\|A(x_0)^{-1}[F(y) - F(x) - A(x)(y - x)]\| \leq b(\|x - y\|)\|x - y\|,$$
$$\text{for all } x, y \in D;$$

- there exist $\eta \geq 0$, $r_0 > \eta$ such that

$$\|A(x_0)^{-1}F(x_0)\| \leq \eta,$$
$$a(r) < 1,$$

and

$$d(r) < 1, \quad \text{for all } r \in (0, r_0],$$

where

$$c(r) = (1 - a(r))^{-1},$$

and

$$d(r) = c(r)b(r);$$

- r_0 is the minimum positive root of equation $h(r) = 0$ on $(0, r_0]$, where

$$h(r) = \frac{\eta}{1 - d(r)} - r;$$

- $\bar{U}(x_0, r_0) \subseteq D$.

Show: sequence $\{x_n\}$ $(n \geq 0)$ generated by Newton-like method

$$x_{n+1} = x_n - A(x_n)^{-1}F(x_n) \quad (n \geq 0)$$

is well defined, remains in $U(x_0, r_0)$ for all $n \geq 0$ and converges to a solution $x^* \in \bar{U}(x_0, r_0)$ of equation $F(x) = 0$.

4.16. (1) Let F be a twice Fréchet differentiable operator defined on a convex subset D of a Hilbert space H with values in H; x_0 be a point in D. Assume:

(a) there exist constants a, b, c, d and η such that
$$\|F''(x) - F''(x_0)\| \leq a\|x - x_0\|,$$
$$\|F''(x_0)\| \leq b,$$
$$\tfrac{1}{c}\|y\|^2 \leq |\langle F'(x)(y), y\rangle|,$$
$$\|F'(x)\| \leq d,$$

and
$$\|F(x_0)\| \leq \eta,$$

for all $x \in D$, $y \in H$;
(b) $p = \tfrac{a}{6}c^3\eta^2 + \tfrac{b}{2}c^2\eta + \sqrt{1 - \tfrac{1}{d^2c^2}} \in [0, 1)$
and
(c) $U(x_0, r^*) \subseteq D$,
where
$$r^* = \tfrac{c\eta}{1-p}.$$

Show: iteration $\{x_n\}$ ($n \geq 0$) generated by
$$x_{n+1} = x_n - \frac{\langle F'(x_n)F(x_n), F(x_n)\rangle}{\|F'(x_n)F(x_n)\|^2}F(x_n) \quad (n \geq 0)$$

is well defined, remains in $U(x_0, r^*)$ for all $n \geq 0$ and converges to a unique solution x^* of equation $F(x) = 0$ in $U(x_0, r^*)$. Moreover, the following error bounds hold for all $n \geq 0$
$$\|F(x_n)\| \leq p^n\eta$$

and
$$\|x_n - x^*\| \leq \tfrac{c\eta}{1-p}p^n$$

If we let $a = 0$, then (1) reduces to Theorem 13.2 in [187, p. 161].

(2) Under the hypothesis of (1), show the same conclusions of (1) hold for iteration
$$x_{n+1} = x_n - \frac{\|F(x_n)\|^2}{\|[F'(x_n)]^*F(x_n)\|^2}[F'(x_n)]^*F(x_n) \quad (n \geq 0)$$

but with p replaced by
$$p_0 = \sqrt{c^2\left[d^2 + b\eta + \tfrac{ac\eta^2}{3}\right] - 1}.$$

(3) Under the hypotheses of (2), show that the conclusions of (1) hold for iteration

$$x_{n+1} = x_n - \frac{\|F(x_n)\|^2}{\langle F'(x_n)F(x_n), F(x_n)\rangle} F(x_n) \quad (n \geq 0).$$

Results (2) and (3) reduce to the corresponding ones in [187, p. 163] for $a = 0$. We give a numerical example to show our results apply, whereas the corresponding ones in [187, pp. 161–163] do not.

(4) Let $H = \mathbb{R}$, $D = [-1, 1]$, $x_0 = 0$ and consider equation

$$F(x) = -\tfrac{1}{6}x^3 - \tfrac{1}{6}x^2 + \tfrac{5}{6}x - \tfrac{1}{3} = 0.$$

The convergence condition in [187, p. 161] using our notation is

$$q = \sqrt{1 - \tfrac{1}{d^2 c^2}} + \tfrac{c^2 L \eta}{2} \in [0, 1) \qquad (\text{i})$$

where L is the Lipschitz constant such that

$$\|F'(x) - F'(y)\| \leq L\|x - y\|, \quad x, y \in D.$$

Show:

$$\eta = \tfrac{1}{3}, \; d = \tfrac{5}{6}, \; c = \tfrac{3}{2}, \; L = \tfrac{4}{3} \text{ and } q = \tfrac{11}{10} > 1.$$

Hence, (i) is violated.

Moreover, show: the condition corresponding to iterations in (2) and (3) in [187, p. 163] is also violated since

$$q_0 = \sqrt{c^2(d^2 + L\eta) - 1} \in [0, 1),$$

gives

$$q_0 = \tfrac{5}{4} > 1.$$

Furthermore show: our conditions hold since for $a = 1$, $b = \tfrac{1}{3}$, we get

$$p = .7875 < 1$$

and

$$p_0 = \sqrt{\tfrac{15}{16}} < 1.$$

Conclude: there is no guarantee that iterations converge to a solution x^* of equation $F(x) = 0$ under the conditions in

[187]. But our results (1)–(3) guarantee convergence for the same iterations.

(5) The results obtained here can easily be extended to include m-times Fréchet differentiable operators, where $m \geq 2$ an integer. Indeed, let us assume there exist constants $a_2, \ldots, a_{m+1}$ such that

$$\|F^{(m)}(x) - F^{(m)}(x_0)\| \leq a_{m+1}\|x - x_0\|$$
$$\|F^{(i)}(x_0)\| \leq a_i, \quad i = 2, \ldots, m.$$

Show:

$$\|F(x_{k+1}) - F(x_k) - F'(x_k)(x_{k+1} - x_k)\| \leq$$
$$\leq \tfrac{a_{m+1}}{(m+1)!}\|x_{k+1} - x_k\|^{m+1}\| + \cdots + \tfrac{a_2}{2!}\|x_{k+1} - x_k\|^2.$$

The results (1)–(3) hold if we replace p, p_0 by

$$p = \sqrt{1 - \tfrac{1}{d^2 c^2}} + \tfrac{a_{m+1}}{(m+1)!}c^{m+1}\eta^n + \cdots + \tfrac{a_2}{2!}c^2\eta,$$

and

$$p_0 = \sqrt{c^2\left[d^2 + \left(\tfrac{2a_{m+1}}{(m+1)!}c^{m-1}\eta^{m-1} + \cdots + a_2\right)\eta\right]} - 1.$$

4.17. (a) Let F be a Fréchet-differentiable operator defined on some closed convex subset D of a Banach space X with values in a Banach space Y; let $A(x) \in L(X, Y)$ $(x \in D)$. Assume: there exists $x_0 \in D$ such that $A(x_0) \in L(X, Y)$, $A(x_0)^{-1} \in L(Y, X)$, and

$$\|A(x_0)^{-1}[F'(y) - A(x)]\| < \varepsilon_0, \quad \text{for all } x, y \in U(x_0, \delta_0).$$

Then, show:

(1) for all $\varepsilon_1 > 0$ there exists $\delta_1 > 0$ such that

$$\|[A(x)^{-1} - A(x_0)^{-1}]A(x_0)\| < \varepsilon_1, \quad \text{for all } x \in U(x_0, \delta_1).$$

Set $\delta = \min\{\delta_0, \delta_1\}$ and $\varepsilon = \max\{\varepsilon_0, \varepsilon_1\}$.

(2) for $\varepsilon > 0$ there exist $\delta > 0$ as defined above such that

$$\|A(x_0)^{-1}[F'(y) - A(x)]\| < \varepsilon$$

and

$$\|[A(x)^{-1} - A(x_0)^{-1}]A(x_0)\| < \varepsilon,$$

for all $x, y \in U(x_0, \delta)$.

(b) Let operators F, A, point $x_0 \in D$, and parameters ε, δ be as in (1). Assume there exist $\eta \geq 0$, $c \geq 0$ such that

$$\|A(x_0)^{-1} F(x_0)\| \leq \eta,$$
$$(1+\varepsilon)\varepsilon \leq c < 1,$$
$$\frac{\eta}{1-c} \leq \delta,$$

and

$$\bar{U}(x_0, \delta) \subseteq D.$$

Show: sequence $\{x_n\}$ ($n \geq 0$) generated by Newton-like method is well defined, remains in $U(x_0, \delta)$ for all $n \geq 0$, and converges to a solution $x^* \in \bar{U}(x_0, \delta)$ of equation $F(x) = 0$. Moreover, if linear operator

$$L = \int_0^1 F'(x + t(y-x))\,dt$$

is invertible for all $x, y \in D$, then x^* is the unique solution of equation $F(x) = 0$ in $\bar{U}(x_0, \delta)$. Furthermore, the following error bounds hold for all $n \geq 0$

$$\|x_{n+1} - x_n\| \leq c^n \|x_1 - x_0\| \leq c^n \eta$$

and

$$\|x_n - x^*\| \leq \tfrac{c^n}{1-c} \|x_1 - x_0\|.$$

(c) Let $X = Y = \mathbb{R}$, $D \supseteq U(0, .3)$, $x_0 = 0$,

$$F(x) = \tfrac{x^2}{2} + x - .04.$$

Set $A(x) = F'(x)$ ($x \in D$), $\delta_3 = \delta_4 = \varepsilon_3 = \varepsilon_4 = .3$. Then we obtain

$$c_3 = \tfrac{6}{7} < 1,$$
$$\tfrac{\eta}{1-c_3} = .28 < \delta = \delta_3.$$

The conclusions of (b) hold and

$$x^* = .039230485 \in U(x_0, \delta).$$

(d) Let $x^* \in D$ be a simple zero of equation $F(x) = 0$. Assume:
$$\|A(x^*)^{-1}[F'(y) - A(x)]\| < \varepsilon_{11}, \quad \text{for all } x \in U(x^*, \delta_{11}),$$
$$\|[A(x)^{-1} - A(x^*)^{-1}]A(x^*)\| < \varepsilon_{12}, \quad \text{for all } x \in U(x^*, \delta_{12}).$$
Set
$$\delta_{13} = \min\{\delta_{11}, \delta_{12}\}, \quad c_8 = (1 + \varepsilon_{12})\varepsilon_{11}.$$
Further, assume:
$$0 < c_8 < 1$$
$$x_0 \in \bar{U}(x^*, \delta_{13}),$$
and
$$\bar{U}(x^*, \delta_{13}) \subseteq D.$$
Show: sequence, $\{x_n\}$ $(n \geq 0)$ generated by Newton-like method is well defined, remains in $U(x^*, \delta_{13})$ for all $n \geq 0$ and converges to x^* with
$$\|x_{n+1} - x^*\| \leq c_8 \|x_n - x^*\|, \quad \text{for all } n \geq 0.$$

Chapter 5

Two-Step Newton Methods and Their Applications

The Kantorovich convergence analysis of Newton methods inaugurated by L.V. Kantorovich (Nobel Prize of Economics) [182]-[183] (see Chapter 4) has enjoyed a very rapid growth especially duing the past three decades. But the study of Kantorovich's analysis for multipoint iterative methods is less developed although the fundamental theory of such methods was developed by Ostrowski and Traub in the early sixties [265]. The main reason is that it is not easy to find a scalar function g such that $\|F(x_n)\| \leq g(t_n)$ $(n \geq 0)$ for some nonnegative sequence $\{t_n\}$ $n \geq 0$ and some Newton method $\{x_n\}$ $n \geq 0$, for multistep Newton methods. Note however that this is a relatively easy task for single step methods (see Chapter 3).

Here we develop a general theory for two-step Newton methods that assists us to overcome this obstacle and control F by g as it happens for single step Newton methods. It is known however that from the efficiency point of view multipoint iterative methods are much better than single Newton methods [66], [99], [265]. The same is true from the complexity theory point of view [265].

In this chapter we use two-step Newton-methods to approximate a locally unique solution of the nonlinear operator equation in a Banach as well as in a POTL-space setting. The order of convergence of our method is also provided that shows superiority of these methods over the ones introduced in Chapter 4.

5.1 Two-Step Newton Methods

In this section, we introduce some new very general ways of constructing fast two-step methods to approximate a locally unique solution of a nonlinear operator equation in a Banach space setting. We provide existence-uniqueness theorems as well as an error analysis for the iterations involved using Newton-Kantorovich-type hypotheses and the majorant method. Our results depend on the existence of a Lipschitz function defined on a closed ball centered at a certain point and of a fixed radius and with values into the positive real axis. Special choices of this function lead to favorable comparisons with results already in the literature. The monotone convergence is also examined in a partially ordered topological space setting. some setting. some applications to the solution of nonlinear integral equations appearing in radiative transfer as well as to the solution of integral equations of Uryson-type are also provided.

Let $x_0, z_0 \in D$ be fixed and define the two-step Newton method for all $n = 0$ by

$$y_n = x_n - F'(x_n)^{-1} F(x_n) \qquad (5.1)$$

$$x_{n+1} = y_n - z_n, \qquad (5.2)$$

Here $z_n \in X$ are points to be determined for all $n \geq 0$. Some choices for z_n can be given by $z_n = 0$ or $z_n = F'(x_n)^{-1} F(y_n)$ or $z_n = F'(y_n)^{-1} F(y_n)$ for all $n \geq 0$. The first choice gives the ordinary Newton method (single step). The second choice leads to a two-step Newton-method, where two function evaluations and one inverse are required at each step. The third choice leads to another two-step method, where two function evaluations and two inverses are required at each step. Here we will give general conditions for the selection of the $z_n's$ $n \geq 0$ that will guarantee that the sequence $\{x_n\}$ $n \geq 0$ generated by the approximations (5.1)-(5.2) converges to a solution x^* of the equation $F(x) = 0$.

Let $x_0, z_0 \in D$ and $R > 0$ be fixed and assume that there exists a function $\overline{\alpha} : \overline{U}^3(x_0, R) \to [0, +\infty)$ such that

$$\left\| F'(x_0)^{-1} [F(y) - F(x) - F'(x)(y - x) + F'(y)(z - y)] \right\| \leq \overline{\alpha}(x, y, z) \qquad (5.3)$$

for all $x, z \in \overline{U}(y, r(x, z)) \subseteq \overline{U}(x_0, R) \subseteq D$, where $r(x, z) = \max\{\|y - x\|, \|y - z\|\} \leq R - \|y - x_0\|$.

Using the majorant theory and condition (5.3) we will show that under certain hypotheses the iteration $\{x_n\}$ $n \geq 0$ generated by (5.1)-(5.2) converges to a locally unique solution x^* of equation $F(x) = 0$. The order of convergence of the iteration $\{x_n\}$ $n \geq 0$ is also examined here. Under special choices of the function α and the points z_n $n \geq 0$ our results can be reduced to the ones obtained already.

We will need to introduce the constants

$$t_0 = 0, \quad s_0 \geq \|y_0 - x_0\|, \quad I_0 \geq \|z_0\|, \tag{5.4}$$

the sequences for all $n \geq 0$

$$s_{n+1} = t_{n+1} + \frac{1}{1 - Lt_{n+1}} h_{n+1}, \quad L = 2L_0 \tag{5.5}$$

$$t_{n+1} = s_n + I_n, \tag{5.6}$$

$$h_{n+1} = \frac{L}{2}(t_{n+1} - s_n)^2 + \alpha_n \tag{5.7}$$

for some fixed $L_0 > 0$ and some given sequences $\{\alpha_n\}$ and $\{I_n\}$ with $\alpha_n \leq 0$, $I_n \leq 0$,

$$\sum_{i=0}^{k+1} \alpha_i \leq \gamma_0 r_0 + \gamma_1 \quad \text{and} \quad \sum_{i=0}^{k+1} I_i \leq \gamma_2 r_0 \tag{5.8}$$

for all nonnegative integers k, some fixed real constants γ_0, γ_1, γ_2 and some fixed $r_0 \in [0, R]$.

Moreover, we define the sequences for all $n \geq 0$

$$\overline{h}_{n+1} = \frac{L}{2}\|x_{n+1} - y_n\|^2 + \overline{\alpha}_n, \tag{5.9}$$

where $\overline{\alpha}_n$ denotes $\overline{\alpha}(x_n, y_n, x_{n+1})$ with

$$\overline{\alpha}_n \leq \alpha_n \quad \text{for all} \quad n \geq 0, \tag{5.10}$$

$$\overline{e}_{n+1} = \left[1 - \frac{L}{2}(\|x^* - x_0\| + \|x_{n+1} - x_0\|)\right]^{-1}, \tag{5.11}$$

$$e_{n+1} = \left[1 - \frac{L}{2}(r_0 + t_{n+1})\right]^{-1}, \tag{5.12}$$

$$p_n = \frac{L}{2}\|x^* - x_n\|^2 \tag{5.13}$$

and the function

$$T(r) = s_0 + \frac{1}{1 - Lr}\left[\frac{L}{2}(r - s_0)^2 + \gamma_0 r + \gamma_1\right] + \gamma_2 r \tag{5.14}$$

on $[0, R]$.

We can now state the result:

Theorem 5.1 Let $F: D \subseteq X \to Y$ be a nonlinear operator whose divided difference $[x, y]$ satisfies

$$\left\| F'(x_0)^{-1} ([x, y] - [u, v]) \right\| \leq L_0 (\|x - u\| + \|y - v\|) \qquad (5.15)$$

for some $L_0 > 0$ and all $x, y, uv \in U(x_0, R)$. Moreover, we assume:

(a) the condition (5.3) is satisfied;
(b) the sequences $\{\alpha_n\}$, $\{I_n\}$ and $\{\bar{\alpha}_n\}$ $n \geq 0$ satisfy conditions (5.8) and (5.10) and $\|z_n\| \leq I_n$ for all $n \geq 0$;
(c) there exists a minimum nonnegative number r_0 such that

$$T(r_0) \leq r_0; \qquad (5.16)$$

(d) the following estimates are true:

$$r_0 \leq R$$

and

$$L_0(r_0 + R) < 1; \qquad (5.17)$$

and
(e) the ball

$$\overline{U}(x_0, R) \subseteq D. \qquad (5.18)$$

Then

(i) the scalar sequence $\{t_n\}$ $n \geq 0$ generated by relations (5.5) – (5.6) is monotonically increasing and bounded above by its limit, which is number r_0;
(ii) the sequence $\{x_n\}$ $n \geq 0$ generated by relations (5.1) – (5.2) is well defined, remains in $\overline{U}(x_0, r_0)$ for all $n \geq 0$, and converges to a solution x^* of the equation $F(x) = 0$, which is unique in $U(x_0, R)$.

Moreover, the following estimates are true for all $n \geq 0$

$$\|y_n - x_n\| \leq s_n - t_n, \tag{5.19}$$

$$\|x_{n+1} - y_n\| \leq t_{n+1} - s_n, \tag{5.20}$$

$$\|x_n - x^*\| \leq r_0 - t_n, \tag{5.21}$$

$$\|y_n - x^*\| \leq r_0 - s_n, \tag{5.22}$$

$$\left\|F'(x_0)^{-1} F(x_{n+1})\right\| \leq \overline{h}_{n+1}, \tag{5.23}$$

$$\|x^* - x_{n+1}\| \leq \overline{e}_{n+1}\overline{h}_{n+1} \leq e_{n+1}h_{n+1} \leq r_0 - t_{n+1} \tag{5.24}$$

and

$$\|y_n - x_n\| \leq \|x^* - x_n\| + \frac{p_n}{1 - L\|x_n - x_0\|}. \tag{5.25}$$

(We will be concerned only with the case $r_0 > 0$, since when $r_0 = 0$, $x_0 = x^*$.)

Proof. (i) Using relations (5.4), (5.5), (5.6), (5.8) and (5.16) we deduce that the scalar sequence $\{t_n\}$ $n \geq 0$ is monotonically increasing, nonnegative, and $t_0 \leq s_0 \leq t_1 \leq s_1 \leq r_0$. Let us assume that $t_k \leq s_k \leq t_{k+1} \leq s_{k+1} \leq r_0$ for $k = 0, 1, 2, ..., n$. Then by relations (5.5), (5.6), (5.8) we can have in turn

$$t_{k+2} = s_{k+1} + I_{k+1} = t_{k+1} + \frac{1}{1 - Lt_{k+1}}\left[\frac{L}{2}(t_{k+1} - s_k)^2 + \alpha_k\right] + I_{k+1}$$

$$\leq t_{k+1} + \frac{1}{1 - Lr_0}\left[\frac{L}{2}(r_0 - s_0)(t_{k+1} - s_k) + \alpha_k\right] + I_{k+1}$$

$$\leq \cdots \leq s_0 + \frac{1}{1 - Lr_0}\left[\frac{L}{2}(r_0 - s_0)\sum_{i=0}^{k+1}(t_{i+1} - s_i) + \sum_{i=0}^{k}\alpha_i\right]$$

$$+ \sum_{i=0}^{k+1} I_i$$

$$\leq s_0 + \frac{1}{1 - Lr_0}\left[\frac{L}{2}(r_0 - s_0)^2 + \gamma_0 r_0 + \gamma_1\right] + \gamma_2 r_0 = T(r_0) \leq r_0$$

by hypothesis (5.16).

Hence, the scalar sequence $\{t_n\}$ $n \geq 0$ is bounded above by r_0. By hypothesis (5.16) the number r_0 is the minimum nonnegative zero of the equation $T(r) - r = 0$ on $[0, r_0]$ and from the above $r_0 = \lim_{n \to \infty} t_n$.

(ii) Using relations (5.1), (5.2), (5.4), (5.5) and (5.6) we deduce that $x_1, x_0 \in \overline{U}(x_0, r_0)$ and that estimates (5.19) and (5.20) are true for $n = 0$. Let us assume that they are true for $k = 0, 1, 2, ..., n-1$. Using the induction hypothesis we can have in turn

$$\begin{aligned}\|x_{k+1} - x_0\| &\leq \|x_{k+1} - y_0\| + \|y_0 - x_0\| \\ &\leq \|x_{k+1} - y_k\| + \|y_k - y_0\| + \|y_0 - x_0\| \\ &\leq \cdots \leq (t_{k+1} - s_k) + (s_k - s_0) + s_0 \leq t_{k+1} \leq r_0,\end{aligned}$$

and

$$\begin{aligned}\|y_{k+1} - x_0\| &\leq \|y_{k+1} - y_0\| + \|y_0 - x_0\| \leq \|y_{k+1} - x_{k+1}\| + \|x_{k+1} - y_k\| \\ &\quad + \|y_k - y_0\| + \|y_0 - x_0\| \\ &\leq \cdots \leq (s_{k+1} - t_{k+1}) + (t_{k+1} - s_k) + (s_k - s_0) + s_0 \\ &= s_{k+1} = r_0.\end{aligned}$$

That is $x_n, y_n \in \overline{U}(x_0 r_0)$ for all $n \geq 0$.

Note that by hypothesis (5.15) $F'(x) = [x, x]$ for all $x \in D$, and for all $x_k \in \overline{U}(x_0, r_0)$

$$\left\|F'(x_0)^{-1}(F'(x_k) - F'(x_0))\right\| \leq L_0(\|x_k - x_0\| + \|x_k - x_0\|)$$
$$= L\|x_k - x_0\| \leq L(t_k - t_0)$$

by hypothesis (5.17). It now follows from the Banach Lemma on invertible operators the linear operator $F'(x_k)$ is invertible, and

$$\left\|F'(x_k)^{-1} F'(x_0)\right\| \leq \frac{1}{1 - L\|x_k - x_0\|} \leq \frac{1}{1 - Lt_k} \quad (5.26)$$

for all k.

We can now have

$$\begin{aligned}F(x_{k+1}) &= [F(x_{k+1}) - F(y_k) - F'(y_k)(x_{k+1} - y_k)] \\ &\quad + [F(y_k) + F'(y_k)(x_{k+1} - y_k)] \\ &= \int_0^1 [F'(y_k + t(x_{k+1} - y_k)) - F'(y_k)](x_{k+1} - y_k) \, dt \\ &\quad + [F(y_k) + F'(y_k)(x_{k+1} - y_k)],\end{aligned}$$

and by using hypotheses (5.3) and (5.15) we obtain

$$\left\| F'(x_0)^{-1} F(x_{k+1}) \right\|$$
$$\leq \int_0^1 \left\| F'(x_0)^{-1} (F'(y_k + t(x_{k+1} - y_k)) - F'(y_k)) \right\| \|x_{k+1} - y_k\| \, dt$$
$$+ \left\| F'(x_0)^{-1} [F(y_k) - F(x_k) - F'(x_k)(y_k - x_k) + F'(y_k)(x_{k+1} - y_k)] \right\|$$
$$\leq \frac{L}{2} \|x_{k+1} - y_k\|^2 + \overline{\alpha}(x_k, y_k, x_{k+1})$$
$$= \frac{L}{2} \|x_{k+1} - y_k\|^2 + \overline{\alpha}_k = \overline{h}_{k+1}$$
$$\leq \frac{L}{2} (t_{k+1} - s_k)^2 + \alpha_k = h'_{k+1}$$

by hypotheses (5.10), (5.20) and relations (5.9) and (5.7). hence, we showed estimate (5.23) for all $k \geq 0$.

Using relations (5.1), (5.5), (5.23) and (5.26) we obtain

$$\|y_{k+1} - x_{k+1}\| \leq \left\| F'(x_{k+1})^{-1} F'(x_0) \right\| \left\| F'(x_0)^{-1} (F(x_{k+1})) \right\|$$
$$\leq \frac{1}{1 - L\|x_{k+1} - x_0\|} \overline{h}_{k+1} \leq \frac{1}{1 - Lt_{k+1}} h_{k+1} = s_{k+1} - t_{k+1},$$

which shows estimate (5.19) for all $n \geq 0$.

similarly, from relations (5.2), (5.6) and hypothesis (ii) we obtain

$$\|x_{k+1} - y_k\| = \|-z_k\| = \|z_k\| \leq I_k = t_{k+1} - s_k,$$

from which it follows that estimate (5.20) is true for all $n \geq 0$.

It now follows from the estimates (5.19) and (5.20) that the sequence $\{x_k\}$ is Cauchy in a Banach space X and as such it converges to some $x^* \in \overline{U}(x_0, r_0)$, which by taking the limit as $k \to \infty$ in (5.1) we obtain $F(x^*) = 0$.

To show uniqueness, we assume that there exists another solution y^* of equation $F(x) = 0$ in $U(x_0, R)$. From hypothesis (5.15) we get

$$\int_0^1 \left\| F'(x_0)^{-1} [F'(x^* + t(y^* - x^*)) - F'(x_0)] \right\| dt$$
$$\leq L \int_0^1 \|x^* + t(y^* - x^*) - x_0\| \, dt$$
$$\leq L \int_0^1 [(1-t)\|y^* - x_0\| + t\|x^* - x_0\|] \, dt$$

$$\leq L_0 (r_0 + R) < 1$$

by hypothesis (5.17).

It now follows that the linear operator

$$\int_0^1 F'(x^* + t(y^* - x^*)) \, dt$$

is invertible, and from the approximation

$$F(y^*) - F(x^*) = \int_0^1 F'(x^* + t(y^* - x^*)) \, dt \, (y^* - x^*),$$

it follows that $x^* = y^*$.

Estimates (5.21) and (5.22) follow easily from estimates (5.19) and (5.20) respectively for all $n \geq 0$.

Finally, using the approximations

$$x_{k+1} - x^* = \left(D_{k+1}^{-1} F'(x_0)\right) \left(F'(x_0)^{-1} F(x_{k+1})\right),$$

$$D_{k+1} = \int_0^1 F'(x^* + t(x_{k+1} - x^*)) \, dt,$$

$$y_k - x_k = x^* - x_k + \left(F'(x_k)^{-1} F'(x_0)\right) \cdot$$

$$\cdot \int_0^1 \left\{ F'(x_0)^{-1} [F'(x_k + t(x^* - x_k)) - F'(x_k)] (x^* - x_k) \right\} dt,$$

and the estimate

$$\int_0^1 \left\| F'(x_0)^{-1} [F'(x^* + t(x_{k+1} - x^*)) - F'(x_0)] \right\| dt$$

$$\leq L \int_0^1 \|x^* + t(x_{k+1} - x^*) - x_0\| \, dt$$

$$\leq L \int_0^1 [(1-t) \|x^* - x_0\| + t \|x_{k+1} - x_0\|] \, dt$$

$$\leq L r_0 < 1 \quad \text{by hypothesis (5.17)}$$

and

$$\left\| D_{k+1}^{-1} F'(x_0) \right\| \leq \bar{e}_{k+1},$$

we can immediately obtain estimates (5.24) and (5.25), where we have also used relations (5.11), (5.12) and (5.13).

That completes the proof of the theorem. □

Remark 5.1 *(a) Theorem was proved by using only the weaker condition*

$$\left\| F'(x_0)^{-1} ([x,x] - [u,u]) \right\| \leq \|x - u\|, \quad L > 0, \quad x, y \in \overline{U}(x_0, R) \quad (5.27)$$

instead of (5.15).

(b) Theorem 5.1 can be further generalized if we assume instead (5.14) that

$$\left\| F'(x_0)^{-1} ([x,x] - [u,u]) \right\| \leq q(r) \|x - y\| \quad (5.28)$$

for all

$$x, u \in \overline{U}(x_0, r), \quad 0 \leq r \leq R$$

where $q(r)$ is a nondecreasing function the interval $[0, R]$.

Let us define the functions

$$w(r) = \int_0^r q(t)\,dt, \quad \varphi(r) = \int_0^r w(t)\,dt - r + s_0, \quad (5.29)$$

$$T(r) = s_0 - \frac{1}{\varphi'(r)} \left[\int_{s_0}^r w(t)\,dt - w(s_0)(r - s_0) + \gamma_0 r + \gamma_1 \right] + \gamma_2 r \quad (5.30)$$

and the iterations for all $n \geq 0$

$$\overline{h}_{n+1} = \int_0^1 [w(\|y_n - x_0\| + t\|x_{n+1} - y_n\|)$$
$$- w(\|y_n - x_0\|)] \|x_{n+1} - y_n\|\,dt + \overline{\alpha}_n,$$
$$h_{n+1} = \varphi(t_{n+1}) - \varphi(s_n) - \varphi'(s_n)(t_{n+1} - s_n) + \alpha_n,$$
$$\overline{e}_{n+1} = (1 - w((1-t)\|x_0 - x^*\| + t\|x_{n+1} - x_0\|))^{-1},$$
$$e_{n+1} = (1 - w((1-t)r_0 + tt_{n+1}))^{-1},$$
$$p_n = \int_0^1 (w(\|x_n - x_0\| + t\|x_n - x^*\|) - w(\|x_n - x_0\|)) \|x_n - x^*\|\,dt.$$

Replace in relations (5.5) and (5.25) the iterations $\frac{1}{1-Lt_{n+1}}$, $\frac{1}{1-L\|x_n-x_0\|}$ by $-\frac{1}{\varphi'(t_{n+1})}$ and $-\frac{1}{\varphi'(\|x_n-x_0\|)}$ respectively. Moreover, replace condition (5.15) with condition (5.28) and condition (5.17) with $w((1-t)r_0 + tR) \leq 1$. Then following the proof of Theorem 5.1 step by step we can produce a more general theorem under exactly the same hypotheses (with the modification introduced above) and the same conclusions. See also the proofs in [68], [99].

In [6], [68], [99] we saw how to choose the function $q(r)$, when it comes to solving nonlinear integral equation of Uryson-type in various spaces.

(c) We will now find the order of convergence for the iterations $\{x_n\}$, $\{t_n\}$ $n \geq 0$, by first showing that conditions (5.16) and (5.17) can somehow correspond to standard Newton-kantorovich-type hypotheses.

Let us assume that there exist nonnegative constants η, h, σ and $K \geq 0$ such that

$$\beta K \leq L, \quad \eta \leq s_0, \quad \sigma \leq 2h = 2K\eta \leq 1, \tag{5.31}$$

$$\sigma_0 \leq 2\eta \quad \text{and} \quad \sigma = 1 - \left(\frac{2\eta}{r_0} - 1\right)^2 \quad \text{provided that } r \neq 0. \tag{5.32}$$

If $r_0 = 0$, we choose $s_0 = \eta = h = 0$, and our conditions reduce to $K \leq L$ only. Moreover we define a scalar function

$$g(t) = \frac{K}{2}t^2 - t + \eta. \tag{5.33}$$

and iterations

$$v_n = w_n - \frac{g(w_n)}{g'(w_n)}, \quad w_0 = 0 \tag{5.34}$$

$$w_{n+1} = v_n - \delta_n, \quad \delta_n \leq 0 \quad \text{for all} \quad n \geq 0 \tag{5.35}$$

$$q_\eta = \frac{2}{K}[g(v_n) - \delta_n g'(v_n)] \tag{5.36}$$

with

$$L_0 q_\eta \leq \alpha_n \quad \text{and} \quad -\delta_n \leq I_n \quad \text{for all} \quad n \geq 0 \tag{5.37}$$

Furthermore, let us define

$$r_1 = \frac{1 - \sqrt{1 - 2h}}{h}\eta, \tag{5.38}$$

$$r_2 = \frac{1 + \sqrt{1 - 2h}}{h}\eta \tag{5.39}$$

and

$$\theta = \frac{r_1}{r_2}. \tag{5.40}$$

Note that the constants given by relations (5.38) and (5.39) are the real solutions of the equation $g(t) = 0$, where g is given by (5.33) (provided that $2h \leq 1$).

We can now prove the following proposition.

Proposition 5.1 *Let us assume:*

(a) *the hypotheses of Theorem 5.1 are true;*
   and that
(b) *the hypotheses (5.31), (5.32) and (5.37) are also true.*

Then the following estimates are true:

$$w_n \leq t_n, \quad v_n \leq s_n, \quad w_{n+1} - v_n \leq t_{n+1} - s_n, \quad v_n - w_n \leq s_n - t_n$$
$$r_0 = r_1 \quad \text{and} \quad r_0 - t_n \leq r_1 - w_n \quad \text{for all} \quad n \geq 0.$$

Proof. We will use induction on n. Using relations (5.5), (5.6), (5.34), (5.35) and the hypotheses for $n = 0$, we get in turn

$$w_1 = v_0 - \delta_0 = \eta - \delta_0 s < s_0 + I_0 = t_1 \Longrightarrow w_1 \leq t_1,$$
$$w_1 - v_0 = -\delta_0 \leq I_0 = t_1 - s_0 \Longrightarrow w_1 - v_0 \leq t_1 - s_0,$$
$$v_1 = w_1 + \frac{K}{2(1 - Kw_1)}\left[(w_1 - v_0)^2 + q_0\right]$$
$$\leq t_1 + \frac{1}{1 - Lt_1}\left[\frac{L}{2}(w_1 - v_0)^2 + L_0 q_0\right]$$
$$\leq t_1 + \frac{1}{1 - Lt_1}\left[\frac{L}{2}(t_1 - s_0)^2 + \alpha_0\right] = s_1 \Longrightarrow v_1 \leq s_1,$$

and from the same arguments $w_1 - s_1 \leq s_1 - t_1$.

That is we have showed all the inequalities for $n = 0$. We now assume that they are true for $k = 0, 1, 2, ..., n - 1$. The induction can now easily be completed if we repeat the proof we gave for $n = 0$, by observing that we can replace the subscript 1 and 0 by $k + 1$ and k respectively.

From hypothesis $\sigma < 2h$ it can easily follow that $r_0 \leq r_1$ and by taking the limit as $n \to \infty$ in the estimate $w_n \leq v_n$, we obtain $r_1 \leq r_0$. Hence, we get $r_0 = r_1$.

The proof of the proposition is now complete. □

Remark 5.2 *From the proof of the Proposition 5.1 and the theorem it can now easily be seen that the uniqueness of the solution x^* can be extended* in the ball $U(x_0, r)$ with $r_1 \leq r \leq r_2$ provided that $\overline{U}(x_0, r) \leq D$ and $L_0(r_1 + r_2) < 1$.

It also follows from the Proposition that the order of convergence of iterations $\{x_n\}$ and $\{t_n\}$ is comparable to the order of convergence of the iteration $\{w_n\}$ $n \geq 0$.

The conditions $\sigma \geq 2h$ and $r_0 \leq 2\eta$ are used only to show that $r_0 \leq r_1$. If these conditions are violated, we can still show that the order of convergence of iterations $\{x_n\}$ and $\{t_n\}$ $n \geq 0$ is "at least asymptotically" equal to the order of convergence of the iteration $\{w_n\}$ $n \geq 0$. See the examples that follow.

Remark 5.3 *Let us examine some special choices for the z_n's. Set*

$$z_n = A_n^{-1} B_n (y_n - x_n)$$

for all $n \geq 0$, some linear operators B_n and A_n with A_n being invertible for all $n \geq 0$. Here the linear operators A_n and B_n may depend on x_n or y_n or both or neither. The condition (5.3) can now be replaced by one of the following three sets of conditions for all $x_n, x_{n+1} \in U(y_n, r(x_n, x_{n+1}))$

$$\left\| F'(x_0)^{-1} \left[\int_0^1 [F'(x_n + t(y_n - x_n)) - (F'(x_n) + B_n)](y_n - x_n)\, dt \right.\right.$$
$$\left.\left. + (F'(y_n) - A_n)(x_{n+1} - y_n) \right] \right\| \leq \bar{a}_n \|x_n - y_n\| + \bar{b}_n \|x_{n+1} - y_n\|$$

or by

$$\left\| F'(x_0)^{-1} \left[\int_0^1 [F'(x_n + t(y_n - x_n)) - (F'(x_n) + B_n)](y_n - x_n)\, dt \right] \right\|$$
$$\leq \bar{a}_n \|x_n - y_n\|.$$

and

$$\left\| F'(x_0)^{-1} (F'(y_n) - A_n)(x_{n+1} - y_n) \right\| \leq \bar{b}_n \|x_{n+1} - y_n\|$$

or by

$$\left\| F'(x_0)^{-1} \left[\int_0^1 [F'(x_n + t(y_n - x_n)) - (F'(x_n) + B_n)]\, dt \right] \right\| \leq \bar{a}_n,$$

and

$$\left\| F'(x_0)^{-1} (F'(y_n) - A_n) \right\| \leq \bar{b}_n.$$

The sequence $\{\bar{a}_n\}$, $\{\bar{b}_n\}$ $n \geq 0$ may differ in each part of conditions but we use the same letters.

The above conditions can be written in the more inclusive form (5.3) if x_n, y_n, x_{n+1}, $\bar{a}_n$, $\bar{b}_n$, B_n, A_n are replaced by x, y, z, $a(x,y,z)$, $b(x,y,z)$, $B(x,y)$, $A(x,y)$ respectively, for all $x, z \in \overline{U}(y, r(x,z)) \subseteq \overline{U}(x_0, R)$. Here $a(x,y,z)$ and $b(x,y,z)$ functions from $\overline{U}^3(x_0, R)$ into $[0, +\infty)$, and $B(x,y)$, $A(x,y)$ denote linear operators for fixed points $x, y \in \overline{U}(x_0, R)$.

However we will use the discrete conditions for our purpose from now on.

Let us now assume that there exist sequences $\{a_n\}$, $\{b_n\}$ and positive numbers a and b which may depend on r_0 such that

$$\bar{a}_n \leq a_n \leq a \quad \text{and} \quad \bar{b}_n \leq b_n \leq b.$$

Then we can set

$$\bar{\alpha}_n = \bar{a}_n \|y_n - x_n\| + \bar{b}_n \|x_{n+1} - y_n\|,$$
$$\alpha_n = a_n(s_n - t_n) + b_n(t_{n+1} - s_n)$$

and we will have that

$$\sum_{i=0}^{k} \alpha_i \leq ar_0 + b(r_0 - s_0) = \gamma_0 r_0 + \gamma_1$$

where $\gamma_0 = a + b$ and $\gamma_1 = -bs_0$ (see also relation (5.8)).

Moreover, let us assume that there exists sequences $\{\bar{c}_n\}$, $\{c_n\}$ and a number c which may depend on r_0 such that

$$\|A_n^{-1} B_n\| \leq \bar{c}_n \leq c_n \leq c \quad \text{for all} \quad n \geq 0.$$

Then we can set

$$I_n = c_n(s_n - t_n)$$

and therefore

$$\sum_{i=0}^{k+1} I_i \leq c \sum_{i=0}^{k+1} (s_n - t_n) \leq cr_0$$

that is by relation (5.8) $\gamma_2 = c$.

From now on we will assume that the function $\bar{\alpha}$ appearing in condition (5.3) can be chosen by any of the three sets of conditions mentioned above. We can now show how to choose the operators A_n and B_n for all $n \geq 0$.

Example 5.1 Let $A_n = F'(x_n)$ and $B_n = [x_n, y_n] - [x_n, x_n]$ for all $n \geq 0$. The iteration $\{x_n\}$ generated by (5.1) – (5.2) with the above choices has been examined in [68]. It is an Euler-Chebysheff-type method [144]-[151], [163], [164], [169]-[176], [181]. We can now set

$$\overline{b}_n = L \|y_n - x_n\| \leq L(s_n - t_n) = b_n \leq Lr_0 = b,$$

$$\overline{a}_n = L_0 \|y_n - x_n\| \leq L_0(s_n - t_n) = a_n \leq L_0 r_0 = a,$$

$$\overline{c}_n = \frac{L_0 \|y_n - x_n\|}{1 - L\|x_n - x_0\|} \leq \frac{L_0(s_n - t_n)}{1 - L(t_n - t_0)} = c_n \leq \frac{L_0 r_0}{1 - L r_0} = c,$$

$$\delta_n = \frac{K(v_n - w_n)^2}{2 \; g'(w_n)}, \quad q_n = -\frac{K(v_n - w_n)^2}{g'(w_n)}$$

and for $h \leq .48528137...$

$$r_1 - w_n = \frac{(1-\theta^2)\eta}{1 - \frac{1}{\sqrt{2}}(\sqrt{2}\theta)^{3^n}} (\sqrt{2}\theta)^{3^n - 1},$$

and the rest of the conclusions of the theorem hold for this method.

Hence, the order of convergence for this method is almost three.

Asymptotic Case.

As mentioned in Remark 5.1, the conditions $c \leq 2h$ and $r_0 \leq 2\eta$ are used only to show that $r_0 \leq r_1$. If these conditions are violated we can reason as follows. for sufficiently large n and since $[x_n, y_n] - [x_n, x_n] = [x_n, y_n, x_n](y_n - x_n)$, the operator $[x_n, y_n, x_n]$ can be "approximated" by $1/2 \; F''(x_n)$. Here $[x_n, y_n, x_n]$ denotes a divided difference of order two for F and $F''(x_n)$ the second Fréchet derivative of F evaluated at $x = x_n$. Replace now the difference operator in the approximation by $1/2 \; F''(x_n)$ and use the same letters for the new iteration $\{x_n\}$ $n = 0$. Moreover, let us assume that $\left\|F'(x_0)^{-1} F''(x)\right\| \leq M$,

$$\left\|F'(x_0)^{-1}(F''(x) - F''(y))\right\| \leq N \|x - y\|$$

for all $x, y \in D$, $\|y_0 - x_0\| \leq \eta$, and $\left(M^2 + \frac{N}{3}\right)^{\frac{1}{2}} \leq K$.

Then the sequence $\{w_n\}$ $n \geq 0$ majorizes the sequence $\{x_n\}$ also and in particular $\|x_n - x^*\| \leq r_1 - w_1$ for all $n \geq 0$. For the proof of this result see the last section of this chapter.

Example 5.2 Let $A_n = F'(x_n)$ and $B_n = (I - E_n)^{-1}([x_n, y_n] - [x_n, x_n])$ with $E_n = -F'(x_n)^{-1}([x_n, y_n] - [x_n, x_n])$ for all $n \geq 0$. The iteration $\{x_n\}$ generated by (5.1) – (5.2) with the above choices has been examined in places mentioned in the previous example. It is a Chebysheff-Halley-type method. We can now set

$$\bar{b}_n = L\,\|y_n - x_n\| \leq L\,(s_n - t_n) = b_n \leq Lr_0 = b,$$

$$\bar{a}_n = \frac{L_0\,\|y_n - x_n\|^2 + (1 - 2L_0\,\|x_n - x\|)\,\|y_n - x_n\|}{1 - L_0\,(2\,\|x_n - x_0\| + \|y_n - x_n\|)},$$

$$\leq \frac{L_0\,(s_n - t_n)^2 + (s_n - t_n)}{1 - L_0\,(s_n - t_n)} = a_n$$

$$\leq \frac{L_0 r_0^2 + r_0}{1 - 2L_0 r_0} = a,$$

$$\bar{c}_n = \frac{L_0\,\|y_n - x_n\|}{1 - L_0\,(2\,\|x_n - x_0\| + \|y_n - x_n\|)} \leq \frac{L_0\,(s_n - t_n)}{1 - L_0\,(s_n + t_n)} = c_n$$

$$\leq \frac{L_0 r_0}{1 - 2L_0 r_0} = c,$$

$$q_n = -\frac{K}{2}\frac{(v_n - w_n)^3\,g'(w_n)^{-1}}{1 + \frac{K}{2}(v_n - w_n)\,g'(w_n)^{-1}}.$$

$$\delta_n = \frac{K}{2}\frac{g'(w_n)^{-1}(v_n - w_n)^2}{1 + \frac{K}{2}g'(w_n)^{-1}(v_n - w_n)}$$

$$r_1 - w_n = \frac{(1 - \theta^2)\,\eta}{1 - \theta^{3^n}}\theta^{3^n - 1} \quad \text{for all} \quad n \geq 0,$$

and the rest of the conclusions of the theorem hold for this method.

Hence, the order of convergence for this method is almost three.

Asymptotic Case.

For the asymptotic case we reason exactly as in Example 5.1. The condition on K becomes

$$\left(3M^2 + \frac{2N}{3}\right)^{\frac{1}{2}} \leq K,$$

but the rest of the hypotheses remain as in Example 5.1. For the proof of this result see the last section of this chapter.

Example 5.3 Let $A_n = F'(x_n)$ and $B_n = \frac{3}{4}(F'(x_n + \frac{2}{3}(y_n - x_n)) - F'(x_n))(I - \frac{3}{2}H_n)$, $H_n = F'(x_n)^{-1}(F'(x_n + \frac{2}{3}(y_n - x_n)) - F'(x_n))$ for all $n \geq 0$. The iteration $\{x_n\}$ generated by (5.1)–(5.2) with the above choices has been examined in [68], [99] see also the last section.

We can now set

$$\bar{b}_n = L \|y_n - x_n\| \le L(s_n - t_n) = b_n \le Lr_0 = b,$$

$$\bar{a}_n = L_0 \left(2 + \tfrac{L\|y_n - x_n\|}{1-L\|x_n-x_0\|}\right) \|y_n - x_n\| \le L_0 \left(2 + \tfrac{L(s_n-t_n)}{1-L(s_n-t_0)}\right) = a_n$$

$$\le L_0 \left(2 + \tfrac{Lr_0}{1-Lr_0}\right) = L_0 \left(\tfrac{2-Lr_0}{1-Lr_0}\right) = a,$$

$$\bar{c}_n = \tfrac{L_0 \|y_n - x_n\|}{1-L\|x_n - x_0\|}\left(1 + \tfrac{L\|y_n - x_n\|}{1-L\|x_n-x_0\|}\right) \le \tfrac{L_0}{1-L(t_n-t_0)}\left(1 + \tfrac{L(s_n-t_n)}{1-L(t_n-t_0)}\right)(s_n - t_n)$$

$$= c_n \le \tfrac{L_0}{1-Lr_0}\left(1 + \tfrac{Lr_0}{1-Lr_0}\right) = \tfrac{L_0 r_0}{(1-Lr_0)^2} = c.$$

$$\delta_n = \tfrac{3}{4} g'(w_n)^{-1} \left(g'\left(v_n + \tfrac{2}{3}(v_n - w_n)\right) - g'(v_n)\right) \cdot$$
$$\cdot \left[I - \tfrac{3}{2} g'(w_n)^{-1} \left(g'\left(v_n + \tfrac{2}{3}(v_n - w_n)\right) - g'(v_n)\right)\right],$$

and for $h \le .46568....$

$$r_1 - w_n = \frac{(1-\theta^2)\eta}{1 - \frac{1}{\sqrt[3]{5}}\left[\sqrt[3]{5\theta}\right]^{4^n}} \left[\sqrt[3]{5\theta}\right]^{4^n - 1}.$$

The rest of the conclusions of the theorem hold for this method. Hence, the order of convergence for this methods is almost four.

Asymptotic Case.

for the asymptotic case we reason exactly as above. The condition on K becomes

$$\left(M^2 + \frac{N}{6}\right)^{\frac{1}{2}} \le K,$$

and the rest of the hypotheses remain as above. For the computational details in the remaining cases see [68].

Example 5.4 Choose $z_n = 0$ for all $n \ge 0$ in approximation (5.2). Then by approximation (5.1) − (5.2) we obtain the single-step Newton-Kantorovich method which is of order two. Note that we can set $\bar{a}_n = 0$, $a_n = I_n = \delta_n = q_n = 0$, $t_n = v_n$ and $s_n = w_n$ for all $n \ge 0$.

Example 5.5 Choose $z_n = M_n^{-1}\varepsilon_n$, $\varepsilon_n \in E_2$ and set $M_n = F'(x_n)$, $\varepsilon_n = F(y_n)$ for all $n \ge 0$. That is we obtain the two-step Newton-Kantorovich method that requires two function evaluations and are inverse at every step. We can then have that

$$r_1 - w_n = \frac{(1-\theta^2)\eta}{1-(d\theta)^{3^n}}(d\theta)^{3^n - 1}, \quad (\bar{a}_n = L \|y_n - x_n\| \le L(s_n - t_n) = \alpha_n \text{ e.t.c.})$$

provided that $h \leq \frac{2d}{(d+1)^2}$, $d = \sqrt[6]{2}$.

Example 5.6 Choose $z_n = M^{-1}\varepsilon_n$, $\varepsilon_n \in E_1$ and set $M_n = F'(y_n)$, $\varepsilon_n = F(y_n)$ for all $n \geq 0$. That is we obtain the two-step Newton-Kantorovich method that requires two function evaluations and two inverse at every step. We can then have that $\bar{a}_n = 0$, $a_n = 0$, $\bar{a}_n = a_n = \bar{b}_n = b_n$ and

$$r_1 - w_n = \frac{(1-\theta^2)\eta}{1-\theta^{4^n}}\theta^{4^n-1} \text{ for all } n \geq 0.$$

Other choices are also possible.

Remark 5.4 *The error estimates (5.19) − (5.25) can be improved even further, if as we did for Newton-like methods, Ptak-type or Zabrejko-Nguen or generalized Zabrejko-Nguen-type conditions are assumed (see also Chapter 4 and the rest of this section).*

Example 5.7 We will provide an example under the case in Example 5.6 that shows how to choose the constants s_0, γ_0, γ_1, γ_2 and the functions $q(r)$, $w(r)$, $\varphi(r)$ and $T(r)$.

Let us assume that $X = Y = C = C[0,1]$ the space of continuous functions on $[0,1]$ equipped with the usual supremum norm. We consider Uryson-type nonlinear integral equations of the form (2.12).

We assume for simplicity that $x_0 = 0$, and make use of the following:

Theorem 5.2 *The Lipschitz condition (5.28) for the Fréchet-derivative F' of the operator (2.12) holds if and only if the second derivative $K''_{uu}(t,s,u)$ exists for all t and almost all s and u, and*

$$\sup_{t \in [0,1]} \int_0^1 \sup_{|u| \leq r} |K''_{uu}(t,s,u)|\, ds < \infty \qquad (5.41)$$

Moreover, the left hand side in relation (5.41) is then the minimal Lipschitz constant

$$\frac{q(r)}{\beta}, \quad \beta = \left\|F'(x_0)^{-1}\right\| \quad \text{in (5.28).}$$

Moreover, the constants s_0 and β are given by

$$s_0 = \sup_{t \in [0,1]} \left| \int_0^1 K(t,s,0)\, ds + \int_0^1 r(t,s) \int_0^1 K(s,p,0)\, dp\, ds \right| \qquad (5.42)$$

and
$$\beta = 1 + \sup_{t \in [0,1]} \int_0^1 |r(t,s)|\, ds, \qquad (5.43)$$

where $r(t,s)$ is the resolvent kernel of the equation

$$h(t) - \int_0^1 K'_u(t,s,0)\, h(s) = -\int_0^1 K(t,s,0)\, ds. \qquad (5.44)$$

Let us consider a simple example. Suppose that $K(t,s,u) = c_1(t) c_2(s) c_3(u)$ with two continuous functions c_1 and c_2, and $c_3 \in C^2$. We set

$$d_1 = \int_0^1 c_2(s)\, ds, \quad d_2 = \int_0^1 c_1(s) c_2(s)\, ds. \qquad (5.45)$$

Then relation (5.44) becomes

$$h(t) = [c'_4 c'_3(0) - d_1 c_3(0)] c_1(t), \qquad (5.46)$$

where

$$c'_4 = \int_0^1 c_2(s) h(s)\, ds. \qquad (5.47)$$

Substituting relation (5.47) into (5.48), one may calculate c'_4 and hence find the resolvent kernel $r(t,s)$ in case $d_2 c'_3(0) < 1$, to get

$$r(t,s) = \frac{c_1(t) c_2(t) c'_3(0)}{1 - d_2 c'_3(0)}. \qquad (5.48)$$

Using relations (5.41)-(5.43), we obtain

$$q(t) = \|c_1\|\, d_1 \sup_{\|u\| \le r} |c''_3(u)|, \qquad (5.49)$$

$$\eta = \frac{d_1 c_3(0)}{1 - d_2 c'_3(0)} \|c_1\| \qquad (5.50)$$

and

$$\beta = 1 + \frac{d_1 c_3(0)}{1 - d_2 c'_3(0)} \|c_1\|. \qquad (5.51)$$

Thus, in this case a complete and explicit computation of the function T given by relation (5.30) is possible. As an example, let us choose

$$c_1(t) = \frac{3}{10} t, \quad c_2(s) = \frac{2}{10} s \quad \text{and} \quad c_3(u) = \frac{1}{3} u^3 + \frac{1}{10} u + 1$$

on $[0, 1]$. Then using relations (5.45), (5.48)-(5.51), (5.30), (5.29) and (5.8), we get

$$d_1 = \tfrac{1}{10}, \quad d_2 = \tfrac{2}{100}, \quad d_2 c'_3(0) = \tfrac{2}{1000} < 1,$$
$$r(t,s) = \tfrac{30}{499} ts, \quad q(r) = \tfrac{6\beta}{100} r, \quad w(r) = \tfrac{3\beta}{100} r^2, \quad \varphi(r) = \tfrac{\beta}{100} r^3 - r + s_0,$$
$$\int_0^1 w(t)\, dt = \tfrac{\beta}{100} r^3, \quad s_0 = \tfrac{15}{499}, \quad \beta = \tfrac{514}{499}, \quad \alpha = \overline{\alpha}_n = \alpha_n = \gamma_0 = \gamma_1 = 0,$$
$$\gamma_2 = \tfrac{\beta(r^2 - 3s_0^2)}{100 - 3\beta r^2}$$

and

$$T(r) - r \leq 0 \text{ if}$$
$$r^3 - .09018036 r^2 - 97.0762904 r + 2.91823361 = 0,$$

and

$$R = 5.688635222 = R_0.$$

That is, the hypotheses of Theorem 5.1 will be satisfied if we choose

$$r_0 = .030061239 \quad \text{and} \quad R = R_0.$$

The conclusions of Theorem 5.1 for the iteration under the case in Example 5.6 can now follow.

Similar work can immediately follow for the rest of the cases mentioned earlier.

5.2 Monotone Convergence

In this section we examine the monotone converges of iteration (5.52)-(5.53). The results that follow can apply for single step methods by just setting $z_n = 0$ in (5.54) for all $n \geq 0$ (see also Chapter 3).

We will assume that the reader is familiar with the notion of a partially ordered topological space, and that X and Y are POTL-spaces (see Chapter 3).

We can now prove the result:

Theorem 5.3 *Let F be a nonlinear operator defined on a convex subset D of a regular POTL-space X with values in a POTL-space Y. Let $\overline{x}_0$ and x_0 be two points of D such that*

$$\overline{x}_0 \leq x_0 \quad \text{and} \quad F(\overline{x}_0) \leq 0 \leq F(x_0). \tag{5.52}$$

Suppose that F has a divided difference of order one on

$$x_0 = \langle \overline{x}_0, x_0 \rangle = \{x \in E_1 \mid \overline{x}_0 \leq x \leq x_0\} \subseteq D \text{ satisfying} \quad (5.53)$$
$$P_0 = [x_0, x_0] \text{ has a continuous nonnegative left subinverse } Q_0 \quad (5.54)$$
$$[x, v] - [x, y] \leq 0 \quad \text{if} \quad v \leq y. \quad (5.55)$$

Consider also the iterations

$$F(x_n) + P_n (y_n - x_n) = 0, \quad P_n = [x_n, x_n], \quad (5.56)$$
$$z_n + x_{n+1} - y_n = 0, \quad (5.57)$$
$$F(\overline{x}_n) + P_n (\overline{y}_n - \overline{x}_n) = 0, \quad (5.58)$$
$$\overline{z}_n + \overline{x}_{n+1} - \overline{y}_n = 0 \quad (5.59)$$

for some sequences $\{z_n\}$, $\{\overline{z}_n\}$ $n \geq 0$ selected so that

$$\overline{z}_n - P_n (x_n - \overline{x}_n) \geq 0, \quad (5.60)$$
$$z_n - P_n (x_n - \overline{x}_n) \leq 0 \quad (5.61)$$
$$(I - B_n [x_n, \overline{x}_n])(x_n - \overline{x}_n) - B_n (z_n - \overline{z}_n) \geq 0 \quad (5.62)$$
$$z_n \geq 0 \quad (5.63)$$

and

$$\overline{z}_n \leq 0 \quad \text{for all} \quad n \geq 0, \quad (5.64)$$

where B_n denote the conditions nonnegative left subinverses of A_n $n = 0$.

Then there exist two sequences $\{\overline{x}_n\}$, $\{x_n\}$ $n \geq 0$ satisfying the approximations (5.56) – (5.59),

$$\overline{x}_0 \leq \overline{y}_0 \leq \overline{x}_1 \leq \cdots \leq \overline{y}_n \leq \overline{x}_{n+1} \leq x_{n+1} \leq y_n \leq \cdots \leq x_1 \leq y_0 \leq x_0, \quad (5.65)$$

$$\lim_{n \to \infty} \overline{x}_n = \overline{x}, \quad \lim_{n \to \infty} x_n = x \quad \text{and} \quad \overline{x}, x \in D_0 \quad \text{with} \quad \overline{x} \leq x. \quad (5.66)$$

Moreover, if the operators $P_n = [x_n, x_n]$ are inverse nonnegative then any aolution u of the equation $F(x) = 0$ in $\langle \overline{x}_0, x_0 \rangle$ belongs to $\langle \overline{x}, x \rangle$.

Proof. Let us define the operator

$$G_1 : \langle 0, x_0 - \overline{x}_0 \rangle \to E_1, \quad G_1(x) = x - Q_0 (F(\overline{x}_0) + P_0(x)).$$

This operator is isotone and continuous. We can have in turn

$$G_1(0) = -Q_0 F(\overline{x}_0) \geq 0, \quad \text{(by (5.52))},$$
$$G_1(x_0 - \overline{x}_0) = x_0 - \overline{x}_0 - Q_0 F(x_0) + Q_0(F(x_0) - F(\overline{x}_0) - P_0(x_0 - \overline{x}_0))$$
$$\leq x_0 - \overline{x}_0 + Q_0([x_0, \overline{x}_0] - [x_0, x_0])(x_0 - \overline{x}_0)$$
$$\leq x_0 - \overline{x}_0 \quad \text{(by (5.55))}.$$

By Kantorovich's Theorem 1.1 the operator G_1 has a fixed point $u_1 \in \langle 0, x_0 - \overline{x}_0 \rangle : G_1(u_1) = u_1$. Set $\overline{y}_0 = \overline{x}_0 + u_1$, then we have the estimates

$$F(\overline{x}_0) + P_0(\overline{y}_0 - \overline{x}_0) = 0,$$
$$F(\overline{y}_0) = F(\overline{y}_0) - F(\overline{x}_0) - P_0(\overline{y}_0 - \overline{x}_0) \leq 0$$

and

$$\overline{x}_0 \leq \overline{y}_0 \leq x_0.$$

Let us now define the operator

$$G_2 : \langle 0, x_0 - \overline{y}_0 \rangle \to E_1, \quad G_2(x) = x + Q_0(F(x_0) - P_0(x)).$$

This operator is isotone and continuous. We have in turn

$$G_2(0) = Q_0 F(x_0) \geq 0 \quad \text{(by (5.52))},$$
$$G_2(x_0 - \overline{y}_0) = x_0 - \overline{y}_0 + Q_0 F(\overline{y}_0) + Q_0(F(x_0) - F(\overline{y}_0) - P_0(x_0 - \overline{y}_0))$$
$$\leq x_0 - \overline{y}_0 + Q_0([x_0, \overline{y}_0] - [x_0, x_0])(x_0 - \overline{y}_0)$$
$$\leq x_0 - \overline{y}_0 \quad \text{(by (5.55))}.$$

By Kantorovich's theorem there exists $u_2 \in \langle 0, x_0 - \overline{y}_0 \rangle$ such that $G_2(u_2) = u_2$.

Set $y_0 = x_0 - u_2$, then we have the estimates

$$F(x_0) + P_0(y_0 - x_0) = 0,$$
$$F(y_0) = F(y_0) - F(x_0) - P_0(y_0 - x_0) \geq 0$$

and

$$\overline{x}_0 \leq \overline{y}_0 \leq y_0 \leq x_0.$$

From the approximations (5.57), (5.59) and estimates (5.60)-(5.64) for $n = 0$ we have respectively that

$$\bar{x}_1 - \bar{y}_0 = -\bar{z}_0 \geq 0 \Longrightarrow \bar{x}_1 \geq \bar{y}_0,$$
$$x_1 - y_0 = -z_0 \leq 0 \Longrightarrow x_1 \leq y_0$$

and

$$\bar{x}_1 \leq x_1.$$

Hence, we obtain $\bar{x}_0 \leq \bar{y}_0 \leq \bar{x}_1 \leq x_1 \leq y_0 \leq x_0$.

By hypothesis (5.55) it follows that the operator P_n has a continuous nonnegative left subinverse Q_n for all $n \geq 0$. Proceeding by induction we can show that there exist two sequences $\{\bar{x}_n\}$, $\{x_n\}$ $n \geq 0$ satisfying relations (5.65) and (5.66) in a regular space X and as such they converge to some $\bar{x}, x \in D_0$ respectively. That is we have

$$\lim_{n \to \infty} \bar{x}_n = \bar{x} \leq x = \lim_{n \to \infty} x_n.$$

If $\bar{x}_0 \leq u \leq x_0$ and $F(u) = 0$, then we can obtain

$$P_0(y_0 - u) = P_0(x_0 - Q_0 F(x_0)) - P_0 u = P_0(x_0 - u)$$
$$- P_0 Q_0 (F(x_0) - F(u))$$
$$= P_0 (I - Q_0 [x_0, u])(x_0 - u) \geq 0,$$

since $Q_0 [x_0, u] \leq Q_0 P_0 \leq I$ by (5.56)

Similarly, we show $P_0(\bar{y}_0 - u) \leq 0$.

If the operator P_0 is inverse nonnegative then it follows from the above that $\bar{y}_0 \leq u \leq y_0$.

Proceeding by induction we deduce that $\bar{y}_n \leq u \leq y_n$, from which it follows that

$$\bar{y}_n \leq \bar{x}_n \leq \bar{y}_{n+1} \leq u \leq y_{n+1} \leq x_n \leq y_n \text{ for all } n \geq 0.$$

That is, we have $\bar{x}_n \leq u \leq x_n$ for all $n \geq 0$. hence, we get $\bar{x} \leq u \leq x$. That completes the proof of the theorem. □

Remark 5.5 *Conditions for $\bar{x} = x$ and $F(\bar{x}) = 0$ can be found at the last case of Section 5 that follows.*

Remark 5.6 *let us now consider the following sets of conditions:*

(C$_1$) *Let $\bar{x}_0$ and x_0 be two points of D such that $\bar{x}_0 \leq x_0$ and $F(\bar{x}_0) \leq 0 \leq F(x_0)$;*

(**C$_2$**) F has a divided difference of order one on
$$D_0 = \langle \bar{x}_0, x_0 \rangle \subseteq D;$$

(**C$_3$**) $P_0 = [x_0, y_0]$ has a continuous nonnegative subinverse Q_0;
(**C$_4$**) $P_n = [x_n, x_n]$ are inverse nonnegative for all $n = 0$;
(**C$_5$**) $N_n = [x_n, y_n]$ are inverse nonnegative for all $n = 0$;
The following conditions are satisfied:
(**C$_6$**) $[x_0, y] \geq 0$ for all $\bar{x}_n \leq y \leq x_0$;
(**C$_7$**) $[u, v] \leq [x, y]$ if $u \leq x$ and $v \leq y$;
(**C$_8$**) $[z, w] + [w, q] - [z, z] - [v, z] \geq 0$ if $v \leq w \leq z$ for some $q \in \langle v, z \rangle$;
(**C$_9$**) $[x, y] + [y, x] + 2[y, y] - 2[x, x] \geq 0$ if $y \leq x$;
(**C$_{10}$**) There exists a positive number c such that
$$[x, y] + [y, x] + 2[y, y] - (c + 2)[x, x] \leq 0,$$
and
$$\frac{c}{2}[[x, y] + [y, x] + 2[y, y]] + [z, x] \leq [p, q]$$
for all $v \leq y \leq p \leq q \leq x$.

We can now provide several results using the following examples. The proofs as identical to the one in the last case of Section 5 are left as exercises.

Example 5.8 We choose $A_n = P_n = [x_n, x_n]$, $B_n = [x_n, y_n] - [x_n, x_n]$, $z_n = -B_n(y_n - x_n)$ and $\bar{z}_n = -B_n(y_n - x_n)$ for all $n \geq 0$. Let us assume that conditions (C$_1$)-(C$_4$), (C$_6$)-(C$_8$) are satisfied. Then the conditions (5.61)-(5.65) are satisfied with the above choices of the sequences $\{z_n\}$, $\{\bar{z}_n\}$ $n \geq 0$. Therefore the conclusions of Theorem 5.3 follow.

Example 5.9 We choose
$$A_n = P_n = [x_n, x_n], \quad B_n = \overline{E}_n([x_n, y_n] - [x_n, x_n]),$$
$$E_n = -Q_n([x_n, y_n] - [x_n, x_n]),$$
$$z_n = -B(y_n - x_n), \quad \bar{x}_n = -B_n(y_n - x_n)$$
and $\overline{E}_n$ denote the conditions nonnegative left

5.3 Exercises

5.1. Prove the claim made in Remark 5.1 (b).
5.2. Verify the claim in Example 5.1.

5.3. Verify the asymptotic case after Example 5.1.
5.4. Verify the claim in Example 5.2.
5.5. Verify the asymptotic case after Example 5.2.
5.6. Verify the claim in Example 5.3.
5.7. Verify the asymptotic case after Example 5.3.
5.8. Verify the claims made in Examples 5.4, 5.5 and 5.6.
5.9. Verify the computations following Theorem 5.2.
5.10. Verify the claims in Remark 5.4 and Example 5.8.
5.11. Follow the case of tangent hyperbolas to produce a similar analysis and comparison for the method of tangent parabolas.
5.19. Consider the problem of approximating a solution x^* of the equation $f(x) = 0$ in the complex plane by the iteration defined by

$$y_n = x_n + \alpha f(x_n)$$

$$z_n = y_n + \frac{\alpha f(y_n)}{1 - f(y_n)/f(x_n)}$$

and

$$x_{n+1} = z_n + \frac{\alpha f(z_n)}{1 - f(y_n)/f(x_n)}, \quad \alpha > 0 \quad (n \geq 0) \quad x_0 \in X \text{ given.}$$

Show:
Let $f : D \subset C \to C$ where C is the complex space and D is a convex open domain. Assume that if f has 2nd order continuous derivatives on D, and satisfies:

$$|f''(x)| \leq k \text{ for all } x \in D, \quad \left|\frac{1}{\alpha} + f'(x_0)\right| \leq \frac{1}{\alpha} - \frac{1}{\beta},$$

$$|y_0 - x_0| \leq \eta, \quad h = k\beta\eta \leq \frac{1}{2},$$

$$\overline{U}(t^* - \eta) \subset D, \quad t^* = \frac{1 - \sqrt{1 - 2h}}{h}\eta.$$

Define also the real function

$$g(t) = \frac{K}{2}t^2 - \frac{1}{\beta}t + \frac{\eta}{\beta}$$

and the iterations

$$s_n = t_n + \alpha f(t_n)$$

$$u_n = s_n + \frac{\alpha f(s_n)}{1 - f(s_n) / f(t_n)}$$

and

$$t_{n+1} = u_n + \frac{af(u_n)}{1 - f(s_n)/f(t_n)}, \quad t_0 = 0.$$

Then the sequence $\{x_n\}$ $n \geq 0$ is well defined, remains in $\overline{U}(z_0, t^* - \eta)$ and converges to a solution x^* of equation $f(x) = 0$. Moreover the following are true:

$$|x_n - x^*| \leq t^* - t_n$$
$$|y_n - x^*| \leq t^* - s_n$$
$$|z_n - x^*| \leq t^* - u_n$$

and

$$|x_n - x^*| \leq t^* - t_n < \frac{(1-\theta^2)\eta}{1-\theta^{3n}}\theta^{3^n-1}$$

for all $n \geq 0$, where

$$\theta = \frac{1 - \sqrt{1-2h}}{1 + \sqrt{1-2h}}.$$

5.13. Consider the problem of approximating a multiple root x^* with multiplicity $m \geq 1$ of the real equation $f(x) = 0$. Assume that f has derivatives as high as we desire. Show that
(i) the iterative function

$$h(x) = g(x) - \frac{f(x)(g(x)-x)}{a_1 f'(g(x))(g(x)-x) + a_2 f(x) + a_3 f'(x)(g(x)-x)}$$

is convergent with order three, where

$$g(x) = x - a\frac{f(x)}{f'(x)},$$
$$b = g'(x^*) = 1 - \frac{a}{m},$$
$$a_1 = -1/mb^m \left[(m+1)b^2 - 2mb + m - 1\right]$$
$$a_2 = (b-1)\left[(m+1)b^2 - 3mb - b + 2m\right] / b\left[(m+1)b^2 - 2mb + m - 1\right]$$

and

$$a_3 = \left[(m+1)b - m\right] / mb\left[(m+1)b^2 - 2mb + m - 1\right].$$

(ii) Choose $a = 1$ or $a = \frac{m}{m+1}$ and consider comparing the resulting iterations with Newton's or the modified Newton method on the example

$$f(x) = (x^*x - 1)(x^*x - 1), \quad m = 2, \quad x^* = 1.$$

Then verify the results

$a = 1$	$a = \frac{m}{m+1}$	(N)	(NM)
$x_0 = .8$	$x_0 = .8$	$x_0 = .8$	$x_0 = .8$
$x_1 = .9995$	$x_1 = 1.0031$	$x_1 = .9125$	$x_1 = 1.025$

5.14. Consider the two-step Newton method of the form

$$y_n = x_n - F'(x_n)^{-1}(F(x_n)),$$
$$x_{n+1} = y_n - F'(y_n)^{-1}(F(y_n)) \quad (n = 0)$$

Assume:

(a) The following conditions are satisfied

$$\|F'(x) - F'(y)\| \leq K \|x - y\|$$
$$\|F'(x) - F'(y)\| \leq q(r) \|x - y\|$$
$$\|F'(x + h)\| \leq D(r, \|h\|)$$

for all $x, y \in \overline{U}(x_0, R)$, $R > 0$ fixed, $0 < \|h\| \leq R - r$, $k > 0$, q is nondecreasing on $[0, R]$, and D is a nonnegative and continuous function of two variable such that if one of the variable is fixed then D is a nondecreasing function of the other on the interval $[0, R]$. Also, the function $\frac{\partial D(0,t)}{\partial t}$ is positive, continuous and nondecreasing on $[0, R - r]$ with $D(0, 0) = 0$. Define the constants:

$$\eta \geq \|y_0 - x_0\| > 0, \quad \|F'(x_0)^{-1}\| \leq \beta,$$
$$t_0 = t_0^1 = t_0^2 = 0, \quad s_0, s_0^1 s_0^2 \geq \eta,$$
$$t_1^1 \geq s_0^1 + p_1, \quad s_1^1 \geq t_1^1 + p_2, \quad t_1^2 \geq s_0^2 + p_3, \quad s_1^2 \geq t_1^2 + p_3,$$
$$p_1 = \frac{\beta}{1 - \beta w(s_0^1)} B(w, t_0^1, s_0^1), \quad p_2 = \frac{\beta}{1 - \beta w(t_1^1)} A(w, t_0^1, s_0^1, t_1^1),$$
$$p_3 = \frac{\beta}{1 - \beta D(0, s_0^2)} \int_{t_0^2}^{s_0^2} D(t_0^2, t) \, dt,$$
$$p_4 = \frac{\beta}{1 - \beta D(0, t_1^2)} \int_{s_0^2}^{t_1^2} D(s_0^2, t) \, dt,$$

the functions

$$g(t) = \tfrac{k}{2}t^2 - \tfrac{1}{\beta}t + \tfrac{\eta}{\beta}, \quad w(r) = \int_0^r q(t)\,dt,$$

$$T_1(r) = s_0^1 + \frac{\beta}{1-\beta w(r)} \int_0^r w(t)\,dt,$$

$$T_2(r) = s_0^2 + \frac{\beta}{1-\beta D(0,r)} \int_0^r D(r,t)\,dt,$$

and for all $n \geq 0$ the scalar iterations

$$s_n = t_n - \frac{g(t_n)}{g'(t_n)}, \quad t_{n+1} = s_n - \frac{g(s_n)}{g'(s_n)},$$

$$s_{n+1}^1 = t_{n+1}^1 + \frac{\beta}{1-\beta w(t_{n+1}^1)} A\left(w, t_n^1, s_n^1, t_{n+1}^1\right),$$

$$t_{n+1}^1 = s_n^1 + \frac{\beta}{1-\beta w(s_n^1)} B\left(w, t_n^1, s_n^1\right),$$

$$A\left(w, t_n^1, s_n^1, t_{n+1}^1\right) = \int_{s_n^1}^{t_{n+1}^1} w(t)\,dt - w\left(s_n^1\right)\left(t_{n+1}^1 - s_n^1\right),$$

$$B(w, t_n, s_n) = \int_{t_n^1}^{s_n^1} w(t)\,dt - w\left(t_n^1\right)\left(s_n^1 - t_n^1\right),$$

$$s_{n+1}^2 = t_{n+1}^2 + \frac{\beta}{1-\beta D(0, t_{n+1}^2)} \int_{s_n^2}^{t_{n+1}^2} D(s_n^2, t)\,dt,$$

$$t_{n+1}^2 = s_n^2 + \frac{\beta}{1-\beta D(0, s_n^2)} \int_{t_n^2}^{s_n^2} D\left(t_n^2, t\right)\,dt.$$

(Note that the supperscripts 1,2 here are not exponents).
(b) The condition

$$h = k\eta\beta \leq \tfrac{1}{2} \text{ is satisfied.}$$

Set $r_2 = \frac{1-\sqrt{1-2h}}{h}\eta$, and $r_1 = \frac{1+\sqrt{1-2h}}{h}\eta$.
(c) There exists a minimum positive number r_1^1 satisfying $r_1^1 \leq R$ and $T_1\left(r_1^1\right) \leq r_1^1$. The number R satisfies $w(R) < \tfrac{1}{\beta}$.
(d) There exists a minimum positive number r_1^2 satisfying $r_1^2 \leq R$ and $T_2\left(r_1^2\right) \leq r_1^2$. The number R satisfies $\beta D(0, R) < 1$.
(e) The following conditions are satisfied:

$$\frac{\beta}{1-\beta w(r)}\left[\int_{s_0^1}^r w(t)\,dt + w(r)r\right] \leq s_1 - t_1, \quad \text{or} \quad \leq -\frac{g(r)}{g'(r)}$$

and

$$\frac{\beta}{1-\beta w(r)}\left[\int_0^z w(t)\,dt + w(r)r\right] \le t_1 - s_0, \quad \text{or} \quad \le -\frac{g(r)}{g'(r)}$$

provided that $s_0 = s_0^1$ for all $r \in [0, R]$.

(f) Moreover the following conditions are also satisfied for all $t \in [0, R]$:

$$D(0,t) \le w(t), \quad D(r_2^2, t) \le w(t)$$

provided that $s_0^1 = s_0^2$.

Then show that:

(i) The sequences $\{t_n\}$, $\{t_n^1\}$, $\{t_n^2\}$ are monotonic ally increasing and bounded above by their limits r_2, r_2^1 and r_2^2 respectively.

(ii) The sequence $\{x_n\}$ $n \ge 0$ is well defined, remains in $\overline{U}(x_0, v)$ for all $n \ge 0$ ($v = r_2$, or r_2^1 or r_2^2) and converges to a unique solution x^* of the equation $F(x) = 0$.

(iii) The following estimates are true:

$$\|y_n - x_n\| \le s_n - t_n, \quad \|x_{n+1} y_n\| \le t_{n+1} - s_n,$$
$$\|x_n - x^*\| \le r_2 - t_n, \tag{5.67}$$
$$\|y_n - x^*\| \le r_2 - s_n, \quad r_2 - t_n = \frac{(1-\theta^2)\eta\theta^{4^n-1}}{1-\theta^{4^n}}, \quad \theta = \frac{r_2}{r_1},$$
$$\|y_n - x_n\| \le s_n^1 - t_n^1, \quad \|x_{n+1} - y_n\| \le t_{n+1}^1 - s_n^1,$$
$$\|x_n - x^*\| \le r_2^1 - t_n^1, \quad \|y_n - x^*\| \le r_2^1 - s_n^1, \tag{5.68}$$
$$\|y_n - x_n\| \le s_n^2 - t_n^2, \quad \|x_{n+1} - y_n\| \le t_{n+1}^2 - s_n^2,$$
$$\|x_n - x^*\| \le r_2^2 - t_n^2, \quad \|y_n - x^*\| \le r_2^2 - s_n^2, \tag{5.69}$$
$$s_n^1 - t_n^1 \le s_n - t_n, \quad t_{n+1} - s_n \le t_{n+1}^1 - s_n^1,$$
$$r_2^1 - t_n^1 \le r_2 - t_n, \quad r_2^1 - s_n^1 \le r_2 - s_n, \tag{5.70}$$

and

$$s_n^2 - t_n^2 \le s_n^1 - t_n^1, \quad t_{n+1}^2 - s_n^2 \le t_{n+1}^1 - s_n^1,$$
$$r_2^2 - t_n^2 \le r_2^1 - s_n^2 \le r_2^1 - s_n^1 \quad (n \ge 0). \tag{5.71}$$

In particular conditions (b), or (c), or (d) or (b), (c) and (e), or (c), (d) and (f) imply the results (5.67), or (5.68), or (5.69) or (5.67), (5.68) and (5.70) or (5.68), (5.69) and (5.71) respectively.

(iv) Conditions (f) can be replaced by the weaker

$$\frac{1}{1-\beta D(0,r)}\int_0^r D(r,t) \le \frac{1}{1-\beta w(r)}\int_0^1 w(t)\,dt$$

for all $r \in [0, R]$.

Note:

1] A choice for the function q can be given by

$$q(r) = \sup_{x,y \in \overline{U}(x_0,r)} \frac{\|F'(x) - F'(y)\|}{\|x-y\|}$$

in which case $q(r) \le k$ for all $x, y \in \overline{U}(x_0, r)$.

2] The function D can be chosen by

$$D(r, \|h\|) = \int_r^{r+\|h\|} q(t)\,dt$$

or

$$D(r, \|h\|) = \sup_{\substack{x,y \in \overline{U}(x_0,r) \\ \|h\| \le R-r}} \|F'(x+h) - F'(x)\|.$$

We will then have with either choice

$$D(r, \|h\|) \le k\|h\|$$

and

$$D(r, \|h\|) \le \int_r^{r+\|h\|} q(t)\,dt$$

for all $0 \le r \le R$ and $0 \le \|h\| \le R - r$.

Finally note that in Chapter 2 we have justified in detail these choices.

5.15. Consider the midpoint method (4.5.264) introduced in 4.5.2 for solving equation $F(x) = 0$.

Define the constants

$$\eta \geq \|y_0 - x_0\|, \quad \beta \geq \left\|F'(x_0)^{-1}\right\|, \quad t_0 = 0, \quad s_0 \geq \eta, \quad t_1 \geq s_0^*,$$

$$s_0^* = s_0 + \frac{\beta\eta\left[w\left(\frac{1}{2}(t_0+s_0)\right) - w(t_0)\right]}{1 - \beta w\left(\frac{1}{2}(t_0+s_0)\right)}, \text{ the scalar iterations}$$

$$s_{n+1} = t_{n+1} + \frac{\beta}{1 - \beta w(t_{n+1})} A(w, t_n, s_n),$$

$$A(w, t_n, s_n) = \int_{t_n}^{t_{n+1}} w(t)\,dt - w(s_n)(t_{n+1} - s_n)$$

$$+ \left[w\left(\tfrac{1}{2}(t_n+s_n)\right) - 2w(t_n)\right](s_n - t_n) + \frac{\beta}{1-\beta w\left(\frac{1}{2}(t_n+s_n)\right)}$$

$$\times \left[w(s_n) - w\left(\tfrac{1}{2}(t_n+s_n)\right)\right]\left[w\left(\tfrac{1}{2}(t_n+s_n)\right) - w(t_n)\right](s_n - t_n),$$

$$t_{n+2} = t_{n+1} + \frac{\beta}{1-\beta w\left(\frac{1}{2}(t_n+s_n)\right)} A(w, t_n, s_n) \quad (n \geq 0)$$

and the functions

$$w(r) = \int_0^r q(t)\,dt,$$

$$T(r) = t_1 + \frac{\beta}{1-\beta w(r)}\left[\int_0^r w(t)\,dt + w(r)r + \frac{\beta}{1-\beta w(r)} w(r)^2 r\right]$$

on $[0, R]$ with $0 \leq r \leq R$, where q is a nondecreasing function on $[0, R]$ such that:

$$\|F'(x_1) - F'(x_2)\| \leq q(r)\|x_1 - x_2\| \quad \text{for all} \quad x_1, x_2 \in \overline{U}(x_0, r).$$

Furthermore, Assume:

(a) there exists a minimum number R_1 satisfying

$$T(R_1) \leq R_1, \quad R_1 \leq R \quad \text{and} \quad \overline{U}(x_0, R_1) \subseteq D.$$

(b) R_1 also satisfies

$$\beta \int_0^1 w((1-t)R + tR_1)\,dt < 1 \quad \text{for all} \quad t \in [0, 1].$$

Then show:
(i) the sequence $\{t_n\}$ $n \geq 0$ is monotonically increasing and bounded above by its limit R_1;
(ii) the iteration $\{x_n\}$ $n \geq 0$, is well defined, remains in $\overline{U}(x_0, R_1)$ for all $n \geq 0$ and converges to a unique zero x^* of equation $F(x) = 0$ in $\overline{U}(x_0, R)$.

(iii) the following estimates are true:
$$\|x_n - x^*\| \leq R_1 - t_n$$
and
$$\|y_n - x^*\| \leq R_1 - s_n \quad (n \geq 0).$$

(iv) Under the hypotheses of Theorem 4.5.11 denote by t_n^1 and s_n^1 the scalar sequences defined there. Assume that

$$\frac{\beta}{1 - \beta w(r)} \left[\int_0^r w(t)\,dt + w(r)\,r + \frac{\beta}{1 - \beta w(r)} w(r)^2 r \right]$$
$$\leq s_1^1 - t_1^1$$

for all $r \in [0, R]$. Then show:

$$s_n - t_n \leq s_n^1 - t_n^1 \quad (n \geq 0).$$

That is under Ptak-type hypotheses the bounds on the distances $\|x_n - y_n\|$ can further be improved.

5.16. Let $F : D \subseteq \mathbf{R} \to \mathbf{R}$ be a function with continuous derivatives of second order on an open interval D, and let $x_0 \in D$ be fixed. Assume:

$$\|F''(x)\| \leq K \quad \text{for all} \quad x \in D, \quad -\frac{1}{\alpha} \leq F'(x_0) \leq -\frac{1}{\beta},$$

$$|y_0 - x_0| \leq \frac{\eta}{\beta}, \quad \bar{U}(y_0, r_1 - \eta) \subseteq D, \quad h = K\beta\eta \leq \frac{1}{2}$$

consider the iterations

$$y_n = x_n + \alpha F(x_n), \quad x_{n+1} = y_n + \frac{\alpha F(y_n)}{1 - \frac{F(y_n)}{F(x_n)}} \quad (n \geq 0)$$

$$s_n = t_n + \alpha \varphi(t_n), \quad t_{n+1} = s_n + \frac{\alpha \varphi(s_n)}{1 - \frac{\varphi(s_n)}{\varphi(t_n)}}, \quad t_0 = 0 \quad (n \geq 0)$$

where

$$\varphi(t) = \frac{1}{2} k t^2 - \frac{1}{\beta} t + \frac{\eta}{\beta}.$$

Then show:

(i) there exists a unique solution x^* of the equation $F(x) = 0$ in $\overline{U}(y_0, r_1 - \eta)$, where r_1 is the small solution of the equation $\varphi(t) = 0$ (r_2 is the large solution).

(ii) Moreover, we have

$$|x_n - x^*| \leq r_1 - t_n < \frac{(1+\theta)\eta}{b_n} \theta^{2^n - 1}, \quad b_n = \sum_{k=0}^{2^n - 1} \theta^k, \quad \theta = \frac{r_1}{r_2},$$

$$|y_n - x^*| \leq r_1 - s_n,$$

$$0 = t_0 < s_0 < t_1 < s_1 < \cdots < r_1,$$

and

$$\lim_{n \to \infty} t_n = \lim_{n \to \infty} s_n = r_1, \quad \lim_{n \to \infty} x_n = x^*.$$

5.17 Let $F : D \subseteq \mathbf{R} \to \mathbf{R}$ be a function with continuous derivatives of third order on an open interval D, and $x_0 \in D$ be fixed. Assume:

$$\left|F'(x_0)^{-1}\right| \leq \beta, \quad |y_0 - x_0| \leq \eta, \quad |F''(x)| \leq M,$$

$$|F'''(x)| \leq N \quad \text{for all} \quad x \in D, \quad \left(M^3 + \frac{2}{3}\frac{NM}{\beta}\right)^{\frac{1}{3}} \leq K,$$

$$h = K\beta\eta \leq \frac{1}{2}, \quad \text{and} \quad 0 \leq \alpha \leq 2.$$

Consider the iterations

$$y_n = x_n - \frac{F(x_n)}{F'(x_n)}, \quad x_{n+1} = y_n - \frac{1 + \alpha \frac{F(y_n)}{F'(x_n)}}{1 - (\alpha - 2)\frac{F(y_n)}{F'(x_n)}} \quad (n \geq 0)$$

$$s_n = t_n - \frac{\varphi(t_n)}{\varphi'(t_n)}, \quad t_0 = 0$$

$$t_{n+1} = s_n - \frac{\varphi(s_n)}{\varphi'(t_n)} \frac{1 + \alpha \frac{\varphi(s_n)}{\varphi'(t_n)}}{1 + (\alpha - 2)\frac{\varphi(s_n)}{\varphi'(t_n)}} \quad (n \geq 0)$$

where

$$\varphi(t) = \frac{1}{2}Kt^2 - \frac{1}{\beta}t + \frac{\eta}{\beta}.$$

Then show:
(i) There exists a unique solution x^* of the equation $F(x) = 0$ in $\overline{U}(x_0, r_1)$, provided that $\overline{U}(x_0, r_1) \subseteq D$, where r_1 is the small solution of the equation $\varphi(t) = 0$ (r_2 is the large solution).
(ii) Moreover, we have

$$|y_n - x^*| \leq r_1 - s_n, \quad |x_{n+1} - x^*| \leq r_1 - t_{n+1} \quad (n \geq 0)$$
$$0 = t_0 < s_0 < t_1 < s_1 < \cdots < t_{n-1} < s_{n-1} < \cdots < r_1$$

and

$$\lim_{n \to \infty} t_n = \lim_{n \to \infty} s_n = r_1, \quad \lim_{n \to \infty} x_n = x^*.$$

Furthermore, we have

$$r_1 - t_n(\alpha) \geq \frac{(1+\theta)\eta}{b_n} \theta^{4^n - 1} = r_1 - t_n(0)$$

where

$$b_n = \sum_{i=0}^{4^n - 1} \theta^i \quad \text{and} \quad \theta = \frac{r_1}{r_2}.$$

(iii) Finally, let us assume that $h = k\beta\eta \leq \frac{2\sqrt[3]{5}}{(1+\sqrt[3]{5})^2}$ then, we have:

$$\frac{(1-\theta^2)\eta}{1-\theta^{4^n}} \theta^{4^n - 1} \leq r_1 - t_n(\alpha)$$
$$\leq \frac{(1-\theta^2)\eta}{1 - \frac{1}{\sqrt[3]{5}}(\sqrt[3]{5\theta})^{4^n}} \left(\sqrt[3]{5\theta}\right)^{4^n - 1} \quad (n \geq 0).$$

5.18. Assume $F \in C^4[a,b]$, $F'(x) \neq 0$, $x^* \in (a,b)$, $F(x^*) = 0$, $H = H(x,y) = F'\left(x - \frac{2}{3}u\right) - F'(x)$. consider also the iterative functions

$$\Phi_\theta(x) = x - u + \frac{3}{4} u \frac{H}{F'(x)} \frac{1 + \frac{\theta H}{F'(x)}}{1 + \frac{\left(\frac{3+\theta}{2}\right)H}{F'(x)}}, \qquad (5.72)$$

$$\Phi(x) = x + \alpha u + \beta u \frac{F'(\varphi(x)) - F'(x)}{F'(x)} \frac{1 + \theta \frac{F'(\varphi(x)) - F'(x)}{F'(x)}}{1 + \delta \frac{F'(\varphi(x)) - F'(x)}{F'(x)}},$$

where
$$u = u(x) = \frac{F(x)}{F'(x)} \quad \text{and} \quad \varphi(x) = x - \lambda u.$$

Then show that (5.72) converges to the solution x^* of the equation $F(x) = 0$ with order four if the parameters α, β, λ and θ are chosen to satisfy the equations

$$1 + \alpha = 0,$$

$$\frac{\alpha}{2} + \beta\lambda = 0,$$

$$\frac{\beta}{2}(\lambda - 2)\lambda - \frac{\alpha}{3} = 0,$$

and

$$\frac{\alpha}{2} + \beta\lambda(2 - \delta\lambda) + \beta\theta\lambda^2 = 0.$$

5.19. Let a and b be real numbers such that $b \geq 0$, $0 \leq a < \frac{2}{3}$. We set $a_0 = 1$, $c_0 = 1$, $b_0 = \frac{a}{2}$, $d_0 = \frac{2}{2-a}$, and, for $n \geq 0$, $a_{n+1} = \frac{a_n}{1-aa_nd_n}$, $c_{n+1} = a_{n+1}\left[\frac{b}{6} + (1-b_n)a_n\frac{a^2}{4}\right]d_n^3$, $b_{n+1} = \frac{a}{2}a_{n+1}c_{n+1}$, $d_{n+1} = \frac{c_{n+1}}{1-b_{n+1}}$, $r_n = d_0 + d_1 + \cdots + d_n$ and $r = \lim_{n\to\infty} r_n$ (if the limit exists !). we denote by $R(a,b) = \{a_n, c_n, b_n, d_n\}$ we say $R(a,b)$ is positive if $a_n \geq 1$ ($n \geq 0$,) stable if there exists a constant $M \geq 1$ such that $a_N \leq M$ for all $n \geq 0$, and convergent if there exists $\lim_{n\to\infty} r_n = r$. Prove the result:

Let X and Y be Banach spaces. Let D be an open convex subset of X. Let F be from D into Y, an operator which is twice Fréchet-differentiable on D. Assume:

(a) there exists a constant k_2 such that
$$\|F''(x)\| \leq k_2 \quad \text{for all} \quad x \in D;$$

(b) there exists a constant k_3 such that
$$\|F''(x) - F'(y)\| \leq k_3 \|x - y\| \quad \text{for all} \quad x, y \in D.$$

(c) Let $x_0 \in D$ be fixed with $\left\|F'(x_0)^{-1}\right\| \leq B$, $\left\|F'(x_0)^{-1} F(x_0)\right\| \leq \eta$ and $F(x_0) \neq 0$, such that if we set $a = k_2 B\eta$ and $b = k_3 B\eta^2$, $R(a,b)$ is positive, convergent and $\overline{U}(x_0, r\eta) \subseteq D$, where $r = \lim_{n\to\infty}(d_0 + \cdots + d_n)$.

Then
(i) the Halley iteration
$$x_{n+1} = x_n - (I - T(x_n))^{-1} F'(x_n)^{-1} F(x_n)$$
where
$$T(x) = \frac{1}{2} F'(x)^{-1} F''(x) F'(x)^{-1} F(x),$$
is well defined and lies in $U(x_0, r\eta)$ for all $n \geq 0$. Moreover the sequence $\{x_n\}$ $n \geq 0$ converges to a point $x^* \in \overline{U}(x_0, r\eta)$, such that $F(x^*) = 0$.

(ii) The following estimates are true:
$$\left\| F'(x_n)^{-1} \right\| \leq a_n B, \quad \left\| F'(x_n)^{-1} F(x_n) \right\| \leq c_n \eta,$$
$$\| T(x_n) \| \leq b_n, \quad \| x_{n+1} - x_n \| \leq d_n \eta,$$
$$\| x^* - x_{n+1} \| \leq (r - r_n)\eta = \sum_{k=n+1} d_k \eta \quad (n \geq 0),$$
and
$$\| x_{n+1} - x_n \| \leq \frac{4M}{3\eta^2} \left(\frac{b}{6} + \frac{5a^2 M}{16} \right) \| x_n - x_{n-1} \|^3 \quad (n \geq 1).$$

(iii) For all $0 < a \leq \frac{1}{2}$ there exist a second degree polynomial F and a point x_0 such that $R(a, 0)$ produces optimal estimates for the Halley method, i.e., for such F and x_0 the above estimates (excluding the last one) hold as equalities.

(iv) For all $0 < a \leq \frac{1}{2}$, $R(a, 0)$ is convergent with
$$\sum_{k=0}^{\infty} d_k = \frac{1 - \sqrt{1 - 2a}}{a}.$$

Chapter 6

The Secant Method

Different treatments of the Secant Method (or Method of Chord) are given here.

6.1 The Modified Secant Method

In this section we study the iterative procedure

$$x_{n+1} = x_n - \delta f(x_{-1}, x_0)^{-1} f(x_n) \tag{6.1}$$

to approximate solutions x^* of the equation

$$f(x) = 0 \tag{6.2}$$

where f is a nonlinear operator between two Banach spaces X and Y, x_{-1} and x_0 are two points in the domain of f, and δf is a consistent approximation of f'.

The Secant method has been known since the time of early Italian algebraists [228] and it was extended to the solution of nonlinear equations in Banach spaces by Sergeev [254] and Schmidt [252], [253].

The iterative procedure (6.1) called the modified Secant method was first considered by Ulm [268].

Here we provide a priori and a posteriori error estimates which are proven to be better than the ones presently in the works mentioned above under the same assumptions.

Finally a simple example is provided where our results are compared favorably with the corresponding results obtained in [235], [252], [263], [268].

In the study of the modified Secant method we shall use the method of nondiscrete mathematical induction. This method was developed by V.

Pták by refining the closed graph theorem [240].

Let T denote either the set of all positive real numbers or an interval of the form $(0,e] = \{x \in \mathbb{R} \mid 0 < x \leq e\}$.

We will need the definitions:

Definition 6.1 A function $w: T \to T$ is called a rate of convergence on T if the series

$$\sigma(r) = \sum_{k=0}^{\infty} w^{(n)}(r) \qquad (6.3)$$

is convergent for each $r \in T$, where the iterates $w^{(n)}$ of w are defined as follows

$$w^{(0)}(r) = r, \quad w^{(n+1)}(r) = w^{(n)}(w(r)), \quad n = 0, 1, 2, \ldots \qquad (6.4)$$

Definition 6.2 Let X and Y be two Banach spaces and let V be a convex and open subset of X. Let $f: V \to Y$ be a nonlinear operator which is Fréchet-differentiable on V. A mapping $\delta f: V \times V \to L(X,Y)$ will be called a consistent approximation of f', if there exists a constant $H > 0$ such that

$$\|\delta f(x,y) - f'(z)\| \leq H(\|x-z\| + \|y-z\|) \quad \text{for all} \quad x, y, z \in \overline{V} \subset V. \qquad (6.5)$$

The above condition implies the Lipschitz continuity of f'. In this case using deduce the following:

$$\|f(u) - f(v) - f'(v)(u-v)\| \leq H \|u-v\|^2; \ u, v \in V \qquad (6.6)$$

and

$$\|f(u) - f(v) - \delta f(x,y)(u-v)\|$$
$$\leq H(\|u-v\| + \|x-v\| + \|y-v\|)\|u-v\|. \qquad (6.7)$$

Let $C(h_0, q_0, r_0)$ be the class of all the triplets (f, x_0, x_{-1}) satisfying the following properties:

(**P$_1$**) f is a nonlinear operator having the domain of definition $V \subset X$ and taking values in Y.

(**P$_2$**) x_0 and x_{-1} are two points of V such that

$$\|x_0 - x_{-1}\| \leq q_0, \quad \|x_0 - x_{-1}\| < \mu, \quad \text{with} \quad q_0, \mu > 0. \qquad (6.8)$$

(**P$_3$**) f is Fréchet-differentiable in the open ball
$$U = U(x_0, \mu) \text{ and continuous on its closure } \overline{U}.$$

(**P$_4$**) There exists a consistent approximation δf of f' such that $D_0 \equiv \delta f(x_{-1}, x_0)$ is invertible and
$$\left\| D_0^{-1} (\delta f(x,y) - f'(z)) \right\|$$
$$\leq h_0 (\|x - z\| + \|y - z\|) \quad \text{for all} \quad x, y, z \in U. \quad (6.9)$$

(**P$_5$**) The following inequalities are satisfied:
$$\left\| D_0^{-1} f(x_0) \right\| \leq r_0, \quad (6.10)$$
$$h_0 q_0 + 2\sqrt{h_0 r_0} \leq 1, \quad (6.11)$$
$$\mu \geq \frac{1}{2h_0} \left(1 - h_0 q_0 - \sqrt{(1 - h_0 q_0)^2 - 4 h_0 r_0} \right) \equiv \mu_0. \quad (6.12)$$

Using the iterative procedure (6.1) Potra showed in [7, Thm. 3] and [8, Thm. 1] that if $(f, x_0, x_{-1}) \in C(h_0, q_0, r_0)$ then the equation $f(x) = 0$ has a locally unique solution x^* and certain error estimates are valid.

In particular he showed:

Theorem 6.1 *If $(f, x_{-1}, x_0) \in C(h_0, q_0, r_0)$, then by the iterative algorithm (6.1) one obtains a sequence $\{x_n\}$, $n = 0, 1, 2, \ldots$ of points belonging to the open ball $U(x_0, \mu_0)$, which converges to a unique root x^* of the equation $f(x) = o$ in $\overline{U}(x_0, \mu_0)$ and the following estimates hold:*

$$\|x_n - x^*\| \leq \sigma_0 \left(w_0^{(n)}(r_0) \right) = s_n - a_0, \quad n = 0, 1, 2, \ldots \quad (6.13)$$

$$\|x_n - x^*\| \leq c(n) \equiv \sqrt{a_0^2 + h_0^{-1} \|x_n - x_{n-1}\|} - a_0, \quad n = 1, 2, \ldots \quad (6.14)$$

$$\|x_{n+1} - x_n\| \leq w_0^{(n)}(t_0) = g(s_n) = s_n - s_{n+1}, \quad n = 0, 1, 2, \ldots \quad (6.15)$$

and

$$\|x_n - x^*\| \leq c_0(n) \equiv b_0 - \|x_n - x_0\|$$
$$- \Big[(b_0 - \|x_n - x_0\|)^2 - (\|x_n - x_0\| + \|x_{n-1} - x_0\|$$
$$+ \|x_0 - x_{-1}\|) \|x_n - x_{n-1}\| \Big]^{1/2}, \quad n = 1, 2, \ldots \quad (6.16)$$

where

$$a_0 = \frac{1}{2h_0}\sqrt{(1-h_0q_0)^2 - 4h_0r_0}, \qquad (6.17)$$

$$b_0 = \frac{1-h_0q_0}{2h_0}, \qquad (6.18)$$

$$w_0(r) = r\left(h_0r + 1 - 2\sqrt{h_0^2 a_0^2 + h_0 r}\right), \qquad (6.19)$$

$$\sigma_0(r) = \sqrt{a_0^2 + h_0^{-1}r} - a_0, \qquad (6.20)$$

$$g(s) = h_0\left(s^2 - a_0^2\right) \qquad (6.21)$$

and the sequence $\{s_n\}$, $n = 0, 1, 2, \ldots$ given by (6.15) with $s_0 = s_0(r) = \sqrt{a_0^2 + h_0^{-1}r}$ is decreasingly convergent to a_0.

We can now improve the error estimates (6.14) of Theorem 6.1 as follows:

Theorem 6.2 *Under the hypotheses of Theorem 6.1 the following inequalities hold for* $n = 1, 2, 3, \ldots$

$$\|x_n - x^*\| \le \sigma_{n-1}\left(w_{n-1}^{(n)}(r_{n-1})\right), \qquad (6.22)$$

$$\|x_n - x^*\| \le c_1(n) \equiv \sqrt{a_{n-1}^2 + h_0^{-1}\|x_n - x_{n-1}\|} - a_{n-1}, \qquad (6.23)$$

$$c_1(n) \le c(n) \qquad (6.24)$$

$$\text{if} \quad q_n \le q_0 \quad \text{and} \quad r_n \le r_0, \quad n = 0, 1, 2, \ldots, \qquad (6.25)$$

where we have denoted

$$w_n(r) = r\left(h_0 r + 1 - 2\sqrt{h_0^2 a_n^2 + h_0 r}\right), \qquad (6.26)$$

$$\sigma_n(r) = \sqrt{a_n^2 + h_0^{-1}r} - a_n, \qquad (6.27)$$

$$a_n = \frac{1}{2h_0}\left[(1-h_0q_n)^2 - 4h_0r_n\right]^{1/2}, \qquad (6.28)$$

$$r_n = \|x_n - x_{n+1}\| \qquad (6.29)$$

and

$$q_n = \|x_n - x_{n-1}\|, \quad \text{for all} \quad n = 0, 1, 2, \ldots. \qquad (6.30)$$

Proof. Let us consider the triplet $(f, x_{-1}, x_0) \in C(h_0, q_0, r_0)$. We shall show that $(f, x_{n-1}, x_n) \in C(h_0, q_n, r_n)$. It suffices to show the inequality

$$h_0 q_n + 2\sqrt{h_0 r_n} \le 1.$$

But, by (6.29), (6.30) and (6.11)

$$h_0 q_n + 2\sqrt{h_0 r_n} \leq h_0 q_0 + 2\sqrt{h_0 r_0} \leq 1, \quad n = 0, 1, 2, \ldots.$$

By applying Theorem 6.1 to the triplet $(f, x_{n-1}, x_n) \in C(h_0, q_n, r_n)$ we deduce (6.22) and (6.23).

Let us define the real function P on $I = [0, q_0] \times [0, r_0]$ by

$$P(q, r) = \frac{1}{h_0} \left[(1 - h_0 q)^2 - 4 h_0 r \right]^{1/2}.$$

The function P is well defined on I since by (6.11).

$$1 - h_0 q \geq 1 - h_0 q_0 \geq 0$$

and

$$(1 - h_0 q)^2 \geq (1 - h_0 q_0)^2 \geq 4 h_0 r_0 \geq 4 h_0 r.$$

It can easily be checked that P is decreasing with respect to each one of the variables q and r (if one variable remains fixed).

Therefore the function P is decreasing in the sense that $q_1 \leq q_2, r_1 \leq r_2$ implies

$$P(q_2, r_2) \leq p(q_1, r_1). \tag{6.31}$$

Indeed we have

$$P(q_1, r_1) \geq P(q_2, r_1) \geq P(q_2, r_2).$$

Using (6.14), (6.17), (6.23), (6.25), (6.28), (6.31) we get

$$a_{n-1} \geq a_0, \quad n = 1, 2, \ldots$$

which implies (6.24) for all $n = 1, 2, \ldots$.

That completes the proof of the theorem. $\square$

Note that inequalities (6.25) will hold after a finite number of steps. Moreover it can easily be seen using induction on n that (6.25) holds for all $n = 0, 1, 2, \ldots$ if $r_0 \leq q_0$.

We can now improve the error estimates (6.13) as follows:

Theorem 6.3 *Assume:*

(a) the triplet $(f, x_{-1}, x_0) \in C(h_0, q_0, r_0^*)$, where $r_0^* = \min(r_0, \bar{r}_0)$ and $\bar{r}_0$ is the minimum nonnegative number such that the real continuous function F_n given by

$$h_0^{-1} F_n(r) = 4h_0 r^3 + h_0 \left(4h_0^2 a_n^2 - 5\right) r^2 + \left(1 - 8h_0^2 a_n^2\right) r \\ + h_0 a_n^2 \left(1 - 4h_0^2 a_n^2\right), \quad n = 0, 1, 2, \ldots \quad (6.32)$$

satisfies

$$F_n(r) \geq 0 \quad \text{for all} \quad r \in [0, \bar{r}_0]; \quad (6.33)$$

(b) Inequalities (6.25) hold for $r_0 = r^*$ and the sequence $\{q_n\}$, $n = 0, 1, 2, \ldots$ is decreasing.

Then

$$\|x_n - x^*\| \leq \sigma_{n-1}\left(w_{n-1}^{(n)}(r_{n-1})\right) \leq \sigma_0\left(w_0^{(n)}(r_0^*)\right), \quad n = 1, 2, \ldots. \quad (6.34)$$

Proof. By (6.4), to show (6.34), it suffices to have

$$w_{n-1}^{(n+k)}(r_{n-1}) \leq w_0^{(n+k)}(r_0^*). \quad (6.35)$$

We shall use double induction. For $k = 0$ and $n = 1$, (6.35) is trivially true. Assume that

$$w_{m-1}^{(m)}(r_{m-1}) \leq w_0^{(m)}(r_0^*), \quad m = 1, 2, \ldots, n. \quad (6.36)$$

We show that (6.36) is true for $m = n + 1$.

By (6.26) we obviously have that w_m an its iterates are decreasing in a_n for fixed r. Moreover it is simple calculus to show that the partial derivatives of w_m and its iterates with respect to r are positive for a_n fixed if (6.33) holds. Note that the existence of $\bar{r}_0$ is guaranteed by the fact that

$$F_n(0) = h_0 a_n^2 \left(1 - 4h_0^2 a_n^2\right) \geq 0$$

and the choice of a_n.

With the above remarks we can get

$$w_m^{(m+1)}(r_m) \leq w_{m-1}^{(m+1)}(r_m) \leq w_{m-1}^{(m+1)}(r_{m-1}) \leq w_{m-1}\left(w_{m-1}^{(m)}(r_{m-1})\right) \\ \leq \left(w_0^{(m)}(r_0^*)\right) = w_0^{(m+1)}(r_0^*),$$

which shows that (6.36) is true for $m = n + 1$.

Moreover assume that

$$w_{n-1}^{(n+i)}(r_{n-1}) \leq w_0^{n+i}(r_0^*) \quad \text{for all} \quad n = 1, 2, ..., \quad i = 0, 1, 2, ..., k-1. \tag{6.37}$$

We shall show that (6.37) is true for $i = k$. Using (6.37), (6.35) and (6.26) we have

$$w_n^{(n+i+1)}(r_n) \leq w_{n-1}^{(n+i+1)}(r_n) \leq w_{n-1}^{(n+i+1)}(r_{n-1}) \leq \left(w_{n-1}\left(w_{n-1}^{(n+1)}(r_{n-1})\right)\right)$$
$$\leq w_0\left(w_{n-1}^{(n+1)}(r_{n-1})\right) \leq w_0\left(w_0^{(n+i)}(r_0^*)\right) = w_0^{(n+i+1)}(r_0^*).$$

That completes the double induction and the result follows. □

We can improve the error estimates (6.16) as follows:

Theorem 6.4 *Assume:*

(a) *the hypotheses of Theorem 6.2 are satisfied;*
(b) *the linear operator $\delta f(x, y)$ is such that*

$$\delta f(x, y)(x - y) = f(x) - f(y), \quad \text{for all} \quad x, y \in V. \tag{6.38}$$

(c) *there exists an integer $N \geq 1$ such that for $n \geq N$*

$$\|x_{n-1} - x^*\| \leq \|x_{n-1} - x_n\|. \tag{6.39}$$

Then, the following are true:

(i) *there exists an integer $\overline{N} \geq N$ such that*

$$\|x_n - x_{n-1}\| + \|x_{-1} - x_{n-1}\| \leq \|x_n - x_0\| + \|x_0 - x_{-1}\| \quad \text{for all} \quad n \geq \overline{N} \tag{6.40}$$

(ii) *For $n \geq \overline{N}$, the following estimates hold:*

$$\|x_n - x^*\| \leq \left[1 - h_0\left(2\|x_0 - x_n\| + \sigma_{n-1}\left(w_{n-1}^{(n)}(r_{n-1})\right)\right)\right]^{-1} \|D_0^{-1} f(x_n)\| \tag{6.41}$$

and

$$\|x_n - x^*\| \leq c_2(n) \equiv \frac{1 - 2h_0\|x_0 - x_n\| - \sqrt{(1 - 2h_0\|x_0 - x_n\|)^2 - 4h_0\|D_0^{-1} f(x_n)\|}}{2h_0}. \tag{6.42}$$

(iii) *The following is true:*

$$\|x_n - x^*\| \leq c_2(n) \leq c_0(n) \quad \text{for all} \quad n \geq \overline{N}. \tag{6.43}$$

Proof. (i) The inequality (6.40) is true for all $n \geq N_1$ for some fixed integer N_1, because if we assume otherwise and let $n \to \infty$

$$\|x_{-1} - x^*\| > \|x_0 - x^*\| + \|x_0 - x_{-1}\|$$

which is a contradiction.

(ii) Let us consider the linear operator D, given by

$$D = \delta f(x^*, x_{n-1}). \tag{6.44}$$

We will show that D is invertible for all $n \geq \overline{N} = \max(N, N_1)$. Indeed we have by (6.9), (6.13), (6.15), (6.21), (6.39), and (6.44) that

$$\left\| D_0^{-1} \left(f'(x_0) - D \right) \right\|$$
$$\leq h_0 \left[\| x_0 - x^* \| + \| x_0 - x_{n-1} \| \right]$$
$$\leq h_0 \left[2 \| x_0 - x_{n-1} \| + \| x_{n-1} - x^* \| \right]$$
$$\leq h_0 \left[2 \| x_0 - x_{n-1} \| + \| x_{n-1} - x_n \| \right]$$
$$\leq h_0 \left[2 \left(\mu_0 - \sigma_0 \left(w_0^{(n-1)}(t_0) \right) \right) + w^{(n-1)}(t_0) \right]$$
$$< h_0 \left[2 \left(\sigma(r_0) - \sigma \left(w^{(n-1)}(t_0) \right) \right) + w^{(n-1)}(t_0) + q_0 \right]$$
$$= \frac{w_0^{(n)}(t_0)}{w^{(n-1)}(t_0)} = \frac{g(s_n)}{g(s_{n-1})} = \frac{s_n^2 - a_0^2}{s_{n-1}^2 - a_0^2} \leq 1$$

since $s_n \leq s_{n-1}$ for all $n = 0, 1, 2, \ldots$.

According to Banach's lemma it follows that the linear operator D is invertible for all $n \geq \overline{N}$ and that

$$\left\| \left(D_0^{-1} D \right)^{-1} \right\| \leq \left[1 - h_0 \left(2 \| x_0 - x_{n-1} \| + \| x_{n-1} - x^* \| \right) \right]^{-1}. \tag{6.45}$$

Using the identity

$$D(x_{n-1} - x^*) = f(x_{n-1})$$

and (6.45), we obtain

$$\| x_{n-1} - x^* \| \leq \left\| \left(D_0^{-1} D \right)^{-1} \right\| \cdot \left\| D_0^{-1} f(x_n) \right\|$$
$$\leq \left[1 - h_0 \left(2 \| x_0 - x_{n-1} \| + \| x_{n-1} - x^* \| \right) \right]^{-1} \left\| D_0^{-1} f(x_{n-1}) \right\|. \tag{6.46}$$

The inequality (6.41) follows from (6.46) and (6.22), whereas the inequality (6.42) follows from (6.46).

That completes the proof of (ii).

(iii) Using the identity

$$f(x_n) = f(x_n) - f(x_{n-1}) + \delta f(x_0, x_{-1})(x_n - x_{n-1}),$$

(6.7) and (6.39) we have

$$\|D_0^{-1} f(x_n)\| \leq h_0 (\|x_n - x_{n-1}\| + \|x_0 - x_{n-1}\| + \|x_{-1} - x_{n-1}\|) \|x_n - x_{n-1}\|$$
$$\leq h_0 (\|x_n - x_0\| + \|x_{n-1} - x_0\| + \|x_0 - x_{-1}\|) \|x_n - x_{n-1}\|,$$
$$n = \overline{N}, \overline{N}+1, \ldots. \qquad (6.47)$$

By (6.18) we see that

$$1 - 2h_0 \|x_0 - x_n\| \geq 2h_0 (b_0 - \|x_n - x_0\|). \qquad (6.48)$$

Using (6.47) and (6.48) it can easily be checked that

$$c_2(n) \leq c_0(n), \quad \text{for all} \quad n \geq \overline{N}.$$

That completes the proof of the theorem. □

A lower bound on $\|x_n - x^*\|$ can be given by the following:

Theorem 6.5 *Under the hypotheses of Theorem 6.1 the following inequality holds for $n = 1, 2, \ldots$*

$$\|x_{n-1} - x^*\| \geq q = q(h_0, \|x_n - x_{n-1}\|, \|x_n - x_0\|),$$

where q is the positive root of the quadratic equation in $\|x_n - x^*\|$ given by,

$$h_0 \|x_n - x^*\|^2 + [1 + h_0 (\|x_n - x_{n-1}\| + \|x_n - x_0\|)] \|x_n - x^*\| - \|x_n - x_{n-1}\| = 0.$$

Proof. Using the identity

$$x_n - x_{n-1} = x^* - x_n + D_0^{-1} [f(x^*) - f(x_n) - \delta f(x_{n-1}, x_0)(x^* - x_n)],$$

(6.7) and the triangle inequality, the result follows immediately. □

Example 6.1 Let us now compare our estimates (6.23) and (6.42) with (6.14) and (6.16) respectively on a very simple example.

We consider the quadratic

$$f(x) = x^2 - 16. \qquad (6.49)$$

Take $x_{-1} = 3$, $x_0 = 3.2$ and $\delta f(x, y)(x - y) = f(x) - f(y)$. Then $h_0 = \frac{10}{62}$, $q_0 = .2$ and $r_0 = .92903225$.

The condition (6.11) is satisfied, since

$$h_0 q_0 + 2\sqrt{\frac{h_0 r_0}{10}} = .806451609 < 1.$$

It is easy to see that $(f, x_{-1}, x_0) \in C(h_0, q_0, r_0)$.
The modified Secant method for (6.49) becomes

$$x_{n+1} = x_n - \frac{x_n^2 - 16}{x_{-1} + x_0}, \quad n = 0, 1, 2, \dots. \qquad (6.50)$$

Using (6.50), (6.28), (6.29) and (6.30) we can compute the following:

$$\begin{aligned}
x_1 &= 4.12903225, & x_6 &= 3.99971974 \\
x_2 &= 3.95985363, & x_7 &= 4.00008134, \\
x_3 &= 4.01139544, & x_8 &= 3.99997638, \\
x_4 &= 3.9966707, & x_9 &= 4.00000685, \\
x_5 &= 4.00096478, & x_{10} &= 3.999998, \\
a_0 &= d_0 = 1.800000012, & d_5 &= 3.099978406, \\
d_1 &= 2.428346725, & d_6 &= 3.099993739, \\
d_2 &= 2.96194909, & d_7 &= 3.099998183, \\
d_3 &= 3.07408061, & d_8 &= 3.099999474, \\
d_4 &= 3.092594588, & d_9 &= 3.099999848,
\end{aligned}$$

and

$$b_0 = 3.$$

Using the above values and noting that $x^* = 4$, we can tabulate the following results:

n	Error (Potra) estimates (6.14)	Error (Argyros) estimates (6.23)	Error (Potra) estimates (6.16)	Error (Argyros) estimates (6.42)
1	1.99999987	1.99999987	.32334261	.25675937
2	.0270967753	.207137151	.07249864	.06930382
3	.086679416	.0534616	.02095686	.02003286
4	.02518311	.014813202	$6.0499141 \cdot 10^{-3}$	$5.78657584 \cdot 10^{-3}$
5	$7.38023 \cdot 10^{-3}$	$4.3013711 \cdot 10^{-1}$	$1.7551266 \cdot 10^{-3}$	$1.67939703 \cdot 10^{-3}$
6	$2.1429599 \cdot 10^{-3}$	$1.2447987 \cdot 10^{-3}$	$5.095243 \cdot 10^{-4}$	$4.8738448 \cdot 10^{-4}$
7	$6.226478 \cdot 10^{-4}$	$3.615796 \cdot 10^{-4}$	$1.47922 \cdot 10^{-4}$	$1.41465214 \cdot 10^{-4}$
8	$1.807553 \cdot 10^{-4}$	$1.039582 \cdot 10^{-4}$	$4.29406 \cdot 10^{-5}$	$4.1069141 \cdot 10^{-5}$
9	$5.24753 \cdot 10^{-5}$	$3.04698 \cdot 10^{-5}$	$1.2466 \cdot 10^{-5}$	$1.190214 \cdot 10^{-5}$
10	$1.52416 \cdot 10^{-5}$	$8.8499 \cdot 10^{-6}$	$3.6205 \cdot 10^{-6}$	$3.451476 \cdot 10^{-6}$

The above table indicates that our estimates (6.23) and (6.42) are better than the corresponding ones given by (6.14) and (6.16) respectively. Note however that the additional information on $\left\|D_0^{-1} f(x_n)\right\|$ is used by (6.42).

Similar favorable comparisons can be made between the lower bound given by Theorem 6.5 and the corresponding one in [235].

All the above strongly exhibit the usefulness of our results in numerical applications.

6.2 Error Bounds for the Secant Method

In this section we study the iterative procedure

$$x_{n+1} = x_n - \delta f(x_{n-1}, x_n)^{-1} f(x_n) \tag{6.51}$$

to approximate solutions x^* of the equation

$$f(x) = 0 \tag{6.52}$$

where f is a nonlinear operator between two Banach spaces X and Y, x_{-1} and x_0 are two points in the domain of f, and δf is a consistent approximation f'.

The iterative procedure (6.51) is called the secant method but it is also known under the name of regular falsi or the method of chords.

Here we provide a priori and a posteriori error estimates which are proven to be eventually better than the ones presently in the literature [252], [254], [268], [235], under the same assumptions.

Finally, a simple example is provided where our results are compared favorably with the corresponding results obtained in the references given above.

using the iterative procedure (6.51) Potra showed in [235, Thm. 3] (see also 6.1) that if $(f, x_0, x_{-1}) \in C(h_0, q_0, r_0)$, then the equation $f(x) = 0$ has a locally unique solution x^*, and certain error estimates are valid.

In particular he showed:

Theorem 6.6 *If $(f, x_0, x_{-1}) \in C(h_0, q_0, r_0)$, then via the iterative procedure (6.51) one obtains a sequence $\{x_n\}$, $n \geq 0$ of points from the open ball $U(x_0, \mu_0)$ which converges to a unique root x^* of the equation $f(x) = 0$ in $\overline{U}(x_0, \mu_0)$ and the following estimates are satisfied:*

$$\|x_n - x^*\| \leq \sigma_0 \left(w_0^{(n)}(t_0) \right), \quad t_0 = (q_0, r_0), \quad n = 0, 1, 2, \ldots \tag{6.53}$$

$$\|x_n - x^*\| \leq c(n)$$
$$\equiv \left[a_0^2 + \|x_n - x_{n-1}\| (\|x_{n-1} - x_{n-2}\| + \|x_n - x_{n-1}\|) \right]^{1/2} - a_0, \tag{6.54}$$

$n = 1, 2, ...$

$$\|x_{n+1} - x_n\| \leq w_0^{(n)}(t_0), \quad n = 0, 1, 2... \tag{6.55}$$

$$\|x_n - x^*\| \leq c_0(n) \equiv s_0 - \|x_n - x_0\| - \Big[(s_0 - \|x_n - x_0\|)^2$$
$$- (\|x_n - x_{n-1}\| + \|x_{n-1} - x_{n-2}\|) \|x_n - x\|\Big]^{1/2},$$
$$n = 1, 2, ..., \tag{6.56}$$

where

$$a_0 = \frac{1}{2h_0}\left[(1 - h_0 q_0)^2 - 4h_0 r_0\right]^{1/2} \tag{6.57}$$

$$s_0 = \frac{1 - q_0 h_0}{2h_0} \tag{6.58}$$

$$w_0(t) = w_0(q, r) = \frac{r(q+r)}{r + 2\sqrt{r(q+r) + a_0^2}} \tag{6.59}$$

and

$$\sigma_0(t) = \sigma_0(q, r) = r - a_0 + \sqrt{r(q+r) + a_0^2}. \tag{6.60}$$

We can prove the following theorem:

Theorem 6.7 *Under the hypothesis of Theorem 6.6 the following inequalities hold for $n = 1, 2, 3, ...$.*

$$\|x_n - x^*\| \leq c_1(n)$$
$$\equiv \left[a_{n-1}^2 + \|x_n - x_{n-1}\| (\|x_{n-1} - x_{n-2}\| + \|x_n - x_{n-1}\|)\right]^{1/2}$$
$$- a_{n-1} \tag{6.61}$$

where

$$h_n = \sup_{x,y,z \in U} \frac{\|D_n^{-1}(\delta f(x, y) - f'(z))\|}{\|x - z\| + \|y - z\|}, \tag{6.62}$$

$$D_n = \delta f(x_{n-1}, x_n), \tag{6.63}$$

$$q_{n+1} = r_n = \|x_n - x_{n+1}\|, \tag{6.64}$$

and

$$a_n(h_n, q_n, r_n) = a_n = \frac{1}{2h_n}\left[(1 - h_n q_n)^2 - 4h_n r_n\right]^{1/2} \quad n = 0, 1, 2, ..., \tag{6.65}$$

The Secant Method

Proof. First let us observe that with the constant a_0 given by (6.57) we have $\sigma_0(r_0) = \mu_0$. hence the closed ball with center x_0 and radius μ_0 is included in U. consider the triplet $(f, x_{-1}, x_0) \in C(h_0, q_0, r_0)$. We will prove that $(f, x_{n-1}, x_n) \in C(h_n, q_n, r_n)$. It suffices to show inequality

$$h_n q_n + 2\sqrt{h_n r_n} \leq 1. \tag{6.66}$$

Using (6.55) and (6.60), we have:

$$\begin{aligned}
&\left\| D_0^{-1}(D_0 - D_n) \right\| \\
&\leq \left\| D_0^{-1}(D_0 - f'(x_{n-1})) \right\| + \left\| D_0^{-1}(f'(x_{n-1}) - D_n) \right\| \\
&\leq h_0 \left[\|x_0 - x_{n-1}\| + \|x_{-1} - x_{n-1}\| + \|x_n - x_{n-1}\| \right] \\
&\leq h_0 \left[2\|x_0 - x_{n-1}\| + \|x_{-1} - x_0\| + \|x_{n-1} - x_n\| \right] \\
&\leq h_0 \left[2\left(\mu_0 - \sigma_0\left(w_0^{(n-1)}(t_0)\right)\right) + q_0 + w_0^{(n-1)}(t_0) \right] \\
&\leq 1 - h_0 \left[w_0^{(n-1)}(t_0) + 2\sqrt{w_0^{(n-1)}(t_0)\left(q_0 + w_0^{(n-1)}(t_0)\right) + a_0^2} \right] < 1.
\end{aligned}$$

According to Banach's lemma this implies that

$$\left\| (D_0^{-1} D_n)^{-1} \right\| \leq \{1 - h_0 [\|x_0 - x_{n-1}\| + \|x_{-1} - x_{n-1}\| + \|x_{n-1} - x_n\|]\}^{-1}. \tag{6.67}$$

From the identity

$$D_n^{-1}(\delta f(x, y) - f'(z)) = (D_0^{-1} D_n)^{-1} D_0^{-1} (\delta f(x, y) - f'(z)),$$

(6.67) we obtain

$$\left\| D_n^{-1}(\delta f(x, y) - f'(z)) \right\| \leq h_0 \left\| (D_0^{-1} D_n)^{-1} \right\| (\|x - z\| + \|y - z\|) \tag{6.68}$$

that is

$$h_n \leq z_n \equiv h_0 \{1 - h_0 [\|x_0 - x_{n-1}\| + \|x_{-1} - x_{n-1}\| + \|x_{n-1} - x_n\|]\}^{-1}. \tag{6.69}$$

By (6.67) we can easily obtain that

$$h_n \leq \left[w_0^{(n-1)}(t_0) + 2\sqrt{w_0^{(n-1)}(t_0)\left(q_0 + w_0^{(n-1)}(t_0)\right) + a_0^2} \right]^{-1}. \tag{6.70}$$

To show (6.66), using (6.64), and (6.70), it suffices to show

$$\frac{w_0^{(n-1)}(t_0)}{w_0^{(n-1)}(t_0) + 2\sqrt{w_0^{(n-1)}(t_0)\left(q_0 + w_0^{(n-1)}(t_0)\right)} + a_0^2}$$

$$+ 2\sqrt{\frac{w_0^{(n)}(t_0)}{w_0^{(n-1)}(t_0) + 2\sqrt{w_0^{(n-1)}(t_0)\left(q_0 + w_0^{(n-1)}(t_0)\right)} + a_0^2}} \leq 1. \quad (6.71)$$

Set $v = w_0^{(n-1)}(t_0)$, then by (6.59)

$$w_0^{(n)}(t_0) = w_0(r_0, v) = \frac{v(r_0 + v)}{v + 2\sqrt{v(r_0 + v)} + a_0^2}.$$

The left hand side of (6.71) now becomes

$$\frac{v + 2\sqrt{v(r_0 + v)}}{v + 2\sqrt{v(r_0 + v)} + a_0^2} \leq 1.$$

By applying Theorem 6.6 to the triplet (f, x_{n-1}, x_n) we deduce (6.61). That completes the proof of the theorem. □

We can now improve the results of Theorem 6.6 through Theorem 6.7 as follows:

Proposition 6.1 *Under the hypotheses of Theorem 6.6 the following are true:*

(a) *For all $n = 1, 2, ...$, the triplet $(f, x_{n-1}, x_n) \in C(z_n, q_n, r_n)$,*

$$\|x_n - x^*\| \leq c_2(n)$$
$$\equiv \left[d_{n-1}^2 + \|x_n - x_{n-1}\|(\|x_{n-1} - x_{n-2}\| + \|x_n - x_{n-1}\|)\right]^{1/2}$$
$$- d_{n-1} \quad (6.72)$$

and if

$$d_n \geq a_0, \quad n = 0, 1, 2, ..., \quad (6.73)$$

then

$$c_2(n) \leq c(n) \quad (6.74)$$

where we have denoted

$$d_n(z_n, q_n, r_n) = d_n = \frac{1}{2z_n}\left[(1 - z_n q_n)^2 - 4z_n r_n\right]^{1/2},$$
$$n = 0, 1, 2, \ldots . \quad (6.75)$$

(b) If

$$a_n \geq a_0, \quad n = 1, 2, \ldots, \quad (6.76)$$

then

$$c_1(n) \leq c(n) \quad \text{for all} \quad n = 1, 2, \ldots . \quad (6.77)$$

(c) Moreover, for all $n = 1, 2, \ldots$, the triplet $(f, x_{n-1}, x_n) \in C(h_0, q_n, r_n)$,

$$\|x_n - x^*\| \leq c_3(n)$$
$$\equiv \left[e_{n-1}^2 + \|x_n - x_{n-1}\|(\|x_{n-1} - x_{n-2}\| + \|x_n - x_{n-1}\|)\right]^{1/2}$$
$$-e_{n-1} \quad (6.78)$$

and if

$$e_n \geq a_0, \quad n = 1, 2, \ldots, \quad (6.79)$$

then

$$c_3(n) \leq c(n) \quad \text{for all} \quad n = 1, 2, \ldots, \quad (6.80)$$

where we have denoted

$$e_n(q_n, r_n) = e_n = \frac{1}{2h_0}\left[(1 - h_0 q_n)^2 - 4h_0 r_n\right]^{1/2}, \quad n = 0, 1, 2, \ldots . \quad (6.81)$$

Proof. (a) By (6.69)-(6.71), it follows that the triplet $(f, x_{n-1}, x_n) \in C(z_n, q_n, r_n)$. By applying Theorem 6.6 to the triplet (f, x_{n-1}, x_n) we obtain (6.72). Using (6.54), (6.72) and (6.73), inequality (6.74) follows immediately.

(b) Using (6.54), (6.61) and (6.70), the result (6.77) follows.

(c) The triplet $(f, x_{n-1}, x_n) \in C(h_0, q_0, r_n)$, since the proof of (6.66) can be repeated with $h_n = h_0$ and h_0 dominated by the right hand side of (6.70). Applying Theorem 6.6 to the triplet $(f.x_{n-1}, x_n)$ we obtain (6.78). Using (6.54), (6.78) and (6.79) the result (6.80) follows.

That completes the proof of the proposition. □

Facts. The functions a_n, d_n and e_n are decreasing with respect to each one of their variables separately. Therefore, they are decreasing in the sense that if P is a function of three variables h, q and r, $h_1 \leq h_2$, $q_1 \leq q_2$ and $r_1 \leq r_2$ implies

$$P(h_2, q_2, r_2) \leq P(h_1, q_1, r_1). \tag{6.82}$$

Indeed, we get

$$P(h_1, q_1, r_1) \geq P(h_2, q_1, r_1) \geq P(h_2, q_2, r_1) \geq P(h_2, q_2, r_2). \tag{6.83}$$

Note that:

(a) Inequality (6.73) holds if

$$q_n \leq q_0, \tag{6.84}$$
$$r_n \leq r_0, \tag{6.85}$$

and

$$z_n \leq z_0, \quad \text{for all} \quad n = 0, 1, 2, \ldots. \tag{6.86}$$

(b) Inequality (6.76) holds if

$$q_n \leq q_0,$$
$$r_n \leq r_0,$$

and

$$h_n s < h_0, \quad \text{for all} \quad n = 0, 1, 2, \ldots. \tag{6.87}$$

(c) Inequality (6.80) holds if

$$r_n \leq r_0$$

and

$$q_n \leq q_0, \quad \text{for all} \quad n = 0, 1, 2, \ldots.$$

Moreover, since the sequence $\{x_n\}$, $n = -1, 0, 1, 2, \ldots$ converges, there exists an integer $N \geq 1$ such that (6.84) and (6.85) hold for all $n \geq N$.

With the exception of the scalar case, the cost of computing h_n may be very high. However, the $d'_n s$ and $z'_n s$ can be computed.

By (6.69) we can easily check that (6.86) is true if

$$\|x_0 - x_{n-1}\| + \|x_{-1} - x_{n-1}\| + \|x_{n-1} - x_n\| \leq 2q_0 \quad \text{for all} \quad n = 0, 1, 2, \ldots. \tag{6.88}$$

It turns out that under certain assumptions the conditions (6.84) and (6.85) are satisfied.

In particular, we can show the following:

Proposition 6.2 *Assume:*

(a) *the hypotheses of Theorem 6.6 are true.*
(b) *The following estimates are true:*

$$r_0 \leq q_0, \tag{6.89}$$

and

$$2h_0(r_0 + q_0) \leq 1. \tag{6.90}$$

Then

$$w_0^{(n)}(t_0) \leq w_0^{(n-1)}(t_0) \leq r_0, \quad \text{for all} \quad n = 1, 2, \ldots, \tag{6.91}$$

$$r_n \leq r_0.$$

and

$$q_n \leq q_0.$$

Proof. It suffices to show (6.91). The rest will follow from (6.54), (6.64) and (6.89). We first show that (6.91) is valid for $n = 1$.

That is

$$w_0^{(0)}(t_0) \leq r_0 \tag{6.92}$$

which is true by (6.59) and (6.100).

Assume now that

$$v_1 = w_0^{(n-1)}(t_0) \leq w^{(n-2)}(t_0) = v_2.$$

We must show that

$$w_0^{(n)}(t_0) \leq w_0^{(n-1)}(t_0)$$

or equivalently
$$\frac{v_1(r_0+v_1)}{v_1+2\sqrt{v_1(r_0+v_1)}+a_0^2} \leq \frac{v_2(r_0+v_2)}{v_2+2\sqrt{v_2(r_0+v_2)}+a_0^2}$$
which is true since the function w_0 is increasing in r.

That completes the proof of the proposition. □

Let us denote by A and B the left hand sides of (6.90) respectively. It can easily be seen that
$$A \leq 1 \nLeftrightarrow B \leq 1,$$
but both can hold at the same time.

Let us take for example:
$$h_0 = 1, \quad q_0 = .5 \quad \text{and} \quad r_0 = \frac{1}{49}; \quad \text{then} \quad A = .785744285$$
$$\text{and} \quad B = 1.040816327$$

or

$$h_0 = 1, \quad q_0 = .29 \quad \text{and} \quad r_0 = .2, \quad \text{then} \quad A = 1.184427191 \quad \text{and} \quad B = .98$$

or

$$h_0 = 1, \quad q_0 = r_0 = .1, \quad \text{then} \quad A = .732455532 \quad \text{and} \quad B = .4.$$

Furthermore, we can produce the following a posteriori error estimates on the distances $\|x_n - x^*\|$.

Theorem 6.8 *Assume:*

(a) *the hypotheses of Theorem 6.6 are true and*

(b) *the linear operator $\delta f(x,y)$ is such that*
$$\delta f(x,y)(x-y) = f(x) - f(y), \quad \text{for all} \quad x, y \in V. \qquad (6.93)$$

Then the following inequalities are true:
$$c_2(n) \leq \|x_{n+1} - x_n\| \qquad (6.94)$$

and

$$\|x_n - x^*\| \leq c_4(n) = \frac{1-2h_0\|x_0-x_n\| - \sqrt{(1-2h_0\|x_0-x_n\|)^2 - 4h_0\|D_0^{-1}f(x_n)\|}}{2h_0},$$
$$n = 1, 2, \ldots . \qquad (6.95)$$

Proof. Using (6.69) and (6.72) it can easily follow that $\lim_{n\to\infty} d_n \geq \frac{1-h_0[\|x_0-x^*\|+\|x_{-1}-x^*\|]}{2h_0} > 0$, which implies (6.94) for sufficiently large n. By reordering the sequence $\{x_n\}$, $n = -1, 0, 1, 2$ we can assume that (6.94) is true for $n = 1, 2, \ldots$. Let us consider the linear operator D, given by

$$D = \delta f(x^*, x_n). \tag{6.96}$$

We will show that D is invertible for all $n \geq N$. Indeed, we have by (6.55), and (6.60) and (6.94) for $n \geq N$,

$$\|D_0^{-1}(f'(x_0) - D)\|$$
$$\leq h_0 [\|x_0 - x^*\| + \|x_0 - x_n\|]$$
$$\leq h_0 [2\|x_0 - x_n\| + \|x_n - x^*\|]$$
$$\leq h_0 [2\|x_0 - x_n\| + c_2(n)]$$
$$\leq h_0 [2\|x_0 - x_n\| + \|x_{n+1} - x_n\|]$$
$$\leq h_0 \left[2\left(\mu_0 - \sigma_0\left(w_0^{(n)}(t_0)\right) + w_0^{(n)}(t_0)\right)\right]$$
$$\leq 1 - h_0 \left[w_0^{(n)}(t_0) + q_0 + 2\sqrt{w_0^{(n)}(t_0)\left(q_0 + w_0^{(n)}(t_0)\right) + a_0^2}\right] < 1.$$

According to Banach's lemma it follows that the linear operator D is invertible for $n \geq N$ and that

$$\left\|\left(D_0^{-1} D\right)^{-1}\right\| \leq [1 - h_0 (2\|x_0 - x_n\| + \|x_n - x^*\|)]^{-1}. \tag{6.97}$$

Using the identity

$$D(x_n - x^*) = f(x_n),$$

(6.96) and (6.97), we obtain

$$\|x_n - x^*\| \leq \left\|\left(D_0^{-1} D\right)^{-1}\right\| \cdot \|D_0^{-1} f(x_n)\|$$
$$\leq [1 - h_0 (2\|x_0 - x_n\| + \|x_n - x^*\|)]^{-1} \|D_0^{-1} f(x_n)\|. \tag{6.98}$$

The inequality (6.95) follows now from (6.98).
That completes the proof of the theorem. □

We now compare the estimates c_4 and c_0.

Proposition 6.3 *Under the hypotheses of Theorem 6.6 the following inequality is true:*

$$c_4(n) \leq c_0(n), \quad n = 1, 2, \ldots \tag{6.99}$$

where c_4 and c_0 are defined by (6.95) and (6.56) respectively.

Proof. Using the identity
$$f(x_n) = f(x_n) - f(x_{n-1}) + \delta f(x_{n-1}, x_{n-2})(x_n - x_{n-1})$$
we obtain
$$\|D_0^{-1} f(x_n)\| \leq h_0 (\|x_n - x_{n-1}\| + \|x_{n-2} - x_{n-1}\|) \|x_n - x_{n-1}\|. \quad (6.100)$$
Moreover, it can easily be seen that
$$1 - 2h_0 \|x_0 - x_n\| \geq 2h_0 (s_0 - \|x_n - x_0\|). \quad (6.101)$$
The estimates $c_4(n)$ and $c_0(n)$ can be written respectively
$$c_4(n) = \|D_0^{-1} f(x_n)\| \left\{ 2 \left[1 - 2h_0 \|x_0 - x_n\| \right. \right.$$
$$\left. \left. + \left((1 - 2h_0 \|x_0 - x_n\|)^2 - 4h_0 \|D_0^{-1} f(x_n)\| \right)^{1/2} \right] \right\}^{-1} \quad (6.102)$$
and
$$c_0(n) = h_0 (\|x_n - x_{n-1}\| + \|x_{n-1} - x_{n-2}\|) \|x_n - x_{n-1}\| \left\{ h_0 [(s_0 - \|x_n - x_0\|)] \right.$$
$$\left. + ((s_0 - \|x_n - x_0\|)^2 - (\|x_n - x_{n-1}\| + \|x_{n-1} - x_{n-2}\|) \|x_n - x_{n-1}\|)^{1/2} \right\}^{-1}. \quad (6.103)$$

Using (6.100) and (6.101) it can easily be seen that the numerator of (6.102) is smaller or equal to the numerator of (6.103). Whereas the denominator of (6.102) is greater or equal to the denominator of (6.103). The estimate (6.99) now follows.

That completes the proof of the proposition. $\square$

Moreover we can show:

Proposition 6.4 *Assume that the set $C_1(h_n, q_n, r_n)$ denoting the class of all triplets $(f, x_{n-1}, x_n) \in C(h_n, q_n, r_n)$ satisfying the estimates (6.86), (6.89) and (6.90).*

Then the following inequalities are true:
$$\|x_n - x^*\| \leq \sigma_k \left(w_k^{(n)}(t_k) \right), \quad \text{with} \quad t_k = (q_k, r_k) \quad (6.104)$$
and
$$\sigma_k \left(w_k^{(n)}(t_k) \right) \leq \sigma_0 \left(w_0^{(n)}(t_0) \right) \quad \text{for all} \quad n = 0, 1, 2, ..., \quad k = 0, 1, 2, ..., n-1$$

where we have denoted:
$$w_k(t) = w_k(q, r) = \frac{r(q+r)}{r + 2\sqrt{r(q+r) + d_k^2}} \qquad (6.105)$$

and
$$\sigma_k(t) = r - d_k + \sqrt{r(q+r) + d_k^2}. \qquad (6.106)$$

Proof. The result (6.104) follows immediately by applying Theorem 6.6 to the triplet $(f, x_{k-1}, x_k) \in C(h_k, q_k, r_k) \subset C(h_k, q_k, r_k)$, $k = 0, 1, 2, ..., n-1$.

By (6.106) we have
$$\sigma_0\left(w_0^{(n)}(t_0)\right) = \sigma_0\left(\left(r_0, w_0^{(n-1)}(t_0)\right)\right) \qquad (6.107)$$

and
$$\sigma_k\left(w_k^{(n)}(t_k)\right) = \sigma_k\left(\left(r_k, w_k^{(n-1)}(t_k)\right)\right), \quad k = 1, 2, ..., n-1. \qquad (6.108)$$

We first show that
$$w_0^{(m)}(t_0) \geq w_k^{(m)}(t_k), \quad m = 0, 1, 2, ..., n-1$$

which is true for $m = 0$.

Assume
$$y_1 = w_0^{(m)}(t_0) \geq w_k^{(m)} = y_2, \quad m = 0, 1, 2, ..., n-2. \qquad (6.109)$$

Then we must show
$$w_0^{(n-1)}(t_0) \geq w_k^{(n-1)}(t_k) \qquad (6.110)$$

or equivalently by (6.105)
$$\frac{y_1(r_0 + y_1)}{y_1 + 2\sqrt{y_1(r_0 + y_1) + a_0^2}} \geq \frac{y_2(r_0 + y_2)}{y_2 + 2\sqrt{y_2(r_0 + y_2) + a_0^2}}$$

which is true since $q_n \geq q_0$, $d_k \geq a_0$, $r_n \leq r_0$ and (6.109) are true.

The induction is now completed.

That completes the proof of the proposition. □

A lower bound on $\|x_n - x^*\|$ can be given by the following:

Proposition 6.5 *Under the hypotheses of Theorem 6.6 the following inequality holds for $n = 1, 2, ...$*

$$\|x_{n-1} - x^*\| \geq q$$

where q is the positive root of the equation

$$h_0 \left\|\left(D_0^{-1}D_{n-1}\right)^{-1}\right\| \|x_{n-1} - x^*\|^2 +$$
$$\left(1 + \left\|\left(D_0^{-1}D_{n-1}\right)^{-1}\right\| \|x_{n-1} - x_{n-2}\| h_0\right) \|x_{n-1} - x^*\| - \|x_{n-1} - x_n\| = 0.$$

Proof. Using the identity

$$x_n - x_{n-1}$$
$$= x^* - x_n + \left(D_0^{-1}D_{n-1}\right)^{-1} D_0^{-1} \left[(f(x^*) - f(x_{n-1})) - D_{n-1}(x^* - x_{n-1})\right],$$

and the triangle inequality, the result follows immediately.

That completes the proof of the proposition. □

Note that q depends on $\left\|\left(D_0^{-1}D_{n-1}\right)^{-1}\right\|$, which in practice can be replaced by the right hand side of (6.106). Denote by $\bar{q}$ the resulting quantity. Then we will certainly have

$$\|x_{n-1} - x^*\| \geq q \geq \bar{q} \quad \text{for all} \quad n = 1, 2, \ldots .$$

Applications

Let us now compare the estimates (6.72) with (6.54) and (6.95) with (6.56), on a very simple example. We consider the quadratic equation

$$f(x) = x^2 - 16. \tag{6.111}$$

Take $x_{-1} = 3$, and $x_0 = 3.2$ and $\sigma f(x,y)(x-y) = f(x) - f(y)$. Then $h_0 = \frac{10}{62}$, $q_0 = .2$ and $r_0 = .92903225$.

The condition (6.66) is now satisfied, since

$$h_0 q_0 + 2\sqrt{h_0 r_0} = .806451609 < 1.$$

It is easy to see that $(f, x_{-1}, x_0) \in C(h_0, q_0, r_0)$.
The secant method for (6.111) becomes

$$x_{n+1} = \frac{x_{n-1}x_n + 16}{x_{n-1} + x_n}, \quad n = 0, 1, 2, \ldots .$$

Note that $x^* = 4$.

We can now compute

$$x_1 = 4.129032258,$$
$$x_2 = 3.985915493,$$
$$x_3 = 3.999776048,$$
$$x_4 = 4.000000395,$$
$$x_5 = 4,$$
$$a_0 = 1.8$$
$$d_0 = a_0$$
$$d_1 = 2.461392145,$$
$$d_2 = 3.014220162,$$
$$d_3 = 3.092844865,$$

and

$$d_4 = 3.099887432.$$

Using the above values, (6.71), (6.54), (6.95) and (6.56), we can tabulate the following results (within a precision of $\frac{1}{2}10^{-8}$).

n	error	error estimates (6.72)	error estimates (6.54)	error estimates (6.95)	error estimates (6.56)
1	.12903226	.27096774	.27096774	.25675941	.27096774
2	.014084507	.030974961	.042129881	.02443182	.03492694
3	$2.23952 \cdot 10^{-4}$	$3.608997 \cdot 10^{-4}$	$6.042855 \cdot 10^{-4}$	$3.8946524 \cdot 10^{-4}$	$4.945036 \cdot 10^{-4}$
4	$3.95 \cdot 10^{-7}$	$5.108 \cdot 10^{-7}$	$8.777498 \cdot 10^{-7}$	$6.8693502 \cdot 10^{-7}$	$7.182 \cdot 10^{-7}$
5	$1.1 \cdot 10^{-11}$	$1.4 \cdot 10^{-11}$	$2.46 \cdot 10^{-8}$	$1.9 \cdot 10^{-11}$	$2.0 \cdot 10^{-11}$

The above table indicates that our estimates (6.72) and (6.95) are better than the corresponding ones given by (6.54) and (6.56) respectively. Note, however, that the additional information (computation) $\left\|D_0^{-1}f(x_n)\right\|$ is used by (6.95).

All the above strongly recommend the usefulness of our estimates in numerical applications.

6.3 Exercises

6.1. Let $F: D \subseteq X \to Y$ and let D be an open set. Assume:

(a) the divided difference $[x, y]$ of F satisfies

$$[x, y](y - x) = F(y) - F(x) \text{ for all } x, y \in D$$
$$\|[x, y] - [y, u]\| \leq I_1 \|x - y\|^p + I_2 \|x - y\|^p + I_2 \|y - u\|^p$$

for all $x, y, u \in D$ where $I_1 \geq 0$, $I_2 \geq 0$ are constants which do not depend on x, y and u, while $p \in (0, 1]$;
(b) $x^* \in D$ is a simple solution of equation $F(x) = 0$;
(c) there exists $\varepsilon > 0$, $b > 0$ such that $\left\|[x, y]^{-1}\right\| \leq b$ for every $x, y \in U(x^*, \varepsilon)$;
(d) there exists a convex set $D_0 \subset D$ such that $x^* \in D_0$, and there exists $\varepsilon_1 > 0$, with $0 < \varepsilon_1 < \varepsilon$ such that $F'(\cdot) \in H_{D_0}(c, p)$ for every $x, y \in D_0$ and $U(x^*, \varepsilon_1) \subset D_0$.
Let $r > 0$ be such that:

$$0 < r < \min\left\{\varepsilon_1 \left(q(p)\right)^{-\frac{1}{p}}\right\}$$

where

$$q(p) = \frac{b}{p+1}\left[2^p (I_1 + I_2)(1 + p) + c\right].$$

Then
i. if $x_0, x_1 \in \bar{U}(x^*, r)$, the secant iterates are well defined, remain in $\bar{U}(x^*, r)$ for all $n \geq 0$, and converge to the unique solution x^* of equation $F(x) = 0$ in $\bar{U}(x^*, r)$. Moreover, the following estimation:

$$\|x_{n+1} - x^*\| \leq \gamma_1 \|x_{n-1} - x^*\|^p \|x_n - x^*\| + \gamma_2 \|x_n - x^*\|^{1+p}$$

holds for sufficiently large n, where

$$\gamma_1 = b(I_1 + I_2)2^p \text{ and } \gamma_2 = \frac{bc}{1+p}.$$

ii. If the above condition hold with the difference that x_0 and x_1 are chosen such that

$$\|x^* - x_0\| \leq ad_0; \ \|x^* - x_1\| \leq \min\left\{ad_0^{t_1}, \|x^* - x_0\|\right\},$$

where $0 < d_0 < 1$, $a = (q(b))^{-\frac{1}{p}}$, while t_1 is the positive root of the equation:

$$t^2 - t - p = 0,$$

then show that for every $n \in N$, $x_n \in U = \{x \in X \mid \|x - x^*\| < a\}$ and

$$\|x_{n+1} - x^*\| \le a d_0^{t_1^{n+1}} \quad (n \ge 0).$$

6.2. Let $F : D \subseteq X \to Y$ and let D be an open set. Assume:
(a) $x_0 \in X$ is fixed, and consider the non-negative real numbers: $B, v, w, p \in (0, 1]$, $\alpha, \beta, q \ge 1$, I_1, I_2 and I_3, where

$$w = B\alpha \left(I_1 B^p + I_2 \beta^p + I_3 B^p \alpha^p \|F(x_0)\|^{p(q-1)} \right)$$

and

$$v = w^{\frac{1}{p+q-1}} \|F(x_0)\|.$$

Denote $r = \max\{B, \beta\}$ and suppose $\bar{U}(x_0, r^*) \subseteq D$, where

$$r^* = \frac{rv}{w^{\frac{1}{p+q-1}} (1 - v^{p+q-1})};$$

(b) Condition (a) of the previous exercise holds with the last I_2 replaced by I_3;
(c) for every $x, y \in \bar{U}(x_0, r^*)$, $[x, y]^{-1}$ exists, and $\left\|[x, y]^{-1}\right\| \le B$;
(d) for every $x \in \bar{U}(x_0, r^*)$, $\|F(g(x))\| \le \alpha \|F(x)\|^q$ where $g : X \to Y$ is an operator having at least one fixed point which coincides with the solution x^* of equation $F(x) = 0$;
(e) for every $x \in \bar{U}(x_0, r^*)$, $\|x - g(x)\| \le \beta \|F(x)\|$;
(f) the number v is such that: $0 < v < 1$.
Then show that the Steffensen-type method

$$x_{n+1} = x_n - [x_n, g(x_n)]^{-1} F(x_n) \quad n \ge 0$$

is well defined, remains in $\bar{U}(x_0, r^*)$ for all $n \ge 0$ and converges to a solution x^* of equation $F(x) = 0$ with

$$\|x^* - x_n\| \le \frac{r w^{(p+q)^n}}{w^{\frac{1}{p+q-1}} (1 - v^{p+q-1})} \quad (n \ge 0).$$

6.3. (a) Consider conditions of the form

$$\|A_0^{-1}([x, y; F] - [z, w; F])\| \le w(\|x - z\|, \|y - w\|), \quad (6.112)$$
$$\|A_0^{-1}([x, y; F] - A_0)\| \le w_0(\|x - x_{-1}\|, \|y - x_0\|) \quad (6.113)$$

for all $x, y, z, w \in D$ provided that $A_0^{-1} \in L(Y, X)$, where $w, w_0 \colon [0, +\infty) \times [0, +\infty) \to [0, +\infty)$ are continuous nondecreasing functions in two variables.

Let $F \colon D \subseteq X \to Y$ be an operator. Assume:
there exists a divided difference of order one such that $[x, y; F] \subseteq L(X, Y)$ for all $x, y \in D$ satisfying (6.112), (6.113); there exist points $x_{-1}, x_0 \in D$ such that $A_0 = [x_{-1}, x_0; F]^{-1} \in L(Y, X)$ and set

$$\|A_0^{-1} F(x_0)\| \leq \eta;$$

equation

$$t = \left[\frac{c_0(t) c_1(t)}{1 - c(t)} + c_0(t) + 1\right] \eta$$

has at least one positive zero. Denote by t^* the smallest such zero;

$$w_0(t^* + \eta_0, t^*) < 1;$$
$$c(t^*) < 1;$$

and

$$\bar{U}(x_0, t^*) \subseteq D.$$

Show: sequence $\{x_n\}$ ($n \geq 0$) generated by the Secant method is well defined, remains in $\bar{U}(x_0, t^*)$ for all $n \geq 0$ and converges to a unique solution x^* of equation $F(x) = 0$ in $\bar{U}(x_0, t^*)$. Moreover the following error bounds hold

$$\|x_2 - x_1\| \leq c_0 \|x_1 - x_0\|$$
$$\|x_{n+1} - x_n\| \leq c \|x_n - x_{n-1}\| \quad (n \geq 3)$$

and

$$\|x_n - x^*\| \leq \frac{c^{n-2}}{1 - c} \|x_3 - x_2\| \quad (n \geq 2),$$

where,

$$c_0 = c_0(t^*), \quad c_1 = c_1(t^*), \quad c = c(t^*).$$

(b) Assume: x^* is a simple zero of operator F such that:

$$A_*^{-1} = F'(x^*)^{-1} \in L(Y,X);$$
$$\|A_*^{-1}([x,y;F] - [x,x^*;F])\| \leq v(\|y - x^*\|),$$
$$\|A_*^{-1}([x,y;F] - F'(x^*))\| \leq v_0(\|x - x^*\|, \|y - x^*\|)$$

for all $x,y \in D$ for some continuous nondecreasing functions $v\colon R_+ \to R_+$ and $v_0\colon R_+ \times R_+ \to R_+$; equation

$$v_0(q,q_0) + v(q_0) = 1$$

where,

$$q_0 = \|x_{-1} - x^*\|, \quad x_{-1} \in D$$

has at least one positive solution. Denote by q^* the minimum positive one;
and

$$\bar{U}(x^*, q^*) \subseteq D.$$

Under the above stated hypotheses, show sequence $\{x_n\}$ ($n \geq 0$) generated by the Secant method is well defined, remains in $\bar{U}(x^*, q^*)$ for all $n \geq 0$ and converges to x^* provided that $x_0 \in U(x^*, q)$ for some $x_{-1} \in D$.
Moreover the following error bounds hold for all $n \geq 0$:

$$\|x_{n+1} - x^*\| \leq \gamma_n \|x_n - x^*\|$$

where,

$$\gamma_n = \frac{v(\|x_{n-1} - x^*\|)}{1 - v_0(\|x_n - x^*\|, \|x_{n-1} - x^*\|)}.$$

6.4. (a) Let $x_0, x_{-1} \in D$ with $x_0 \neq x_{-1}$. It is convenient to define the parameters α, n by

$$\alpha = \|x_0 - x_{-1}\|,$$
$$\|L_0^{-1} F(x_0)\| \leq n,$$

and functions a, b, L_n by

$$a(r) = \frac{w(\alpha, r)}{1 - w(\alpha, r)},$$

$$b(r) = \frac{2w(\alpha + r, r)}{1 - w(\alpha + r, r)},$$

$$L_n = [x_{n-1}, x_n; F] \quad (n \geq 0).$$

We can now state and prove the following semilocal convergence theorem for the secant method.

Let F be a nonlinear operator defined on an open convex subset D of a Banach space X with values in a Banach space Y. Assume:
(1) there exist distinct points x_0, x_{-1} such that $L_0^{-1} \in L(Y, X)$;
(2) condition

$$\|[x_{-1}, x_0; F]^{-1}([x, y; F] - [x_{-1}, x_0; F])\| \leq w(\|x - x_{-1}\|, \|y - x_0\|),$$

holds for all $x, y \in D$;
(3) there exists a minimum positive zero denoted by r^* such that:

$$r \geq \left[\frac{a(r)b(r)}{1 - b(r)} + a(r) + 1\right]\eta \quad \text{for all } r \in (0, r^*];$$

(4)

$$w(\alpha + r^*, r^*) < 1,$$
$$b(r^*) < 1$$

and

$$U(x_0, r^*) \subseteq D.$$

Show: sequence $\{x_n\}$ $n \geq -1$ generated by secant method is well defined, remains in $\bar{U}(x_0, r^*)$ for all $n \geq -1$ and converges to a solution x^* of equation $F(x) = 0$, which is unique in $U(x_0, r^*)$.
(b) Let us consider the two boundary value problem:

$$\begin{matrix} y'' + y^{1+p} = 0 \\ y(0) = y(1) = 0' \end{matrix}, p \in [0, 1] \qquad (6.114)$$

considered first in [68]. As in [68] we divide the interval $[0, 1]$ into m subintervals and let $h = \frac{1}{m}$. We denote the points of subdivision by $t_i = ih$, and $y(t_i) = y_i$. We replace y'' by the standard

approximations
$$y''(t) \cong [y(t+h) - 2y(t) + y(t-h)]/h^2$$
$$y''(t_i) \cong (y_{i+1} - 2y_i + y_{i-1})/h^2, \quad i = 1, 2, \ldots, m-1.$$

System (6.114) becomes
$$2y_1 - h^2 y_1^{1+p} - y_2 = 0,$$
$$-y_{i-1} + 2y_i - h^2 y_i^{1+p} - y_{i+1} = 0$$
$$-y_{m-2} + 2y_{m-1} - h^2 y_{m-1}^{1+p} = 0, \quad i = 2, 3, \ldots, m-2.$$

Define operator $F \colon \mathbb{R}^{m-1} \to \mathbb{R}^{m-1}$ by
$$F(y) = H(y) - h^2 g(y),$$

where
$$y = (y_1, y_2, \ldots, y_{m-1})^t, \quad g(y) = (y_1^{1+p}, y_2^{1+p}, \ldots, y_{m-1}^{1+p})^t,$$

and
$$H = \begin{bmatrix} -2 & -1 & 0 & \cdots & 0 \\ -1 & 2 & -1 & \cdots & 0 \\ 0 & -1 & 2 & \cdots & 0 \\ \cdots & \cdots & \cdots & \cdots & \cdots \\ 0 & 0 & 0 & \cdots & 2 \end{bmatrix}.$$

We apply our Theorem to approximate a solution y^* of equation
$$F(y) = 0. \tag{6.115}$$

Let $x \in \mathbb{R}^{m-1}$, and choose the norm $\|x\| = \max_{1 \leq i \leq m-1} |x_i|$. The corresponding matrix $M \in \mathbb{R}^{m-1} \times \mathbb{R}^{m-1}$ is
$$\|M\| = \max_{1 \leq i \leq m-1} \sum_{j=1}^{m-1} |m_{ij}|.$$

A standard divided difference at the points $x, y \in \mathbb{R}^{m-1}$ is defined by the matrix whose entries are
$$[x, y; F]_{ij} = \tfrac{1}{x_i - y_i} \big[F_i(x_1, \ldots, x; y_{j+1}, \ldots, y_k) - F_i(x_1, \ldots, x_{j-1}, y; , \ldots, y_k) \big], \quad k = m-1.$$

We can set $[x, y; F] = \int_0^1 F'[x + t(y - x)]dt$.
Let $x, v \in \mathbb{R}^{m-1}$ with $|x_i| > 0$, $|v_i| > 0$, $i = 1, 2, \ldots, m - 1$. Using the max-norm we obtain

$$\begin{aligned}
\|F'(x) - F'(v)\| &= \|diag\{h^2(1+p)(v_i^p - x_i^p)\}\| \\
&= \max_{1 \leq i \leq m-1} |h^2(1+p)(v_i^p - x_i^p)| \\
&\leq (1+p)h^2 \max_{1 \leq i \leq m-1} |v_i^p - x_i^p| \\
&\leq (1+p)h^2 |v_i - x_i|^p = (1+p)h^2 \|v - x\|^p.
\end{aligned}$$

Hence, we get

$$\begin{aligned}
\|[x, y; F] &- [v, w; F]\| \\
&\leq \int_0^1 \|F'(x + t(y - x)) - F'(v + t(w - v))\| \, dt \\
&\leq h^2 \int_0^1 (1+p)\|(1-t)(x-v) + t(y-w)\|^p \, dt \\
&\leq h^2(1+p) \int_0^1 [(1-t)^p \|x - v\|^p + t^p \|y - w\|^p] \, dt \\
&= h^2(\|x - v\|^p + \|y - w\|^p).
\end{aligned}$$

Define the function w by

$$w(r_1, r_2) = \|[y_{-1}, y_0; F]^{-1}\| h^2 (r_1^p + r_2^p),$$

where y_{-1}, y_0 will be the starting points for the secant method

$$y_{n+1} = y_n - [y_{n-1}, y_n; F]^{-1} F(y_n) \quad (n \geq 0)$$

applied to equation $F(y) = 0$ to approximate a solution y^*. Choose $p = \frac{1}{2}$ and $m = 10$, then (6.115) gives 9 equations. Since a solution of (6.115) vanishes at the end points and is positive in the interior, a reasonable initial approximation seems to be $135 \sin nt$. This choice gives the following vector

$$z_{-1} = \begin{bmatrix} 41.7172942406179 \\ 79.35100905948387 \\ 109.2172942406179 \\ 128.3926296998458 \\ 135.0000000000000 \\ 128.3926296998458 \\ 109.2172942406179 \\ 79.35100905948387 \\ 41.7172942406179 \end{bmatrix}.$$

Choose y_0 by setting $z_0(t_i) = z_{-1}(t_i) - 10^{-5}$, $i = 1, 2, \ldots, 9$. Using secant method, we obtain after 3 iterations

$$z_2 = \begin{bmatrix} 33.64838334335734 \\ 65.34766285832966 \\ 91.77113354118937 \\ 109.4133887062593 \\ 115.6232519796117 \\ 109.4133887062593 \\ 91.77113354118937 \\ 65.34766285832964 \\ 33.64838334335733 \end{bmatrix}$$

and

$$z_3 = \begin{bmatrix} 33.57498274928053 \\ 65.204528678501265 \\ 91.56893412724006 \\ 109.1710943553677 \\ 115.3666988182897 \\ 109.1710943553677 \\ 91.56893412724006 \\ 65.20452867501265 \\ 33.57498274928053 \end{bmatrix}.$$

Set $y_{-1} = z_2$ and $y_0 = z_3$. Show:
We obtain

$$\alpha = .256553, \quad \eta = .00365901.$$

Moreover, show:

$$\|[y_{-1}, y_0; F]^{-1}\| \leq 26.5446,$$

$r^* = .0047$, $w(\alpha+r^*, r^*) = .153875247 < 1$ and $b(r^*) = .363717635$. All hypotheses are satisfied. Hence, equation has a unique solution $y^* \in U(y_0, r^*)$. Note that in [175] they found $r^* = .0043494$.

Chapter 7

Newton-Like Methods

Efficient Newton-like methods are discussed in this Chapter

7.1 Stirling's Method

Let F be a nonlinear operator defined on a closed and convex subset D of a Banach space X and with values into itself.

Consider Stirling's method [258]:

$$x_{n+1} = x_n - \left(I - F'(F(x_n))\right)^{-1} (x_n - F(x_n))), \quad n = 0, 1, 2, \ldots \quad (7.1)$$

for approximating a fixed point x^* of the equation

$$F(x) = x. \quad (7.2)$$

Iteration methods of the form (7.1) have been studied extensively. Bartle [102] studied (7.1) under the assumption that a modulus of continuity is known for $F'(x)$ as a function of x. Rall [244], gave general convergence results under the assumption that the Fréchet-derivative F' of F is uniformly bounded. Relevant work can be found in [68], [99] and the references there.

Stirling's method (7.1) can be viewed as a combination of the method of successive substitutions and Newton's method. It is consequently reasonable to examine the convergence of (7.1) under conditions which guarantee the convergence of the method of successive substitutions. In terms of computational effort, iteration (7.1) and Newton's method require essentially the same labor per step, as each requires the evaluation of F, F', and the solution of a linear equation (or the inversion of a linear operator) independently. Example where iteration (7.1) can be applied where Newton's method fails can be easily be constructed (see e.x. [244]).

One has remarked that from the view of the numerical efficiency it is not advantageous to change the operator $(I - F'(F(x_n)))^{-1}$ at each step of the iteration. Optimal recepts can be prescribed according to the dimension of the space (see [252]).

In this section will study the iteration (7.1) in case for each n, the linear operator T_n can be such that

$$T_n \in \left\{ (I - F'(F(x_{p_n})))^{-1} \right\}, \quad n = 0, 1, 2, \ldots \quad (7.3)$$

where $\{p_n\}$, $n = 0, 1, 2, \ldots$ is a nondecreasing sequence of integers satisfying the conditions

$$p_0 = 0, \quad 0 < p_n \leq n, \quad n = 1, 2, 3\ldots \quad (7.4)$$

The iteration

$$x_{n+1} = x_n - T_n(x_n - F(x_n)), \quad n = 0, 1, 2, \ldots \quad (7.5)$$

can be viewed as a generalization of (7.1) for $p_n = n$, $n = 0, 1, 2, \ldots$.

In what follows we will give sufficient conditions for the convergence of (7.5) to a locally unique fixed point x^* of equation (7.2). The error estimates obtained here are better than the ones previously obtained.

Finally an example is also provided.

We will need the definitions:

Definition 7.1 Let F be as in the introduction and let $x_0 \in D$. The Fréchet-derivative F' of F is said to be locally bounded if a constant $\alpha > 0$ exists such that

$$\|F'(x)\| \leq \alpha, \quad (7.6)$$

for all $x \in \overline{U}(x_0, R) \subset D$ with $R > 0$ and sufficiently large.

Definition 7.2 The Fréchet-derivative F' of F is said to be Lipschitz continuous if a constant $\gamma > 0$ exists such that

$$\|F'(x) - F'(y)\| \leq \gamma \|x - y\| \quad \text{for all} \quad x, y \in D. \quad (7.7)$$

In what follows we assume that F' satisfies (7.6) and (7.7) on D.

Definition 7.3 Let $x_0 \in D$ be such that the linear operator $I - F'(F(x_0))$ is invertible and $\left\| (I - F'(F(x_0)))^{-1} \right\| \leq p$, for some $p > 0$.

For $0 < \alpha < 1$, we define the quantities R_i, δ, δ_i, c, d, $i = 0, 1, ...5$ by

$$R_0 = \frac{1}{\gamma \alpha p},$$

$$R_1 = \frac{1 - \delta - \sqrt{\delta^2 - 2\delta(1 + p + 4\alpha p) + 1}}{p\gamma(1 + 4\alpha)},$$

$$R_2 = \frac{1 - \delta + \sqrt{\delta^2 - 2\delta(1 + p + 4\alpha p) + 1}}{p\gamma(1 + 4\alpha)},$$

$$R_3 = \frac{\|x_0 - F(x_0)\|}{1 - 2},$$

$$R_4 = \max(R_1, R_3),$$

$$R_5 = \frac{2 - \delta}{p\gamma(1 + 5\alpha)},$$

where

$$\delta = \delta(x_0) = p\gamma \|x_0 - F(x_0)\|, \tag{7.8}$$

$$\delta_0 = \frac{1}{\alpha p},$$

$$\delta_1 = 1 + p + 4\alpha p - \sqrt{(1 + p + 4\alpha p)^2 - 1},$$

$$\delta_2 = \frac{2(1 - \alpha p \gamma R_4)}{1 + 2\alpha},$$

$$\delta_3 = \frac{1 - \alpha}{2 + 3\alpha},$$

$$\delta_4 = 2,$$

$$\delta_5 = \frac{1 + p + 4\alpha p + \alpha + 3\alpha^2 - \sqrt{\Delta}}{24\alpha^2 + 10\alpha + 1},$$

$$c = \frac{p\gamma[(1 + 3\alpha) R_1 + \|x_0 - F(x_0)\|]}{2(1 - \alpha p \gamma R_1)},$$

and

$$d = \frac{p\gamma(1 + 2\alpha)}{2(1 - \alpha p \gamma R_4)},$$

where

$$\Delta = \Delta(\alpha) = (1 + P + 4\alpha P + \alpha + 3\alpha^2)^2 - 4\alpha(1 + 4\alpha)(24\alpha^2 + 10\alpha + 1).$$

The following can easily be verified:

(i) if
$$\delta < \min(\delta_0, \delta_1) \quad (7.9)$$
then
$$0 \leq R_1 \leq R_2 \leq R_0. \quad (7.10)$$

(ii) if
$$\delta < \min(\delta_0, \delta_1, \delta_2) \quad (7.11)$$
then
$$R_4 \leq R_2 \quad \text{and} \quad 0 < dp\,\|x_0 - F(x_0)\| < 1. \quad (7.12)$$

(iii) if
$$\delta < \min(\delta_0, \delta_1, \delta_2, \delta_3) \quad (7.13)$$
then
$$R_3 \leq R_2. \quad (7.14)$$

(iv) if
$$\delta < \min(\delta_0, \delta_1, \delta_2, \delta_4, \delta_5) \quad (7.15)$$
then
$$R_2 \leq R_5 \quad \text{and} \quad 0 < c < 1. \quad (7.16)$$

Finally by the form of Δ and, since
$$\Delta(0) = (p+1)^2 > 0$$
there exists α with $0 < \alpha < 1$ such that
$$\Delta(\alpha) > 0.$$

We can now prove the following theorem concerning the convergence of (7.5) to a fixed point x^* of equation (7.2).

Theorem 7.1 *Let $F : D \subset X \to X$ be a nonlinear operator and $x_0 \in D$ be such that the linear operator $I - F'(F(x_0))$ is invertible and*
$$\left\|(I - F'(F(x_0)))^{-1}\right\| \leq p, \quad \text{for some} \quad p > 0.$$

The following are true:

(a) if (7.9) is true, then the sequence $\{x_n\}$ generated by (7.5) is well defined for all $n = 0, 1, 2, ...$ and remains in $\overline{U}(x_0, R) \subset D$ with $R_1 \leq R \leq R_2$.

(b) If (7.13) is true and $p_n = n$, $n = 0, 1, 2, ...$, then the sequence $\{x_n\}$, $n = 0, 1, 2, ...$ converges to a unique fixed point $x^* \in D$ in $\overline{U}(x_0, R_4)$. Moreover x^* is unique in $\overline{U}(x_0, R_2)$.
The error is given by

$$\|x_n - x^*\| \leq \frac{1}{d} \sum_{j=n}^{\infty} (d \|x_0 - x_1\|)^{2^j}$$

$$\leq \frac{1}{d} \sum_{j=n}^{\infty} (dp \|x_0 - F(x_0)\|)^{2^j},$$

$$\leq b - \frac{1}{2} \sum_{j=0}^{n-1} (dp \|x_0 - F(x_0)\|)^{2^j},$$

$$b = \frac{1}{d} \sum_{j=0}^{\infty} (dp \|x_0 - F(x_0)\|)^{2^j}, n = 0, 1, 2, ... \quad (7.17)$$

(c) if (7.15) is true and $0 \leq p_n < n$, $n = 0, 1, 2, ...$, then the sequence $\{x_n\}$, $n = 0, 1, 2, ...$ converges to a unique fixed point $x^* \in D$ in $\overline{U}(x_0, R_1)$. Moreover, x^* is unique in $\overline{U}(x_0, R_2)$.
The error is given by

$$\|x_n - x^*\| \leq \frac{\|x_1 - x_0\|}{1 - c} c^n \leq \frac{p\|x_0 - F(x_0)\|}{1 - c} c^n, \quad n = 0, 1, 2, ... \quad (7.18)$$

(d) Under the hypotheses of (b) or (c) the following estimates is true:

$$\|x_n - x^*\| \geq \left[1 + \frac{\gamma p (3 + 2\alpha)}{1 - \alpha p \gamma \|x_{p_n} - x_0\|} \|x_{p_n} - x^*\|\right]^{-1} \|x_{n+1} - x_n\|. \quad (7.19)$$

and if

$$p\gamma \left[2\|x_{p_n} - x_0\| + \alpha \|x^* - x_{p_n}\| + \|x_0 - F(x_0)\|\right] < 2,$$
$$n, p_n = 0, 1, 2, ... \quad (7.20)$$

then

$$\|x_n - x^*\| \leq \frac{2p}{2 - p\gamma(2\|x_{p_n} - x_0\| + \alpha\|x_{p_n} - x^*\| + \|x_0 - F(x_0)\|)} \|x_n - F(x_n)\|. \quad (7.21)$$

Proof. (a) First let us observe that the linear operator $I - F'(F(x))$ is invertible for all $x \in \overline{U}(x_0, R)$ with all R such that $0 < R, R_0$.
Indeed form (7.6) and (7.7) it follows that

$$\left\| I - (I - f'(F(x_0)))^{-1} (I - F'(F(x))) \right\|$$
$$= \left\| (I - F'(F(x_0)))^{-1} (F'(F(x)) - F'(F(x_0))) \right\|$$
$$\leq p\alpha\gamma \|x - x_0\| \leq p\alpha\gamma R_0 < 1$$

so that according to Banach's lemma $I - F'(F(x))$ is invertible,

$$\left\| (I - F'(x))^{-1} \right\| \leq \frac{p}{1 - p\alpha\gamma \|x_0 - x_0\|}, \qquad (7.22)$$

and

$$\left\| \left[(I - F'(F(x_0)))^{-1} (I - F'(F(x))) \right]^{-1} \right\| \leq \frac{1}{1 - p\alpha\gamma \|x - x_0\|}. \qquad (7.23)$$

We will show that $x_n \in \overline{U}(x_0, R)$ with $R_1 \leq R \leq R_2$ for all $n = 0, 1, 2, \ldots$. By the choice of R, it follows that $x_1 \in \overline{U}(x_0, R)$. Let us assume that $x_{k+1} \in \overline{U}(x_0, R)$. Then T_{k+1} and consequently x_{k+2} are well defined. we will show that $x_{k+2} \in \overline{U}(x_0, R)$.
We have

$$\|x_{k+2} - x_0\| = \|x_{k+1} - x_0 - T_{k+1}(x_{k+1} - F(x_{k+1}))\|$$
$$= \|(I - F'(F(x_{p_{k+1}})))^{-1}[(I - F'(F(x_{p_{k+1}})))(x_{k+1} - x_0)$$
$$- (x_{k+1} - F(x_{k+1}))]\|$$
$$= \|(I - F'(F(x_{p_{k+1}})))^{-1}[F(x_{k+1}) - F(x_0)$$
$$- F'(F(x_{p_{k+1}}))(x_{k+1} - x_0) + (F(x_0) - x_0)]\|. \qquad (7.24)$$

Using the identity

$$F(x) - F(y) - F'(w)(x - y)$$
$$= \int_0^1 [F'(\theta x + (1-\theta)y) - F'(\theta w + (1-\theta)w)](x - y)\, d\theta \qquad (7.25)$$

for $x = x_{k+1}$, $y = x_0$, $w = F(x_{p_{k+1}})$, (7.22) and (7.7), the identity (7.24) becomes

$$\|x_{k+2} - x_0\|$$
$$\leq \frac{p\gamma}{2(1 - p\alpha\gamma R)} \left[\|x_{k+1} - F(x_{p_{k+1}})\| + \|x_0 - F(x_{p_{k+1}})\|\right] \|x_{k+1} - x_0\|$$
$$+ \frac{p}{1 - p\alpha\gamma R} \|F(x_0) - x_0\|. \qquad (7.26)$$

But

$$\|x_{k+1} - F(x_{p_{k+1}})\|$$
$$= \|(x_{k+1} - x_0) + (x_0 - F(x_0)) + (F(x_0) - F(x_{p_{k+1}}))\|$$
$$\leq R + \|x_0 - F(x_0)\| + \alpha R = R(\alpha + 1) + \|x_0 - F(x_0)\| \qquad (7.27)$$

and

$$\|x_0 - F(x_{p_{k+1}})\| = \|(x_0 - F(x_0)) + (F(x_0) - F(x_{p_{k+1}}))\|$$
$$\leq \|x_0 - F(x_0)\| + \alpha R. \qquad (7.28)$$

Therefore by (7.26), (7.27) and (7.28), $\|x_{k+2} - x_0\| \leq R$ if

$$\frac{p\gamma}{2(1 - p\alpha\gamma R)} \left[R(2\alpha + 1) + 2\|x_0 - F(x_0)\|\right] R$$
$$+ \frac{p}{1 - p\alpha\gamma R} \|x_0 - F(x_0)\| \leq R$$

or

$$p\gamma(4\alpha + 1)R^2 + 2[\gamma p \|x_0 - F(x_0)\| - 1] + 2p\|x_0 - F(x_0)\| \leq 0$$

which is true by the choice of R and δ. That is $x_n \in \overline{U}(x_0, R)$, $R_1 \leq R \leq R_2$.

(b) For $R_4 \leq R \leq R_2$ and $\delta < \min(\delta_0, \delta_1, \delta_2, \delta_3)$, using the identity

$$x_{n+1} - x_{n+2}$$
$$= (I - F'(F(x_{p_{n+1}})))^{-1} \left[F(x_n) - F(x_{n+1}) - F'(F(x_{p_{n+1}}))(x_n - x_{n+1})\right] \qquad (7.29)$$

with $p_{n+1} = n+1$, inequality (7.22) and the fact that $x_{n+1} \in \overline{U}(x_0, R)$ we obtain

$$\|x_{n+2} - x_{n+1}\| \leq \frac{p}{1 - \gamma\alpha p \|x_0 - x_{n+1}\|}$$
$$\cdot \frac{1}{2}\gamma [\|x_n - F(x_n)\| + \|F(x_n)\|] \|x_n - x_{n+1}\| \quad (7.30)$$
$$\leq \frac{\gamma p}{2(1 - \gamma\alpha p R_4)}[(1+\alpha)\|x_n - x_{n+1}\|$$
$$+\alpha \|x_n - x_{n+1}\|] \|x_n - x_{n+1}\|$$
$$\leq d \|x_n - x_{n+1}\|^2, \quad n = 0, 1, 2, \ldots . \quad (7.31)$$

But,

$$\|x_{n+2} - x_{n+1}\| \leq d \|x_n - x_{n+1}\|^2 \leq d \cdot d^2 \|x_{n-1} - x_n\|^2$$
$$\leq \cdots \leq d^{2^0 + 2^1 + \cdots + 2^n} \|x_0 - x_1\|^{2^{n+1}}$$
$$\leq \frac{1}{d}(d\|x_0 - x_1\|)^{2^{n+1}}$$
$$\leq \frac{1}{d}(dp\|x_0 - F(x_0)\|)^{2^{n+1}}.$$

Therefore, for $q > 1$

$$\|x_{n+1} - x_{n+q}\|$$
$$= \|(x_{n+1} - x_{n+2}) + (x_{n+2} - x_{n+3}) + \cdots + (x_{n+q-1} - x_{n+q})\|$$
$$\leq \frac{1}{d} \sum_{j=k+1}^{k+q-1} (dp\|x_0 - F(x_0)\|)^{2^j}. \quad (7.32)$$

Since, $dp\|x_0 - F(x_0)\| < 1$ by the choice of R and δ, it follows that the sequence $\{x_n\}$, $n = 0, 1, 2, \ldots$ generated by (7.1) is a Cauchy sequence in a Banach space, and as such it converges to some $x^* \in D$. By continuity, it follows from (7.2) that $x^* = F(x^*)$.

Moreover, letting $q \to \infty$ in (7.32) we obtain (7.17) which proves (b).

(c) For $R_1 \leq R \leq R_2$ and $\delta < \min(\delta_0, \delta_1, \delta_2, \delta_4, \delta_5)$, using (7.30), (7.7) and (7.9) we obtain

$$\|x_{n+1} - x_{n+2}\|$$
$$\leq \frac{p\gamma}{2[1-\gamma\alpha p\|x_0-x_{p_{n+1}}\|]}\left[\|x_n - F(x_{p_{n+1}})\|\right.$$
$$\left. + \|F(x_{p_{n+1}}) - F(x_{n+1})\|\right]\|x_n - x_{n+1}\|$$
$$\leq \frac{p\gamma}{2(1-\gamma\alpha pR_1)}\left[R_1(3\alpha+1) + \|x_0 - F(x_0)\|\right]\|x_n - x_{n+1}\|$$
$$= c\|x_n - x_{n+1}\|.$$

Therefore, for $q > 1$ we easily obtain that

$$\|x_{n+q} - x_n\| \leq c^n \frac{1-c^q}{1-c}\|x_1 - x_0\|. \tag{7.33}$$

But $0 < c < 1$, that is the sequence generated by (7.5), is a Cauchy sequence in a Banach space and as such it converges to an element $x^* \in D$. By continuity it follows from (7.5) that $x^* = F(x^*)$.

Moreover, letting $q \to \infty$ in (7.33) we obtain (7.18) which proves (c).

(d) Denote by L the linear operator given by

$$L = \int_0^1 F'(\theta x_{p_n} + (1-\theta)x^*)\,d\theta.$$

The linear operator $I - L$ is invertible if (7.20) holds. Indeed from (7.7), (7.20) and the identity

$$I - (I - F'(F(x_0)))^{-1}(I - L) = (I - F'(F(x_0)))^{-1}[L - F'(F(x_0))],$$

it follows that

$$\left\|I - (I - F'(F(x_0)))^{-1}(I - L)\right\|$$
$$\leq p\left\|\int_0^1 [F'(\theta x_n + (1-\theta)x^*) - F'(F(x_0))]\,d\theta\right\|$$
$$\leq p\gamma \int_0^1 [\|x_{p_n} - x_0\|\theta + \|x^* - x_0\|\alpha(1-\theta) + \|x_0 - F(x_0)\|\theta]\,d\theta$$
$$= \frac{1}{2}p\gamma\left[2\|x_{p_n} - x_0\| + \alpha\|x^* - x_{p_n}\| + \|x_0 - F(x_0)\|\right] < 1$$

so that according to Banach's lemma $I - L$ is invertible and

$$\left\|[I - F'(F(x_0)))^{-1}(I - L)]^{-1}\right\| \leq \frac{2}{2-p\gamma[2\|x_{p_n}-x_0\|+\alpha\|x^*-x_{p_n}\|+\|x_0-F(x_0)\|]}. \tag{7.34}$$

The result (7.21) now follows from (7.34) and the identity

$$x_n - x^* = ((I - F'(F(x_0)))^{-1}(I - L))^{-1}(I - F'(F(x_0)))^{-1}(x_n - F(x_n)). \tag{7.35}$$

Finally, using the identity

$$x_{n+1} - x_n = x^* - x_n + \left[(I - F'(F(x_0)))^{-1}(I - F'(F(x_{p_n})))\right]^{-1} \cdot$$
$$\cdot (I - F'(F(x_0)))^{-1}[F(x^*) - F(x_n) - F'(F(x_{p_n}))(x^* - x_n)] \tag{7.36}$$

and the estimate

$$\|x_{p_n} - F(x_{p_n})\| \leq (1 + \alpha)\|x_{p_n} - x^*\| \tag{7.37}$$

we obtain (7.19).

Finally it is routine to show that in either case (b) or (c) the fixed point x^* of equation (7.2) is unique in the corresponding balls and that completes the proof of the theorem. □

Remark 7.1 *(i) Note that (7.17) and (7.18) indicate quadratic and linear convergence respectively.*

(ii) The condition (7.20) in practice may be replaced by the condition

$$p\gamma[R_6(1 + \alpha) + \|x_0 - F(x_0)\|] < 2$$

where

$$R_6 = R_4 \text{ in case (b)}$$

or

$$R_6 = R_1 \text{ in case (c)}.$$

Moreover, the inequality

$$\|x_{p_n} - x^*\| \leq \|x_{p_n} - x_0\| + \|x_0 - x^*\| \leq \|x_{p_n} - x_0\| + R_6$$

can be used to replace $\|x_{p_n} - x^*\|$ on the right hand side of (7.21) or (7.19).

(iii) By definition δ_2 depends on R_4 which in turn depends on δ. In practice we may like to have δ_2 independent of R_4.

It can easily be checked that if

$$0 < p \leq \frac{22}{9} \quad \text{and} \quad \delta < \delta_6 \tag{7.38}$$

where

$$\delta_6 = \frac{2(1-\alpha)\left[4\alpha^2 p - (2+3p)\alpha - (3+p)\right]}{8\alpha^2 + 14\alpha + 3}$$

then $R_3 \geq R_1$. That is $R_4 = \max\{R_1, R_3\} = R_3$.

Define the number $\bar{\delta}_2 = \frac{2(1-\alpha p\gamma R_3)}{1+2\alpha}$. The condition $\delta < \bar{\delta}_2$ is now easier to check than $\delta < \delta_2$. The number R_3 can replace R_4 in the previous theorem if $\bar{\delta}_2$ replaces δ_2 and (7.38) holds.

Theorem 7.2 *Let $F : D \subset X \to X$ be a nonlinear operator and $x_0 \in D$ be such that the linear operator $I - F'(F(x_0))$ is invertible and*

$$\left\|(I - F'(F(x_0)))^{-1}\right\| \leq p, \quad \text{for some} \quad p > 0.$$

Assume:

(a) *the estimate*

$$\delta < \min(\delta_0, \delta_1, \delta_3), \quad \text{is true} \tag{7.39}$$

and

(b) *the sequence $\{t_n\}$, $n = 0, 1, 2, \ldots$ generated by*

$$t_{n+2} = t_{n+1} - \frac{p\gamma(1+2\alpha)(t_n - t_{n+1})^2}{2[1 - \alpha p\gamma(t_0 - t_{n+1})]}, \quad n = 0, 1, 2, \ldots \tag{7.40}$$

$$t_1 = t_0 - p\|x_0 - F(x_0)\|$$

is positive, and converges to some $t^* > 0$ for a particular $t_0 > 0$. Then

(i) *the sequence $\{t_n\}$, $n = 0, 1, 2, \ldots$ is decreasing*

(ii) *the sequence $\{x_n\}$ generated by (7.5) for $p_n = 0, 1, 2, \ldots$ is well defined, remains in $\overline{U}(x_0, R_4) \subset D$ and converges to a unique fixed point x^* of equation (7.2) in $\overline{U}(x_0, R_4)$.*

The following estimates are true:

$$\|x_n - x^*\| \leq t_n - t^* \tag{7.41}$$

and

$$\|x_n - x^*\| \geq \left[1 + \frac{(3+2\alpha)p\gamma}{1 - \alpha p\gamma(t_0 - t_{p_n})}(t_{p_n} - t^*)\right]^{-1}(t_n - t_{n+1}) \tag{7.42}$$

Proof. By (a) above and part (a) of Theorem 7.1, the iteration $\{x_n\}$, $n = 0, 1, 2, \ldots$ is well defined and remains in $\overline{U}(x_0, R_4)$. We shall now prove that

$$\|x_n - x_{n+1}\| \le t_n - t_{n+1}, \quad n = 0, 1, 2, \ldots . \tag{7.43}$$

By the choice of t_0 and t_1 it is easy to see that (7.43) is true for $n - 0$. Let us assume that (7.43) is true for $n = 0, 1, 2, \ldots, k$, where $k \ge 0$. Using (7.6), (7.7) and (7.30) we obtain as before that

$$\|x_{n+2} - x_{n+1}\| \le \frac{p\gamma(1 + 2\alpha)\|x_n - x_{n+1}\|^2}{2(1 - \alpha p\gamma\|x_0 - x_{n+1}\|)}$$

$$\le \frac{p\gamma(1 + 2\alpha)(t_n - t_{n+1})^2}{2[1 - \alpha p\gamma(t_0 - t_{n+1})]} = t_{n+1} - t_{n+2}.$$

We thus have proved that (7.43) holds for all n. By hypothesis the sequence $\{t_n\}$ converges to t^* and X is a Banach space, therefore there exists a point $x^* \in D$ such that $\lim_{n \to \infty} x_n = x^*$ and

$$\|x_n - x^*\| \le t_n - t^*.$$

By continuity, x^* is a fixed point of the equation (7.2) and that proves (7.41). It is routine to show that x^* is the unique fixed point of equation (7.2) in $\overline{U}(x_0, R_4)$.

Finally, the estimate (7.42) can be proved using (7.36), (7.37) and (7.43). That completes the proof of the theorem. □

We now cover the case when $0 \le p_n < n$, $n = 0, 1, 2, \ldots$. The proof of the following theorem is omitted as similar to Theorem 7.2.

Theorem 7.3 *Let F, x_0 be as in Theorem 7.1 and $0 \le p_n < n$, $n = 0, 1, 2, \ldots$.*

Assume:

(a) *the hypothesis (7.9) is true and*

(b) *the sequence $\{\bar{t}_n\}$, $n = 0, 1, 2, \ldots$ generated by*

$$\bar{t}_{n+2} = \bar{t}_{n+1} - \frac{p\gamma}{2[1 - \alpha p\gamma(\bar{t}_0 - \bar{t}_{p_{n+1}})]}$$
$$\times \left[\bar{t}_n + (1 + 2\alpha)\bar{t}_{p_{n+1}} - 2\bar{t}_{n+1} - (1 + \alpha)\bar{t}_{p_{n+2}}\right](\bar{t}_n - \bar{t}_{n+1}),$$
$$n = 0, 1, 2, \ldots$$
$$\bar{t}_1 = \bar{t}_0 - p\|x_0 - F(x_0)\|$$

is positive, decreasing and converges to some t^* for a particular $\bar{t}_0 > 0$.

Then the sequence $\{x_n\}$ generated by (7.5) for $0 \le p_n < n$, $n = 0, 1, 2, ...$ is well defined, remains in $\overline{U}(x_0, R_1)$ and converges to a unique fixed point x^* of equation (7.2) in $\overline{U}(x_0, R_1) \subset D$.

The following estimates are true:

$$\|x_n - x^*\| \le \bar{t}_n - t^*$$

and

$$\|x_n - x^*\| \ge \left[1 + \frac{(3 + 2\alpha) p\gamma}{1 - \alpha p\gamma (\bar{t}_0 - \bar{t}_{p_n})} (\bar{t}_{p_n} - t^*)\right]^{-1} (\bar{t}_n - \bar{t}_{n+1}), \quad n = 0, 1, 2, ...$$

We now proved some sufficient conditions for the convergence of the sequence $\{t_n\}$ given by (7.40), $n = 0, 1, 2, ...$ for a special choice of t_0. Similar conditions can be given for the sequence $\{\bar{t}_n\}$, $n = 0, 1, 2, ...$

Proposition 7.1 *Let F, x_0 be as in Theorem 7.1. If the real number δ given by (7.8) satisfies:*

$$\delta_g \le \delta < \min\left(\frac{\delta_0}{r+1}, \delta_1, \delta_3, \delta_7\right) \quad \text{if} \quad 1 - e + r \ne 0$$

or

$$\delta_9 \le \delta < \min\left(\frac{\delta_0}{r+1}, \delta_1, \delta_3, \delta_8\right) \quad \text{if} \quad 1 - e + r = 0 \text{ for some (finite) } r > 0,$$

where

$$\delta_7 = \frac{2 + r - \sqrt{r^2 + 4e}}{2(1 - e + r)\alpha p},$$

$$\delta_8 = \frac{1}{(2 + r)\alpha p},$$

$$\delta_9 = \frac{\gamma t_0 \left(1 + 2\sqrt{2e(2e - 1)}\right)}{4er}$$

and

$$e = \frac{1 + 2\alpha}{2\alpha}.$$

Then

(i) the real sequence $\{t_n\}$, $n = 0, 1, 2, \dots$ given by (7.40) for $t_0 = \frac{1}{\alpha\gamma p}$ is decreasing converging to some $t^* = t^*(r) \geq rp\|x_0 - F(x_0)\|$. The following are true:

$$t_{n+1} - t_{n+2} \leq s(t_n - t_{n+1})^2 \tag{7.44}$$

$$t_n - t^* \leq b^* - \frac{1}{s}\sum_{j=0}^{n-1}[s(t_0 - t_1)]^{2^j}$$

$$\leq b_1^* - \frac{1}{s}\sum_{j=0}^{n-1}[sp\|x_0 - F(x_0)\|]^{2^j}, \quad n = 0, 1, 2, \dots \tag{7.45}$$

and

$$e < r \tag{7.46}$$

where

$$s = \frac{e}{rp\|x_0 - F(x_0)\|}$$

and

$$b^* = \frac{1}{s}\sum_{j=0}^{\infty}[s(t_0 - t_1)]^{2^j} \quad \text{and} \quad b_1^* = \frac{1}{s}\sum_{j=0}^{\infty}[sp\|x_0 - F(x_0)\|]^{2^j}.$$

(ii) Moreover, the sequence $\{x_n\}$, $n = 0, 1, 2, \dots$ generated by (7.5) for $p_n = n$, $n = 0, 1, 2, \dots$ is well defined, remains in $\overline{U}(x_0, R_4)$ and converges to a unique fixed point x^* of equation (7.2) in $\overline{U}(x_0, R_4) \subset D$.
The following estimates are also true:

$$\|x_{n+1} - x_{n+2}\| \leq t_{n+1} - t_{n+2} \tag{7.47}$$

and

$$\|x_n - x^*\| \leq t_n - t^*, \quad n = 0, 1, 2, \dots. \tag{7.48}$$

Proof. We will first show that

$$rp\|x_0 - F(x_0)\| \leq t_{n+1} \leq t_n \leq t_0 = \frac{1}{\alpha\gamma p}, \quad \text{for all} \quad n = 0, 1, 2, \dots. \tag{7.49}$$

The sequence (7.40) now becomes

$$t_0 = \frac{1}{\alpha \gamma p},$$
$$t_1 = t_0 - p \|x_0 - F(x_0)\|$$

and

$$t_{n+2} = t_{n+1} - \frac{e}{t_{n+1}} (t_n - t_{n+1})^2, \quad n = 0, 1, 2, \ldots. \qquad (7.50)$$

It can easily be seen that

$$t_0 \geq rp \|x_0 - F(x_0)\| \quad \text{if} \quad \delta \leq \frac{\delta_0}{r}$$
$$t_1 \geq rp \|x_0 - F(x_0)\| \quad \text{if} \quad \delta \leq \frac{\delta_0}{r+1}$$

and

$$t_2 \geq rp \|x_0 - F(x_0)\| \quad \text{if} \quad \begin{cases} \delta \leq \delta_7 & \text{if } 1-e+r \neq 0 \\ \delta \leq \delta_8 & \text{if } 1-e+r = 0. \end{cases}$$

We now assume that

$$t_k \geq rp \|x_0 - F(x_0)\|, \quad k = 0, 1, 2, \ldots, n+1.$$

To show

$$t_{k+2} \geq rp \|x_0 - F(x_0)\|,$$

by (7.40) it suffices to show

$$t_{k+1} - \frac{e}{t_{k+1}} (t_k - t_{k+1})^2 \geq rp \|x_0 - F(x_0)\|$$

or

$$(e-1) t_{k+1}^2 - (2et_k - rp \|x_0 - F(x_0)\|) t_{k+1} + et_k^2 \leq 0. \qquad (7.51)$$

But, since

$$t_k \leq t_0, \quad k = 0, 1, 2, \ldots, n+1$$

we have

$$(e-a) t_{k+1}^2 - (2et_k - rp \|x_0 - F(x_0)\|) t_{k+1} + et_k^2$$
$$\leq (e-1) t_0^2 - 2e (rp \|x_0 - F(x_0)\|)^2 + rp \|x_0 - F(x_0)\| t_0 + et_0^2.$$

To show (7.51) it suffices to show

$$(e-1)t_0^2 - 2e(rp\|x_0 - F(x_0)\|)^2 + rp\|x_0 - F(x_0)\|t_0 + et_0^2 \leq 0$$

which is true if $\delta \geq \delta_9$.

The induction is now completed. It follows now from (7.49) that there exists $t^* = t^*(r) \geq rp\|x_0 - F(x_0)\|$, such that

$$\lim_{n \to 0} t_n = t^* \leq t_0.$$

Moreover, the sequence $\{t_n\}$, $n = 0, 1, 2, \ldots$ is decreasing and by (7.50)

$$\begin{aligned}
t_{n+1} - t_{n+2} &= \frac{e}{t_{n+1}}(t_n - t_{n+1})^2 \\
&\leq s(t_n - t_{n+1})^2 \\
&\leq s[s(t_0 - t_1)]^{2^{n+1}} \leq s[sp\|x_0 - F(x_0)\|]^{2^{n+1}}.
\end{aligned}$$

From the above inequality it follows that

$$sp\|x_0 - F(x_0)\| < 1$$

or

$$e < r$$

since the sequence $\{t_n\}$, $n = 0, 1, 2, \ldots$ is a Cauchy sequence (as convergent to t^*). The rest of the proof follows as the proof of Theorem 7.1 part (b). □

Note that r is not difficult to be determined, in practice and due to (7.46), the following is certainly true

$$r > \frac{3}{2}.$$

It can easily be seen by referring to the papers mentioned in the introduction that our results are better than those given in the literature for different particular cases. They are also obtained under weaker assumptions in general.

For simplicity we only compare our results with the ones obtained by Rall [244, Theorem 4], which are the most recent ones concerning iteration (7.1).

Assume that $X = \mathbb{R}$ and consider the interval $[-2.2, 1.8]$ on which (7.6) is true, the number $x_0 = -.2$ and the equation

$$F(x) = x \qquad (7.52)$$

where

$$F(x) = \frac{1}{8}(x^2 - 2).$$

Using the definitions 7.1, 7.2 and (7.3) we can easily calculate the following numbers

$$\alpha = .55,$$
$$\gamma = .25$$
$$p = .942285041,$$
$$\delta = .010600706,$$
$$\delta_0 = 1.909090909,$$
$$\delta_1 = .12651649,$$
$$\delta_2 = .633848081,$$
$$\delta_3 = .123287671,$$
$$\min(\delta_0, \delta_1, \delta_3) = .633848081.$$

According to Theorem 7.1, part (b) the condition (7.13) is satisfied, since

$$\delta < \min(\delta_0, \delta_1, \delta_2, \delta_3).$$

Therefore iteration (7.1) converges to the fixed point $x^* = -.24264087$ of equation (7.52).

Our error estimate (7.17) becomes

$$\|x_n - x^*\| \leq \frac{1}{d}\sum_{j=n}^{\infty}(dp\,\|x_0 - F(x_0)\|)$$

$$\leq (2.690685107)\sum_{j=n}^{\infty}(.015759119)^{2^j}, \quad n = 0, 1, 2, \ldots. \qquad (7.53)$$

The corresponding error estimate (3.10) of Rall becomes:

$$\|x_n - x^*\| \leq (h_s)^{2^n-1} \frac{\|x_0 - F(x_0)\|}{1-\alpha}$$
$$\leq (.1)(.05833333)^{2^n-1}, \quad n = 0,1,2,\ldots. \qquad (7.54)$$

Moreover, the iteration (7.1) becomes

$$x_{n+1} = \frac{4x_n^2 - x_n^3 + 2x_n - 3}{34 - x_n^2}, \quad n = 0,1,2,\ldots.$$

We can now tabulate the following results:

$x^* = -.242640687$		Results by Rall (7.54) $\|x_n - x^*\|$	Results by Argyros (7.53) $\|x_n - x^*\|$
	$x_0 = -.2$	.1	.043071219
	$x_1 = -.2424028$	.00583333	.000668393

The above table clearly indicates the superiority of the estimates (7.53) or (7.54) in this case.

7.2 Convergence for a Certain Class of Newton-Like Methods

Let F be a nonlinear operator mapping some subset D of a real Banach space E into a subset of a real Banach space $\hat{E}$. we are concerned with finding solutions $x^* \in D$ of the equation

$$F(x) = 0. \qquad (7.55)$$

a number of very interesting iterations have been introduced to approximate a solution $x^* \in D$ of (7.1). Most of them can be described, through the so called Newton-like methods of the form

$$x_{n+1} = x_n - A_n^{-1} F(x_n), \quad n = 0,1,2,\ldots \text{ for some } x_0 \in D. \qquad (7.56)$$

The sequence $\{A_n\}$ denotes invertible linear operators that constitute a conscious approximation of the first Fréchet-derivative $F'(x_n)$ of F evaluated at x_n.

Note that for $A_n = F'(x_n)$, one obtains the Newton method, whereas for $A_n = \delta F(x_{n-1}, x_n)$, one can obtain the secant method.

Relevant work on (7.56) (or special cases of (7.56)) can be found in the investigations of Kantorovich [183], Ortega and Rheinboldt [227], Gragg and Tapia [160], Dennis [132], Potra and Ptak [160], [68], [99] and the references there.

In particular, Dennis in [132], [239] did work of the following type. Given an operator F and a rule A which assigns to each x an approximate Jacobian $A(x)$, a simple technique is given to build a scalar function f and a scalar rule $a(x)$ such that if the corresponding Newton-like method for f converges, then so does the vector method for F. He then presented an error analysis based on the majorant theory.

Here, under the same assumptions, and using the same information, we show that our lower and upper bounds on the distances $\|x_n - x^*\|$, $\|x_{n+1} - x_n\|$, $n = 0, 1, 2, ...$ are sharper than the corresponding ones in [132].

An example is also provided.

We will need the definition:

Definition 7.4 Denote by $F'(x)$ the first Fréchet-derivative of F evaluated at $x \in D$. We say that $F' \in Lip_{k_0} D$ if

$$\|A_0^{-1}(F'(x) - F'(y))\| \leq k_0 \|x - y\| \quad \text{for all} \quad x, y \in D, \qquad (7.57)$$

provided that the inverse of the linear operator A_0 exists.

We state the following version of Theorem 2.6 in [132, p. 435]:

Theorem 7.4 *Let $F' \in Lip_{k_0} D_0$, where D_0 is the closure of an open convex set with $D_0 \subset D$.*
Assume:

(a) *for all $n = 0, 1, 2, ...$, there exist positive numbers b, a, a_n, such that*

$$a_0 = 1, \qquad (7.58)$$
$$b \leq a_n \leq a, \qquad (7.59)$$

and

$$\left\|\left(A_0^{-1} A_n\right)^{-1}\right\| \leq a_n^{-1}, \quad n = 0, 1, 2, ..., \qquad (7.60)$$

where $A_n \in L\left(E, \hat{E}\right)$ is a boundedly invertible operator for all $n = 0, 1, 2, ...$.

(b) For $\sigma \geq 1$ and $\Delta > 0$ both independent of n the following estimate holds

$$\|A_0^{-1}(F'(x_n) - A_n)\| \leq a_n + \sigma k_0 \sum_{j=1}^{n} \|x_j - x_{j-1}\| - \Delta,$$

for all $n = 0, 1, 2, \ldots$. \hfill (7.61)

(c) Denote by h_0, $s_0^{(0)}$, and r_0 the following:

$$h_0 = \frac{\sigma k_0 r_0 a_0}{\Delta^2}, \tag{7.62}$$

$$s_0^{(0)} = \frac{1 - \sqrt{1 - 2h_0}}{\sigma k_0} \Delta, \tag{7.63}$$

$$r_0 = \|A_0^{-1} F(x_0)\| \tag{7.64}$$

and assume that

$$h_0 \leq \frac{1}{2} \tag{7.65}$$

and

$$U\left(x_0, s_0^{(0)}\right) \subset D_0. \tag{7.66}$$

Denote by $f_0(t)$ and $t_n^{(0)}$ the following

$$f_0(t) = \frac{1}{2}\sigma k_0 t^2 - \Delta t + a_0 r_0 \tag{7.67}$$

and

$$t_{n+1}^{(0)} = t_n^{(0)} + \frac{f_0(t_n^{()})}{a_n}, \quad t_0^{(0)} = 0. \tag{7.68}$$

Then the sequence generated by (7.56) remains in $U\left(x_0, s_0^{(0)}\right)$ and converges to a unique solution x^* of the equation $F(x) = 0$ in $\overline{U}\left(x_0, s_1^{(0)}\right)$ with

$$s_1^{(0)} \leq s_0^{(0)} \tag{7.69}$$

where

$$s_1^{(0)} = \frac{1 - \sqrt{1 - 2h_0}}{\sigma k_0} \left(1 - \delta^{(0)}\right) a_0, \tag{7.70}$$

$$h_0' = \frac{k_0 r_0}{a_0 \left(1 - \delta^{(0)}\right)^2} \tag{7.71}$$

and

$$\delta^{(0)} = \frac{\left\| A_0^{-1} \left(F'(x_0) - A_0\right) \right\|}{a_0}. \tag{7.72}$$

Moreover, the following estimates are true:

$$\|x_{n+1} - x_n\| \leq t_{n+1}^{(0)} - t_n^{(0)}, \tag{7.73}$$

$$\|x_{n+1} - x^*\| \leq s_0^{(0)} - t_{n+1}^{(0)}, \tag{7.74}$$

$$\frac{1}{2} \|x_{n+1} - x_n\| \leq \|x_n - x^*\| \quad \text{for all} \quad n = 0, 1, 2, ..., \tag{7.75}$$

and

$$\sum_{j=0}^{n} \|x_j - x_{j-1}\| \leq t_n^{(0)} \leq s_0^{(0)} = \lim_{n \to \infty} t_{n+1}^{(0)}, \quad f_0\left(s_0^{(0)}\right) = 0. \tag{7.76}$$

Note that the linear operator A_0^{-1} does not appear in the estimates (7.60) and (7.61) of the original theorem. However, it can easily be seen by just repeating the proof of the original theorem [132, p. 435] that Theorem 7.4 above stands as it is.

Let us denote by $C(k_0, h_0, r_0)$ the class of all pairs (F, x_0) satisfying the hypotheses of 7.4.

Then we can improve the above theorem with what follows. But firstly we must define certain quantities.

Denote by h_m, $s_0^{(m)}$, r_m, k_m, f_m and $t_n^{(m)}$ the following quantities

$$h_m = \frac{\sigma k_m r_m a_m}{\Delta^2}, \tag{7.77}$$

$$s_0^{(m)} = \frac{1 - \sqrt{1 - 2h_m}}{\sigma k_m} \Delta, \tag{7.78}$$

$$r_m = \|x_{m+1} - x_m\|, \tag{7.79}$$

$$k_m = \sup_{x,y \in D_0} \frac{\|A_m^{-1}(F'(x) - F'(y))\|}{\|x - y\|}, \tag{7.80}$$

$$f_m(t) = \frac{1}{2}\sigma k_m t^2 - \Delta t + a_m r_m, \tag{7.81}$$

and

$$t_{n+1}^{(m)} = t_n^{(m)} + \frac{f_m\left(t_n^{(m)}\right)}{a_n}, \tag{7.82}$$

for each fixed m, $m = 0, 1, 2, \ldots$ and all $n = m, m+1, \ldots$ with

$$t_m^{(m)} = 0. \tag{7.83}$$

Theorem 7.5 Let $F' \in Lip_{k_0} D_0$ is the closure of an open convex set with $D_0 \subset D$.
Assume:

(a) the pair $(F, x_0) \in C(k_0, h_0, r_0)$;
(b) the sequence $\{a_n\}$, $n = 0, .1, 2, \ldots$ is increasing;
(c) the following are true for each fixed m, $m = 0, 1, 2, \ldots$

$$U\left(x_m, s_0^{(m)}\right) \subset D_0, \tag{7.84}$$

$$b \geq 1, \tag{7.85}$$

$$r_m \leq \frac{r_0}{a}, \tag{7.86}$$

$$\frac{1}{2}\left[\sigma k_0 \left(t_n^{(0)} + t_n^{(m)}\right) - 2\Delta + 2\sigma k_0 \left(s_0^{(m)} - t_n^{(m)}\right)\right] \cdot \left(t_n^{(0)} - t_n^{(m)}\right) + a_0 r_0 - a_m r_m + \frac{1}{2}\sigma(k_0 - k_m)\left(s_0^{(m)}\right)^2 \geq 0, \tag{7.87}$$

$$a_n + \frac{1}{2}\sigma k_0 \left(t_n^{(m)} + t_n^{(0)}\right) - \Delta \geq 0, \tag{7.88}$$

and
$$h_0 \leq \frac{(4\sigma + 1)^2 - 1}{2(4\sigma + 1)^2}. \quad (7.89)$$

Then for each fixed m, $m = 0, 1, 2, ...$, $(F, x_m) \in C(k_m, h_m, r_m)$, the sequence generated by (7.56) remains in $U\left(x_m, s_0^{(m)}\right)$ and converges to a unique solution x^* of the equation $F(x) = 0$ in $\overline{U}\left(x_m, s_1^{(m)}\right)$, with

$$s_1^{(m)} \leq s_0^{(m)} \quad (7.90)$$

where

$$s_1^{(m)} = \frac{1 - \sqrt{1 - 2h'_m}}{\sigma k_m}\left(1 - \delta^{(m)}\right) \quad (7.91)$$

$$h'_m = \frac{k_m r_m}{a_m \left(1 - \delta^{(m)}\right)^2} \quad (7.92)$$

and

$$\delta^{(m)} = \frac{\left\|A_0^{-1}\left(F'(x_m)0A_m\right)\right\|}{a_m}. \quad (7.93)$$

Moreover for each fixed m, $m = 0, 1, 2, ...$ and all $n = m, m+1, ...$, the following estimates are true:

$$\|x_{n+1} - x_n\| \leq t_{n+1}^{(m)} - t_n^{(m)}, \quad (7.94)$$
$$\|x_{n+1} - x^*\| \leq s_0^{(m)} - t_{n+1}^{(m)}, \quad (7.95)$$
$$c(n) \leq \|x_n - x^*\|, \quad (7.96)$$

$$\sum_{j=m}^{n} \|x_j - x_{j-1}\| \leq t_n^{(m)} \leq s_0^{(m)} = \lim_{n \to \infty} t_{n+1}^{(m)}, \quad f_m\left(s_0^{(m)}\right) = 0,$$
$$(7.97)$$

$$s_0^{(m)} \leq s_0^{(0)}, \quad (7.98)$$
$$t_n^{(m)} \leq t_n^{(0)}, \quad (7.99)$$
$$t_{n+1}^{(m)} - t_n^{(m)} \leq t_{n+1}^{(0)} - t_n^{(0)}, \quad (7.100)$$
$$s_0^{(m)} - t_n^{(m)} \leq s_0^{(0)} - t_n^{(0)} \quad (7.101)$$

and
$$c(n) \geq \frac{1}{2} \qquad (7.102)$$

where we have denoted

$$c(n) = 2\left\{\left[\left[2 + a_n^{-1}\left(\sigma k_0 \sum_{j=1}^{n} \|x_j - x_{j-1}\| - \Delta\right)\right]^2 \right.\right.$$
$$\left. + 2a_n^{-1} k_0 \|x_n - x_{n+1}\|\right]^{1/2}$$
$$\left. + \left[2 + a_n^{-1}\left(\sigma k_0 \sum_{j=1}^{n} \|x_j - x_{j-1}\| - \Delta\right)\right]\right\}^{-1}, \quad n = 0, 1, 2, \dots \quad (7.103)$$

Proof. We shall first show that

$$r_n \leq t_{n+1}^{(0)} - t_n^{(0)} \leq r_0, \quad n = 0, 1, 2, \dots \quad (7.104)$$

By (7.73), (7.79), and (a) to show the relation (7.104) it suffices:

$$t_{n+1}^{(0)} - t_n^{(0)} \leq r_0 \quad \text{for all} \quad n = 0, 1, 2, \dots \quad (7.105)$$

The above inequality is true as equality for $n = 0$ by (7.67), (7.68), (7.64) and (7.56). Let us assume that

$$t_n^{(0)} - t_{n-1}^{(0)} \leq t_1^{(0)} - t_0^{(0)} = r_0, \quad n = 1, 2, \dots \quad (7.106)$$

We must show that relation (7.105) holds for all $n = 0, 1, 2, \dots$.

The function f_0 given by (7.67) is decreasing on $\left[0, s_0^{(0)}\right]$ and the sequence $\left\{t_n^{(0)}\right\}$ is increasing. Using (b) and (7.68) we get

$$t_{n+1}^{(0)} - t_n^{(0)} = \frac{f_0\left(t_{n-1}^{(0)}\right)}{a_{n-1}} = t_n^{(0)} - t_{n-1}^{(0)}. \qquad (7.107)$$

The relation (7.105) now follows from (7.107) and (7.106).

We shall show that the pair $(F, x_m) \in C(k_m, h_m, r_m)$ for each fixed m, $m = 0, 1, 2, \dots$ by showing that

$$h_m \leq \frac{1}{2} \qquad (7.108)$$

and

$$U\left(x_m, s_0^{(m)}\right) \subset D_0. \qquad (7.109)$$

The results (7.94), (7.95) and (7.97) will then follow by applying Theorem 7.4 to the pair (F, x_m).

By (7.80) and (7.57) we get

$$k_m = \sup_{x,y \in D_0} \frac{\left\|(A_0^{-1} A_m)^{-1} A_0^{-1}(F'(x) - F'(y))\right\|}{\|x-y\|} \left\|(A_0^{-1} A_m)^{-1}\right\| k_0 \leq a_m^{-1} k_0. \tag{7.110}$$

By (7.65), (7.77), (7.105) and (7.110) we obtain

$$h_m = \frac{\sigma k_m r_m a_m}{\Delta^2} \leq \frac{\sigma a_m^{-1} k_0 r_0 a_m}{\Delta^2} = h_0 \leq \frac{1}{2}. \tag{7.111}$$

That shows (7.108) and (7.109) follows immediately from (7.84).

using the identities

$$x_{n+1} - x_n = x^* - x_n + \left(A_0^{-1} A_n\right)^{-1} A_0^{-1} \left[F(x^*) - F(x_n) - A_n(x^* - x_n)\right], \tag{7.112}$$

$$F(x^*) - F(x_n) - A_n(x^* - x_n) = \\ F(x^*) - F(x_n) - F'(x_n)(x^* - x_n) + (F'(x_n) - A_n)(x^* - x_n), \tag{7.113}$$

the triangle inequality on (7.112) and (7.113), (7.57), (7.60), (7.61) the inequality (7.96) follows immediately.

To show relation (7.98) we must first prove that

$$f_m(t) \leq f(t) \quad \text{for each fixed} \quad m, m = 0, 1, 2, \ldots \text{ and all } t \geq 0. \tag{7.114}$$

But by (7.67), (7.81), (7.85), (7.86) and (7.110) we easily get

$$f_m(t) = \frac{1}{2}\sigma k_m t^2 - \Delta t + a_m r_m \leq \frac{1}{2}\sigma k_0 t^2 - \Delta t - a_0 r_0 = f_0(t).$$

Let us assume that $s_0^{(0)} \in \left[0, s_0^{(m)}\right)$ then

$$f_m\left(s_0^{(0)}\right) > 0 \quad \text{for each fixed} \quad m, m = 0, 1, 2, \ldots. \tag{7.115}$$

Set $t = s_0^{(0)}$ in (7.114) and using the second relation in (7.76) we get

$$f_m\left(s_0^{(0)}\right) \leq f_0\left(s_0^{(0)}\right) = 0 \tag{7.116}$$

contradicting relation (7.115). The result (7.98) follows immediately.

We shall show the relation (7.99) using induction on $n = m, m+1, \ldots$ for every fixed m, $m = 0, 1, 2, \ldots$. The relation (7.99) is true for $n = m$, since by (7.83) and (7.68)

$$t_m^{(m)} = 0 \le t_n^{(0)}.$$

Assuming that (7.99) holds for every fixed m, $m = 0, 1, 2...$ and all $n = m, m+1, ...$, we shall show

$$t_{n+1}^{(m)} \le t_{n+1}^{(0)}. \tag{7.117}$$

By (7.67), (7.68), (7.81), (7.82), (7.99) and (7.114), the relation (7.117) is true if

$$t_{n+1}^{(m)} \le t_n^{(m)} + \frac{f_m\left(t_n^{(m)}\right)}{a_n} \le t_n^{(m)} + \frac{f_0\left(t_n^{(m)}\right)}{a_n} \le t_n^{(0)} + \frac{f_0\left(t_n^{(0)}\right)}{a_n}$$

or if

$$a_n\left(t_n^{(m)} - t_n^{(0)}\right) + \frac{1}{2}\sigma k_0 \left(t_n^{(m)} - t_n^{(0)}\right)\left(t_n^{(m)} + t_n^{(0)}\right) - \Delta\left(t_n^{(m)} - t_n^{(0)}\right) \le 0. \tag{7.118}$$

The above inequality is true by (7.99), (7.88) and the first relations of (7.76) and (7.97) respectively.

To show (7.100), it suffices by (7.68) and (7.82) to prove

$$f_m\left(t_n^{(m)}\right) \le f_0\left(t_n^{(0)}\right) \quad \text{for each fixed} \quad m, m = 0, 1, 2, ... \tag{7.119}$$

and all $n = m, m+1, ...$.

By (7.67) and (7.81) the relation (7.119) is true if

$$\frac{1}{2}\sigma k_0 \left[\left(t_n^{(m)}\right)^2 - \left(t_n^{(0)}\right)^2\right] - \Delta\left(t_n^{(m)} - t_n^{(0)}\right) + a_m r_m - a_0 r_0 \le 0 \tag{7.120}$$

since (7.85) and (7.110), $k_m \le k_0$.

Moreover by (7.86)

$$a_m r_m \le a_0 r_0. \tag{7.121}$$

But the relation (7.120) is certainly true if (7.121), (7.86) and (7.99) hold. This argument shows the relation (7.100).

Let us now assume contrary to (7.101) that

$$s_0^{(m)} - t_n^{(m)} + t_n^{(0)} > s_0^{(0)} \tag{7.122}$$

for each fixed m, $m = 0, 1, 2, ...$ and all $n = m, m+1, ...$.

The function f_0 is decreasing on $\left[0, s_0^{(0)}\right]$ and

$$f_0(t) \geq 0 \quad \text{for all} \quad t \in \left[0, s_0^{(0)}\right]. \tag{7.123}$$

Therefore by (7.122) we deduce that

$$f_0\left(s_0^{(m)} + t_n^{(0)} - t_n^{(m)}\right) < 0. \tag{7.124}$$

But by (7.67) and the second relation of (7.97),

$$f_0\left(s_0^{(m)} + t_n^{(0)} - t_n^{(m)}\right) = \left(\frac{1}{2}\sigma k_m \left(s_0^{(m)}\right)^2 - \Delta s_0^{(m)} + a_m r_m\right)$$
$$+ \frac{1}{2}\left[\sigma k_0\left(t_n^{(0)} + t_n^{(m)}\right) - 2\Delta + 2\sigma k_0\left(s_0^{(m)} - t_n^{(m)}\right)\right]\left(t_n^{(0)} - t_n^{(m)}\right)$$
$$+ a_0 r_0 - a_m r_m + \frac{1}{2}\sigma(k_0 - k_m)\left(s_0^{(m)}\right)^2$$
$$\geq 0 \tag{7.125}$$

by hypothesis (7.87), contracting (7.122) and (7.124). The relation (7.101) now follows.

Finally, we shall show relation (7.102). Using (7.103), (7.63), the first relation of (7.76) we can easily deduce that (7.102) holds if

$$\|x_n - x_{n+1}\| \leq \frac{4\Delta\sqrt{1 - 2h_0}}{k_0}. \tag{7.126}$$

But, by the first relation of (7.76)

$$\|x_n - x_{n+1}\| \leq t_{n+1} \leq s_0^{(0)}. \tag{7.127}$$

By (7.127) and (7.63), to show (7.126) it suffices

$$s_0^{(0)} \leq \frac{4\sqrt{1 - 2h_0}}{k_0},$$

which is true by (7.89).

That shows relation (7.102) and completes the proof of the theorem. □

The estimates on the distances $\|x_n - x_{n+1}\|$ and $\|x_{n+1} - x^*\|$ given in Theorem 7.5 are sharper than the ones given in [132, p. 435, 453]. However, the computation of the sequences $\left\{t_n^{(m)}\right\}$ involves the evaluation of the sequence $\{k_m\}$ given by (7.80). The cost of such computation may be very high in interesting cases (excluding the scalar case).

In the theorem that follows we show how to overcome this difficulty and still obtain sharper estimates on the distances $\|x_n - x_{n+1}\|$ and $\|x_{n+1} - x^*\|$ than the corresponding ones in Theorem 7.4.

Let us first denote by $\bar{h}_m, \bar{k}_m, v_0^{(m)}, g_m, v_{n+1}^{(m)}, v_1^{(m)}$, and ℓ_m the following quantities for each fixed m, $m = 0, 1, 2, \ldots$ and all $n = m, m+1, \ldots$

$$\bar{h}_m = \frac{\sigma \bar{k}_m r_m a_m}{\Delta^2},$$

$$\bar{k}_m = a_m^{-1} k_0,$$

$$g_m(t) = \frac{1}{2}\sigma \bar{k}_m t^2 - \Delta t + a_m r_m,$$

$$w_{n+1}^{(m)} = w_n^{(m)} + \frac{g_m\left(w_n^{(m)}\right)}{a_n}, \quad w_m^{(m)} = 0,$$

$$v_1^{(m)} = \frac{1 - \sqrt{1 - 2\ell}}{\sigma \bar{k}_m}\left(1 - \delta^{(m)}\right), \quad v_0^{(m)} = \lim_{n \to \infty} w_{n+1}^{(m)}$$

and

$$\ell_m = \frac{\bar{k}_m r_m}{a_m \left(1 - \delta^{(m)}\right)^2}.$$

Then we can show:

Theorem 7.6 *Let $F' \in Lip_{k_0} D_0$, where D_0 is the closure of an open convex set with $D_0 \subset D$ and assume that the hypotheses of Theorem 7.5 are true.*

Then with the notation introduced above, for each fixed m, $m = 0, 1, 2, \ldots$ the pair $(F, x_m) \in C\left(\bar{k}_m, \bar{h}_m, r_m\right)$, the sequence generated by (7.56) remains in $U\left(x_m, v_0^{(m)}\right)$ and converges to a unique solution x^* of the equation $F(x) = 0$ in $\overline{U}\left(x_m, v_1^{(m)}\right)$.

Moreover, for each fixed $m, m = 0, 1, 2, \ldots$ and all $n = m, m+1, \ldots$ the following estimates are true:

$$\|x_{n+1} - x_n\| \leq w_{n+1}^{(m)} - w_n^{(m)}, \tag{7.128}$$

$$\|x_{n+1} - x^*\| \leq v_0^{(m)} - w_{n+1}^{(m)}, \tag{7.129}$$

$$\sum_{j=m}^{n} \|x_j - x_{j-1}\| \leq w_{n+1}^{(m)} \leq v_0^{(m)}, \quad g_m\left(v_0^{(m)}\right) = 0,$$

$$s_0^{(m)} \leq v_0^{(m)} \leq s_0^{(0)},$$
$$t_n^{(m)} \leq w_n^{(m)} \leq t_n^{(0)},$$
$$t_{n+1}^{(m)} - t_n^{(m)} \leq w_{n+1}^{(m)} - w_n^{(m)} \leq t_{n+1}^{(0)} - t_n^{(0)}, \tag{7.130}$$

and

$$s_0^{(m)} - t_0^{(m)} \leq v_0^{(m)} - w_n^{(m)} \leq s_0^{(0)} - t_n^{(0)}. \tag{7.131}$$

Proof. The proof is identical to Theorem 7.5 with h_m, $s_0^{(m)}$, k_m, f_m, $t_{n+1}^{(m)}$, $s_1^{(m)}$, h_m' replaced respectively by $\overline{h}_m$, $v_0^{(m)}$, $\overline{k}_m$, g_m, $w_{n+1}^{(m)}$, $v_1^{(m)}$ and ℓ_m for each fixed m, $m = 0, 1, 2, ...$ and all $n = m, m+1, ...$. □

Remark 7.2 (a) *If inequality (7.61) holds for $n = 0$ and $\sigma > 1$ then Theorem 7.4 gives*

$$x^* \in \overline{U}\left(x_0, s_1^{(0)}\right) \subset U\left(x_0, s_0^{(0)}\right)$$

with

$$s_1^{(0)} < s_0^{(0)}. \tag{7.132}$$

Then the relation

$$\|x_m - x_0\| + s_0^{(m)} \leq s_0^{(0)} \tag{7.133}$$

for each fixed m, $m = 0, 1, 2, ...$, implies that

$$x^* \in U\left(x_m, s_0^{(m)}\right) \subset U\left(x_0, s_0^{(0)}\right) \subset D_0 \tag{7.134}$$

by hypothesis (7.66). The relation (7.132) holds for sufficiently large m. Indeed otherwise since $\lim_{m\to\infty} s_0^{(m)} = 0$, there exists M such that if $m \geq M$, then by (7.133),

$$s_1^{(0)} \geq \|x^* - x_0\| \geq s_0^{(0)}$$

contradicting (7.132).

Due to the conservative nature of our estimates one may expect that the relation (7.133) holds for sufficiently small m. If this is the case due to (7.133), Theorems 7.5 and 7.6 provide better information on the location of the solution x^* Theorem 7.4.

(b) The relations (7.95), (7.96), (7.100), (7.101), (7.128), (7.129), (7.130) and (7.131) indicate that our estimates on the distances $\|x_{n+1} - x^*\|$ are sharper than the corresponding ones in (7.74) and (7.75). A similar observation can be made for the distances $\|x_{n+1} - x_n\|$. Moreover, these results are obtained with almost the same hypotheses as the ones in Theorem 7.4.

Note, however, that only the estimates (7.96), (7.128), (7.129), (7.130) and (7.131) are always computable in the most interesting cases (excluding the scalar case).

(c) It is often the case that does not wish to recalculate the approximate derivative at every iteration but will instead use A_n in place of $A_{n+1}, ..., A_{n+k}$ and then calculate A_{n+k+1} and use it for q iterations. Let $\{p_n\}$ be a nondecreasing sequence of nonnegative real numbers such that

$$p_0 = 0, \quad \text{and} \quad p_n = p_{n-1} \quad \text{or} \quad p_n = n.$$

Then the general iteration referred to above can be written as

$$x_{n+1} = x_n - A_{p_n}^{-1} F(x_n), \quad n = 0, 1, 2, ... \tag{7.135}$$

The conclusions of Theorem 7.4 hold for iteration (7.135) [132, Th.2.6, p. 435]. Therefore, by just repeating the proofs of Theorem 7.5 and 7.6 and interchanging the role of n with p_n we can easily produce the corresponding results (7.195) − (7.102) and (7.128) − (7.131) of Theorems 7.5 and 7.6 respectively for iteration (7.135).

(d) By following arguments similar to the ones introduced in the proofs of Theorems 7.5 and 7.6 one can easily produce bounds on the distances $\|x_{n+1} - x^*\|$ and $\|x_{n+1} - x_n\|$ that are sharper than the corresponding ones given in [132, p. 457] for Newton's or the secant method.

Note that in [132, p. 457] by appropriate choices of A_n, a_n, σ and Δ the iteration 2 can be reduced to Newton's or the secant method.

The details of this approach are left to the motivated reader.

(e) The condition (7.87), due to (7.125) can be replaced in practice by the equivalent relation

$$f_0\left(s_0^{(m)} + t_n^{(0)} - t_n^{(m)}\right) \geq 0 \quad \text{for every fixed} \quad m, \; m = 0, 1, 2, ...$$
$$\text{and all} \quad n = m, m+1, ...$$

We now prove a more general theorem which at first seems to reduce to a minimum the assumptions necessary to apply the majorant technique.

Second it gives an easy semilocal convergence theorem for the class of methods using "consistent derivative approximations" [68].

Theorem 7.7 *Let $F' \in Lip_{p_{k_0}} D_0$, where D_0 is the closure of an open convex set with $D_0 \subset D$ and assume:*

(a) *there exist nonnegative real numbers δ_n, δ and γ such that for every n for which $x_0, ..., x_n$ as defined by (7.56), are in D_0, with*

$$\left\| A_0^{-1} \left(A_n - F'(x_n) \right) \right\| \leq \delta_n + \gamma \sum_{j=1}^{n} \|x_j - x_{j-1}\|; \qquad (7.136)$$

and

$$\delta_n \leq \delta \quad for \quad n = 1, 2, ...\;.$$

(b) *Let $x_0 \in D_0$ and A_0 be an invertible element of $L\left(E, \hat{E}\right)$ with*

$$\left\| A_0^{-1} F(x_0) \right\| \leq r_0.$$

(c) *The following conditions are satisfied:*

$$a_0 > \delta_0 + 2\delta, \qquad (7.137)$$

$$\frac{1}{2} \geq d_0 = \frac{(2\gamma + k_0) a_0 r_0}{(a_0 - \delta_0 - 2\delta)}, \qquad (7.138)$$

and

$$U(x_0, e_0) \subset D_0, \quad where \quad e_0 = \frac{1 - \sqrt{1 - 2d_0}}{(2\gamma + k_0)} (a_0 - \delta_0 - 2\delta).$$

Then
(i) *the sequence $\{x_n\}$, $n = 0, 1, 2, ...$ generated by (7.56) exists in $U(x_0, e_0)$ and converges to a unique solution x^* of the equation $F(x) = 0$ in $\overline{U}(x_0, e'_0)$, where we have denoted*

$$e'_0 = \frac{1 - \sqrt{1 - 2d'_0}}{k_0} (a_0 - \delta_0)$$

and

$$d'_0 = \frac{a_0 k_0 r_0}{(a_0 - \delta)^2}.$$

(ii) *If*
$$d_0' < \frac{1}{2}$$
then the solution x^* *is unique in* $D_0 \cap U(x_0, e_0'')$ *where we have denoted*
$$e_0'' = \frac{1 + \sqrt{1 - d_0'}}{k_0}(a_0 - \delta_0).$$

(iii) *For all* $n \geq 1$ *and with* $t_1^{(0)} = r_0$ *the following estimate is true*
$$\|x_{n+1} - x^*\| \leq e_0 - t_{n+1}^{(0)}$$
where the sequence $\left\{t_{n+1}^{(0)}\right\}$, $n = 0, 1, 2, \ldots$ *is as defined in* (7.68) *with*
$$\sigma = \frac{k_0 + 2\gamma}{k_0},$$
$$\Delta = a_0 - \delta_0 - 2\delta,$$
$$a_0 = 1$$
and
$$a_n = \left[1 - \left(\delta_0 + \delta_n + (k_0 + \gamma)\sum_{j=1}^{n}\|x_j - x_{j-1}\|\right)\right], \quad n = 1, 2, \ldots . \tag{7.139}$$

(iv) *Let us denote by* $C_1(k_0, d_0, r_0)$ *the class of all pairs* (F, x_0) *satisfying the hypotheses of this theorem. If for each fixed* m, $m = 0, 1, 2, \ldots$, $r_m \leq r_0$, $k_m \leq k_0$ *and* $U(x_m, e_m) \subset D_0$ *then the pair* $(F, x_m) \in C_1(k_m, d_m, r_m)$ *where*
$$d_m = \frac{(2\gamma + k_m)a_m r_m}{(a_0 - \delta_0 - 2\delta)^2}, \tag{7.140}$$
$$e_m = \frac{1 - \sqrt{1 - 2d_m}}{(2\gamma + k_m)}(a_0 - \delta_0 - 2\delta),$$

and k_m *and* r_m *are as defined before.*

(v) *The sequence generated by* (7.56) *remains in* $U(x_m, e_m)$ *and converges to a unique solution* x^* *of equation* $f(x) = 0$ *in* $\overline{U}(x_m, e_m')$. *If*

$d'_m < \frac{1}{2}$ then the solution x^* is unique in $D_0 \cap (x_m, e''_m)$, where we have denoted

$$e'_m = \frac{1 - \sqrt{1 - d'_m}}{k_m}(a_0 - \delta_0),$$

$$d'_m = \frac{a_m k_m r_m}{(a_0 - \delta)^2}$$

and

$$e''_m = \frac{1 + \sqrt{1 - d'}}{k_m}(a_0 - \delta_0),$$

and for each fixed m, $m = 0, 1, 2, \ldots$.

(vi) *Moreover for each fixed m, $m = 0, 1, 2, \ldots$ and all $n = m, m+1, \ldots$ the following estimates are true:*

$$\|x_{n+1} - x_n\| \leq t_{n+1}^{(m)} - t_n^{(m)},$$

$$\|x_{n+1} - x^*\| \leq e_m - w_{n+1}^{(m)},$$

$$\sum_{j=m}^{n} \|x_j - x_{j-1}\| \leq t_{n+1}^{(m)} \leq e_m = \lim_{n \to \infty} t_{n+1}^{(m)}, \quad f_m(e_m) = 0.$$

Furthermore, if the rest of the hypotheses of Theorem 7.5 are satisfied excluding (7.85) and (7.86), then

$$e_m \leq e_0,$$
$$t_n^{(m)} \leq t_n^{(0)},$$
$$t_{n+1}^{(m)} - t_n^{(m)} \leq t_{n+1}^{(0)} - t_n^{(0)}$$

and

$$e_m - t_n^{(m)} \leq e_0 - t_n^{(0)}$$

where the sequence $\left\{t_n^{(m)}\right\}$ is as defined in Theorem 7.6 with σ, Δ and a_n as defined in (iii) above.

Proof. We will make use of Theorem 7.4. Let us assume that $x_0, \ldots, x_n \in U(x_0, e_0)$ and $e_0 \geq \sum_{j=1}^{n} \|x_j - x_{j-1}\|$.

Using (7.136) and (7.57) we get

$\|A_0^{-1}(A_n - A_0)\|$
$\leq \|A_0^{-1}(A_n - F'(x_n))\| + \|A_0^{-1}(F'(x_n) - F'(x_0))\| + \|A_0^{-1}(F'(x_0) - A_0)\|$
$\leq \delta_n + \gamma \sum_{j=1}^{n} \|x_j - x_{j-1}\| + k_0 \|x_n - x_0\| + \delta_0$
$\leq \delta_0 + \delta_n + (k_0 + \gamma) \sum_{j=1}^{n} \|x_j - x_{j-1}\|$
$< \delta_0 + \delta + (k_0 + \gamma) e_0$
$\leq 1,$

by the choice of e_0.

By the Banach lemma on invertible operators the inverse of the linear operator A_n, $n = 1, 2, ...$ exists and is bounded. In particular

$$\left\|\left(A_0^{-1} A_n\right)^{-1}\right\| \leq a_n^{-1}, \quad n = 0, 1, 2, ...$$

where the sequence $\{a_n\}$, $n = 0, 1, 2, ...$ is as defined in (7.139).

With the above choices of σ, Δ and a_n it can easily be seen that for all $n = 0, 1, 2, ...,$

$$\left[1 - \left(\delta_0 + \delta_n + (k_0 + \gamma) \sum_{j=1}^{n} \|x_j - x_{j-1}\|\right)\right] + \sigma k_0 \sum_{j=1}^{n} \|x_j - x_{j-1}\| - \Delta$$
$$\geq \delta_n + \gamma \sum_{j=1}^{n} \|x_j - x_{j-1}\|.$$

We now observe that (7.136) is (7.61) and (7.138) is (7.65). The results (i)-(iii) follow now immediately from Theorem 7.4.

We now show that $(F, x_m) \in C_1(k_m, d_m, r_m)$ for each fixed m, $m = 0, 1, 2, ...$. We only need to show

$$d_m \leq \frac{1}{2}. \tag{7.141}$$

Using the hypothesis $r_m \leq r_0$, (7.138) and (7.139) we easily deduce

$$d_m \leq d_0. \tag{7.142}$$

By (7.137) and (7.141) we can obtain (7.140). The results (iv)-(vi) now follow immediately through Theorem 7.6.

That completes the proof of the theorem. □

Remark 7.3 *(a) Note that by defining e'_0 (or e'_m) earlier in the statement of this theorem we could have given a less restrictive existence statement based on Theorem 7.4. This can be a very useful type of result (see the discussion on p. 441 in [132] and the reference [109] for an application).*

But we wanted to leave our main concern, the convergence of (7.56) as uncluttered as possible.

(b) Using (7.139) we can easily see that the sequence $\{a_n\}$, $n = 0, 1, 2, ...$ is increasing if

$$\|x_{n+1} - x_n\| \leq \frac{\delta_n - \delta_{n+1}}{k_0 + \gamma}$$

provided that

$$\delta_{n+1} \leq \delta_n \quad \text{for all} \quad n = 0, 1, 2, ... \;.$$

We shall now compare the estimates (7.74) and (7.95) on a simple scalar example. Let us consider the operator F defined on the real line by

$$F(x) = \frac{1}{2}(x^2 - 16). \tag{7.143}$$

Let us choose

$$A_n = \left(\frac{2n+1}{n+1}\right) x_n,$$

$$a_n = \left(\frac{2n+1}{n+1}\right) \frac{x_n}{x_0}, \tag{7.144}$$

$$\sigma = \Delta = 1.$$

It can easily be seen that

$$k_m = a_m^{-1}$$

$$h_0 = .312162162.$$

The iteration (7.56) for solving (7.143) with the above values becomes

$$x_{n+1} = \frac{(3n+1)x_n^2 + 16(n+1)}{2(2n+1)x_n}. \tag{7.145}$$

Let us choose $x_0 = 3.7$. Using (7.144) and (7.145) we obtain

$$x_1 = 4.012162162, \ a_1 = 1.626552228$$
$$x_2 = 4.004066343, \ a_2 = 1.803633488$$
$$x_3 = 4.001627776, \ a_3 = 1.892661786$$
$$x_4 = 4.000697807, \ a_4 = 1.94628542$$
$$x_5 = 4.00031017, \ a_5 = 1.98213567$$
$$x_6 = 4.000140992, \ a_6 = 2.007792776$$
$$x_7 = 4.000065075, \ a_7 = 2.027060004$$
$$x_8 = 4.000030368, \ a_8 = 2.042057545$$
$$x_9 = 4.000014291, \ a_9 = 2.054061393$$
$$x_{10} = 4.000006769 \ a_{10} = 2.063885556.$$

We can now set $b = 1$ and $a = 2.162162162$. From (7.63) and (7.78) for $m = 5$ we obtain

$$s_0^{(0)} = .38707612$$

and

$$s_0^{(5)} = 3.35362132.10^{-4}.$$

Using (7.68) and (7.82) we obtain

$$t_0^{(0)} = 0, \qquad t_5^{(5)} = 0$$
$$t_1^{(0)} = .312162162, \ t_6^{(5)} = 1.69178.10^{-4}$$
$$t_2^{(0)} = .342116692, \ t_7^{(5)} = 2.530086635.10^{-4}$$
$$t_3^{(0)} = .357955489, \ t_8^{(5)} = 2.945504381.10^{-4}$$
$$t_4^{(0)} = .367610007, \ t_9^{(5)} = 3.151367587.10^{-4}$$
$$t_5^{(0)} = .37383762, \quad \text{and}$$
$$t_6^{(0)} = .377975492, \ t_{10}^{(5)} = 3.25339009.10^{-4}.$$
$$t_7^{(0)} = .38077424,$$
$$t_8^{(0)} = .382689577,$$
$$t_9^{(0)} = .384010911,$$
$$t_{10}^{(0)} = .384927846,$$

It can easily be seen that the hypotheses of Theorem 7.4 and 7.5 are satisfied with the above values and by observing that $x^* = 4$, we can tabulate the following results:

n	Actual error estimates	Error estimates (7.74)	Error estimates (7.95)
1	$1.2162162.10^{-2}$	$7.4913958.10^{-2}$	–
2	$4.066343.10^{-3}$	$4.4959428.10^{-2}$	–
3	$1.627776.10^{-3}$	$2.9120631.10^{-2}$	–
4	$6.97807.10^{-4}$	$1.9466113.10^{-2}$	–
5	$3.1017.10^{-4}$	$1.32385.10^{-2}$	–
6	$1.40992.10^{-4}$	$9.100628.10^{-3}$	$1.66184132.10^{-4}$
7	$6.5075.10^{-5}$	$6.30183.10^{-3}$	$8.2354.10^{-5}$
8	$3.0368.10^{-5}$	$4.386543.10^{-3}$	$4.0812.10^{-5}$
9	$1.4291.10^{-5}$	$3.065209.10^{-3}$	$3.8485.10^{-5}$
10	$6.769.10^{-6}$	$2.148274.10^{-3}$	$1.00224908.10^{-5}$

The above table indicates that our estimates are sharper that the corresponding ones obtained in [132]. Note, however, that both approaches use the same information.

All the above strongly recommend the usefulness of our results in numerical applications.

7.3 Newton-Like Methods Under Mild Differentiability Conditions

Consider an equation

$$F(x) = 0 \qquad (7.146)$$

where F is a nonlinear operator between two Banach space X, Y. A Newton-like method can be defined as any iterative method of the form

$$x_{n+1} = x_n - L_n^{-1} F(x_n), \quad n = 0, 1, 2, ...; \quad x_0 \text{ pre-chosen} \qquad (7.147)$$

for generating approximate solutions to (7.146). The $\{L_n\}$ denotes a sequence of invertible linear operators. This is plainly too general and what is really implicit in the title is that L_n should be a conscious approximation to $F'(x_n)$, since when $L_n = F'(x_n)$, the method reduces to the Newton-Kantorovich method. The convergence of (7.147) to a solution of (7.146) has been described already in 7.2 and the references there. The basic assumption made is that F is twice Fréchet-differentiable in some ball around the initial iterate. We relax this requirement to operators that are only once Fréchet-differentiable. An error analysis is also provided.

Our results can be compared with the ones obtained in [132], [160], [183], [227] when $L_n = F'(x_n)$, $n = 0, 1, 2, ...$. But, even then, they are proved to be stronger.

From now on we assume that F is once Fréchet-differentiable at a point $x \in X$ and note that $F'(x) \in L(X,Y)$, the space of bounded linear operators from X to Y.

Definition 7.5 We say that the Fréchet-derivative $F'(x)$ is Hölder continuous over a domain D if for some $c > 0$, $p \in [0,1]$

$$\|F'(x) - F'(y)\| \leq c \|x - y\|^p, \quad \text{for all} \quad x, y \in D. \tag{7.148}$$

We then say that $f'(\cdot) \in H_D(c,p)$.

Definition 7.6 Let t_0 and t' be non-negative real numbers and let g be a continuously differentiable real function on $[t_0, t_0 + t']$ and P be a continuously Fréchet-differentiable operator on

$$\overline{U}(x_0, t') \subset X$$

into Y. Then the equation

$$t = g(t)$$

will be said to majorize the equation

$$x = P(x) \quad \text{on} \quad U(x_0, t')$$

if

$$\|P(x_0) - x_0\| \leq g(t_0) - t_0$$

and

$$\|P'(x)\| \leq g'(t) \quad \text{for} \quad \|x - x_0\| \leq t - t_0 < t'.$$

We will need the following results whose proofs can be found in [183] and [227] respectively.

Lemma 7.1 *Let $\{x_k\}$, $k = 0, 1, 2, \dots$ be a sequence in X and $\{t_k\}$, $k = 0, 1, 2, \dots$ a sequence of non-negative real numbers such that*

$$\|x_{k+1} - x_k\| \leq t_{k+1} - t_k, \quad k = 0, 1, 2, \dots$$

and

$$t_k \to t^* < \infty \quad as \quad k \to \infty.$$

Then there exists a point $x^* \in X$ such that

$$x_k \to x^* \quad as \quad k \to \infty$$

and

$$\|x^* - x_k\| \leq t^* - t_k, \quad k = 0, 1, 2, \dots.$$

Lemma 7.2 *Let $F : X \to X$ and $D \subset X$. assume D is open and that $F'(\cdot)$ exists for every $x \in D$. Let D_0 be a convex set with $D_0 \subseteq D$ such that $F'(\cdot) \in H_{D_0}(c, p)$, then*

$$\|F(x) - F(y) - F'(x)(x - y)\| \leq \frac{c}{1 + p} \quad \text{for all} \quad x, y \in D_0.$$

We can now prove the following:

Proposition 7.2 *Let $F'(\cdot) \in H_{D_0}(c, p)$, where D_0 is the closure of an open convex set and $D_0 \subset D$. Assume that for every n with $\{x_k\} \subset D_0$, $k = 0, 1, 2, \dots, n$, there exists an invertible operator $L_n \in L(X, Y)$ and a positive real number d_n such that:*

$$\|L_n^{-1}\| \leq d_n^{-1}. \tag{7.149}$$

For a and $b > 0$, both independent of n with

$$a \geq \frac{2}{p(p+1)} \quad \text{if} \quad p \neq 0$$

and

$$a \geq 1 \quad \text{if} \quad p = 0,$$

assume:

$$\|F'(x_n) - L_n\| \leq d_n + ap \left(\sum_{j=1}^{n} \|x_j - x_{j-1}\| \right)^p - b, \quad n = 0, 1, 2, \dots. \tag{7.150}$$

Set

$$f(t) = \frac{ca}{p+1} t^{p+1} - bt + d_0 \|L_0^{-1} F(x_0)\|, \quad t \in [0, \infty) \tag{7.151}$$

and

$$t_{n+1} = t_n + \frac{f(t_n)}{d_n}; \quad t_0 = 0. \tag{7.152}$$

Then if $\{x_n\} \subset D_0$, (7.152) majorizes iteration (7.147).

Proof. We will use induction on n and Definitions 7.5 and 7.6. Note:

$$\|x_1 - x_0\| = \|L_0^{-1} F(x_0)\| = t_1 - t_0.$$

and assume that:

$$\{x_k\} \subset D_0, \quad k = 0, 1, 2, ..., n$$

and

$$\|x_j - x_{j-1}\| \leq t_j - t_{j-1} \quad \text{for} \quad j = 1, ..., n.$$

The iterate x_{n+1} is well defined since $F(x_n)$ and L_n^{-1} are. we will use the obvious estimate

$$\sum_{j=1}^{n} \|x_j - x_{j-1}\| \leq t_n$$

to compute:

$$\|x_{n+1} - x_n\| \leq \|L_n^{-1}\| \|F(x_n)\|$$
$$\leq d_n^{-1} [\|F(x_n) - F(x_{n-1}) - F'(x_{n-1})(x_n - x_{n-1})\|$$
$$+ \|L_{n-1} - F'(x_{n-1})\| \|x_n - x_{n-1}\|]$$
$$\geq d_n^{-1} \left[\frac{c}{p+1} \|x_n - x_{n-1}\|^{p+1} + (d_{n-1} + act_{n-1}^p) \|x_n - x_{n-1}\| \right]$$
$$\leq d_n^{-1} \left[\frac{c}{p+1} (t_n - t_{n-1})^{p+1} + (act_{n-1}^p - b)(t_n - t_{n-1}) + d_{n-1}(t_n - t_{n-1}) \right]$$
$$\leq d_n^{-1} \left[\frac{1}{2} f''(t_{n-1})(t_n - t_{n-1})^2 + f'(t_{n-1})(t_n - t_{n-1}) + f(t_{n-1}) \right].$$

(7.153)

But

$$f(t_n) = f(t_{n-1}) + f'(t_{n-1})(t_n - t_{n-1}) + \frac{1}{2} f''(t_{n-1})(t_n - t_{n-1})^2$$
$$+ \frac{1}{6} f'''(\tilde{t}_n)(t_n - t_{n-1})^3$$
$$\geq f(t_{n-1}) + f'(t_{n-1})(t_n - t_{n-1}) + \frac{1}{2} f''(t_{n-1})(t_n - t_{n-1})^2$$

since
$$f'''(\tilde{t}) = p(p-1)ca(\tilde{t}_n)^{p-2} \geq 0 \quad \text{for some} \quad \tilde{t}_n \in [t_{n-1}, t_n] \subset [0, \infty)$$
and
$$t_n \geq t_{n-1}$$
by the induction hypothesis.

Therefore (7.153) becomes
$$\|x_{n+1} - x_n\| \leq d_n^{-1} f(t_n) = t_{n+1} - t_n$$
and the induction is completed. $\square$

Proposition 7.3 *Let* $F'(\cdot) \in H_{D_0}(c, p)$, *where* D_0 *is the closure of an open convex set and* $D_0 \subset D$.

Assume:

(i) *Inequality (7.150) holds for* $n = 0$;

(ii)
$$\frac{\|F'(x_0) - L_0\|}{a_0} \leq \delta^1 < 1;$$

and

(iii) *the function* $\overline{f}(t)$ *defined by*

$$\overline{f}(t) = \frac{c}{p+1} t^{p+1} + (\delta^1 - 1) d_0 t + d_0 \|L_0^{-1} F(x_0)\|, \quad t \in [0, \infty) \tag{7.154}$$

has a minimum positive solution r_0' *such that* $U(x_0, r_0') \subset D_0$.

Then, F *has a unique solution* $x^* \in \overline{U}(x_0, r_0')$. *If* r_0' *is the unique fixed point of the equation*

$$g(t) = t + \frac{\overline{f}(t)}{d_0}$$

on some interval $[r_0', r_1']$, $r_0' \leq r_1'$, *then* x^* *is also unique in* $D_0 \cap U(x_0, r_1')$.

Moreover:

(a) *Iteration*

$$x_{n+1}' = x_n' - L_0^{-1} F(x_n');$$

converges to x^* *for* $\|x_0' - x_0\| < r_2 \leq r_1'$ *and* $U(x_0, r_2) \subset D_0$.

(b) *The following estimate is true:*

$$\|x'_n - x^*\| \leq |r'_0 - t'_n|$$

where $\{t'_n\}$ is generated by

$$t'_{n+1} = t'_n + \frac{\overline{f}(t'_n)}{d_0}.$$

Proof. define the nonlinear operator P on D_0 by

$$P(x) = x - L_0^{-1} F(x).$$

We will show that if $t' \in [r'_0, r'_1)$, then $g(t)$ majorizes $P(x)$ on $\overline{U}(x_0, t') \cap D_0$. We have

$$\|P(x_0) - x_0\| = \|L_0^{-1} F(x_0)\| = g(0) - 0.$$

Let x, t be such that $x \in \overline{U}(x_0, t') \cap D_0$ and $\|x - x_0\| \leq t < t'$. Then

$$\|P'(x)\| = \|I - L_0^{-1} P'(x)\| = \|L_0^{-1}((L_0 - F'(x_0)) + (F'(x_0) - F'(x)))\|$$
$$\leq \|L_0^{-1}\| (\|F'(x) - F'(x_0)\| + \|F'(x_0) - L_0\|)$$
$$\leq \delta^1 + c \frac{t^p}{d_0} = g'(t).$$

By hypothesis r'_0 is the unique fixed point of $g(t)$ in $[0, t']$ and $g(t) \leq t'$ with equality holding if and only if $t' = r'_0$.

The results now follows from the well known classical theorem on the existence and uniqueness of solutions of equations (7.146) via majorizing sequences given in Kantorovich [5, p. 697] or in Theorem 3.2.4.

We remark that if $\{t_n\}$ converges to t^*, then t^* is the least upper bound for $\sum_{j=1}^{n} \|x_j - x_{j-1}\|$, independent of n. Therefore, if we assume that $U(x_0, t^*) \subset D_0$, using Lemma 7.1 we obtain that $\{x_n\}$ exists and converges to a solution x^* of (7.146).

Usually we not wish to calculate the derivative of each L_n but instead use L_n in place of $L_{n+1}, ..., L_{n+q}$ and then calculate L_{n+q+1} and use it for $\bar{q}$ calculations. That is why, as in [5], we find it useful to define a nondecreasing sequence of non-negative real numbers $\{e_n\}$ such that

$$e_0 = 0$$

and

$$e_n = e_{n-1} \quad \text{or} \quad e_n = n.$$

We then replace (7.147) by the iteration

$$x_{n+1} = x_n - L_{e_n}^{-1} F(x_n), \quad n = 0, 1, 2, \ldots. \tag{7.155}$$

We can now prove a basic result.

Theorem 7.8 *Assume:*

(i) *The hypotheses of Proposition 7.2 hold;*
(ii) *the sequence $\{d_n\}$ is uniformly bounded above and*
(iii) *the hypothesis (iii) of Proposition 7.3 is true.*

Then (7.155) converges to a solution x^* of (7.146) according to

$$\|x_{n+1} - x^*\| \leq r_0 - t_n - d_{e_n}^{-1}(f(t_n)); \quad t_0 = 0, \quad n = 0, 1, 2, \ldots. \tag{7.156}$$

Moreover if

(iv) *the hypothesis on r_1' in Proposition 7.3 holds then x^* is a unique solution of (7.146) in $U(x_0, r_1') \cap D_0$.*

Proof. Let us define $C_n = L_{e_n}$ and $c_n = d_{e_n}$, $n = 0, 1, 2, \ldots$. The proof will be a consequence of the following steps. □

Step 1. We will show that $\{x_n\} \subset U(x_0, r_0) \subset D_0$ and that the rest of the hypotheses of Proposition 7.3 hold.

We easily note:

(i) (7.149) holds for C_n and c_n, $n = 0, 1, 2, \ldots$;
(ii) (7.150) holds by the choice of a and $d_n \leq n$.

we now estimate

$$\begin{aligned}
\|C_n - F'(x_n)\| &= \|L_{e_n} - F'(x_n)\| \\
&= \|(L_{e_n} - F'(x_{e_n})) + (F'(x_{e_n}) - F'(x_n))\| \\
&\leq \|L_{e_n} - F'(x_{e_n})\| + \|F'(x_{e_n}) - F'(x_n)\| \\
&\leq d_n + ap \left(\sum_{j=1}^{e_n} \|x_j - x_{j-1}\| \right)^p - b + c \|x_{e_n} - x_n\|^p \\
&\leq d_n + ap \left(\sum_{j=1}^{n} \|x_j - x_{j-1}\| \right)^p - b.
\end{aligned}$$

Let f be defined by Proposition 7.2, r_0 is the smallest positive zero of f. By Proposition 7.2

$$\|x_{n+1} - x_n\| \leq g_n(t_n) - t_n,$$

where the function $g_n(t)$ is defined on $[0, \infty)$ by

$$g_n(t) = t + \frac{f(t)}{c_n}, \quad n = 0, 1, 2, \ldots.$$

Assume that

$$t_n < r_0.$$

Then via the mean value theorem we can find $z_n \in (t_n, r_0)$ such that

$$r_0 - t_{n+1} = g_n(r_0) - g_n(t_n) = g'_n(z_n)(r_0 - t_n)$$
$$= \left[1 + \frac{f'(z_n)}{c_n}\right](r_0 - t_n)$$
$$= c_n^{-1}[c_n + caz_n^p - b](r_0 - t_n).$$

Using (7.150) we easily get

$$0 \leq c_n^{-1}[c_n + caz_n^p - b](r_0 - t_n) < r_0 - t_{n+1}$$
$$< c_n^{-1}[c_n + car_0^p - b](r_0 - t_n) \leq r_0 - t_n.$$

Therefore $\{t_n\}$ is bounded and convergent to some $t^* \leq r_0$. The estimate,

$$0 = \lim_{n \to \infty}(t_{n+1} - t_n) = \lim_{n \to \infty}\frac{f(t_n)}{c_n} \geq \lim_{n \to \infty}\frac{f(t_n)}{e}$$

where e denotes the uniform upper bound on $\{d_n\}$, implies that

$$f(t^*) = 0,$$

that is

$$t^* = r_0$$

and (7.156) holds.

Step 2. We show that $x^* = \lim_{n\to\infty} x_n$ is a solution of F. We have

$$\|C_n\| \leq \|F'(x_n)\| + c_n - b + ap\left(\sum_{j=1}^{n} \|x_j - x_{j-1}\|\right)^p$$
$$\leq \|F'(x_0)\| + c\|x_0 - x_n\|^p + c_n - b + apr_0^p$$
$$\leq \|F'(x_0)\| + (c + ap)r_0^p - b + e \equiv B.$$

Therefore the inequality

$$\|F(x_n)\| \leq \|C_n(x_{n+1} - x_n)\|$$
$$\leq \|C_n\| \|x_{n+1} - x_n\|$$
$$\leq B\|x_{n+1} - x_n\| \to 0 \quad \text{as} \quad n \to \infty$$

implies that $F(x^*) = 0$.

The unicity results will now hold if (ii) of Proposition 7.3 is satisfied and the hypothesis (iii) and (iv) of the theorem hold.

for $n = 0$ in (7.150) we obtain

$$0 < b \leq d_0 - \|F'(x_0) - L_0\|$$

that is

$$0 < \frac{b}{d_0} < 1 - \delta^1,$$

so (ii) of Proposition 7.3 is satisfied. It can easily be checked that $r_0' \leq r_0 \leq r_1'$ and the proof of the theorem is now completed.

We now state a theorem which seems to reduce to a minimum of the assumptions necessary to apply the majorant technique.

Theorem 7.9 *Let $F'(\cdot) \in H_{D_0}(c, p)$, where D_0 is the closure of an open convex set $D_0 \subset D$.*
Assume:

(i) *If $x_0 \in D$, let $L_0 \in L(X, Y)$ be an invertible operator with*

$$\|L_0^{-1} F(x_0)\| \leq \alpha$$

and

$$\|L_0^{-1}\| \leq \beta.$$

(ii) *There exist real numbers δ and γ such that if $\{x_k\} \subset D_0$, $k = 0, 1, 2, ..., n$ then*

$$\|L_{e_n} - F'(x_n)\| \leq \delta_n + \gamma \left(\sum_{j=1}^{n} \|x_j - x_{j-1}\| \right)^p ;$$

$$\delta_n \leq \delta_0 \equiv \delta, \quad n = 0, 1, 2, ...$$

(iii) *The following estimate holds:*

$$3\beta\delta < 1.$$

(iv) *There exists an interval $[0, \bar{r}_0]$ such that for $r \in [0, \bar{r}_0]$*

$$2\beta\delta + \beta(\gamma + c) r^p < 1,$$

and $U(x_0, r) \subset D_0$.

(v) *Assume that there exists a nonempty interval $[r_3, r_4] = [0, \bar{r}_0] \cap [r_0, r'_1]$ where r_0 is the small positive solution of (7.151).*
Then the following are true:

(a) *for $\bar{a}, \bar{b}$ such that:*

$$\bar{a} \geq \max\left(\frac{2\gamma + c}{p}, \frac{2}{p(p+1)} \right)$$

and

$$0 < \bar{b} \leq \frac{1 - 3\beta\delta}{\beta}.$$

$$\delta_n + \gamma \left(\sum_{j=1}^{n} \|x_j - x_{j-1}\| \right)^p \leq \bar{d}_n + \bar{a}p \left(\sum_{j=1}^{n} \|x_j - x_{j-1}\| \right)^p - \bar{b}$$

(7.157)

where

$$\bar{d}_n^{-1} = \beta \left[1 - \beta(\delta + \delta_n) - \beta(\gamma + c) \left(\sum_{j=1}^{n} \|x_j - x_{j-1}\| \right)^p \right]^{-1},$$

$n = 0, 1, 2, ...$ *and $\bar{d}_n \leq d_0$.*

(b) The sequence $\{x_n\}$ given by (7.155) exists in $U(x_0,r)$, $r_3 \leq r \leq r_4$ and converges to a unique solution x^* of (7.146) in $U(x_0,r_3)$. Moreover, the solution x^* is unique in $U(x_0,r_4)$.

(c) The following estimate holds if $t_1 = \alpha$

$$\|x_{n+1} - x^*\| \leq \overline{r}_0 - t_n - d_{e_n}^{-1}(f(t_n))$$

where f is given by (7.151) with

$$a = \overline{a}, \ b = \overline{b} \quad \text{and} \quad d_n = \overline{d}_n, \ n = 0, 1, 2, \ldots .$$

Proof. Assume $\{x_k\} \subset U(x_0, \overline{r}_0)$, $k = 0, 1, 2, \ldots, n$ and $\sum_{j=1}^{n} \|x_j - x_{j-1}\| < r$ with $r_3 \leq rr \leq r_4$.

We have

$$\|L_n - L_0\| \leq \|L_n - F'(x_n)\| + \|F'(x_n) - F'(x_0)\| + \|F'(x_0) - L_n\|$$

$$\leq \delta_n + \gamma \left(\sum_{j=1}^{n} \|x_j - x_{j-1}\|\right)^p + c\|x_n - x_0\|^p + \delta$$

$$\leq \delta_0 + \delta_n + \gamma(r)^p + cr^p$$

$$\leq 2\delta + (\gamma + c)r^p < \frac{1}{\beta}.$$

Therefore,

$$\|L_0^{-1} L_n - I\| \leq \beta(\delta + \delta_n) + \beta(\gamma + c)\left(\sum_{j=1}^{n} \|x_j - x_{j-1}\|\right)^p.$$

The Banach lemma can now be used to show that L_n^{-1} exists and is bounded in norm by $(\overline{d}_n)^{-1}$, $n = 0, 1, 2, \ldots$. Moreover, $\{\overline{d}_n\}$ is uniformly bounded by $\overline{d}_0$.

It is now easy to check that (7.66) is satisfied by the choice of $\overline{d}_n$, $\overline{a}$ and $\overline{b}$.

The rest of the theorem now follows from Theorem 7.8. □

When we solve equation (7.146) numerically using iteration (7.155) we generate instead of the sequence $\{x_n\}$ the perturbed sequence $\{z_n\}$ given by

$$z_{n+1} = z_n + \left[L_{e_n} + \overline{L}_{z_n}\right]^{-1} [F(z_n) + a_{z_n}] - q_{z_n}, \quad n = 0, 1, 2, \ldots,$$

assuming $z_0 = x_0$ and $\left[L_{e_n} + \overline{L}_{z_n}\right]^{-1}$ exists for $n = 0, 1, 2, \ldots$.

The problem of estimating the bound on $\|x_n - z_n\|$ when $L_{e_n} = F'(x_n)$ and under certain assumptions, basically on the norm of the linear operator L_{z_n} and on the norm of the elements of $\overline{L}_{z_n}$, a_{z_n} and q_{z_n} has been solved in [250].

Here we can easily prove the analog of Lemma 2 and Theorem 3 in [250] for the more general iteration (7.155). However, we leave that to the motivated reader and we show that the order of convergence of (7.147) when $L_n = F'(x_n)$ to a solution x^* of (7.146) is $1 + p$.

We then show that iteration (7.155) under appropriate choice of the L_{e_n} s converges to x^* with order $1 + 2p$.

This improves the results in [250] where the order of convergence is not given. If the second Fréchet derivative of F is bounded and $p = 1$, our results coincide with the one's in [243].

Proposition 7.4 *Let $L_{e_n} = F'(x_n)$ in (7.155). Then under the hypotheses of Theorem 7.9, the obtained solution x^* of (7.146) via iteration (7.155) is such that*

$$\|x_{n+1} - x^*\| \le k \|x_n - x^*\|^{1+p}, \quad n = 0, 1, 2, \ldots \quad (7.158)$$

where

$$k = \frac{c\overline{d}_0}{(p+1)^2}.$$

Proof. We have

$$x_{n+1} - x^* = x_n - x^* - F'(x_n)^{-1} F(x_n)$$
$$= (F'(x_n))^{-1} [F'(x_n)(x_n - x^*) - (F(x_n) - F(x^*))].$$

By taking norms in the above identity we obtain

$$\|x_{n+1} - x^*\|$$
$$\le \left\| F'(x_n)^{-1} \right\| \left\| \int_0^1 (F'(x_n) - F'(x_n + t(x^* - x_n))) \, dt \right\| \|x_n - x^*\|$$
$$\le \frac{c\overline{d}_0}{p+1} \|x_n - x^*\|^{p+1} \int_0^1 t^p \, dt$$
$$\le k \|x_n - x^*\|^{p+1}, \quad n = 0, 1, 2, \ldots.$$

□

Proposition 7.5 *Consider the iteration (7.155) for the solution of (7.146) given in the form*

$$\left.\begin{array}{l} y_n = x_n - F'(x_n)^{-1} F(x_n), \quad n = 0, 1, 2, ... \\ x_{n+1} = y_n - F'(x_n)^{-1} F(y_n), \quad n = 0, 1, 2, ... \end{array}\right\} \quad (7.159)$$

with x_0 pre-chosen.

Then under the hypotheses of Theorem 7.9 the obtained solution x^* of (7.146) via iteration (7.159) is such that

$$\|x_{n+1} - x^*\| \leq k_1 \|x_n - x^*\|^{1+2p}, \quad n = 0, 1, 2, ... \quad (7.160)$$

where,

$$k_1 = \frac{2^p \left(c\overline{d}_0\right)^2}{(p+1)^3}.$$

Proof. We have

$$\begin{aligned} x_{n+1} - x^* &= y_n - x^* - F'(x_n)^{-1} F(y_n) \\ &= F'(x_n)^{-1} \left[F'(x_n)(y_n - x^*) - (F(y_n) - F(x^*))\right]. \end{aligned}$$

By taking norms in the above identity we obtain

$$\begin{aligned} \|x_{n+1} - x^*\| &\leq \left\|F'(x_n)^{-1}\right\| \left\|\int_0^1 (F'(x_n) - F'(x^* + t(y_n - x^*))) \, dt\right\| \\ &\quad \cdot \|y_n - x^*\| \\ &\leq \frac{c\overline{d}_0}{p+1} \|(x_n - x^*) + t(y_n - x^*)\|^p \|y_n - x^*\| \\ &\leq \frac{c\overline{d}_0}{p+1} (\|x_n - x^*\| + \|y_n - x^*\|)^p \|y_n - x^*\| \\ &\leq \frac{2^p c\overline{d}_0}{p+1} \|x_n - x^*\|^p \|y_n - x^*\| \quad (\text{since } \|y_n - x^*\| \leq \|x_n - x^*\|). \end{aligned}$$

$$(7.161)$$

similarly,

$$\begin{aligned} y_n - x^* &= x_n - x^* - F'(x_n)^{-1} F(x_n) \\ &= F'(x_n)^{-1} \left[F'(x_n)(x_n - x^*) - (F(x_n) - F(x^*))\right]. \end{aligned}$$

Therefore,

$$\|y_n - x^*\| \leq \left\|F'(x_n)^{-1}\right\| \left\|\int_0^1 (F'(x_n) - F'(x^* + t(x_n - x^*)))\, dt\right\|$$
$$\cdot \|x_n - x^*\|$$
$$\leq \frac{c\overline{d}_0}{p+1} \left\|\int_0^1 (1-t)^p\, dt\right\| \cdot \|x_n - x^*\|^{p+1}$$
$$\leq \frac{c\overline{d}_0}{(p+1)^2} \|x_n - x^*\|^{p+1}. \qquad (7.162)$$

Finally, by (7.161) and (7.162) we obtain

$$\|x_{n+1} - x^*\| \leq \frac{2^p c\overline{d}_0}{p+1} \|x_n - x^*\|^p \frac{c\overline{d}_0}{(p+1)^2} \|x_n - x^*\|^{p+1}$$
$$\leq k_1 \|x_n - x^*\|^{1+2p}. \qquad \square$$

Definition 7.7 Define the efficiency X of an iteration $\{x_n\}$ for solving (7.146) by

$$X = \frac{\ell_n k}{T}, \qquad (7.163)$$

where k is the order of convergence of $\{x_n\}$ and T denotes "time per step". That is, the number of function evaluations required to compute each iterate x_n, $n = 1, 2, \ldots$.

Let E_1, E_2 denote the efficiencies of iterations (7.147) and (7.159) respectively. Take $p = \frac{1}{2}$ then

$$E_1 = \frac{\ell_n \frac{3}{2}}{2},$$
$$E_2 = \frac{\ell_n 2}{3}$$

and

$$E_1 < E_2.$$

7.4 Perturbed Newton-Like Methods

Let F be a nonlinear operator mapping some subset D of a real Banach space X into a subset of a real Banach space Y. The most popular methods for approximating solutions x^* of the equation

$$F(x) = 0 \qquad (7.164)$$

are the so-called Newton-like methods of the form

$$x_{n+1} = x_n - A(x_n)^{-1} F(x_n), \quad n = 0, 1, 2, \dots . \qquad (7.165)$$

Here $x_0 \in D$ is given and $\{A(x_n)\}$, $n = 0, 1, 2, \dots$ denotes a sequence of invertible linear operators.

Yamamoto [284] has unified the study of finding sharp error bounds for Newton-like methods of the form (7.165) under Kantorovich type assumptions. He obtains results that improve error bounds obtained before by Rheinboldt [248], Dennis [132], Miel [202], Moret [209], Potra [235]–[240]. Relevant and interesting results for the local and semilocal case can be found in Cătinaș [110]–[112].

The results obtained by the above authors however have great theoretical but little practical value, since the sequence generated by (7.165) can rarely be computed exactly.

In this section we find it useful to consider that the iterative procedure (7.165) is perturbed. We suppose that all the elements contained in the construction of these procedures are known only approximately. Moreover we suppose that at each step the solution of the respective linear system is also performed approximately.

In particular, we consider the iterative procedures corresponding to (7.165) to be of the form

$$\widetilde{x}_{n+1} = \widetilde{x}_n - (A(\widetilde{x}_n) + L_n)^{-1}(F(\widetilde{x}_n) + y_n) + z_n, \quad \widetilde{x}_0 = x_0, \quad n = 1, 2, \dots \qquad (7.166)$$

$$\widetilde{x}_{n+1} = \widetilde{x}_n - (F'(x_0) + L_0)^{-1}(F(\widetilde{x}_n) + y_n) + z_n, \quad \widetilde{x}_0 = x_0, \quad n = 0, 1, 2, \dots \qquad (7.167)$$

where

$$L_n \in L(X, Y), \quad y_n \in Y \quad \text{and} \quad z_n \in X.$$

We provide upper bounds on the distances $\|x_n - \widetilde{x}_n\|$ and $\|\widetilde{x}_n - x^*\|$.

Finally, our results are applied to an "ill conditioned" scalar equation considered also in [192].

To make the section self-contained we will reproduce some of the results obtained in [284] to fit our purposes.

Let F, D and x_0 be defined and consider the iterative procedure (7.165). According to Dennis [132], Schmidt [252] and Yamamoto [284], we assume the following:

$$\left\|A(x_0)^{-1}(F'(x) - F'(y))\right\| \le K \|x - y\|, \quad x, y \in D, \quad K > 0, \quad (7.168)$$

$$\left\|A(x_0)^{-1}(A(x) - A(x_0))\right\| \le L \|x - x_0\| + \ell, \quad x \in D, \quad L \ge 0, \quad \ell \ge 0, \quad (7.169)$$

$$\left\|A(x_0)^{-1}(F'(x) - A(x))\right\| \le M \|x - x_0\| + m,$$
$$x \in D, \quad M \ge 0, \quad m \ge 0, \quad (7.170)$$

$$\ell + m < 1, \quad \sigma = \max\left(1, \frac{L+M}{K}\right), \quad F(x_0) \ne 0, \quad (7.171)$$

$$\eta = \left\|A(x_0)^{-1} F(x_0)\right\|, \quad h = \sigma K \eta / (1 - \ell - m)^2 \le \frac{1}{2}, \quad (7.172)$$

$$t^* = (1 - \ell - m)\left(1 - \sqrt{1 - 2h}\right) / (\sigma K), \quad (7.173)$$

$$t^{**} = \left(1 - m + \sqrt{(1 - m)^2 - 2K\eta/K,}\right), \quad (7.174)$$

$$\overline{S} = \overline{S}(x_1, t^* - \eta) \subseteq D. \quad (7.175)$$

Under these assumptions, define the sequence $\{t_n\}$ by

$$t_0 = 0, \quad t_{n+1} = t_n + f(t_n)/g(t_n), \quad n = 0, 1, 2, \dots \quad (7.176)$$

where

$$f(t) = \frac{1}{2}\sigma K t^2 - (1 - \ell - m)t + \eta \quad (7.177)$$

and

$$g(t) = 1 - \ell - Lt. \quad (7.178)$$

We can now state the following result [284, Th. 4.1].

Theorem 7.10 *With the above notation and assumptions, we have the following:*

(a) *The iterative procedure (7.165) is well defined for every $n \ge 0$, $x_n \in S$ for $n \ge 1$ and $\{x_n\}$ converges to a solution $x^* \in \overline{S}$ of the equation (7.164).*

(b) *The solution x^* is unique in*

$$\overline{S} = \begin{cases} S(x_0, t^{**}) \cap D & (if \ 2K\eta < (1-m)^2) \\ \overline{S}(x_0, t^{**}) \cap D & (if \ 2K\eta < (1-m)^2). \end{cases} \quad (7.179)$$

Moreover, the following estimates are true:

$$\|x_n - x^*\| \leq t^* - t_n, \quad n = 0, 1, 2, \ldots \quad (7.180)$$

where the nonnegative sequence $\{t_n\}$, $n = 0, 1, 2, \ldots$ is increasingly converging to t^*.

We will finally need the result [189, Cor. 4.1.1].

Theorem 7.11 *Consider the modified Newton method*

$$x_{n+1} = x_n - F'(x_0)^{-1} F(x_n), \quad n = 0, 1, 2, \ldots \quad (7.181)$$

where we assume the following:

$$x_0 \in D, \quad F'(x_0)^{-1} \ exists, \quad (7.182)$$

$$\left\| F'(x_0)^{-1} (F'(x) - F'(x_0)) \right\| \leq K \|x - x_0\|, \quad x \in D, \quad (7.183)$$

$$\eta = \left\| F'(x_0)^{-1} F(x_0) \right\| > 0, \quad h = K\eta \leq \frac{1}{2} \quad (7.184)$$

$$\widetilde{t}^* = \left(1 - \sqrt{1 - 2h}\right)/K, \quad \widetilde{t}^{**} = \left(1 + \sqrt{1 - 2h}\right)/K, \quad (7.185)$$

$$\overline{S}_1 = \overline{S}_1\left(x_1, \widetilde{t}^* - \eta\right) \subseteq D. \quad (7.186)$$

Then:

(a) *the iterative procedure (7.181) is well defined for every $n \geq 0$, $x_n \in S_1$ for $n \geq 1$ and $\{x_n\}$ converges to a solution of equation x^* (7.1).*

(b) *The solution x^* is unique in*

$$\widetilde{S} = \begin{cases} S_1(x_0, \widetilde{t}^{**}) \cap D & (if \ 2h < 1) \\ \widetilde{S}_1(x_0, \widetilde{t}^{**}) \cap D & (if \ 2h = 1). \end{cases} \quad (7.187)$$

Moreover, the following estimates are true

$$\|x_n - x^*\| \leq \widetilde{t}^* - \widetilde{t}_n, \quad n = 0, 1, 2, \ldots \quad (7.188)$$

where the nonnegative sequence $\{\tilde{t}_n\}$, $n = 0, 1, 2, \ldots$ is given by

$$\tilde{t}_0 = 0, \quad \tilde{t}_{n+1} = \frac{1}{2}K\tilde{t}_n^2 + \eta, \quad n = 0, 1, 2, \ldots \quad (7.189)$$

and is increasingly converging to $\tilde{t}^*$.

In what follows we shall suppose that there exist three possible numbers ϵ_1, ϵ_2, and ϵ_3 such that

$$\|y_n\| \leq \epsilon_1, \quad \|L_n\| \leq \epsilon_2, \quad \|z_n\| \leq \epsilon_3 \quad \text{for all} \quad n \in N. \quad (7.190)$$

In Theorems 7.10 and 7.11 we have seen that the sequences produced by the iterative procedures (7.165) and (7.181) remain in the open balls S and S_1 respectively and consequently in D. However, in the perturbed case we have to suppose that F is defined on the balls $S^* = S^*(x_0, r)$ and $S_1^* = S^*(x_0, r_1)$ respectively with $r > t^{**}$, $r_1 > \tilde{t}^{**}$ and $S^*, S_1^* \subset D$. Set $S_2 = S^* \cup S_1^*$.

In the perturbed case it is more convenient to suppose that the following conditions are satisfied for all $x, y \in S_2$:

$$\left\|A(x)^{-1}(F'(x) - F'(y))\right\| \leq K \|x - y\|, \quad (7.191)$$

$$\left\|A(x)^{-1}(A(x) - A(x_0))\right\| \leq L \|x - x_0\| + \ell, \quad (7.192)$$

and

$$\left\|A(x)^{-1}(F'(y) - A(y))\right\| \leq M \|y - x_0\| + m. \quad (7.193)$$

These conditions are more restrictive than conditions (7.168), (7.169) and (7.170) respectively but they are satisfied by the usual examples of approximation.

In order to assure the invertibility of $A(\tilde{x}_n) + X_n$ for all $n = 0, 1, 2, \ldots$ we shall suppose that $A(x)$ is invertible for all $x \in S_2$ and that the norms $\left\|A(x)^{-1}\right\|$ are bounded. More precisely, in the perturbed case we shall impose one, or both, of the following conditions:

(**C$_1$**) The open ball S_2 is included into the domain of definition of F and conditions (7.191)-(7.193) hold for all $x, y \in S_2$.

(**C$_2$**) The linear operator $A(x)$ is invertible for all $x \in S_2$ and there exists a positive number a such that

$$a^{-1} \geq \sup\left\{\left\|A(x)^{-1}\right\|; x \in S_2\right\}. \quad (7.194)$$

We can now prove the following theorem concerning the iterative procedure (7.166).

Theorem 7.12 *Assume:*

(a) *The hypotheses of Theorem 7.10 are satisfied and let $\{x_n\}$, $n = 0, 1, 2, \ldots$ be the sequence generated by (7.165);*
(b) *the conditions (C_1) and (C_2) are satisfied; and*
(c) *the inequalities*

$$(1-\beta)^2 - 4\alpha\gamma \geq 0, \quad a > \epsilon_2, \tag{7.195}$$

$$\beta \leq 1, \tag{7.196}$$

$$\delta = \frac{1 - \beta - \sqrt{(1-\beta)^2 - 4\alpha\gamma}}{2\alpha} \leq r - t^* + \eta. \tag{7.197}$$

are satisfied, where

$$\alpha = \frac{a(K + 2L)}{2(a - \epsilon_2)}, \tag{7.198}$$

$$\beta = [(K+L)\epsilon_0 + m + \ell + \epsilon_1^* + \epsilon_2^* + L\epsilon_0] \frac{a}{a - \epsilon_2}, \quad \epsilon_0 = t^* - t_0,$$

$$\epsilon_1^* = \frac{\epsilon_1}{a}, \quad \epsilon_2^* = \frac{\epsilon_2}{a}, \tag{7.199}$$

and

$$\gamma = \frac{a}{a - \epsilon_2}(2L + \epsilon_0 + \ell + \epsilon_2^*)\epsilon_0 + \epsilon_3. \tag{7.200}$$

Then the iterative procedure (7.166) is well defined and for each $n \in N$ we shall have the estimates

$$\|x_n - \tilde{x}_n\| \leq w_n \leq \delta, \tag{7.201}$$

where the real sequence $\{w_n\}$ is given by:

$$w_{n+1} = \alpha_{n+1} w_n^2 + \beta_{n+1} w_n + \gamma_{n+1}, \quad w_0 = 0, \quad n = 0, 1, 2, \ldots \tag{7.202}$$

with

$$\alpha_n = \frac{a(2L + K)}{2(a - \epsilon_2)}, \tag{7.203}$$

$$\beta_n = [(K+L)(t_n - t_0) + m + \ell + \epsilon_2^* + L(t_{n+1} - t_n)] \frac{a}{1 - \epsilon_2}, \quad (7.204)$$

and

$$\gamma_n = (2L(t_n - t_0) + \epsilon_1^* + \epsilon_2^* + \ell)(t_{n+1} - t_n)\frac{a}{a - \epsilon_2} + \epsilon_3, \quad n = 1, 2, \ldots. \quad (7.205)$$

Proof. For $n=0$ the inequalities (7.201) are trivially satisfied as equalities. Suppose they are satisfied for $n = 0, 1, 2, \ldots, k, k \geq 0$. From (7.201) it follows that $\widetilde{x}_k \in S_2$. In this case, condition (C_2) implies, according to the Banach lemma on invertible operators that the linear operator $A(\widetilde{x}_k) + X_k$ is invertible and

$$7.206 \left\| (A(\widetilde{x}_k) + X_k)^{-1} \right\| = \left\| \left(I + A(\widetilde{x}_k)^{-1} L_k \right)^{-1} A(\widetilde{x}_k)^{-1} \right\|$$

$$\leq \left\| \left(I + A(\widetilde{x}_k)^{-1} L_k \right)^{-1} \right\| \left\| A(\widetilde{x}_k)^{-1} \right\|$$

$$\leq (a - \epsilon_2)^{-1}. \quad (7.206)$$

From (7.165) and (7.166) we obtain the identity

$$\widetilde{x}_{k+1} - x_{k+1}$$
$$= \left(I + A(\widetilde{x}_k)^{-1} L_k \right)^{-1} A(\widetilde{x}_k)^{-1} \{ [F(x_k) - F(\widetilde{x}_k) - F'(x_k)(x_k - \widetilde{x}_k)]$$
$$+ [(F'(x_k) - A(\widetilde{x}_k))(x_k - \widetilde{x}_k)] + y_k + [L_k(\widetilde{x}_k - x_k)]$$
$$+ \left[(A(\widetilde{x}_k) - A(x_k)) A(x_k)^{-1} F(x_k) \right] + \left[L_k A(x_k)^{-1} F(x_k) \right] \} + z_k.$$
$$(7.207)$$

Using (7.191), (7.201) and (7.206) we obtain

$$\left\| A(\widetilde{x}_k)^{-1} [F(x_k) - F(\widetilde{x}_k) - F'(x_k)(x_k - \widetilde{x}_k)] \right\| \leq \tfrac{1}{2} K \left\| x_k - \widetilde{x}_k \right\|$$
$$\leq \tfrac{1}{2} K w_k^2. \quad (7.208)$$

By (7.191)-(7.193) and (7.201) we get

$$\left\|A(\widetilde{x}_k)^{-1}(F'(x_k)-A(\widetilde{x}_k))\right\|$$
$$=\|[(F'(x_k)-F'(x_0))+(F'(x_0)-A(x_0))$$
$$+(A(x_0)-A(\widetilde{x}_k))](x_k-\widetilde{x}_k)\|$$
$$\leq [K\|x_k-x_0\|+L\|\widetilde{x}_k-x_0\|+\ell+m]\|x_k-\widetilde{x}_k\|$$
$$\leq [K(t_k-t_0)+Lw_k+L(t_k-t_0)+m+\ell]w_k$$
$$\leq (K\epsilon_0+Lw_k+L\epsilon_0+m+\ell)w_k. \qquad (7.209)$$

From (7.190) and (7.201) we get

$$\left\|A(\widetilde{x}_k)^{-1}L_k(\widetilde{x}_k-x_k)\right\| \leq \epsilon_2^*\|\widetilde{x}_k-x_k\| \leq \epsilon_2^*w_k. \qquad (7.210)$$

Using (7.165), (7.192) we obtain

$$\left\|A(\widetilde{x}_k)^{-1}(A(\widetilde{x}_k)-A(x_k))\right\|$$
$$=\left\|A(\widetilde{x}_k)^{-1}[(A(\widetilde{x}_k)-A(x_0))+(A(x_0)-A(x_k))](x_k-x_{k+1})\right\|$$
$$\leq (L(\|\widetilde{x}_k-x_0\|+\|x_k-x_0\|)+2\ell)\|x_k-x_{k+1}\|$$
$$\leq (Lw_k+2L(t_k-t_0)+2\ell)(t_{k+1}-t_k)$$

(since, by Theorem 7.10, $\|x_k-x_{k+1}\| \leq t_{k+1}-t_k \leq t^*-t_0 = \epsilon_0$)

$$\leq (Lw_k+2L(t^*-t_0)+2\ell)(t^*-t_0). \qquad (7.211)$$

Finally, by (7.175) and (7.190)

$$\left\|A(\widetilde{x}_k)^{-1}L_kA(x_k)^{-1}F(x_k)\right\| \leq \epsilon_2^*\|x_k-x_{k+1}\| \leq \epsilon_2^*(t_{k+1}-t_k)$$
$$\leq \epsilon_2^*\epsilon_0. \qquad (7.212)$$

With these majorizations in (7.207), using (7.206), (7.190), (7.195)-(7.200), and (7.202) we can easily obtain that

$$\|\widetilde{x}_{k+1}-x_{k+1}\| \leq w_{k+1} \leq \alpha\delta^2+\beta\delta+\gamma = \delta. \qquad (7.213)$$

That completes the induction and the proof of the theorem. □

Concerning the perturbed iteration (7.167) we have:

Theorem 7.13 *Assume:*

(a) The hypotheses of Theorem 7.11 are satisfied and let $\{x_n\}$, $n = 0, 1, 2, \ldots$ be the sequence generated by (7.181);
(b) the condition (C_1) is satisfied and (7.191) with $A(x) = F'(x)$ and $y = x_0$.
(c) The inequalities

$$d > e_2, \text{ with } 0 < d \leq \left\|F'(x_0)^{-1}\right\|^{-1}, \qquad (7.214)$$

$$(1 - \beta_1)^2 - 4\alpha_1\gamma_1 \geq 0, \qquad (7.215)$$

$$\beta_1 \leq 1, \qquad (7.216)$$

$$\delta_1 = \frac{1 - \beta_1 - \sqrt{(1 - \beta_1)^2 - 4\alpha_1\gamma_1}}{2\alpha_1} \leq r_1 - \widetilde{t}^* + \eta \qquad (7.217)$$

are satisfied, where

$$g\alpha_1 = \frac{dK}{2(d - \epsilon_2)}, \qquad (7.218)$$

$$\beta_1 = \frac{\epsilon_0 dK + \epsilon_2}{d - \epsilon_2}, \qquad (7.219)$$

and

$$\gamma_1 = \frac{\epsilon_2\epsilon_0 + \epsilon_1 + \epsilon_3(d - \epsilon_2)}{d - \epsilon_2}. \qquad (7.220)$$

Then the iterative procedure (7.181) is well defined and for each $n \in \mathbb{N}$ we shall have the estimates

$$\|x_n - \widetilde{x}_n\| \leq S_n \leq \delta_1, \qquad (7.221)$$

where the real sequence $\{S_n\}$ is given by:

$$S_{n+1} = \alpha^*_{n+1} S_n^2 + \beta^*_{n+1} S_n + \gamma^*_{n+1}, \quad S_0 = 0, \quad n = 0, 1, 2, \ldots \qquad (7.222)$$

with

$$\alpha^*_n = \frac{dK}{2(d - \epsilon_2)}, \qquad (7.223)$$

$$\beta^*_n = \frac{dK(t_n - t_0) + \epsilon_2}{d - \epsilon_2}, \qquad (7.224)$$

and

$$\gamma^*_n = \frac{\epsilon_2^*(t_{n+1} - t_n) + \epsilon_1}{d - \epsilon_2} + \epsilon_3, \quad n = 1, 2, \ldots . \qquad (7.225)$$

Proof. By the Banach lemma on invertible operators it follows that the linear operator $F'(x_0) + L_0$ is invertible and

$$\left\|(F'(x_0) + L_0)^{-1}\right\| = \left\|\left(I + F'(x_0)^{-1}\right)^{-1} F'(x_0)\right\|$$
$$\leq \left\|\left(I + F'(x_0)^{-1} L_0\right)^{-1}\right\| \left\|F'(x_0)^{-1}\right\| \leq (d - \epsilon_2)^{-1}. \qquad (7.226)$$

This fact, together with the remark that (7.217) and (7.221) imply $\widetilde{x}_n \in S_2$, shows us that if (7.221) is satisfied, then the iterative procedure (7.181) makes sense.

We will show that (7.221) holds for all $n = 0, 1, 2, \ldots$. For $n = 0$ the inequalities (7.221) are trivially satisfied as equalities. Suppose they are satisfied for $n = 0, 1, \ldots, k$. We shall prove that they hold for $n = k+1$ too.

From (7.167) and (7.181) we obtain the identity

$$x_{k+1} - \widetilde{x}_{k+1}$$
$$= \left(I + F'(x_0)^{-1} L_0\right)^{-1} F'(x_0)^{-1} \{[F(\widetilde{x}_k) - F(x_k) - F'(x_0)(\widetilde{x}_k - x_k)]$$
$$+ y_k + L_0(x_n - \widetilde{x}_n) - L_0 F'(x_0)^{-1} F(x_n)]\} - z_n. \qquad (7.227)$$

By taking norms in the above inequality and using (7.191), (7.227), (7.190), (7.181), (7.188), (7.189), (7.214)-(7.225), we obtain as in (7.213) that

$$\|x_{k+1} - \widetilde{x}_{k+1}\| \leq \frac{dK}{2(d - \epsilon_2)} \left(\|x_0 - x_k\| + \|x_0 - \widetilde{x}_k\|\right) \|x_k - \widetilde{x}_k\|$$
$$+ \frac{1}{2 - \epsilon_2} \left[\epsilon_2 \|x_k - \widetilde{x}_k\| + \epsilon_2 \|x_k - x_{k+1}\| + \epsilon_1\right] + \epsilon_3$$
$$\leq \frac{dK}{2(d - \epsilon_2)} \left(2\|x_0 - x_k\| + \|x_k - \widetilde{x}_k\|\right) \|x_k - \widetilde{x}_k\|$$
$$+ \frac{1}{d - \epsilon_2} \left[\epsilon_2 \|x_k - \widetilde{x}_k\| + \epsilon_2 \|x_k - x_{k+1}\| + \epsilon_1\right] + \epsilon_3$$
$$\leq \frac{dK}{2(d - \epsilon_2)} \left[2\left(\widetilde{t}_k - \widetilde{t}_0\right) + S_k\right] S_k + \frac{1}{d - \epsilon_2} \left[S_k + \left(\widetilde{t}_{k+1} - \widetilde{t}_k\right) + \epsilon_1\right] + \epsilon_3$$
$$= S_{k+1} \leq \frac{dK}{2(d - \epsilon_2)} \left[2\left(\widetilde{t}^* - \widetilde{t}_0\right) + \delta_1\right] \delta_1 + \frac{1}{d - \epsilon_2} \left(\delta_1 + \widetilde{t}^* - \widetilde{t}_0 + \epsilon_1\right) + \epsilon_3$$
$$= \delta_1. \qquad (7.228)$$

That completes the induction and the proof of the theorem. $\square$

The following result follows immediately from Theorems 7.10-7.13.

Corollary 7.1 *Under the hypotheses of Theorems 7.12 and 7.13 the following estimates hold for $n = 0, 1, 2, ...$*

$$\|\tilde{x}_n - x^*\| \leq w_n + t^* - t_n \tag{7.229}$$

and

$$\|\tilde{x}_n - x^*\| \leq S_n + \tilde{t}^* - \tilde{t}_n \tag{7.230}$$

for iterations (7.166) and (7.167) respectively.

Finally, we remark that the approach employed here applies for the rest of the error bounds obtained in [284, pp. 550, 555].

We shall apply Theorem 7.13 to an "ill conditioned" example proposed by Wilkinson [280] and considered also by Lancaster [192].

Example 7.1 Consider solving iteratively the quadratic equation

$$x^2 - 2.028888800x + 1.02876900 = 0 \tag{7.231}$$

using a computer characterized by the accuracy $\epsilon_1 = \epsilon_2 = \epsilon_3 = .5 \times 10^{-9}$. Starting with $z_0 = 1.032567321$ and using (7.181) we get

$$z_1 = 1.032567323$$
$$z_2 = 1.032567326$$
$$z_3 = 1.032567329$$
$$z_n = z_3, \quad n \geq 3.$$

If we take $x_0 = z_2$ and $\max(r, r_1) = .03624585$, then we can easily obtain from (7.214)-(7.225) $K = 55.17872776$, $\eta = 2.75893638.10^{-9}$, $h = 1.52234599.10^{-7}$, $\tilde{t}^* = 2.75903425.10^{-9}$, $\tilde{t}^{**} = .0362445849$, $\epsilon_0 = \tilde{t}^*$, $d = .036245852$, $\alpha_1 = 27.5893654$, $\beta_1 = 1.660346909.10^{-7}$, $\gamma_1 = 1.42946827.10^{-8}$ and $\delta_1 = 11.3630739.10^{-9}$.

We want to find an estimate for the distance $|z_3 - x^*|$. The hypotheses of Theorem 7.13 being satisfied we can use the Corollary.

From (7.230) we get $|z_3 - x^*| \leq 21 \times 10^{-9}$.

Taking advantage of the fact that we know that the sequence $\{z_n\}$ becomes constant beginning with $n = 3$, we easily obtain that $|z_3 - x^*| \leq 11.3630739.10^{-9}$. This is very close to reality because $x^* = 1.032567332$ is the solution of equation (7.231).

7.5 Projection Methods and Inexact Newton-like Iterations

We consider the inexact Newton-like method

$$x_{n+1} = x_n + y_n, \quad PA(x_n)y_n = -(F(x_n) + G(x_n)) + r_n, \quad n \geq 0 \quad (7.232)$$

for some $x_0 \in U(x_0, R)$, $R > 0$, to approximate a solution x^* of the equation

$$F(x) + G(x) = 0, \quad \text{in } \overline{U}(x_0, R). \quad (7.233)$$

Here $A(x)$, F, G denote operators defined on the closed ball $\overline{U}(x_0, R)$ with center x_0 and radius R, of a Banach space X with values in a Banach space Y, whereas r_n are suitable points in Y. The operator $A(x)(\cdot)$ is linear and approximates the Fréchet derivatives of F at $x \in U(x_0, R)$. P is a projection operator in X such that $P^2 = P$. we will assume that for any $x, y \in \overline{U}(x_0, R)$ with $0 \leq \|x - y\| s < R - r$,

$$\|P(F'x + t(x - y)) - A(x)\| \leq B_1(r, \|x - x_0\| + t\|y - x\|), \quad t \in [0, 1] \quad (7.234)$$

and

$$\|P(G(x) - G(y))\| \leq B_2(r, \|x - y\|). \quad (7.235)$$

The functions $B_1(r, r')$ and $B_2(r, r')$ defined on $[0, R] \times [0, R]$ and $[0, R] \times [0, R - r]$ are respectively nonnegative, continuous and nondecreasing functions of two variables. Moreover B_2 is linear in the second variable.

Note that Newton method, the modified Newton method and the secant method are special cases of (7.232) with $A(x_n) = F'(x_n)$, $A(x_n) = F'(x_0)$ and $A(x_n) = S(x_n, x_{n-1})$ respectively (when $P = I$ the identity operator on X, or not).

If we take

$$w(r') + c \quad (7.236)$$

and

$$e(r'), \quad (7.237)$$

where w, e are nonnegative, nondecreasing functions on $[0, R - r]$, to be the right hand sides of (7.234) and (7.235) respectively, then we obtain conditions similar but not identical to the Zabrejko-type assumptions [295] considered by Chen and Yamamoto [116]. They provided sufficient conditions for the convergence of the sequence $\{x_n\}$, $n \geq 0$ generated by (7.232) to solution x^* of equation (7.233), when $r_n = 0$, $n \geq 0$ and $P = I$.

Moret [212] also studied (7.232), when $G = 0$ and condition (7.236) is satisfied.

In this section we will derive a criterion for controlling the residuals r_n in such a way that the convergence of the sequence $\{x_n\}$, $n \geq 0$ to a solution x^* of equation (7.233) is ensured.

We believe that conditions of the form (7.234)-(7.235) are useful not only because we can treat a wider range of problems than before, but it turns out that under natural assumptions we can find better error bounds on the distances $\|x_n - x^*\|$, $n \geq 0$.

The iterates $\{x_n\}$ generated by (7.232) when $P = I$ can rarely be computed in infinite dimensional spaces. It may be difficult or impossible to compute the inverses $A(x_n)$ at each step. It is easy to see however that the solution of equation (7.232) reduces to solving certain operator equations in the space E_p. If, moreover E_p is a finite dimensional space of dimension N, we obtain a system of linear algebraic equations of at most order N.

We will need the following proposition.

Proposition 7.6 Let $a \geq 1$, $\sigma > 0$, $0 \leq \mu < 1$, $0 \leq \rho < R$, $s > 0$ be real constants such that the equation

$$\varphi(t) := \partial\sigma \left[\int_0^t B_1(R, \rho + \theta) \, d\theta + B_2(R, t) \right] - t(1-\mu) + s = 0 \quad (7.238)$$

has solutions in the interval $[0, R)$ and let us denote by t^* the least of them. Let $v > 0$, $\mu^1 \geq 0$ such that

$$v(1-\mu) - (1 - \mu^1) \leq 0. \quad (7.239)$$

Then, for every s^1 satisfying

$$0 < s^1 \leq v \left[\sigma \left(\int_0^s B_1(R, \rho + \theta) \, d\theta + B_2(R, s) \right) + s\mu \right] \quad (7.240)$$

and for every ρ^1 such that

$$0 \leq \rho^1 \leq \rho + s, \quad (7.241)$$

the equation

$$\varphi^1(t) := av\sigma \left[\int_0^t B_1\left(R, \rho^1 + \theta\right) d\theta + B_2(R, t) \right] - t\left(1 - \mu^1\right) + s^1 = 0 \quad (7.242)$$

has nonnegative solutions and at least one of them, denoted by t^{**}, lies in the interval $\left[s^1, t^* - s\right]$.

Proof. We first observe that since $\varphi(t^*) = 0$ and $0 \leq \mu < 1$, we obtain from (7.238) that $s \leq t^*$. We will show that

$$\varphi^1(t^* - s) \leq 0. \quad (7.243)$$

Using (7.238)-(7.242), we obtain

$$\varphi^1(t^* - s) = av\sigma \left[\int_0^{t^* - s} B_1(R, \rho^1 + \theta) d\theta + B_2(R, t^* - s) \right]$$
$$- (t^* - s)(1 - \mu^1) + s^1$$
$$\leq v \left[a\sigma \left(\int_s^{t^*} B_1(R, \rho + \theta) d\theta + B_2(R, t^*) - B_2(R, s) \right) \right.$$
$$+ \sigma \left(\int_0^s B_1(R, \rho + \theta) d\theta + B_2(R, s) \right) + s\mu - \frac{(t^* - s)}{v}(1 - \mu^1) \right]$$
$$\leq v \left[a\sigma \left(\int_0^{t^*} B_1(R, \rho + \theta) d\theta + B_2(R, t^*) \right) - t^*(1 - \mu) + s \right.$$
$$+ t^*(1 - \mu) - s + s\mu - \frac{(t^* - s)}{v}(1 - \mu^1) \right]$$
$$\leq v(t^* - s) \left[(1 - \mu) - \frac{(1 - \mu^1)}{v} \right] \leq 0, \quad \text{by (7.239)}.$$

Hence, $\varphi^1(t)$ has nonnegative real roots and for the least of them t^{**}, it is

$$s^1 \leq t^{**} \leq t^* - s.$$

Moreover, from (7.242) we get $\mu^1 < 1$.
That completes the proof of the proposition. □

We can now prove the following result.

Theorem 7.14 *Let* $\{s_n\}, \{\mu_n\}, \{\sigma_n\}, n \geq 0$ *be real sequence, with*

$s_n > 0$, $\mu_n \geq 0$, $\sigma_n > 0$. Let $\{\rho_n\}$ be a sequence on $[0, R)$, with $\rho_0 = 0$ and

$$\rho_{n+1} \leq \sum_{j=0,1,2,\ldots,n} s_j, \quad n \geq 0. \tag{7.244}$$

Suppose that $1 - \mu_0 > 0$ and that, for a given constant $a \geq 1$, the function

$$\varphi_0(t) := a\sigma_0 \left[\int_0^t B_1(R, \rho_0 + \theta) \, d\theta + B_2(R, t) \right] - t(1 - \mu_0) + s_0 \tag{7.245}$$

has roots on $[0, R)$.

Assume that for every $n \geq 0$ the following conditions are satisfied

$$s_{n+1} \leq v_n \left[\sigma_n \left(\int_0^{s_n} B_1(R, \rho_n + \theta) \, d\theta + B_2(R, s_n) \right) + s_n, \mu_n \right], \tag{7.246}$$

$$v_n(1 - \mu_n) - (1 - \mu_n) \leq 0, \tag{7.247}$$

where $v_n = \frac{\sigma_{n+1}}{\sigma_n}$.

Then,

(a) for every $n \geq 0$, the equation

$$\varphi_n(t) := av_n\sigma_n \left[\int_0^t B_1(R, \rho_n + \theta) \, d\theta + B_2(R, t) \right] - t(1 - \mu_n) + s_n \tag{7.248}$$

has solutions in $[0, R)$ and, denoting by t_n^* the least of them, we have

$$\sum_{j=n,\ldots,\infty} s_j \leq t_n^*. \tag{7.249}$$

(b) Let $\{x_n\}$, $n \geq 0$ be a sequence in a Banach space such that $\|x_{n+1} - x_n\| \leq s_n$. Then, it converges and denoting its limit by x^*, the error bounds

$$\|x^* - x_n\| \leq t_n^* \tag{7.250}$$

and

$$\|x^* - x_{n+1}\| \leq t_n^* - s_n \tag{7.251}$$

are true for all $n \geq 0$.

(c) *If there exists $h_0 \in [0, R)$ such that*
$$\varphi_0(h_0) \leq 0, \qquad (7.252)$$
then $\varphi_0(t)$ has roots on $[0, R)$.

Proof. (a) We use induction on n. Let us assume that for some $n \geq 0$, $1 - \mu_n > 0$, $\varphi_n(t)$ has roots on $[0, R)$ and t_n^* is the least of them. This is true for $n = 0$. Then, by (7.244), (7.246), (7.247) and the proposition, by setting $s = s_n$, $s^1 = s_{n+1}$, $\mu = \mu_n$, $\mu^1 = \mu_{n+1}$ and $v = v_n$, it follows that t_{n+1}^* exists, with
$$s_{n+1} \leq t_{n+1}^* \leq t_n^* - s_n$$
and $1 - \mu_{n+1} > 0$.

That completes the induction and proves (a).

(b) This part follows easily from part (a).

(c) Using (7.252), we deduce immediately that $\varphi_0(t)$ has roots on $[0, R)$. That completes the proof of theorem. $\square$

We can now prove the main result.

Theorem 7.15 *Consider the method (7.232). Assume that for $s_0 > 0$, $\sigma_0 > 0$, $0 \leq \mu_0 < 1$ and $a \geq 1$, (7.252) is true. Then, the function $\varphi_0(t)$ defined by (7.245) has roots on $[0, R)$. Denote by t_0^* the least of them and suppose that*
$$t_0^* < R_0 \leq R. \qquad (7.253)$$

Let $s_n > 0$, $\mu_n \geq 0$, $\sigma_n > 0$, $n \geq 0$ be such that $\liminf \sigma_n > 0$ as $n \to \infty$ and condition (7.246) is true for all $n \geq 0$.

Theorem 7.16 *Assume that, for all $n \geq 0$,*
$$\|y_n\| \leq s_n \leq \sigma_n \|P(F(x_n) + G(x_n))\| \qquad (7.254)$$
and
$$\|\mathrm{Pr}_n\| \leq \frac{\mu_n s_n}{\sigma_n}. \qquad (7.255)$$

Then the sequence $\{x_n\}$, $n \geq 0$ generated by (7.232) remains in $U(x_0, t_0^*)$ and converges to a solution x^* of equation (7.233). Moreover, the error bounds (7.250) and (7.251) are true for all $n \geq 0$, where t_n^* is the least root in $[0, R)$ of the function $\varphi_n(t)$ defined by (7.248), with $\rho_n = \|x_n - x_0\|$, $n \geq 0$.

Proof. The existence of t_0^* is guaranteed by (7.252). Let us assume that $x_n, x_{n+1} \in U(x_0, t_0^*)$. We will show that for every $n \geq 0$, condition (7.246) is true. Since $\|y_0\| \leq s_0$, this is true for $n = 0$.

Using the identity

$$P(F(x_{n+1}) + G(x_{n+1}))$$
$$= \int_0^1 P[F'(x_n + t(x_{n+1} - x_n)) - A(x_n)](x_{n+1} - x_n) dt$$
$$+ P(G(x_{n+1}) - G(x_n)) + \Pr_n,$$

(7.234), (7.235), (7.254), (7.255), setting $\rho_n = \|x_n - x_0\|$ and by taking norms in the above identity we get

$$s_{n+1} \leq \sigma_{n+1} \|P(F(x_{n+1}) + G(x_{n+1}))\|$$
$$\leq v_n \left[\sigma_n \left(\int_0^{s_n} B_1(R, \rho_n + \theta) d\theta + B_2(R, s_n) \right) + s_n \mu_n \right]$$

which shows (7.246) for all $n \geq 0$.

The hypothesis (b) of Theorem 7.14 can now easily be verified by induction and thus, by (7.249) and (7.254), the sequence $\{x_n\}$, $n \geq 0$ remains in $U(x_0, t_0^*)$, converges to x^* and (7.250) and (7.251) hold. Finally, from the inequality

$$\|P(F(x_n) + G(x_n))\| \leq \|P(A(x_n) - F'(x_0))\| \|y_n\|$$
$$+ \|PF'(x_0)\| \|y_n\| + \|\Pr_n\|,$$

(7.234), (7.255) and the continuity of F and G, as $\liminf \sigma_n > 0$ and $s_n \to 0$, as $n \to \infty$ it follows that $P(F(x^*) + G(x^*)) = 0$. Hence $F(x^*) + G(x^*) = 0$.

That completes the proof of the theorem. □

Remark 7.4 (a) In the special case when B_1 and B_2 are given (7.236) and (7.237) respectively, then our results can be reduced to the ones obtained by Moret [5, p. 359] (when $G = 0$ and $P = I$).

(b) Let $G = 0$, $P = I$, $A(x) = F'(x)$ and define the functions $\overline{\varphi}_0(t)$, $\overline{\varphi}_n(t)$ by

$$\overline{\varphi}_0(t) = a\sigma_0 \int_0^t (t - \theta) k(\theta) d\theta - t(1 - \mu_0) + s_0,$$

$$\overline{\varphi}_n(t) = av_n \sigma_n \int_0^t (t - \theta) k(\rho_n + \theta) d\theta - t(1 - \mu_n) + s_n,$$

where k is a nondecreasing function on $[0, R]$ such that

$$\|F'(x) - F'(y)\| \leq k(r) \|x - y\|, \quad x, y \in \overline{U}(x_0, r) \quad (r < R_0).$$

Assume that B_1 can be chosen in such a way that

$$\varphi_n(t) \leq \overline{\varphi}_n(t), \quad n \geq 0. \tag{7.256}$$

Then under the hypotheses of Theorem 7.15 above and Proposition 1 in [212, p. 359], using (7.256) we can show

$$\|x^* - x_n\| \leq t_n^* \leq m_m^*, \quad n \geq 0$$

and

$$\|x^* - x_{n+1}\| \leq t_n^* - s_n \leq m_n^* - s_n, \quad n \geq 0$$

where by m_n^*, we denote the least solutions of the equations

$$\overline{\varphi}_n(t) = 0, \quad n \geq 0 \quad in \quad [0, R).$$

7.6 Exercises

7.1. Consider the Stirling method

$$z_{n+1} = z_n - [I - F'(F(z_n))]^{-1} [z_n - F(z_n)]$$

for approximating a fixed point x^* of the equation $x = F(x)$ in a Banach space X.

Show:
(i) If $\|F'(x)\| \leq \alpha < \frac{1}{3}$, then the sequence $\{x_n\}$ $(n \geq 0)$ converges to the unique fixed point x^* of equation $x = F(x)$ for any $x_0 \in X$. Moreover show that:

$$\|x^* - x_n\| \leq \left(\frac{2\alpha}{1-\alpha}\right)^n \frac{\|x_0 - F(z_0)\|}{1-\alpha} \quad (n \geq 0).$$

(ii) If F' is Lipschitz continuous with constant K and $\|F'(x)\| \leq \alpha < 1$, then Newton's method converges to x^* for any $x_0 \times X$ such that

$$h_N = \frac{1}{2} K \frac{\|x_0 - F(x_0)\|}{(1-\alpha)^2} < 1.$$

and

$$\|x_n - x^*\| \le (h_N)^{2^n-1}\frac{\|x_0 - F(x_0)\|}{1-\alpha} \quad (n \ge 0).$$

(iii) If F' is Lipschitz continuous with constant K and $\|F'(x)\| \le \alpha < 1$, then $\{z_n\}$ $(n \ge 0)$ converges to x^* for any $z_0 \in X$ such that

$$h_s = \frac{K}{2}\frac{1+2\alpha}{1-\alpha}\frac{\|x_0 - F(z_0)\|}{1-\alpha} \quad (n \ge 0).$$

7.2. Let H be a real Hilbert space and consider the nonlinear operator equation $P(x) = 0$ where $P: U(x_0, r) \subseteq H \to H$. Let P be differentiable in $U(x_0, r)$ and set $F(x) = \|P(x)\|^2$. Then $P(x) = 0$ reduces to $F(x) = 0$. Define the iteration

$$x_{n+1} = x_n - \frac{\|P(x_n)\|^2}{2\|Q(x_n)\|^2}Q(x_n) \quad (n \ge 0)$$

where $Q(x) = P'(x)P(x)$, and the linear operator $P'(x)$ is the adjoint of $P'(x)$. Show that if:
(a) there exist two positive constants B and K such that

$$B^2 K < 4;$$

(b) $\|P'(x)y\| \ge B^{-1}\|y\|$ for all $y \in H$, $x \in u(x_0, r)$
(c) $\|Q'(x)\| \le K$ for all $x \in U(x_0, r)$;
(d) $\|x_1 - x_0\| < \eta_0$ and $r = \frac{2\eta_0}{2 - B\sqrt{K}}$.
The equation $P(x) = 0$ has a solution $x^* \in U(x_0, r)$ and the sequence $\{x_n\}$ $(n \ge 0)$ converges to x^* with

$$\|x_n - x^*\| \le \eta_0 \frac{\alpha^n}{1-\alpha}$$

where

$$\alpha = \frac{1}{2}B\sqrt{K}.$$

7.3. Consider the equation

$$x = T(x)$$

in a Banach space X, where $T: D \subset X \to X$ and D is convex. Let $T_1(x)$ be another nonlinear continuous operator acting from X into X, and let P be a projection operator in X. Then the operator

$PT_1(x)$ will be assumed to be Fréchet differentiable on D. consider the iteration

$$x_{n+1} = T(x_n) + PT_1'(x_n)(x_{n+1} - x_n) \ (n \geq 0).$$

Assume:
(a) $\left\|[I - PT_1'(x_0)]^{-1}(x_0 - T(x_0))\right\| \leq \eta$,
(b) $\Gamma(x) = \Gamma = [I - PT_1'(x)]^{-1}$ exists for all $x \in D$ and $\|\Gamma\| \leq b$,
(c) $PT_1'(x)$, $QT_1(x)$ $(Q = I - P)$ and $T(x) - T_1$ satisfy a Lipschitz condition on D with respective constants M, q and f,
(d) $\bar{U}(x_0, H\eta) \subseteq D$, where

$$H = 1 + \sum_{j=1}^{\infty} \prod_{i=1}^{j} J_i, \ J_1 = b + \frac{h}{2},$$

$$J_i = b + \frac{h}{2} J_1 \cdots J_{i-1}, \ i \geq 2, \ J_0 = \eta,$$

(e) $h = BM\eta < 2(1-b)$, $b = B(q+f) < 1$. Then show that the equation $x = T(x)$ has a solution $x^* \in \bar{U}(x_0, H\eta)$ and the sequence $\{x_n\} \ (n \geq 0)$ converges to x^* with

$$\|x_n - x^*\| \subseteq H\eta \prod_{i=1}^{n} J_i.$$

7.4. Let H be a real separable Hilbert space. An operator F on H is said to be weakly closed is
(a) x_n converges weakly to x, and
(b) $F(x_n)$ converges weakly to y imply that

$$F(x) = y.$$

Let F be a weakly closed operator defined on $\bar{U}(x_0, r)$ with values in H. Suppose that F maps $\overline{U}(x_0, r)$ into a bounded set in H provided the following conditions is satisfied:

$$(F(x), x) \leq (x, x) \text{ for all } x \in S$$

where $S = \{x \in H \mid \|x\| = r\}$.
Then show that there exists $x^* \in U(x_0, r)$ such that

$$F(x^*) = x^*.$$

7.5. Let X be a Banach space, $LB(X)$ the Banach space of continuous linear operators on X equipped with the uniform norm, B_1 the unit ball. Recall that a nonlinear operator K on X is compact if it maps every bounded set into a set with compact closure. We shall say a family H of operators on X is collectively compact if and only if every bounded set $B \subset X$, $\bigcup_{P \in H} H(B)$ has compact closure.

Show:

(i) If

(a) H is a collectively compact family of operators on X,

(b) K is in the pointwise closure of H,

then K is compact

(ii) If

(a) H is a collectively compact family on X,

(b) H is equidifferentiable on $D \subset X$.

Then for every $x \in D$, the family $\{P'(x) \mid P \in H\}$ is collectively compact.

7.6. Consider the Newton-like method. Let $A : D \to L(X,Y)$, $x_0 \in D$, $M_{-1} \in L(X,Y)$, $X \subset Y$, $L_{-1} \in L(X,X)$. For $n \geq 0$ choose $N_n \in L(X,X)$ and define $M_n = M_{n-1} N_n + A(x_n) L_{n-1}$, $L_n = L_{n-1} + L_{n-1} N_n$, $x_{n+1} = x_n + L_n(y_n)$, y_n being a solution of $M_n(y_n) = -[F(x_n) + z_n]$ for a suitable $z_n \in y$.

Assume:

(a) F is Fréchet-differentiable on D.

(b) There exist non-negative numbers α, α_0 and nondecreasing functions $w, w_0 : \mathbb{R}^+ \to \mathbb{R}^+$ with $w(0) = w_0(0) = 0$ such that

$$\|F(x_0)\| \leq \alpha_0, \quad \|R_0(y_0)\| \leq \alpha,$$
$$\|A(x) - A(x_0)\| \leq w_0(\|x - x_0\|)$$

and

$$\|F'(x + t(y - x)) - A(x)\| \leq w(\|x - x_0\| + t\|x - y\|)$$

for all $x, y \in \bar{U}(x_0, R)$ and $t \in [0, 1]$.

(c) Let M_{-1} and L_{-1} be such that M_{-1} is invertible,

$$\|M_{-1}^{-1}\| \leq \beta, \quad \|L_{-1}\| \leq \gamma \text{ and } \|M_{-1} - A(x_0) L_{-1}\| \leq \delta.$$

(d) There exist non-negative sequence $\{a_n\}$, $\{\bar{a}_n\}$, $\{b_n\}$ and $\{c_n\}$ such that for all $n \geq 0$

$$\|N_n\| \leq a_n, \quad \|I + N_n\| \leq \bar{a}_n, \quad \|M_{-1}^{-1}\| \cdot \|M_{-1} - M_n\| \leq b_n < 1$$

and
$$\|z_n\| \le c_n \|F(x_n)\|.$$

(e) The scalar sequence $\{t_n\}$ $(n \ge 0)$ given by

$$t_{n+1} = t_{n+1} + e_{n+1}d_{n+1}(1 + c_{n+1})$$
$$\left[I_n + \sum_{i=1}^{n} h_i w(t_i)(t_i - t_{i-1}) + w(t_{n+1})(t_{n+1} - t_n) \right]$$

$(n \ge)$, $t_0 = 0$, $t_1 = \alpha$ is bounded above by a t_0^* with $0 < t_0^* \le R$, where

$$e_0 = \gamma \bar{a}_0, \ e_n = I_{n-1}\bar{a}_n \ (n \ge 1), \ d_n = \frac{\beta}{1 - d_n} \ (n \ge 0)$$
$$I_n = \varepsilon_n \varepsilon_{n-1} \ldots \varepsilon_0 \alpha_0 \ (n \ge 0) \ \varepsilon_n = p_n d_n (1 + c_n) + c_n,$$
$$p_n = q_{n-1} a_n \ (n \ge 1), \ p_0 = \delta a_0,$$
$$q_n = p_n + w_0(t_{n+1}) e_n \ (n \ge 1)$$

and

$$h_i = \prod_{m=i}^{n} \varepsilon_m \ (i \le n).$$

(f) The following estimate is true $\varepsilon_n \le \varepsilon < 1$ $(n \ge 0)$.
Then show:
i. the scalar sequence $\{t_n\}$ $(n \ge 0)$ is nondecreasing and convergence to a t^* with $0 < t^* \le t_0^*$ as $n \to \infty$.
ii. The Newton-like method is well defined, remains in $\bar{U}(x_0, t^*)$ and converges to a solution x^* of equation $F(x) = 0$.
iii. The following estimates are true:

$$\|x_{n+1} - x_n\| \le t_{n+1} - t_n$$

and

$$\|x_n - x^*\| \le t^* - t_n \ (n \ge 0).$$

7.7 (Application of Newton's method to differential equations). Consider the boundary value problem

$$\frac{dx}{dt} - g(t, x) = 0, \ x(0) = 0.$$

We want to find a solution $x = x(t)$ in $C'[0,c]$. The operator $F : C'[0,c] \to C[0,c]$ given by

$$F(x) = \frac{dx}{dt} - g(t,x)$$

corresponds to the given equation, assuming that g is continuous for $0 \le t \le c$, and $x \in C'[0,c]$. If g is differentiable with respect to x, then at $x_0 = x_0(t)$, we have that the derivative of F is

$$F'(x_0) = \frac{d}{dt} - g'_2(t, x_0(t)) I,$$

a linear differential operator from $C'[0,c]$ into $C[0,c]$, where

$$g'_2(t, x_0(t)) = \left.\frac{\partial g(t,x)}{\partial x}\right|_{x=x_0(t)}, \quad 0 \le t \le c.$$

Set $x_{n+1} = x_{n+1}(t)$ $(0 \le t \le c)$, $x_{n+1}(0) = 0$,

$$a_n(t) = x_{n+1}(t) - x_n(t),$$
$$b_n(t) = -g'_2(t, x_n(t)),$$
$$c_n(t) = -F(x_n) = -\frac{dx_n(t)}{dt} + g(t, x_n(t)),$$
$$d_n(t) = \int_0^t b_n(s)\,ds, \quad 0 \le t \le c.$$

Show that Newton's method can now be written as

$$x_{n+1}(t) = x_n(t) + \int_0^t e^{d_n(s) - d_n(t)} c_n(s)\,ds, \quad 0 \le t \le c, \eta \ge 0,$$

(with $x_0(0) = 0$).

7.8. Establish Euler's identity

$$\frac{1}{1-c} = 1 + c + c^2 + \cdots = \prod_{n=0}^{\infty}\left(1 + c^{2^n}\right)$$

for scalar c with $|c| < 1$. Extend this to the Neumann series of L^{-1}, assuming that $\|I - ML\| < 1$ (see earlier exercise). Compare the rate of convergence of the partial sums of the Neumann series with the rate of convergence of the corresponding partial products.

7.9. (a) Let $F\colon D \subseteq X \to Y$ be a Fréchet differentiable operator and $A(x) \in L(X,Y)$ ($x \in D$). Assume there exists a point $x_0 \in D$, $\eta \geq 0$ and nonnegative continuous functions a, b, c such that

$$A(x_0)^{-1} \in L(Y,X),$$
$$\|A(x_0)^{-1}F(x_0)\| \leq \eta,$$
$$\|A(x_0)^{-1}[F'(x) - F'(x_0)]\| \leq a(\|x - x_0\|),$$
$$\|A(x_0)^{-1}[F'(x_0) - A(x)]\| \leq b(\|x - x_0\|),$$
$$\|A(x_0)^{-1}[A(x) - A(x_0)]\| \leq c(\|x - x_0\|)$$

for all $x \in D$; equation

$$\int_0^1 a[(1-t)r]r\,dt + [b(r) + c(r) - 1]r + \eta = 0$$

has nonnegative solutions. Denote by r_0 the smallest. Point r_0 satisfies

$$a(r_0) + b(r_0) + c(r_0) < 1,$$

and

$$\bar{U}(x_0, r_0) = \subseteq D.$$

Then show sequence $\{x_n\}$ ($n \geq 0$) generated by Newton-like method is well defined, remains in $\bar{U}(x_0, r_0)$ for all $n \geq 0$ and converges to a unique solution $x^* \in \bar{U}(x_0, r_0)$ of equation $F(x) = 0$. Moroever the following error bounds hold for all $n \geq 0$

$$\|x_{n+2} - x_{n+1}\| \leq q\|x_{n+1} - x_n\|$$

and

$$\|x_n - x^*\| \leq \frac{\eta}{1-q}q^{n+1}$$

where,

$$q = \frac{a(r_0) + b(r_0)}{1 - c(r_0)}.$$

Furthermore x^* is unique in $U(x_0, R)$ for $R > t^*$ and

$$U(x_0, R) \subseteq D$$

if

$$\int_0^1 a[(1-t)r_0 + tR]dt + b(0) \leq 1.$$

(b) Let $F\colon D \subseteq X \subseteq Y$ be a Fréchet-differentiable operator and $A(x) \in L(X,Y)$. Assume: there exist a simple zero x^* of F and nonnegative continuous functions α, β, γ such that

$$A(x^*)^{-1} \in L(Y,X),$$

$$\|A(x^*)^{-1}[F'(x) - F'(x^*)]\| \leq \alpha(\|x - x^*\|),$$
$$\|A(x^*)^{-1}[F'(x^*) - A(x)]\| \leq \beta(\|x - x^*\|),$$
$$\|A(x^*)^{-1}[A(x) - A(x^*)]\| \leq \gamma(\|x - x^*\|)$$

for all $x \in D$;
equation

$$\int_0^1 \alpha[(1-t)r]dt + \beta(r) + \gamma(r) = 1$$

has nonnegative solutions. Denote by r^* the smallest; and

$$\bar{U}(x^*, r^*) \subseteq D.$$

Show: Under the above stated hypotheses: sequence $\{x_n\}$ ($n \geq 0$) generated by the Newton-like method is well defined, remains in $\bar{U}(x^*, r^*)$ for all $n \geq 0$ and converges to x^*, provided $x_0 \in U(x^*, r^*)$.

Moreover the following error bounds hold for all $n \geq 0$:

$$\|x_{n+1} - x^*\| \leq \delta_n \|x_n - x^*\|,$$

where

$$\delta_n = \frac{\int_0^1 \alpha[(1-t)\|x_n - x^*\|]dt + \beta(\|x_n - x^*\|)}{1 - \gamma(\|x_n - x^*\|)}.$$

7.10. (a) Assume:
there exist parameters $K \geq 0$, $M \geq 0$, $L \geq 0$, $\ell \geq 0$, $\mu \geq 0$, $\eta \geq 0$, $\lambda_1, \lambda_2, \lambda_3 \in [0,1]$, $\delta \in [0,2)$ such that:

$$h_q = K\eta^{\lambda_1} + (1+\lambda_1)\left[M\left(\tfrac{\eta}{1-q}\right)^{\lambda_2} + \mu\right] + \left[\ell + L\left(\tfrac{\eta}{1-q}\right)^{\lambda_3}\right]\delta \leq \delta,$$

and
$$\ell + L\left(\frac{\eta}{1-q}\right)^{\lambda_3} \leq 1,$$
where,
$$q = \frac{\delta}{1+\lambda_1}.$$
Then, show: iteration $\{t_n\}$ $(n \geq 0)$ given by
$$t_0 = 0, \ t_1 = \eta,$$
$$t_{n+2} = t_{n+1} + \frac{K(t_{n+1}-t_n)^{\lambda_1}+(1+\lambda_1)[Mt_n^{\lambda_2}+\mu]}{(1+\lambda_1)[1-\ell-Lt_{n+1}^{\lambda_3}]} \cdot (t_{n+1} - t_n) \ (n \geq 0)$$
is non-decreasing, bounded above by
$$t^{**} = \frac{\eta}{1-q},$$
and converges to some t^* such that
$$0 \leq t^* \leq t^{**}.$$
Moreover, the following error bounds hold for all $n \geq 0$
$$0 \leq t_{n+2} - t_{n+1} \leq q(t_{n+1} - t_n) \leq q^{n+1}\eta.$$

(b) Let $\lambda_1 = \lambda_2 = \lambda_3 = 1$. Assume: there exist parameters $K \geq 0$, $M \geq 0$, $L \geq 0$, $\ell \geq 0$, $\mu \geq 0$, $\eta \geq 0$, $\delta \in [0,1]$ such that:
$$h_\delta = \left(K + L\delta + \frac{4M}{2-\delta}\right)\eta + \delta\ell + 2\mu \leq \delta,$$
$$\ell + \frac{2L\eta}{2-\delta} \leq 1,$$
$$L \leq K,$$
and
$$\ell + 2\mu < 1,$$
then, show: iteration $\{t_n\}$ $(n \geq 0)$ is non-decreasing, bounded above
$$t^{**} = \frac{2\eta}{2-\delta}$$
and converges to some t^* such that
$$0 \leq t^* \leq t^{**}.$$

Moreover the following error bounds hold for all $n \geq 0$

$$0 \leq t_{n+2} - t_{n+1} \leq \tfrac{\delta}{2}(t_{n+1} - t_n) \leq \left(\tfrac{\delta}{2}\right)^{n+1} \eta.$$

(c) Let $F: D \subseteq X \to Y$ be a Fréchet-differentiable operator. Assume:
(1) there exist an approximation $A(x) \in L(X,Y)$ of $F'(x)$, an open convex subset D_0 of D, $x_0 \in D_0$, parameters $\eta \geq 0$, $K \geq 0$, $M \geq 0$, $L > 0$, $\mu \geq 0$, $\ell \geq 0$, $\lambda_1 \in [0,1]$, $\lambda_2 \in [0,1]$, $\lambda_3 \in [0,1]$ such that:

$$A(x_0)^{-1} \in L(Y,X), \quad \|A(x_0)^{-1} F(x_0)\| \leq \eta,$$
$$\|A(x_0)^{-1}[F'(x) - F'(y)]\| \leq K\|x-y\|^{\lambda_1},$$
$$\|A(x_0)^{-1}[F'(x) - A(x)]\| \leq M\|x-x_0\|^{\lambda_2} + \mu,$$

and

$$\|A(x_0)^{-1}[A(x) - A(x_0)]\| \leq L\|x-x_0\|^{\lambda_3} + \ell \quad \text{for all } x,y \in D_0;$$

(2) hypotheses of (a) or (b) hold;
(3)
$$\bar{U}(x_0, t^*) \subseteq D_0.$$

Then, show sequence $\{x_n\}$ ($n \geq 0$) generated by Newton's method is well defined, remains in $\bar{U}(x_0, t^*)$ for all $n \geq 0$ and converges to a solution $x^* \in \bar{U}(x_0, t^*)$ of equation $F(x) = 0$.
Moreover the following error bounds hold for all $n \geq 0$:

$$\|x_{n+1} - x_n\| \leq t_{n+1} - t_n$$

and

$$\|x_n - x^*\| \leq t^* - t_n.$$

Furthermore the solution x^* is unique in $\bar{U}(x_0, t^*)$ if

$$\frac{1}{1 - \ell - L(t^*)^{\lambda_3}} \left[\frac{K}{1+\lambda_1}(t^*)^{1+\lambda_1} + M(t^*)^{\lambda_2} + \mu\right] < 1,$$

or in $U(x_0, R_0)$ if $R_0 > t^*$, $U(x_0, R_0) \subseteq D_0$, and

$$\frac{1}{1 - \ell - L(t^*)^{\lambda_3}} \left[\frac{K}{1+\lambda_1}(R + t^*)^{1+\lambda_1} + M(t^*)^{\lambda_2} + \mu\right] \leq 1.$$

(d) Let $F: D \subseteq X \to Y$ be a Fréchet-differentiable operator. Assume:

(a) there exist an approximation $A(x) \in L(X,Y)$ of $F'(x)$, a simple solution $x^* \in D$ of equation $F(x) = 0$, a bounded inverse $A(x^*)$ and parameters $\bar{K}, \bar{L}, \bar{M}, \bar{\mu}, \bar{\ell} \geq 0$, $\lambda_4, \lambda_5, \lambda_6 \in [0,1]$ such that:

$$\|A(x^*)^{-1}[F'(x) - F'(y)]\| \leq \bar{K}\|x-y\|^{\lambda_4},$$
$$\|A(x^*)^{-1}[F'(x) - A(x)]\| \leq \bar{M}\|x-x^*\|^{\lambda_5} + \bar{\mu},$$

and

$$\|A(x^*)^{-1}[A(x) - A(x^*)]\| \leq \bar{L}\|x-x^*\|^{\lambda_6} + \bar{\ell}$$

for all $x, y \in D$;

(b) equation

$$\frac{\bar{K}}{1+\lambda_4}r^{\lambda_4} + \bar{L}r^{\lambda_6} + \bar{M}r^{\lambda_5} + \bar{\mu} + \bar{\ell} - 1 = 0$$

has a minimal positive zero r_0 which also satisfies:

$$\bar{L}r_0^{\lambda_6} + \bar{\ell} < 1$$

and

$$U(x^*, r_0) \subseteq D.$$

Then, show: sequence $\{x_n\}$ $(n \geq 0)$ generated by Newton's method is well defined, remains in $U(x^*, r_0)$ for all $n \geq 0$, and converges to x^* provided that $x_0 \in U(x^*, r_0)$. Moreover the following error bounds hold for all $n \geq 0$:

$$\|x_{n+1} - x^*\| \leq \frac{1}{1-\bar{L}\|x_n-x^*\|^{\lambda_6}-\bar{\ell}}\left[\frac{\bar{K}}{1+\lambda_4}\|x_n-x^*\|^{\lambda_4}\right.$$
$$\left. + \bar{M}\|x_n-x^*\|^{\lambda_5} + \bar{\mu}\right]\|x_n-x^*\|.$$

7.11. Let F be a nonlinear operator defined on an open convex subset D of a Banach space X with values in a Banach space Y and let $A(x) \in L(X,Y)$ $(x \in D)$. Assume:

(a) there exists $x_0 \in D$ such that $A(x_0)^{-1} \in L(Y,X)$;

(b) there exist non-decreasing, non-negative functions a, b such that:

$$\|A(x_0)^{-1}[A(x) - A(x_0)]\| \leq a(\|x-x_0\|),$$
$$\|A(x_0)^{-1}[F(y) - F(x) - A(x)(y-x)]\| \leq b(\|x-y\|)\|x-y\|$$

for all $x, y \in D$;

(c) there exist $\eta \geq 0$, $r_0 > \eta$ such that

$$\|A(x_0)^{-1}F(x_0)\| \leq \eta,$$
$$a(r) < 1,$$

and

$$d(r) < 1 \quad \text{for all} \quad r \in (0, r_0],$$

where

$$c(r) = (1 - a(r))^{-1},$$

and

$$d(r) = c^2(r)b(r);$$

(d) r_0 is the minimum positive root of equation $h(r) = 0$ on $(0, r_0]$ where,

$$h(r) = \frac{\eta}{1 - d(r)} - r.$$

(e) $\bar{U}(x_0, r_0) \subseteq D$.

Then show: sequence $\{x_n\}$ ($n \geq 0$) generated by Newton-like method is well defined, remains in $U(x_0, r_0)$ for all $n \geq 0$ and converges to a solution $x^* \in \bar{U}(x_0, r_0)$ of equation $F(x) = 0$.

7.12. (a) Let $F : D \subseteq X \to Y$ be differentiable. Assume: There exist functions $f_1 : [0,1] \times [0,\infty)^2 \to [0,\infty)$, $f_2, f_3 : [0,\infty) \to [0,\infty)$, nondecreasing on $[0,\infty)^2, [0,\infty), [0,\infty)$ such that

$$\left\|A(x_0)^{-1}[F'(x + t(y - x)) - F'(x)]\right\|$$
$$\leq f_1(t, \|x - y\|, \|x - x_0\|, \|y - x_0\|),$$
$$\left\|A(x_0)^{-1}[F'(x) - A(x)]\right\| \leq f_2(\|x - x_0\|),$$
$$\left\|A(x_0)^{-1}[A(x) - A(x_0)]\right\| \leq f_3(\|x - x_0\|),$$

hold for all $t \in [0,1]$ and $x, y \in D$;

For $\left\|A(x_0)^{-1}F(x_0)\right\| \leq \eta$, equation

$$\eta + b_0\eta + \frac{b_0 b_1 \eta}{1 - b(r)} = r$$

has non-negative solutions, and denote by r_0 the smallest one. In addition, r_0 satisfies:

$$\bar{U}(x_0, r_0) \subseteq D,$$

and

$$\int_0^1 f_1(t, b_1 b_0 r_0, r_0, r_0) \, dt + f_2(r_0) + f_3(r_0) < 1,$$

where

$$b_0 = \frac{\int_0^1 f_1(t, \eta, 0, \eta) \, dt + f_2(0)}{1 - f_3(\eta)},$$

$$b_1 = \frac{\int_0^1 f_1(t, b_0 \eta, \eta, \eta + b_0 \eta) \, dt + f_2(\eta)}{1 - f_3(\eta + b_0 \eta)},$$

and

$$b = b(r) = \frac{\int_0^1 f_1(t, b_1 b_0 \eta, r, r) \, dt + f_2(r)}{1 - f_3(r)},$$

Then, show iteration $\{x_n\}$ $(n \geq 0)$ generated by Newton-like method is well defined, remains in $\bar{U}(x_0, r_0)$ for all $n \geq 0$ and converges to a solution $x^* \in \bar{U}(x_0, r_0)$ of equation. Moreover, the following error bounds hold

$$\|x_2 - x_1\| \leq b_0 \eta$$
$$\|x_3 - x_2\| \leq b_1 \|x_2 - x_1\|.$$
$$\|x_{n+1} - x^*\| \leq b_2 \|x_n - x_{n-1}\|, \ (n \geq 3)$$

and

$$\|x^* - x_n\| \leq \frac{b_0 b_1 b_2^{n-2} \eta}{1 - b_2}, \ (n \geq 3),$$

$$\|x^* - x_n\| \leq \frac{b_2^n}{1 - b_2}, \ (n \geq 0)$$

where $b_2 = b(r_0)$.
Furthermore if r_0 satisfies

$$\int_0^1 f_1(t, 2r_0, r_0, r_0) \, dt + f_2(r_0) + f_3(r_0) < 1,$$

x^* is the unique solution of equation $F(x) = 0$ in $\overline{U}(x_0, r_0)$.

Finally if there exists a minimum non-negative number R satisfying equation

$$\int_0^1 f_1(t, r + r_0, r_0, r)\, dt + f_2(r_0) + f_3(r_0) = 1,$$

such that $U(x_0, R) \subseteq D$, then the solution x^* is unique in $U(x_0, R)$.

(b) There exist a simple zero x^* of F and continuous functions $f_4 : [0,1] \times [0, \infty) \to [0, \infty)$, $f_5, f_6 : [0, \infty) \to [0, \infty)$, non-decreasing on $[0, \infty)$ such that

$$\left\| A(x^*)^{-1} [F'(x + t(x^* - x)) - F'(x)] \right\| \le f_4(t, \|x^* - x\|),$$

$$\left\| A(x^*)^{-1} [F'(x) - A(x)] \right\| \le f_5(\|x^* - x\|),$$

and

$$\left\| A(x^*)^{-1} [A(x) - A(x^*)] \right\| \le f_6(\|x^* - x\|),$$

hold for all $t \in [0,1]$ and $x \in D$;

Equation

$$\int_0^1 f_4(t, r)\, dt + f_5(r) + f_6(r) = 1$$

has a minimum positive zero r^*, and

$$U(x^*, r^*) \subseteq D.$$

Then, show iteration $\{x_n\}$ $(n \ge 0)$ generated by Newton-like sequence is well defined, remains in $U(x^*, r^*)$ for all $n \ge 0$ and converges to x^* provided that $x_0 \in U(x^*, r^*)$. Moreover, the following error bounds hold for all $n \ge 0$

$$\|x_{n+1} - x^*\| \le \frac{\int_0^1 f_4(t, \|x_n - x^*\|)\, dt + f_5(\|x_n - x^*\|)}{1 - f_6(\|x_n - x^*\|)} \|x_n - x^*\|$$

$$\le \gamma \|x_n - x^*\|,$$

where

$$\gamma = \frac{\int_0^1 f_4(t, \|x_0 - x^*\|)\, dt + f_5(\|x_0 - x^*\|)}{1 - f_6(\|x_0 - x^*\|)} < 1.$$

(c) Let $X = Y = \mathbb{R}$, $D = (-1, 1)$, $x^* = 0$ and define function F on D by

$$F(x) = x + \frac{x^{p+1}}{p+1}, \quad p > 1.$$

For the case $A = F'$ show

$$r_R = \frac{2}{3p} < r^* = \frac{2}{2+p},$$

where r_R stands for Rheinboldt's radius (see section 4.1) where r_R is the convergence radius given by Rheinboldt's [247].

(d) Let $X = Y = C[0, 1]$, the space of continuous functions defined on $[0, 1]$ equipped with the max-norm. Let $D = \{\phi \in C[0, 1]; \|\phi\| \leq 1\}$ and F defined on D by

$$F(\phi)(x) = \phi(x) - 5 \int_0^1 xt\phi(t)^3 \, dt$$

with a solution $\phi^*(x) = 0$ for all $x \in [0, 1]$.

In this case, for each $\phi \in D$, $F'(\phi)$ is a linear operator defined on D by the following expression:

$$F'(\phi)[\nu](x) = \nu(x) - 15 \int_0^1 xt\phi(t)^2 \nu(t) \, dt, \quad \nu \in D.$$

In this case and by considering again $A = F'$,

$$r_R = \frac{2}{45} < r^* = \frac{1}{15}.$$

Chapter 8

Two-Point Newton-Like Methods

We discuss two Point-Newton like methods in this Chapter.

8.1 Two-Point Newton-Like Methods in Banach Space

In this section we are concerned with the problem of approximating a locally unique solution x^* of the nonlinear equation

$$F(x) + G(x) = 0, \qquad (8.1)$$

where F, G are operators defined on a closed ball $\overline{U}(w, R)$ centered at point w and of radius $R \geq 0$, which is a subset of a Banach space X with values in a Banach space Y. F is Fréchet-differentiable on $\overline{U}(w, R)$, while the differentiability of the operator G is not assumed.

We use the two-point Newton method

$$y_{-1}, y_0 \in \overline{U}(w, R), \ y_{n+1} = y_n - A(y_{n-1}, y_n)^{-1}[F(y_n) + G(y_n)] \quad (n \geq 0) \qquad (8.2)$$

to generate a sequence converging to x^*. Here $A(x, y) \in L(X, Y)$, the space of bounded linear operators from X into Y for each fixed $x, y \in \overline{U}(w, R)$. We provide a local as well as a semilocal convergence analysis for method (8.2) under very general Lipschitz-type hypotheses (see (8.25), (8.26)).

Our new idea is to use center-Lipschitz conditions instead of Lipschitz conditions for the upper bounds on the inverses of the linear operators involved. It turns out that this way we obtain more precise majorizing sequences. Moreover, despite the fact that our conditions are more general than related ones already in the literature, we can provide weaker sufficient convergence conditions, and finer error bounds on the distances involved.

We note that our analysis is also useful in particular in the numerical solution of problems appearing in visco-elasticity. However we leave the details to the motivated reader. Finally we mention that our approach compares favorably with the classical and elegant work of J.W. Schmidt on the Secant method (see [68], [99], [252], [253], and our Example 8.2).

Several applications are provided: e.g., in the semilocal case we show that the famous Newton–Kantorovich hypothesis (for its simplicity and transparency, see (8.5)) is weakened (see (8.20)), whereas in the local case we can provide a larger convergence radius using the same information (see (8.134) and (8.135)).

Part A: Motivation

Deuflhard and Heindl have proved the following affine invariant form of the Newton–Kantorovich theorem which is the motivation for this study (see also Theorem 4.2.4 for the nonaffine invariant form)

Theorem 8.1 *Let $F: D \subseteq X \to Y$ be a Fréchet-differentiable operator on an open convex set D. Suppose that $d_0 \in D$ is such that $F'(d_0)^{-1}$ exists and*

$$\|F'(d_0)^{-1} F(d_0)\| \leq \eta, \tag{8.3}$$

$$\|F'(d_0)^{-1}[F'(x) - F'(y)]\| \leq \gamma_1 \|x - y\| \quad \text{for all } x, y \in D \text{ and } \gamma_1 > 0, \tag{8.4}$$

$$h = 2\gamma_1 \eta \leq 1, \tag{8.5}$$

$$\overline{U}(d_0, d^1) \subseteq D, \tag{8.6}$$

where,

$$d^1 = \frac{1 - \sqrt{1-h}}{\gamma_1}. \tag{8.7}$$

Then, sequence $\{d_n\}$ $(n \geq 0)$ generated by Newton's method

$$d_{n+1} = d_n - F'(d_n)^{-1} F(d_n) \quad (n \geq 0), \tag{8.8}$$

is well defined, remains in $\overline{U}(d_0, d^1)$ for all $n \geq 0$ and converges to a unique solution d^* of equation

$$F(d) = 0 \tag{8.9}$$

in $\overline{U}(d_0, d^1) \cup (D \cap U(d_0, d^2))$, where

$$d^2 = \frac{1 + \sqrt{1-h}}{\gamma_1}. \tag{8.10}$$

Moreover the following error bounds hold:

$$\|d_{n+1} - d_n\| \leq \overline{d}_{n+1} - \overline{d}_n \tag{8.11}$$

$$\|d_n - d^*\| \leq d^1 - \overline{d}_n, \quad d^1 = \lim_{n \to \infty} \overline{d}_n \tag{8.12}$$

where sequence $\{\overline{d}_n\}$ ($n \geq 0$) is given by

$$\overline{d}_0 = 0, \quad \overline{d}_1 = \eta, \quad \overline{d}_{n+2} = \overline{d}_{n+1} + \frac{\gamma_1(\overline{d}_{n+1} - \overline{d}_n)^2}{2(1 - \gamma_1 \overline{d}_{n+1})}. \tag{8.13}$$

Condition (8.5) is the famous Newton–Kantorovich hypothesis which is the essential sufficient convergence condition for the semilocal convergence of Newton's method (8.8). However Newton's method may converge to a solution of equation (8.9) even when (8.5) is violated.

Example 8.1 Let $X = Y = R$, $d_0 = 1$, $D = [p, 2-p]$, $p \in [0, \frac{1}{2})$, and define F on D by

$$F(d) = d^3 - p. \tag{8.14}$$

Using (8.3), (8.4) and (8.14) we get

$$\eta = \frac{1}{3}(1-p), \quad \gamma_1 = 2(2-p), \tag{8.15}$$

which imply

$$h = \frac{4}{3}(1-p)(2-p) > 1 \quad \text{for all } p \in \left[0, \frac{1}{2}\right). \tag{8.16}$$

That is there is no guarantee that method (8.8) converges since (8.5) is violated. However one can find values of p in $[0, \frac{1}{2})$ such that method (8.8) converges.

For example if $p = .48$, then using (8.8) we find $d^* = \sqrt[3]{.48}$. Hence, we wonder if (8.5) can be weakened. Hypothesis (8.5) is used to show that majorizing sequence $\{\overline{d}_n\}$ is monotonically increasing and converges to d^1. We have noticed that sequence $\{\overline{\overline{d}}_n\}$ ($n \geq 0$) given by

$$\overline{\overline{d}}_n = 0, \quad \overline{\overline{d}}_1 = \eta, \quad \overline{\overline{d}}_{n+2} = \overline{\overline{d}}_{n+1} + \frac{\gamma_1(\overline{\overline{d}}_{n+1} - \overline{\overline{d}}_n)^2}{2(1 - \gamma_0 \overline{\overline{d}}_{n+1})} \tag{8.17}$$

is also a more precise majorizing sequence for Newton's method (8.8) than (8.13), where γ_0 is the center-Lipschitz constant such that

$$\|F'(d_0)^{-1}[F'(x) - F'(d_0)]\| \leq \gamma_0 \|x - d_0\| \quad \text{for all } x \in D. \tag{8.18}$$

In general the inequality

$$\gamma_0 \leq \gamma_1 \qquad (8.19)$$

holds. Note also that in practice finding constant γ_1 requires the computation of γ_0. Hence no additional computational effort is required to compute (γ_0, γ_1) instead of γ_1. As it is shown in a more general setting in what follows (see Application 8.2) in this case (8.5) can be replaced by

$$h_1 = (\gamma_0 + \gamma_1)\eta \leq 1 \qquad (8.20)$$

(see also (8.72)). By comparing (8.5) and (8.20) we get $h_1 \geq \frac{1}{2}h$. Moreover note that:

$$h \leq 1 \Rightarrow h_1 \leq 1 \qquad (8.21)$$

but not vice versa unless if $\gamma_0 = \gamma_1$. Futhermore as it is shown in a more general setting for all $n \geq 0$ ($\gamma_0 < \gamma_1$),

$$\|d_{n+1} - d_n\| \leq \overline{\overline{d}}_{n+1} - \overline{\overline{d}}_n < \overline{d}_{n+1} - \overline{d}_n, \qquad (8.22)$$

$$\|d_n - d^*\| \leq d^3 - \overline{\overline{d}}_n \leq d_1 - \overline{d}_n, \qquad (8.23)$$

and

$$d^3 = \lim_{n \to \infty} \overline{\overline{d}}_n \leq d_1 \qquad (8.24)$$

(see Remark 8.5). Hence we also obtain finer error bounds and at least as precise information on the location of the solution d^* as in Theorem 8.1. Returning back to Example 8.1, since $\gamma_0 = 3 - p$ we find that (8.20) holds if $p \in \left[\frac{5-\sqrt{13}}{3}, \frac{1}{2}\right)$, which improves Theorem 8.1.

Part B: Main Results

In order for us to show that these observations hold in a more general setting we first need to introduce the following assumptions:

Let $R \geq 0$ be given. Assume there exist $v, w \in X$ such that $A(v, w)^{-1} \in L(Y, X)$, and for any $x, y, z \in \overline{U}(w, r) \subseteq \overline{U}(w, R)$, $t \in [0, 1]$, the following hold:

$$\|A(v, w)^{-1}[A(x, y) - A(v, w)]\| \leq h_0(\|x - v\|, \|y - w\|) + a, \qquad (8.25)$$

and

$$\|A(v,w)^{-1}\{[F'(y+t(z-y))-A(x,y)](z-y)+G(z)-G(y)\}\| \quad (8.26)$$
$$\leq [h_1(\|y-w\|+t\|z-y\|)-h_2(\|y-w\|)+h_3(\|z-x\|)+b]\|z-y\|,$$

where, $h_0(r,s)$, $h_1(r+\overline{r})-h_2(r)$ $(\overline{r} \geq 0)$, $h_2(r)$, $h_3(r)$ are monotonically increasing functions for all r,s on $[0,R]^2$, $[0,R]^2$, $[0,R]$, $[0,R]$ respectively with $h_0(0,0) = h_1(0) = h_2(0) = h_3(0) = 0$, and the constants a,b satisfy $a \geq 0$, $b \geq 0$. Given y_{-1}, y_0, v, w in X, define parameters c_{-1}, c, c_1 by

$$\|y_{-1}-v\| \leq c_{-1}, \quad \|y_{-1}-y_0\| \leq c, \quad \|v-w\| \leq c_1. \quad (8.27)$$

Remark 8.1 *Conditions similar to (8.25), (8.26) but less flexible were considered by Chen and Yamamoto in [116] in the special case when $A(x,y) = A(x)$ for all $x,y \in \overline{U}(w,R)$ $(A(x) \in L(X,Y))$ (see also Theorem 8.4). Operator $A(x)$ is intended there to be an approximation to the Fréchet-derivative $F'(x)$ of F. However we also want the choice of operator A to be more flexible, and be related to the difference $G(z) - G(y)$ for all $y,z \in \overline{U}(w,R)$ (see also Application 8.1). Note also that if we choose:*

$$A(x,y) = F'(x), \; G(x) = 0, \; w = d_0, \; h_0(r,r) = \gamma_0 r,$$
$$h_1(r) = h_2(r) = \gamma_1 r, \; h_3(r) = 0, \quad (8.28)$$

for all $x,y \in \overline{U}(w,R)$, $r \in [0,R]$, and $a = b = 0$ then conditions (8.25), (8.26) reduce to (8.18) and (8.4) respectively. Other choices of operators, functions and constants appearing in (8.25) and (8.26) can be found in the applications that follow.

With the above choices, we show the following result on majorizing sequences for method (8.2).

Lemma 8.1 *Assume:*
there exist parameters $\eta \geq 0$, $a \geq 0$, $b \geq 0$, $c_{-1} \geq 0$, $c \geq 0$, $\delta \in [0,2)$, $r_0 \in [0,R]$ such that:

$$2\left[\int_0^1 h_1(r_0+\theta\eta)d\theta - h_2(r_0) + b + h_3(c+\eta)\right]$$
$$+ [a + h_0(c+c_{-1}, \eta+r_0)]\delta \leq \delta, \quad (8.29)$$

$$\frac{2\eta}{2-\delta} + r_0 + c \leq R, \quad (8.30)$$

$$h_0\left[\frac{1-\left(\frac{\delta}{2}\right)^{n+1}}{1-\frac{\delta}{2}}\eta + c + c_{-1}, \frac{1-\left(\frac{\delta}{2}\right)^{n+2}}{1-\frac{\delta}{2}}\eta + r_0\right] + a < 1, \quad (8.31)$$

and

$$2\int_0^1 h_1\left[\frac{1-\left(\frac{\delta}{2}\right)^{n+1}}{1-\frac{\delta}{2}}\eta + \theta\left(\frac{\delta}{2}\right)^{n+1}\eta + r_0\right]d\theta - 2h_2\left[\frac{1-\left(\frac{\delta}{2}\right)^{n+1}}{1-\frac{\delta}{2}}\eta + r_0\right]$$

$$+ 2h_3\left[\left(\frac{\delta}{2}\right)^n\left(1+\frac{\delta}{2}\right)\eta\right] + \delta h_0\left[\frac{1-\left(\frac{\delta}{2}\right)^{n+1}}{1-\frac{\delta}{2}}\eta + c + c_{-1}, \frac{1-\left(\frac{\delta}{2}\right)^{n+2}}{1-\frac{\delta}{2}}\eta + r_0\right]$$

$$+ 2b + \delta a \leq \delta \tag{8.32}$$

for all $n \geq 0$.

Then, iteration $\{t_n\}$ $(n \geq -1)$ given by

$$t_{-1} = r_0, \quad t_0 = c + r_0, \quad t_1 = c + r_0 + \eta, \quad t_{n+2} = t_{n+1} \tag{8.33}$$

$$+ \frac{\{\int_0^1[h_1(t_n-t_0+r_0+\theta(t_{n+1}-t_n))-h_2(t_n-t_0+r_0)+b]d\theta+h_3(t_{n+1}-t_{n-1})\}(t_{n+1}-t_n)}{1-a-h_0(t_n-t_{-1}+c_{-1},t_{n+1}-t_0+r_0)}$$

is monotonically increasing, bounded above by

$$t^{**} = \frac{2\eta}{2-\delta} + r_0 + c, \tag{8.34}$$

and converges to some t^* such that

$$0 \leq t^* \leq t^{**} \leq R. \tag{8.35}$$

Moreover the following error bounds hold for all $n \geq 0$:

$$0 \leq t_{n+2} - t_{n+1} \leq \frac{\delta}{2}(t_{n+1} - t_n) \leq \left(\frac{\delta}{2}\right)^{n+1}\eta. \tag{8.36}$$

Proof. We must show:

$$2\left\{\int_0^1[h_1(t_k - t_0 + r_0 + \theta(t_{k+1} - t_k)) - h_2(t_k - t_0 + r_0) + b]d\theta + h_3(t_{k+1} - t_{k-1})\right\}$$
$$+ \delta[a + h_0(t_k - t_{-1} + c_{-1}, t_{k+1} - t_0 + r_0)] \leq \delta, \tag{8.37}$$

$$0 \leq t_{k+1} - t_k, \tag{8.38}$$

and

$$h_0(t_k - t_{-1} + c_{-1}, t_{k+1} - t_0 + r_0) + a < 1 \tag{8.39}$$

for all $k \geq 0$.

Estimate (8.36) can then follow from (8.37)–(8.39) and (8.33).

Using induction on the integer $k \geq 0$, first for $k = 0$ in (8.37)–(8.39) we must show:

$$2\left[\int_0^1 h_1(r_0 + \theta\eta)d\theta - h_2(r_0) + b + h_3(c+\eta)\right] + \delta[a + h_0(c + c_{-1}, \eta + r_0)] \leq \delta,$$

$$0 \leq t_1 - t_0,$$

$$h_0(c + c_{-1}, \eta + r_0) + a < 1,$$

which hold by (8.29) and the definition of t_1.

By (8.33) we get

$$0 \leq t_2 - t_1 \leq \frac{\delta}{2}(t_1 - t_0).$$

Assume that (8.37)–(8.39) hold for all $k \leq n+1$. Using (8.37)–(8.39) we obtain in turn

$$2\left\{\int_0^1 [h_1(t_{k+1} - t_0 + r_0 + \theta(t_{k+2} - t_{k+1}))\right.$$
$$\left. - h_2(t_{k+1} - t_0 + r_0) + b]d\theta + h_3(t_{k+2} - t_k)\right\}$$
$$+ \delta[a + h_0(t_{k+1} - t_{-1} + c_{-1}, t_{k+2} - t_0 + r_0)]$$
$$\leq 2\left\{\int_0^1 h_1\left[\left(\frac{1 - (\frac{\delta}{2})^{k+1}}{1 - \frac{\delta}{2}}\right) + \theta\left(\frac{\delta}{2}\right)^{k+1}\right)\eta + r_0\right]\right.$$
$$\left. - h_2\left[\frac{1 - (\frac{\delta}{2})^{k+1}}{1 - \frac{\delta}{2}}\eta + r_0\right] + b + h_3\left[\left(\frac{\delta}{2}\right)^{k+1}\eta + \left(\frac{\delta}{2}\right)^k\eta\right]\right\}$$
$$+ \delta\left[a + h_0\left(\frac{1 - (\frac{\delta}{2})^{k+1}}{1 - \frac{\delta}{2}}\eta + c + c_{-1}, \frac{1 - (\frac{\delta}{2})^{k+2}}{1 - \frac{\delta}{2}}\eta + r_0\right)\right]$$
$$\leq \delta$$

by (8.29) and (8.32). Hence we showed (8.37) holds for $k = n+2$. Moreover, we must show:

$$t_k \leq t^{**} \tag{8.40}$$

$$t_{-1} = r_0 \leq t^{**}, \quad t_0 = r_0 + c \leq t^{**}, \quad t_1 = c + r_0 + \eta \leq t^{**},$$
$$t_2 \leq c + r_0 + \eta + \frac{\delta}{2}\eta = \frac{2+\delta}{2}\eta + r_0 + c \leq t^{**}.$$

Assume that (8.40) holds for all $k \leq n+1$. It follows from (8.33), (8.37)–(8.39):

$$t_{k+2} \leq t_{k+1} + \frac{\delta}{2}(t_{k+1} - t_k) \leq t_k + \frac{\delta}{2}(t_k - t_{k-1}) + \frac{\delta}{2}(t_{k+1} - t_k)$$

$$\leq \cdots \leq c + r_0 + \eta + \frac{\delta}{2}\eta + \left(\frac{\delta}{2}\right)^2 \eta + \cdots + \left(\frac{\delta}{2}\right)^{k+1} \eta$$

$$= \frac{1 - \left(\frac{\delta}{2}\right)^{k+2}}{1 - \frac{\delta}{2}} \eta + r_0 + c \leq \frac{2\eta}{2-\delta} + r_0 + c = t^{**}. \tag{8.41}$$

Hence, sequence $\{t_n\}$ $(n \geq -1)$ is bounded above by t^{**}. Inequality (8.39) holds for $k = n+2$ by (8.30) and (8.31). Moreover (8.38) holds for $k = n+2$ by (41) and since (8.37) and (8.39) also hold for $k = n+2$. Furthermore, sequence $\{t_n\}$ $(n \geq 0)$ is monotonically increasing by (8.38) and as such it converges to some t^* satisfying (8.35).

That completes the proof of Lemma 8.1. □

We provide the main result on the semilocal convergence of method (8.2) using majorizing sequence (8.33).

Theorem 8.2 *Assume:*
hypotheses of Lemma 8.1 hold, and there exist

$$y_{-1} \in \overline{U}(w, R), \quad y_0 \in \overline{U}(w, r_0) \tag{8.42}$$

such that:

$$\|A(y_{-1}, y_0)^{-1}[F(y_0) + G(y_0)]\| \leq \eta. \tag{8.43}$$

Then, sequence $\{y_n\}$ $(n \geq -1)$ generated by Newton-like method (8.2) is well defined, remains in $\overline{U}(w, t^*)$ for all $n \geq -1$, and converges to a solution x^* of equation $F(x) + G(x) = 0$. Moreover the following error bounds hold for all $n \geq -1$:

$$\|y_{n+1} - y_n\| \leq t_{n+1} - t_n \tag{8.44}$$

and

$$\|y_n - x^*\| \leq t^* - t_n. \tag{8.45}$$

Furthermore the solution x^* is unique in $\overline{U}(w, t^*)$ if

$$\int_0^1 h_1((1+t)t^*)dt - h_2(t^*) + h_3(t^*) + h_0(t^* + c_1, t^*) + a + b < 1, \tag{8.46}$$

and in $\overline{U}(w, R_0)$ for $R_0 \in (t^*, R]$ if

$$\int_0^1 h_1(t^* + tR_0)dt - h_2(t^*) + h_3(R_0) + h_0(t^* + c_1, t^*) + a + b < 1, \quad (8.47)$$

provided that $y_{-1} = v$ and $y_0 = w$.

Proof. We first show estimate (8.44), and $y_n \in \overline{U}(w, t^*)$ for all $n \geq -1$. For $n = -1, 0$, (8.44) follows from (8.27), (8.33) and (8.43). Suppose (8.44) holds for all $n = 0, 1, \ldots, k+1$; this implies in particular (using (8.27), (8.42))

$$\|y_{k+1} - w\| \leq \|y_{k+1} - y_k\| + \|y_k - y_{k-1}\| + \cdots + \|y_1 - y_0\| + \|y_0 - w\|$$
$$\leq (t_{k+1} - t_k) + (t_k - t_{k-1}) + \cdots + (t_1 - t_0) + r_0$$
$$= t_{k+1} - t_0 + r_0 \leq t_{k+1} \leq t^*.$$

That is, $y_{k+1} \in \overline{U}(w, t^*)$.

We show (8.44) holds for $n = k+2$. By (8.25) and (8.33) we obtain for all $x, y \in \overline{U}(w, t^*)$

$$\|A(v, w)^{-1}[A(x, y) - A(v, w)]\| \leq h_0(\|x - v\|, \|y - w\|) + a. \quad (8.48)$$

In particular for $x = y_k$ and $y = y_{k+1}$ we get using (8.25), (8.27),

$$\|A(v, w)^{-1}[A(y_k, y_{k+1}) - A(v, w)]\| \leq h_0(\|y_k - v\|, \|y_{k+1} - w\|) + a$$
$$\leq h_0(\|y_k - y_{-1}\| + \|y_{-1} - v\|, \|y_{k+1} - x_0\| + \|y_0 - w\|) + a$$
$$\leq h_0(t_k - t_{-1} + c_{-1}, t_{k+1} - t_0 + r_0) + a$$
$$\leq h_0\left[\frac{1 - \left(\frac{\delta}{2}\right)^k}{1 - \frac{\delta}{2}}\eta + c + c_{-1}, \frac{1 - \left(\frac{\delta}{2}\right)^{k+1}}{1 - \frac{\delta}{2}}\eta + r_0\right] + a < 1, \quad \text{(by (31))}.$$
(8.49)

It follows from (8.49) and the Banach Lemma on invertible operators that $A(y_k, y_{k+1})^{-1}$ exists, and

$$\|A(y_k, y_{k+1})^{-1} A(v, w)\| \leq [1 - a - h_0(t_k - t_{-1} + c_{-1}, t_{k+1} - t_0 + r_0)]^{-1}. \quad (8.50)$$

Using (8.2), (8.26), (8.33), (8.50) we obtain in turn

$$\|y_{k+2} - y_{k+1}\| = \|A(y_k, y_{k+1})^{-1}[F(y_{k+1}) + G(y_{k+1})]\|$$
$$= \|A(y_k, y_{k+1})^{-1}[F(y_{k+1}) + G(y_{k+1}) - A(y_{k-1}, y_k)(y_{k+1} - y_k) - F(y_k) - G(y_k)]\|$$
$$\leq \|A(y_k, y_{k+1})^{-1} A(v, w)\| \, \|A(v, w)^{-1}[F(y_{k+1}) - F(y_k)$$
$$\quad - A(y_{k-1}, y_k)(y_{k+1} - y_k) + G(y_{k+1}) - G(y_k)]\|$$
$$\leq \frac{\{\int_0^1 [h_1(\|y_k - w\| + t\|y_{k+1} - y_k\|) - h_2(\|y_k - w\|) + b] dt + h_3(\|y_{k+1} - y_{k-1}\|)\} \|y_{k+1} - y_k\|}{1 - a - h_0(t_k - t_{-1} + c_{-1}, t_{k+1} - t_0 + r_0)}$$
$$\leq \frac{\{\int_0^1 [h_1(t_k - t_0 + r_0 + t(t_{k+1} - t_k)) - h_2(t_k - t_0 + r_0) + b] dt + h_3(t_{k+1} - t_{k-1})\}(t_{k+1} - t_k)}{1 - a - h_0(t_k - t_{-1} + c_{-1}, t_{k+1} - t_0 + r_0)}$$
$$= t_{k+2} - t_{k+1}, \tag{8.51}$$

which shows (8.44) for all $n \geq 0$.

Note also that

$$\|y_{k+2} - y_{k+1}\| \leq \|y_{k+2} - y_{k+1}\| + \|y_{k+1} - z\|$$
$$\leq t_{k+2} - t_{k+1} + t_{k+1} - t_0 + r_0 = t_{k+2} - t_0 + r_0 \leq t_{k+2} \leq t^*. \tag{8.52}$$

That is, $y_{k+2} \in \overline{U}(z, t^*)$.

It follows from (8.44) that $\{y_n\}$ $(n \geq -1)$ is a Cauchy sequence in a Banach space X, and as such it converges to some $x^* \in X \in \overline{U}(w, t^*)$ (since $\overline{U}(w, t^*)$ is a closed set). By letting $k \to \infty$ in (8.51) we obtain $F(x^*) + G(x^*) = 0$. Estimate (8.45) follows from (8.44) by using standard majorization techniques (see Chapter 4 or [68], [99]).

To show uniqueness in $\overline{U}(w, t^*)$, let y^* be a solution of equation (8.1) in $\overline{U}(w, t^*)$. We define Newton-like iteration $\{x_n\}$ $(n \geq -1)$ by

$$x_{-1} = v, \quad x_0 = w, \quad x_{n+1} = x_n - A(x_{n-1}, x_n)^{-1}[F(x_n) + G(x_n)] \quad (n \geq 0) \tag{8.53}$$

Iteration $\{x_n\}$ $(n \geq -1)$ is a special case of $\{y_n\}$ $(n \geq -1)$. Hence, we have

$$\|x_{k+1} - x_k\| \leq \bar{t}_{k+1} - \bar{t}_k, \quad \lim_{n \to \infty} x_n = x^*$$

and

$$\|x^* - x_k\| \leq t^* - \bar{t}_k, \quad \lim \bar{t}_k = t^*, \tag{8.54}$$

where $\{\bar{t}_n\}$ is $\{t_n\}$ $(n \geq -1)$ for $r_0 = 0$.

We shall show:
$$\|y^* - x_k\| \leq t^* - \bar{t}_k. \tag{8.55}$$

For $k = 0$ (8.54) holds since $y^* \in \overline{U}(w, t^*)$. Suppose (8.54) holds for all $n \leq k$. Then as in (8.51) we obtain the identity:

$$\begin{aligned}
y^* - x_{k+1} &= y^* - x_k + A(x_{k-1}, x_k)^{-1}(F(x_k) + G(x_k)) \\
&\quad - A(x_{k-1}, x_k)^{-1}(F(y^*) + G(y^*)) \\
&= [A(x_{k-1}, x_k)^{-1} A(x_{-1}, x_0)] A(x_{-1}, x_0)^{-1} [F(x_k) - F(y^*) \\
&\quad - A(x_{k-1}, x_k)(y^* - x_k) + G(x_k) - F(y^*)].
\end{aligned} \tag{8.56}$$

Using (8.56) we obtain in turn:

$$\begin{aligned}
\|y^* - x_{k+1}\| &\leq \frac{[\int_0^1 h_1(\|x_k - x_0\| + t\|y^* - x_k\|)dt - h_2(\|x_k - x_0\|) + h_3(\|y^* - x_{k-1}\|) + b]\|y^* - x_k\|}{1 - a - h_0(\|x_{k-1} - x_{-1}\|, \|x_n - x_0\|)} \\
&\leq \frac{[\int_0^1 h_1((1+t)t^*)dt - h_2(t^*) + h_3(t^*) + b]}{1 - a - h_0(t^* + c_1, t^*)} \|y^* - x_k\| \\
&< \|y^* - x_k\| \leq t^* - \bar{t}_k \to 0 \quad \text{as } k \to \infty.
\end{aligned} \tag{8.57}$$

That is, $x^* = y^*$.

If $y^* \in \overline{U}(x_0, R_0)$ then as in (8.57) we get

$$\|y^* - x_{k+1}\| \leq \frac{[\int_0^1 h_1(t^* + tR_0)dt - h_2(t^*) + h_3(R_0) + b]}{1 - a - h_0(t^* + c_1, t^*)} \|y^* - x_k\|$$
$$< \|y^* - x_k\|. \tag{8.58}$$

Hence, again we get $x^* = y*$.

That completes the proof of Theorem 8.2. □

Remark 8.2 *Conditions (8.31), (8.32) can be replaced by the stronger but easier to check*

$$h_0 \left[\tfrac{2\eta}{2-\delta} + c + c_{-1}, \tfrac{2\eta}{2-\delta} + r_0 \right] + a \leq 1, \tag{8.59}$$

and

$$\begin{aligned}
&2 \int_0^1 h_1 \left[\tfrac{2\eta}{2-\delta} + \theta\tfrac{\delta}{2} + r_0 \right] d\theta - 2h_2 \left[\tfrac{2\eta}{2-\delta} + r_0 \right] \\
&\quad + 2h_3 \left[\left(1 + \tfrac{\delta}{2}\right) \eta \right] + \delta h_0 \left[\tfrac{2\eta}{2-\delta} + c + c_{-1}, \tfrac{2\eta}{2-\delta} + r_0 \right] + 2b + \delta a \\
&\leq \delta
\end{aligned} \tag{8.60}$$

respectively. Note also that conditions (8.29) − (8.32), (8.59), (8.60) are of the Newton–Kantorovich-type hypotheses (see also (8.5)) which are always present in the study of Newton-like methods.

Application 8.1 Let us consider some special choices of operator A, functions h_i $i = 0, 1, 2, 3$, parameters a, b and points v, w.

Define

$$A(x,y) = F'(y) + [x,y;G], \qquad (8.61)$$

$$v = y_{-1}, \quad w = y_0, \qquad (8.62)$$

and set

$$r_0 = 0, \qquad (8.63)$$

where F', $[\cdot,\cdot;G]$ denote the Fréchet-derivative of F and the divided difference of order one for operator G respectively. Hence, we consider Newton-like method (8.2) in the form

$$y_{n+1} = y_n - (F'(y_n) + [y_{n-1}, y_n; G])^{-1}(F(y_n) + G(y_n)) \quad (n \geq 0). \qquad (8.64)$$

The method was studied in [112], [68], [99]. It is shown to be of order $\frac{1+\sqrt{5}}{2} \approx 1.618\ldots$ (same as the order of Chord), but higher than the order of

$$z_{n+1} = z_n - F'(z_n)^{-1}(F(z_n) + G(z_n)) \quad (n \geq 0) \qquad (8.65)$$

and

$$w_{n+1} = w_n - A(w_n)^{-1}(F(w_n) + G(w_n)) \quad (n \geq 0), \qquad (8.66)$$

where $A(\cdot)$ is an operator approximating F'. Assume:

$$\|A(y_{-1}, y_0)^{-1}[F'(y) - F'(y_0)]\| \leq \gamma_2 \|y - y_0\|, \qquad (8.67)$$

$$\|A(y_{-1}, y_0)^{-1}[F'(x) - F'(y)]\| \leq \gamma_3 \|x - y\|, \qquad (8.68)$$

$$\|A(y_{-1}, y_0)^{-1}([x,y;G] - [y_{-1}, y_0; G])\| \leq \gamma_4(\|x - y_{-1}\| + \|y - y_0\|), \qquad (8.69)$$

and

$$\|A(y_{-1}, y_0)^{-1}([x,y;G] - [z,x;G])\| \leq \gamma_5 \|z - y\| \qquad (8.70)$$

for some non-negative parameters γ_i, $i = 2, 3, 4, 5$ and all $x, y \in \overline{U}(y_0, r) \subseteq \overline{U}(y_0, R)$.

Then we can define

$$a = b = 0, \quad h_1 = h_2, \quad h_1(q) = \gamma_3 q, \quad h_3(q) = \gamma_5 q \quad \text{and}$$
$$h_0(q_1, q_2) = \gamma_4 q_1 + (\gamma_2 + \gamma_4) q_2. \tag{8.71}$$

If the hypotheses of Theorem 8.2 hold for the above choices, the conclusions follow.

Note that conditions (8.67)–(8.70) are weaker than the corresponding ones in [109, pp. 48–49]. Indeed, conditions $\|F'(x) - F'(y)\| \leq \gamma_6 \|x - y\|$, $\|A(x, y)^{-1}\| \leq \gamma_7$, $\|[x, y, z; G]\| \leq \gamma_8$, and

$$\|[x, y; G] - [z, w; G]\| \leq \gamma_9 (\|x - z\| + \|y - w\|)$$

for all $x, y, z, w \in \overline{U}(y_0, r)$ are used there instead of (8.67)–(8.70), where $[x, y, z; G]$ denotes a second order divided difference of G at (x, y, z), and γ_i, $i = 6, 7, 8, 9$ are non-negative parameters.

Let us provide an example for this case:

Example 8.2 Let $X = Y = (\mathbf{R}^2, \|\cdot\|_\infty)$. Consider the system

$$3x^2 y + y^2 - 1 + |x - 1| = 0$$
$$x^4 + xy^3 - 1 + |y| = 0.$$

Set $\|x\|_\infty = \|(x', x'')\|_\infty = \max\{|x'|, |x''|\}$, $F = (F_1, F_2)$, $G = (G_1, G_2)$. For $x = (x', x'') \in \mathbf{R}^2$ we take $F_1(x', x'') = 3(x')^2 x'' + (x'')^2 - 1$, $F_2(x', x'') = (x')^4 + x'(x'')^3 - 1$, $G_1(x', x'') = |x' - 1|$, $G_2(x', x'') = |x''|$. We shall take $[x, y; G] \in M_{2 \times 2}(\mathbf{R})$ as $[x, y; G]_{i,1} = \frac{G_i(y', y'') - G_i(x', y'')}{y' - x'}$, $[x, y; G]_{i,2} = \frac{G_i(x', y'') - G_i(x', x'')}{y'' - x''}$, $i = 1, 2$.

Using method (8.65) with $z_0 = (1, 0)$ we obtain

n	$z_n^{(1)}$	$z_n^{(2)}$	$\|z_n - z_{n-1}\|$
0	1	0	
1	1	0.333333333333333	$3.333E-1$
2	0.906550218340611	0.354002911208151	$9.344E-2$
3	0.885328400663412	0.338027276361322	$2.122E-2$
4	0.891329556832800	0.326613976593566	$1.141E-2$
5	0.895238815463844	0.326406852843625	$3.909E-3$
6	0.895154671372635	0.327730334045043	$1.323E-3$
7	0.894673743471137	0.327979154372032	$4.809E-4$
8	0.894598908977448	0.327865059348755	$1.140E-4$
9	0.894643228355865	0.327815039208286	$5.002E-5$
10	0.894659993615645	0.327819889264891	$1.676E-5$
11	0.894657640195329	0.327826728208560	$6.838E-6$
12	0.894655219565091	0.327827351826856	$2.420E-6$
13	0.894655074977661	0.327826643198819	7.086E-7
...			
39	0.894655373334687	0.3278826521746298	$5.149E-19$

Using the method of chord (i.e. (8.66) with $A(w_n) = [w_{n-1}, w_1; G]$) with $w_{-1} = (5,5)$, $w_0 = (1,0)$, we obtain

n	$w_n^{(1)}$	$w_n^{(2)}$	$\|w_n - w_{n-1}\|$
0	5	5	
1	1	0	5.000E+00
2	0.989800874210782	0.012627489072365	1.262E–02
3	0.921814765493287	0.307939916152262	2.953E–01
4	0.900073765669214	0.325927010697792	2.174E–02
5	0.894939851625105	0.327725437396226	5.133E–03
6	0.894658420586013	0.327825363500783	2.814E–04
7	0.894655375077418	0.327826521051833	3.045E–04
8	0.894655373334698	0.327826521746293	1.742E–09
9	0.894655373334687	0.327826521746298	1.076E–14
10	0.894655373334687	0.327826521746298	5.421E–20

Example 8.3 Using our method (8.64) with $y_{-1} = (5,5)$, $y_0 = (1,0)$, we obtain

n	$y_n^{(1)}$	$y_n^{(2)}$	$\|y_n - y_{n-1}\|$
0	5	5	
1	1	0	5
2	0.909090909090909	0.363636363636364	3.636E−01
3	0.894886945874111	0.329098638203090	3.453E−02
4	0.894655531991499	0.327827544745569	1.271E−03
5	0.894655373334793	0.327826521746906	1.022E−06
6	0.894655373334687	0.327826521746298	6.089E−13
7	0.894655373334687	0.327826521746298	2.710E−20

Example 8.4 We did not verify the hypotheses of Theorem 8.3 for the above starting points. However, it is clear that the hypotheses of Theorem 8.3 are satisfied for all three methods for starting points closer to the solution

$$x^* = (.894655373334687, .327826521746298)$$

chosen from the lists of the tables displayed above.

Hence method (8.2) (i.e. method (8.64) in this case) converges faster than (8.65) suggested in Chen and Yamamoto [116], Zabrejko and Nguen [295] in this case and the method of chord.

In the application that follows we show that the famous Newton–Kantorovich hypothesis is weakened under the same hypotheses/information.

Application 8.2 Returning back to Remark 8.1 and (8.28), iteration (8.2) reduces to the famous Newton–Kantorovich method (8.8).

Condition (8.29) reduces to:

$$h_\delta = (\gamma_1 + \delta\gamma_0)\eta \leq \delta. \tag{8.72}$$

Case 8.1 Let us restrict $\delta \in [0,1]$. Hypothesis (8.32) now becomes

$$2\int_0^1 \gamma_1 \left[\tfrac{2\eta}{2-\delta}\left(1-\left(\tfrac{\delta}{2}\right)^{k+1}\right) + \theta\left(\tfrac{\delta}{2}\right)^{k+1}\eta\right]d\theta$$
$$- 2\gamma_1\left[\tfrac{2\eta}{2-\delta}\left(1-\left(\tfrac{\delta}{2}\right)^{k+1}\right)\right] + \delta\gamma_0\left[\tfrac{2\eta}{2-\delta}\left(1-\left(\tfrac{\delta}{2}\right)^{k+1}\right)\right]$$
$$\leq 2\gamma_1\int_0^1 \theta\eta d\theta + \delta\gamma_0\eta$$

or
$$\left[\tfrac{\gamma_0\delta^2}{2-\delta}-\gamma_1\right]\left[1-\left(\tfrac{\delta}{2}\right)^{k+1}\right]\le 1,$$

which is true for all $k \ge 0$ by the choice of δ. Furthermore (8.31) gives
$$\tfrac{2\gamma_0\eta}{2-\delta} \le 2\gamma_0\eta < 1.$$

Hence in this case conditions (8.29), (8.31) and (8.32) reduce only to (8.72) provided $\delta \in [0,1]$. Condition (8.72) for say $\delta = 1$ reduces to (8.20).

Case 8.2 It follows from Case 1 that (8.29), (8.31) and (8.32) reduce to (8.72),
$$\tfrac{2\gamma_0\eta}{2-\delta} \le 1 \tag{8.73}$$

and
$$\tfrac{\gamma_0\delta^2}{2-\delta} \le \gamma_1 \tag{8.74}$$

respectively provided $\delta \in [0,2)$.

Case 8.3 It turns out that the range for δ can be extended (see also Example 8.5). Introduce conditions:
$$\gamma_0\eta \le 1 - \tfrac{1}{2}\delta, \quad \text{for } \delta \in [\delta_0, 2),$$

where,
$$\delta_0 = \tfrac{-b+\sqrt{b^2+8b}}{2}, \quad b = \tfrac{\gamma_1}{\gamma_0} \text{ and } \gamma_0 \ne 0.$$

Indeed the proof of Theorem 8.2 goes through if instead we show the weaker condition
$$\gamma_1\left(\tfrac{\delta}{2}\right)^{k+1} + \tfrac{2\gamma_0\delta}{2-\delta}\left[1-\left(\tfrac{\delta}{2}\right)^{k+2}\right] \le \delta,$$

or
$$\left(b - \tfrac{b\delta}{2} - \tfrac{\delta^2}{2}\right)\left(\tfrac{\delta}{2}\right)^{k+1} \le 0,$$

or
$$\delta \ge \delta_0,$$

which is true by the choice of δ_0.

Example 8.5 Returning back to Example 8.1 but using Case 8.3 we can do better. Indeed, choose

$$p = p_0 = .4505 < \frac{5-\sqrt{13}}{3} = .464816242\ldots.$$

Then we get

$$\eta = .183166\ldots, \quad \gamma_0 = 2.5495, \quad \gamma_1 = 3.099, \quad \text{and} \quad \delta_0 = 1.0656867.$$

Choose $\delta = \delta_0$. Then we get

$$\gamma_0 \eta = .466983415 < 1 - \frac{\delta_0}{2} = .46715665.$$

That is the interval $\left[\frac{5-\sqrt{13}}{2}, \frac{1}{2}\right)$ can be extended to at least $\left[p_0, \frac{1}{2}\right)$.

In the example that follows we show that $\frac{\gamma_1}{\gamma_0}$ can be arbitrarily large. Indeed:

Example 8.6 Let $X = Y = \mathbf{R}$, $d_0 = 0$ and define functions F, G on R by

$$F(x) = c_0 x + c_1 + c_2 \sin e^{c_3 x}, \quad G(x) = 0, \quad (8.75)$$

where, c_i, $i = 0, 1, 2, 3$ are given parameters. Using (8.75) it can easily be seen that for c_3 large and c_2 sufficiently small $\frac{\gamma_1}{\gamma_0}$ can be arbitrarily large.

Part C: Specialization to One-step Methods

In order to compare with earlier results, we consider the case when $x = y$ and $v = w$ (single step methods). We can then prove along the same lines to Lemma 8.1 and Theorem 8.2 respectively the following results by assuming:

there exists $w \in X$ such that $A(w)^{-1} \in L(Y, X)$, for any $x, y \in \overline{U}(w, r) \subseteq \overline{U}(w, R)$, $t \in [0, 1]$:

$$\|A(w)^{-1}[A(x) - A(w)]\| \leq g_0(\|x - w\|) + \alpha \quad (8.76)$$

and

$$\|A(w)^{-1}\{[F(x + t(y - x)) - A(x)](y - x) + G(y) - G(x)\}\|$$
$$\leq [g_1(\|x - w\| + t\|y - x\|) - g_2(\|x - w\|) + g_3(r) + \beta]\|y - x\|, \quad (8.77)$$

where g_0, g_1, g_2, g_3, α, β are as h_0, (one variable) h_1, h_2, h_3, a, and b respectively.

Then we can show the following result on majorizing sequences.

Lemma 8.2 *Assume:*
there exist $\eta \geq 0$, $\alpha \geq 0$, $\beta \geq 0$, $\delta \in [0, 2)$, $r_0 \in [0, R]$ such that:

$$\overline{h}_\delta = 2\left[\int_0^1 g_1(r_0 + \theta\eta)d\theta - g_2(r_0) + g_3(r_0 + \eta) + \beta\right] \\ + \delta[\alpha + g_0(r_0 + \eta)] \leq \delta, \tag{8.78}$$

$$\tfrac{2\eta}{2-\delta} + r_0 \leq R, \tag{8.79}$$

$$g_0\left[\tfrac{2\eta}{2-\delta}\left(1 - \left(\tfrac{\delta}{2}\right)^{n+1}\right) + r_0\right] + \alpha < 1, \tag{8.80}$$

$$2\int_0^1 g_1\left[\tfrac{2\eta}{2-\delta}\left(1 - \left(\tfrac{\delta}{2}\right)^{n+1}\right) + r_0 + \theta\left(\tfrac{\delta}{2}\right)^{n+1}\eta\right]d\theta \\ - 2g_2\left[\tfrac{2\eta}{2-\delta}\left(1 - \left(\tfrac{\delta}{2}\right)^{n+1}\right) + r_0\right] + 2g_3\left[\tfrac{2\eta}{2-\delta}\left(1 - \left(\tfrac{\delta}{2}\right)^{n+1}\right) + r_0\right] \\ + \delta g_0\left[\tfrac{2\eta}{2-\delta}\left(1 - \left(\tfrac{\delta}{2}\right)^{n+1}\right) + r_0\right] + 2\beta + \delta\alpha \\ \leq \delta \tag{8.81}$$

for all $n \geq 0$.
Then, iteration $\{s_n\}$ ($n \geq 0$) given by

$$s_0 = r_0, \quad s_1 = r_0 + \eta, \quad s_{n+2} = s_{n+1} \\ + \frac{\int_0^1\{g_1(s_n + \theta(s_{n+1}-s_n)) - g_2(s_n) + \beta\}d\theta(s_{n+1}-s_n) + \int_{s_n}^{s_{n+1}} g_3(\theta)d\theta}{1 - \alpha - g_0(s_{n+1})} \tag{8.82}$$

is monotonically increasing, bounded above by

$$s^{**} = \frac{2\eta}{2-\delta} + r_0, \tag{8.83}$$

and converges to some s^* such that

$$0 \leq s^* \leq s^{**}. \tag{8.84}$$

Moreover the following error bounds hold for all $n \geq 0$

$$0 \leq s_{n+2} - s_{n+1} \leq \tfrac{\delta}{2}(s_{n+1} - s_n) \leq \left(\tfrac{\delta}{2}\right)^{n+1}\eta. \tag{8.85}$$

Theorem 8.3 *Assume:*
hypotheses of Lemma 8.2 hold and there exists $y_0 \in \overline{U}(w, r_0)$ *such that*

$$\|A(y_0)^{-1}[F(y_0) + G(y_0)]\| \leq \eta. \tag{8.86}$$

Then, sequence $\{w_n\}$ $(n \geq 0)$ *generated by Newton-like method* (8.66) *is well-defined, remains in* $\overline{U}(w, s^*)$ *for all* $n \geq 0$, *and converges to a solution* x^* *of equation* $F(x) + G(x) = 0$. *Moreover the following error bounds hold for all* $n \geq 0$

$$\|w_{n+1} - w_n\| \leq s_{n+1} - s_n \tag{8.87}$$

and

$$\|w_n - x^*\| \leq s^* - s_n. \tag{8.88}$$

Furthermore the solution x^* *is unique in* $\overline{U}(w, s^*)$ *if*

$$\int_0^1 [g_1(s^* + \theta s^*) - g_2(s^*)]d\theta + g_3(s^*) + g_0(s^*) + \alpha + \beta < 1, \tag{8.89}$$

or in $\overline{U}(w, R_0)$ *if* $s^* < R_0 \leq R$, *and*

$$\int_0^1 [g_1(s^* + \theta R_0) - g_2(s^*)]d\theta + g_3(s^* + R_0) + g_0(s^*) + \alpha + \beta < 1, \tag{8.90}$$

provided that $w_0 = w$.

We state the relevant results due to Chen and Yamamoto [116, pp. 40]. We assume: $A(w)^{-1}$ exists, and for any $x, y \in \overline{U}(w, r) \subseteq \overline{U}(z, R)$:

$$0 < \|A(w)^{-1}(F(w) + G(w))\| \leq \overline{\eta}, \tag{8.91}$$

$$\|A(w)^{-1}(A(x) - A(w))\| \leq \overline{g}_0(\|x - w\|) + \overline{\alpha}, \tag{8.92}$$

$$\|A(w)^{-1}[F'(x + t(y - x)) - A(x)]\| \leq \overline{g}_1(\|x - w\|) + t\|y - x\|)$$
$$- \overline{g}_0(\|x - w\|) + \overline{\beta}, \quad t \in [0, 1], \tag{8.93}$$

$$\|A(w)^{-1}[G(x) - G(y)]\| \leq g_3(r)\|x - y\|, \tag{8.94}$$

where $\overline{g}_0, \overline{g}_1, \overline{\alpha}, \overline{\beta}$ are as g_0, g_1, α, β respectively, but $\overline{g}_0$ is also differentiable with $\overline{g}'_0(r) > 0$, $r \in [0, R]$, and $\overline{\alpha} + \overline{\beta} < 1$.

As in [116] set:

$$\varphi(r) = \overline{\eta} - r + \int_0^r \overline{g}_1(t)dt, \quad \psi(r) = \int_0^r g_3(t)dt, \qquad (8.95)$$

$$\chi(r) = \phi(r) + \psi(r) + (\overline{\alpha} + \overline{\beta})r. \qquad (8.96)$$

Denote the minimal value of $\chi(r)$ on $[0, R]$ by χ^*, and the minimal point by r^*. If $\chi(R) \leq 0$, denote the unique zero of χ by $r_0^* \in (0, r^*]$. Define scalar sequence $\{r_n\}$ $(n \geq 0)$ by

$$r_0 \in [0, R], \quad r_{n+1} = r_n + \frac{u(r_n)}{g(r_n)} \quad (n \geq 0) \qquad (8.97)$$

where

$$u(r) = \chi(r) - x^* \qquad (8.98)$$

and

$$g(r) = 1 - \overline{g}_0(r) - \overline{\alpha}. \qquad (8.99)$$

With the above notation they showed:

Theorem 8.4 *[116, pp. 40] Suppose $\chi(R) \leq 0$. Then equation (8.1) has a solution $x^* \in \overline{U}(w, r_0^*)$, which is unique in*

$$\tilde{U} = \begin{cases} \overline{U}(w, R) & \text{if } \chi(R) = 0 \text{ or } \psi(R) = 0, \text{ and } r_0^* < R. \\ U(w, R) & \text{if } \chi(R) = 0 \text{ and } r_0^* < R. \end{cases} \qquad (8.100)$$

Let

$$D^* = \overline{U}_{r \in [0, r^*)} \left\{ y \in \overline{U}(w, r) \mid \|A(y)^{-1}[F(y) + G(y)]\| \leq \frac{u(r)}{g(r)} \right\}. \qquad (8.101)$$

Then, for any $y_0 \in D$, sequence $\{y_n\}$ $(n \geq 0)$ generated by Newton-like method (8.66) is well defined, remains in $\overline{U}(w, r^*)$ and satisfies

$$\|y_{n+1} - y_n\| \leq r_{n+1} - r_n, \qquad (8.102)$$

and

$$\|y_n - x^*\| \leq r^* - r_n \qquad (8.103)$$

provided that r_0 is chosen as in (8.97) so that $r_0 \in Ry_0$, where for $y \in D^*$

$$R_y = \left\{ r \in [0, r^*) \mid \|A(y)^{-1}(F(y) + G(y))\| \leq \frac{u(r)}{y(r)}, \|y - z\| \leq r \right\}. \qquad (8.104)$$

Remark 8.3 *(a) Hypothesis on $\bar{g}_0$ is stronger than the corresponding one on g_0.*

(b) Iteration (8.97) converges to r^* (even if $r_0 = 0$) not r_0^*.

(c) Choices of y_{-1}, y_0 other than the ones in Theorems 8.2, 8.3 can be given by (8.101) and (8.102).

Remark 8.4 *The conclusions of Theorem 8.4 hold (i.e. the results in [116] were improved) if the more general conditions (8.76), (8.77) replace (8.92) − (8.94), and*

$$\bar{g}_0(r) \leq g_2(r), \quad r \in [0, R], \tag{8.105}$$

is satisfied. Moreover if strict inequality holds in (8.105) we obtain more precise error bounds. Indeed, define the sequence $\{\bar{r}_n\}$ ($n \geq 0$), using (8.77), g_2 instead of (8.93), $\bar{g}_0$ respectively (with $\bar{g}_1 = g_1$, $\alpha = \bar{\alpha}$, $\beta = \bar{\beta}$) by

$$\bar{r}_0 = r_0, \ \bar{r}_1 = r_1, \ \bar{r}_{n+1} - \bar{r}_n = \frac{u(\bar{r}_n) - u(\bar{r}_{n-1}) + (1 - g_2(\bar{r}_{n-1}) - \bar{\alpha})(\bar{r}_n - \bar{r}_{n-1})}{g(\bar{r}_n)} \ (n \geq 1) \tag{8.106}$$

It can easily be seen using induction on n (see also the proof of Proposition 8.1 that follows) that

$$\bar{r}_{n+1} - \bar{r}_n < r_{n+1} - r_n, \tag{8.107}$$

$$\bar{r}_n < r_n, \tag{8.108}$$

$$\bar{r}^* - \bar{r}_n \leq r^* - r_n, \ \bar{r}^* = \lim_{n \to \infty} \bar{r}_n, \tag{8.109}$$

and

$$\bar{r}^* \leq r^*. \tag{8.110}$$

Furthermore condition (8.77) allows us more flexibility in choosing functions and constants.

Remark 8.5 *Returing back to Newton's method (8.8) (see also (8.28)), the iterations corresponding to (8.97) and (8.106) are (8.13) and (8.17) respectively. Moreover condition (8.105) reduces to (8.19), and in case $\gamma_0 < \gamma_1$, estimates (8.22) − (8.24) hold.*

Remark 8.6 *Our error bounds (8.87), (8.88) are finer than the corresponding ones (8.102) and (8.103) respectively in many interesting cases. Let us choose:*

$$\alpha = \bar{\alpha}, \quad \beta = \bar{\beta}, \quad g_0(r) = \bar{g}_0(r), \quad g_1(r) = g_2(r) = \bar{g}_1(r), \quad \text{and}$$
$$\bar{g}_3(r) = g_3(r) \quad \text{for all } r \in [0, R].$$

Then we can show:

Proposition 8.1 *Under the hypotheses of Theorems 8.3 and 8.4, further assume:*

$$s_1 < r_1. \tag{8.111}$$

Then, the following hold:

$$s_n < r_n \quad (n \geq 1), \tag{8.112}$$
$$s_{n+1} - s_n < r_{n+1} - r_n \quad (n \geq 0), \tag{8.113}$$
$$s^* - s_n \leq r^* - r_n \quad (n \geq 0), \tag{8.114}$$

and

$$s^* \leq r^*. \tag{8.115}$$

Proof. It suffices to show (8.112) and (8.113), since then (8.114) and (8.115) respectively can easily follow. Inequality (8.112) holds for $n = 1$ by (8.111). By (8.82) and (8.97) we get in turn

$$s_2 - s_1 = \frac{\int_0^1 \{g_1(s_0 + \theta(s_1 - s_0))d\theta - g_2(s_0) + \alpha\}(s_1 - s_0) + \int_{s_0}^{s_1} g_3(\theta)d\theta}{1 - \beta - g_0(s_1)}$$

$$< \frac{\int_0^1 \{\overline{g}_1(r_0 + \theta(r_1 - r_0))d\theta - \overline{g}_2(r_0) + \overline{\alpha}\}(r_1 - r_0) + \int_{r_0}^{r_1} \overline{g}_3(\theta)d\theta}{1 - \overline{\beta} - \overline{g}_0(r_1)}$$

$$= \frac{u(r_1) - u(r_0) + g(r_0)(r_1 - r_0)}{1 - \overline{\beta} - \overline{g}_0(r_1)} = \frac{u(r_1)}{g(r_1)} = r_2 - r_1. \tag{8.116}$$

Assume:

$$s_{k+1} < r_{k+1}, \tag{8.117}$$

and

$$s_{k+1} - s_k < r_{k+1} - r_k \tag{8.118}$$

hold for all $k \leq n$.

Using (8.82), (8.88), and (8.118) we obtain

$$s_{k+2} - s_{k+1}$$
$$= \frac{\int_0^1 \{g_1[s_k + \theta(s_{k+1} - s_k)]d\theta - g_2(s_k) + \alpha\}(s_{k+1} - s_k) + \int_{s_k}^{s_{k+1}} g_3(\theta)d\theta}{1 - \beta - g_0(s_{k+1})}$$

$$< \frac{\int_0^1 \{\overline{g}_1[r_k + \theta(r_{k+1} - r_k)]d\theta - \overline{g}_2(r_k) + \overline{\alpha}\}(r_{k+1} - r_k) + \int_{r_k}^{r_{k+1}} \overline{g}_3(\theta)d\theta}{1 - \overline{\beta} - \overline{g}_0(r_{k+1})}$$

$$= \frac{u(r_{k+1}) - u(r_k) + g(r_k)(r_{k+1} - r_k)}{g(r_{k+1})} = \frac{u(r_{k+1})}{g(r_{k+1})} = r_{k+2} - r_{k+1}.$$

That completes the proof of Proposition 8.1. □

In order for us to include a case where operator G is nontrivial, we consider the following example for Theorem 8.2 (or Theorem 8.3):

Example 8.7 Let $X = Y = C[0,1]$ the space of continuous functions on $[0,1]$ equipped with the sup-norm. Consider the integral equation on $\overline{U}(x_0, \frac{R}{2})$ given by

$$x(t) = \int_0^1 k(t, s, x(s))ds, \qquad (8.119)$$

where the kernel $k(t, s, x(s))$ with $(t, s) \in [0, 1] \times [0, 1]$ is a nondifferentiable operator on $\overline{U}(x_0, \frac{R}{2})$. Define operators F, G on $\overline{U}(x_0, \frac{R}{2})$ by

$$F(x)(t) = Ix(t) \quad (I \text{ the identity operator}) \qquad (8.120)$$

$$G(x)(t) = -\int_0^1 k(t, s, x(s))ds. \qquad (8.121)$$

Choose $x_0 = 0$, and assume there exists a constant $\theta_0 \in [0, 1)$, a real function $\theta_1(t, s)$ such that

$$\|k(t, s, x) - k(t, s, y)\| \leq \theta_1(t, s)\|x - y\| \qquad (8.122)$$

and

$$\sup_{t \in [0,1]} \int_0^1 \theta_1(t, s)ds \leq \theta_0 \qquad (8.123)$$

for all $t, s \in [0, 1]$, $x, y \in \overline{U}(x_0, \frac{R}{2})$.

Moreover choose in Theorem 8.3: $r_0 = 0$, $y_0 = y_{-1}$, $A(x, y) = A(x) = I(x)$, I the identity operator on X, $g_0(r) = r$, $\alpha = \beta = 0$, $g_1(r) = g_2(r) = 0$, and $g_3(r) = \theta_0$ for all $x, y \in \overline{U}(x_0, \frac{R}{2})$, $r, s \in [0, 1]$ (similar choices for Theorem 8.3). It can easily be seen that the conditions of Theorem 8.2 hold if

$$t^* = \frac{\eta}{1 - \theta_0} \leq \frac{R}{2}. \qquad (8.124)$$

Local Convergence.

In order to cover the local case, let us assume x^* is a zero of equation (8.1), $A(x^*, x^*)^{-1}$ exists and for any $x, y \in \overline{U}(x^*, r) \subseteq \overline{U}(x^*, R)$, $t \in [0, 1]$:

$$\|A(x^*, x^*)^{-1}[A(x, y) - A(x^*, x^*)]\| \leq \overline{h}_0(\|x - x^*\|, \|y - x^*\|) + \overline{a}, \quad (8.125)$$

and

$$\|A(x^*, x^*)^{-1}(F'(x^* + t(y - x^*)) - A(x, y))(y - x^*) + G(y) - G(x^*)]\|$$
$$\leq [\overline{h}_1(\|y - x^*\|(1 + t)) - \overline{h}_2(\|y - x^*\|) + \overline{h}_3(\|x - x^*\|) + \overline{b}]\|y - x^*\|,$$
$$(8.126)$$

where, $\overline{h}_0$, $\overline{h}_1$, $\overline{h}_2$, $\overline{h}_3$, $\overline{a}$, $\overline{b}$ are as h_0, h_1, h_2, h_3, a, b respectively. Then exactly as in (8.56) but using (8.125), (8.126), instead of (8.25), (8.26) we can show the following local result for method (8.2).

Theorem 8.5 *Assume:*
there exists a solution of equation

$$f(\lambda) = 0, \quad (8.127)$$

in $[0, R]$ where

$$f(\lambda) = \int_0^1 [\overline{h}_1((1+t)\lambda) - \overline{h}_2(\lambda)]dt + \overline{h}_3(\lambda) + \overline{h}_0(\lambda, \lambda) + \overline{a} + \overline{b} - 1. \quad (8.128)$$

Denote by λ_0 the smallest of the solutions in $[0, R]$. Then, sequence $\{x_n\}$ ($n \geq -1$) generated by Newton-like method is well defined, remains in $\overline{U}(x^*, \lambda_0)$ for all $n \geq 0$ and converges to x^* provided that $x_{-1}, x_0 \in \overline{U}(x^*, \lambda_0)$.

Moreover the following error bounds hold for all $n \geq 0$:

$$\|x^* - x_{n+1}\| \leq p_n, \quad (8.129)$$

where,

$$p_n = \frac{\{\int_0^1 [\overline{h}_1((1+t)\|x_n - x^*\|) - \overline{h}_2(\|x_n - x^*\|)]dt + \overline{a} + \overline{h}_3(\|x_{n-1} - x^*\|)\}}{1 - \overline{b} - \overline{h}_0(\|x_n - x^*\|)} \|x_n - x^*\|.$$
$$(8.130)$$

Application 8.3 Let us again consider Newton's method, i.e., $F'(x) = A(x, y)$, $G(x) = 0$, and assume:

$$\|F'(x^*)^{-1}[F'(x) - F'(x^*)]\| \leq \lambda_1 \|x - x^*\|, \quad (8.131)$$

and

$$\|F'(x^*)^{-1}[F'(x) - F'(y)]\| \leq \lambda_2 \|x - y\| \quad (8.132)$$

for all $x, y \in \overline{U}(x^*, r) \subseteq \overline{U}(x^*, R)$. Then we can set:

$$\overline{a} = \overline{b} = 0, \overline{h}_3 = 0, \overline{h}_1(r) = \overline{h}_2(r) = \lambda_2 r, \text{ and } \overline{h}_0(r, r) = \lambda_1 r \text{ for all } r \in [0, R]. \tag{8.133}$$

Using (8.131), (8.132) we get:

$$\lambda_0 = \frac{2}{2\lambda_1 + \lambda_2}. \tag{8.134}$$

Local results were not given in [116], [295]. However Rheinboldt in [247] showed that under only (8.132) the convergence radius is given by

$$\lambda_3 = \frac{2}{3\lambda_2}. \tag{8.135}$$

But in general

$$\lambda_1 \leq \lambda_2. \tag{8.136}$$

Hence we conclude:

$$\lambda_3 \leq \lambda_0. \tag{8.137}$$

The corresponding error bounds become:

$$\|x_{n+1} - x^*\| \leq e_n, \tag{8.138}$$
$$\|x_{n+1} - x^*\| \leq e_n^1, \tag{8.139}$$

where,

$$e_n = \frac{\lambda_2 \|x_n - x^*\|^2}{2[1 - \lambda_1 \|x_n - x^*\|]} \tag{8.140}$$

and

$$e_n^1 = \frac{\lambda_2 \|x_n - x^*\|^2}{2[1 - \lambda_2 \|x_n - x^*\|]}. \tag{8.141}$$

That is

$$e_n \leq e_n^1 \quad (n \geq 0). \tag{8.142}$$

If strict inequality holds in (8.136) then (8.137) and (8.142) hold as strict inequalities also (see also Example 8.6).

Remark 8.7 As noted in [2], [68], [99], [108], [287]-[291] the local results obtained here can be used for projection methods such as Arnoldi's, the generalized minimum residual method (GMRES), the generalized conjugate residual method (GCR), for combined Newton/finite projection methods and in connection with the mesh independence principle to develop the cheapest and most efficient mesh refinement strategies.

Remark 8.8 The local results can also be used to solve equations of the form $F(x) = 0$, where F' satisfies the autonomous differential equation [68], [99], [166]:

$$F'(x) = P(F(x)), \tag{8.143}$$

where, $P: Y \to X$ is a known continuous operator. Since $F'(x^*) = P(F(x^*)) = P(0)$, we can apply our results without actually knowing the solution x^* of equation (8.1).

Example 8.8 Let $X = Y = \mathbf{R}$, $\overline{U}(x^*, R) = \overline{U}(0, 1)$, $G = 0$, $A(x, y) = F'(x)$, and define function F on $\overline{U}(0, 1)$ by

$$F(x) = e^x - 1. \tag{8.144}$$

Then we can set $P(x) = x + 1$ in (8.143). Using (8.132) we get $\lambda_2 = e$. Moreover by (144) we get

$$F'(x) - F'(x^*) = e^x - 1 = x + \frac{x^2}{2!} + \cdots + \frac{x^n}{n!} + \cdots$$

$$= \left(1 + \frac{x}{2!} + \cdots + \frac{x^{n-1}}{n!} + \cdots\right)(x - x^*) \tag{8.145}$$

and

$$F'(x^*)^{-1}[\|F'(x) - F'(x^*)\|] \leq (e-1)\|x - x^*\|.$$

That is $\lambda_1 = e - 1$. By (8.134) and (8.135) we get

$$\lambda_3 = .245252961 \tag{8.146}$$

and

$$\lambda_0 = .254028662. \tag{8.147}$$

That is our convergence radius λ_0 is larger than the corresponding one λ_3 due to Rheinboldt and our error bounds (8.140) are also finer than (8.141) so that (8.142) holds as a strict inequality. Finally note that all these improvements are made using the same hypotheses/information as in the earlier results. This observation is important in computational mathematics, since a wider choice of initial guesses x_0 becomes available (see also Remark 8.7).

The results obtained here can be extended to m-point methods ($m > 2$ an integer) [68], [99] and can be used in the solution of variational inequalities [267], [274]-[277].

8.2 A Fast Convergent Method

In this section we are concerned with the problem of approximating a solution x^* of the nonlinear equation

$$F(x) = 0, \qquad (8.148)$$

where F is a Fréchet-differentiable operator defined on an open subset D of a Banach space X with values in a Banach space Y.

The secant method is the most popular iterative procedure using two previous iterates and divided differences of order one for approximating x^*. The order of the Secant method is $1.618\ldots$. In the elegant paper of F.A. Potra [236] a three point method was used and divided differences of order one (see also (8.26) and (8.27)). This method is of order $1.839\ldots$. More recently Secant-like methods of order between $1.618\ldots$ and $1.839\ldots$ were introduced in the works of Hernandez, Gutierrez et al. [144]-[151], [169]-[176]. The question arises if it is then possible to realize an iterative method using two previous iterates and divided differences of only order one with at least quadratic convergence.

It turns out that this is possible. Indeed we introduce the method

$$x_{n+1} = x_n - [2x_n - x_{n-1}, x_{n-1}]^{-1} F(x_n) \quad (x_{-1}, x_0 \in D) \quad (n \geq 0) \quad (8.149)$$

for approximating x^*. Here, a linear operator from X into Y, denoted by $[x, y; F]$ or simply $[x, y]$ which satisfies the condition

$$[x, y](x - y) = F(x) - F(y), \qquad (8.150)$$

is called a divided difference of order one. Iteration (8.149) has a geometrical interpretation similar to the Secant method in the scalar case.

In Section 2 we provide a local and semilocal convergence analysis for method (8.149) using Lipschitz-type conditions and the majorant principle. The monotone convergence of method (8.149) is examined on partially ordered topological spaces in Section 3.

We can show the following local convergence result for method (8.149).

Theorem 8.6 *Let F be a nonlinear operator defined on an open subset D of a Banach space X with values in a Banach space Y. Assume:*

equation $F(x) = 0$ has a solution $x^* \in D$ at which the Fréchet derivative $F'(x^*)$ exists, and is invertible;

operator F is Fréchet-differentiable with divided difference of order one on $D_0 \subseteq D$ satisfying the Lipschitz conditions:

$$\|F'(x^*)^{-1}[F'(x) - F'(x^*)]\| \le a\|x - x^*\|, \tag{8.151}$$
$$\|F'(x^*)^{-1}([x,y] - [x,x^*])\| \le b\|y - x^*\| \tag{8.152}$$

and

$$\|F'(x^*)^{-1}([y,y] - [2y - x, x])\| \le c\|y - x\|^2; \tag{8.153}$$

the ball

$$U^* = U(x^*, r^*) \subseteq D_0, \tag{8.154}$$

where,

$$r^* = \frac{4}{a + b + \sqrt{(a+b)^2 + 32c}}; \tag{8.155}$$

$$F'(x) = [x, x] \quad \text{for all} \quad x \in D; \tag{8.156}$$

$$\text{for all } x, y \in D_0 \Rightarrow 2y - x \in D. \tag{8.157}$$

Then, sequence $\{x_n\}$ $(n \ge 0)$ generated by method (8.149) is well defined, remains in $U(x^*, r^*)$ for all $n \ge 0$ and converges to x^* provided that x_{-1}, x_0 belong in $U(x^*, r^*)$. Moreover the following error bounds hold for all $n \ge 0$:

$$\|x_{n+1} - x^*\| \le \frac{b\|x_n - x^*\| + c\|x_{n-1} - x_n\|^2}{1 - a\|x_n - x^*\| - c\|x_{n-1} - x_n\|^2}\|x_n - x^*\|. \tag{8.158}$$

Proof. Let us denote by $L = L(x, y)$ the linear operator

$$L = [2y - x, x]. \tag{8.159}$$

Assume $x, y \in U(x^*, r^*)$. We shall show L is invertible on $U(x^*, r^*)$, and

$$\|L^{-1}F'(x^*)\| \le [1 - a\|y - x^*\| - c\|x - y\|^2]^{-1} \le [1 - ar^* - 4c(r^*)^2]^{-1}. \tag{8.160}$$

Using (8.149), (8.151)–(8.157), we obtain in turn:

$$\|F'(x^*)^{-1}[F'(x^*) - L]\|$$
$$= \|F'(x^*)^{-1}[([x^*, x^*] - [y, y]) + ([y, y] - [2y - x, x])]\|$$
$$\leq a\|y - x^*\| + c\|y - x\|^2$$
$$\leq ar^* + c[\|y - x^*\| + \|x^* - x\|]^2$$
$$\leq ar^* + 4c(r^*)^2 < 1 \qquad (8.161)$$

by the choice of r^*.

It follows from the Banach lemma on invertible operators and (8.161) that L^{-1} exists on $U(x^*, r^*)$, so that estimate (8.160) holds. We can also have:

$$\|F'(x^*)^{-1}([y, x^*] - L)\|$$
$$= \|F'(x^*)^{-1}[([y, x^*] - [y, y]) + ([y, y] - L)]\|$$
$$\leq \|F'(x^*)^{-1}([y, x^*] - [y, y])\| + \|F'(x^*)^{-1}([y, y] - L)\|$$
$$\leq b\|y - x^*\| + c\|y - x\|^2$$
$$\leq br^* + 4c(r^*)^2. \qquad (8.162)$$

Moreover by (8.149) we get for $y = x_n$, $x = x_{n-1}$

$$\|x_{n+1} - x^*\| = \| - L_n^{-1}([x_n, x^*] - L_n)(x_n - x^*)\|$$
$$\leq \|L_n^{-1} F'(x^*)\| \cdot \|F'(x^*)^{-1}([x_n, x^*] - L_n)\| \cdot \|x_n - x^*\|. \qquad (8.163)$$

Estimate (8.158) now follows from (8.161)–(8.163). Furthermore from (8.158), (8.161) and (8.162) we get

$$\|x_{n+1} - x^*\| < \|x_n - x^*\| < r^* \quad (n \geq 0). \qquad (8.164)$$

Hence, sequence $\{x_n\}$ $(n \geq -1)$ is well defined, remains in $U(x^*, r^*)$ for all $n \geq -1$ and converges to x^*.

That completes the proof of Theorem 8.6. □

Let $x, y, z \in D_0$, and define the divided difference of order two of operator F at the points x, y and z denoted by $[x, y, z]$ by

$$[x, y, z](y - z) = [x, y] - [x, z]. \qquad (8.165)$$

Remark 8.9 In order for us to compare method (8.149) with others [236] using divided differences of order one, consider the condition

$$\|F'(x^*)^{-1}([x,y] - [u,v])\| \leq \overline{a}(\|x-u\| + \|y-v\|) \tag{8.166}$$

instead of (8.151) and (8.152). Note that (8.166) implies (8.151) and (8.152). Moreover we have:

$$a \leq 2\overline{a} \tag{8.167}$$

and

$$b \leq 2\overline{a}. \tag{8.168}$$

Therefore stronger condition (8.166) can replace (8.151), (8.153) and (8.156) in Theorem 8.6.

Assuming F has divided differences of order two, condition (8.153) can be replaced by the stronger

$$\|F'(x^*)^{-1}([y,x,y] - [2y-x,x,y])(y-x)\| \leq \overline{c}\|y-x\|, \tag{8.169}$$

or the even stronger

$$\|F'(x^*)^{-1}([u,x,y] - [v,x,y])(y-x)\| \leq \overline{\overline{c}}\|u-v\|. \tag{8.170}$$

Note also that

$$\overline{c} \leq \overline{\overline{c}} \tag{8.171}$$

and we can set

$$c = \overline{c} \tag{8.172}$$

despite the fact that $\overline{c}$ is more difficult to compute since we use divided differences of order two (instead of one). Conditions (8.166) and (8.170) were used in [236] to show method

$$y_{n+1} = y_n - ([y_n, y_{n-1}] + [y_{n-2}, y_n] - [y_{n-2}, y_{n-1}])^{-1} F(y_n) \quad (n \geq 0) \tag{8.173}$$

converges to x^* with order $1.839\ldots$ which is the solution of the scalar equation

$$t^3 - t^2 - t - 1 = 0. \tag{8.174}$$

Potra in [236] has also shown how to compute the Lipschitz constants appearing here in some cases. It follows from (8.158) that there exist c_0, N sufficiently large such that:

$$\|x_{n+1} - x^*\| \leq c_0 \|x_n - x^*\|^2 \quad \text{for all } n \geq N. \tag{8.175}$$

Hence the order of convergence for method (8.149) is essentially at least two, which is higher than $1.839\ldots$. Note also that the radius of convergence r^* given by (8.156) is larger than the corresponding one given in [[236], estimate (8.169)]. This observation is very important since it allows a wider choice of initial guesses x_{-1} and x_0.

It turns out that our convergence radius r^* given by (8.156) can even be larger than the one given by Rheinboldt [247] (see, e.g., Remark 4.2 in [236]) for Newton's method. Indeed under condition (8.166) radius r_R^* is given by

$$r_R^* = \frac{1}{3\overline{a}}. \tag{8.176}$$

We showed in 8.1 that $\frac{\overline{a}}{a}$ (or $\frac{\overline{a}}{\overline{b}}$) can be arbitrarily large. Hence we can have:

$$r_R^* < r^*. \tag{8.177}$$

In 8.1 we also showed that r_R^* is enlarged under the same hypotheses and computational cost as in [247].

We note that condition (8.157) suffices to hold only for x, y being iterates of method (8.149) (see, e.g., Example 8.9).

Condition (8.171) can be removed if $D = X$. In this case (8.154) is also satisfied.

Delicate condition (8.156) can also be replaced by stronger but more practical one which we decided not to introduce originally in Theorem 8.6, so we can leave the result as uncluttered-general as possible.

Indeed, define ball U_1 by

$$U_1 = U(x^*, R^*) \quad \text{with} \quad R^* = 3r^*.$$

If $x_{n-1}, x_n \in U^*$ ($n \geq 0$) then we conclude $2x_n - x_{n-1} \in U_1$ ($n \geq 0$). This is true since it follows from the estimates

$$\|2x_n - x_{n-1} - x^*\| \leq \|x_n - x^*\| + \|x_n - x_{n-1}\|$$
$$\leq 2\|x_n - x^*\| + \|x_{n-1} - x^*\| < 3r^* = R^* \quad (n \geq 0).$$

Hence the proof of Theorem 8.6 goes through if both conditions (8.154), (8.156) are replaced by

$$U_1 \subseteq D_0. \tag{8.2.7'}$$

We can show the following result for the semilocal convergence of method (8.149).

Theorem 8.7 *Let F be a nonlinear operator defined on an open set D of a Banach space X with values in a Banach space Y. Assume:*
operator F has divided differences of order one and two on $D_0 \subseteq D$; there exist points x_{-1}, x_0 in D_0 such that $2x_0 - x_{-1} \in D_0$ and $A_0 = [2x_0 - x_{-1}, x_{-1}]$ is invertible on D;
Set $A_n = [2x_n - x_{n-1}, x_{n-1}]$ $(n \geq 0)$.
there exist constants α, β such that:

$$\|A_0^{-1}([x,y] - [u,v])\| \leq \alpha(\|x - u\| + \|y - v\|), \tag{8.178}$$
$$\|A_0^{-1}([y,x,y] - [2y - x, x, y])\| \leq \beta\|x - y\| \tag{8.179}$$

for all $x, y, u, v \in D$, and condition (8.157) holds;
Define constants γ, δ by

$$\|x_0 - x_{-1}\| \leq \gamma, \tag{8.180}$$
$$\|A_0^{-1} F(x_0)\| \leq \delta, \tag{8.181}$$
$$2\beta\gamma^2 \leq 1; \tag{8.182}$$

define θ, r, h by

$$\theta = \left\{(\alpha + \beta\gamma)^2 + 3\beta(1 - \beta\gamma^2)\right\}^{1/2}, \tag{8.183}$$
$$r = \frac{1 - \beta\gamma^2}{\alpha + \beta\gamma + \theta}, \tag{8.184}$$

and

$$h(t) = -\beta t^3 - (\alpha + \beta\gamma)t^2 + (1 - \beta\gamma^2)t, \tag{8.185}$$
$$\delta \leq h(r) = \frac{1}{3}\frac{\alpha + \beta\gamma + 2\theta}{1 - 2\beta\gamma^2}r^2; \tag{8.186}$$
$$U_0 = U(x_0, r_0) \subseteq D_0, \tag{8.187}$$

where $r_0 \in (0, r]$ is the unique solution of equation

$$h(t) = (1 - 2\beta\gamma^2)\delta \tag{8.188}$$

on interval $(0, r]$.

Then sequence $\{x_n\}$ ($n \geq -1$) generated by method (8.149) is well defined, remains in $U(x_0, r_0)$ for all $n \geq -1$ and converges to a solution x^* of equation $F(x) = 0$.

Moreover the following error bounds hold for all $n \geq -1$

$$\|x_{n+1} - x_n\| \leq t_n - t_{n+1}, \tag{8.189}$$

and

$$\|x_n - x^*\| \leq t_n, \tag{8.190}$$

where,

$$t_{-1} = r_0 + \gamma, \quad t_0 = r_0, \tag{8.191}$$

$$\gamma_0 = \alpha + 3\beta r_0 + \beta\gamma, \quad \gamma_1 = 3\beta r_0^2 - 2\gamma_0 r_0 - \beta\gamma^2 + 1, \tag{8.192}$$

and for $n \geq 0$

$$t_{n+1} = \frac{\gamma_0 t_n - (t_n - t_{n-1})^2 \beta - 2\beta t_n^2}{\gamma_1 + 2\gamma_0 t_n - (t_n - t_{n-1})^2 - 3\beta t_n^2} \cdot t_n. \tag{8.193}$$

Furthermore if D is a convex set and

$$2\alpha(\gamma + 2r_0) < 1, \tag{8.194}$$

x^* is the unique solution of equation (8.148) in $\overline{U}(x_0, r_0)$.

Proof. Sequence $\{t_n\}$ ($n \geq -1$) generated by (8.191)–(8.193) can be obtained if method (8.149) is applied to the scalar polynomial

$$f(t) = -\beta t^3 + \gamma_0 t^2 + \gamma_1 t. \tag{8.195}$$

It is simple calculus to show sequence $\{t_n\}$ ($n \geq -1$) converges monotonically to zero (decreasingly).

We can have:

$$t_{n+1} - t_{n+2}$$

$$= \frac{2(t_{n+1} - t_n)}{f(2t_{n+1} - t_n) - f(t_n)} f(t_{n+1}) \tag{8.196}$$

$$= \frac{\{[\gamma_0 - (2t_n + t_{n+1})\beta](t_n - t_{n+1}) + (t_n - t_{n-1})^2 \beta\}(t_n - t_{n+1})}{1 - \beta\gamma^2 - 2(t_0 - t_{n+1})\alpha - [3(t_0 - t_{n+1})(3t_0 + t_{n+1}) - (t_n - t_{n+1})^2]\beta}$$

$$\leq \frac{(t_n - t_{n+1})\alpha + (t_{n-1} - t_n)^2 \beta}{1 - 2(t_0 - t_{n+1})\alpha - \beta\gamma^2}(t_n - t_{n+1}). \tag{8.197}$$

We show (8.189) holds for all $n \geq -1$. Using (8.180)–(8.185) and

$$t_0 - t_1 = \left[1 - \frac{\gamma_0 t_0 - (t_0 - t_{-1})^2 \beta - 2\beta t_0^2}{\gamma_1 + 2\gamma_0 t_0 - (t_0 - t_{-1})^2 \beta - 3\beta t_0^2}\right] t_0 = \frac{h(r_0)}{1 - 2\beta\gamma^2} = c \qquad (8.198)$$

we conclude that (8.189) holds for $n = -1, 0$. Assume (8.189) holds for all $n \leq k$ and $x_k \in U(x_0, r_0)$. By (8.157) and (8.189) $x_{k+1} \in U(x_0, r_0)$. By (8.178), (8.179) and (8.189)

$$\|A_0^{-1}(A_0 - A_{k+1})\|$$
$$= \|A_0^{-1}([2x_0 - x_{-1}, x_{-1}] - [x_0, x_{-1}] + [x_0, x_{-1}] - [x_0, x_0]$$
$$+ [x_0, x_0] - [x_{k+1}, x_0] + [x_{k+1}, x_0] - [x_{k+1}, x_k]$$
$$+ [x_{k+1}, x_k] - [2x_{k+1} - x_k, x_k])\|$$
$$= \|A_0^{-1}(([2x_0 - x_{-1}, x_{-1}, x_0] - [x_0, x_{-1}, x_0])(x_0 - x_{-1})$$
$$+ ([x_0, x_0] - [x_{k+1}, x_0])$$
$$+ ([x_{k+1}, x_0] - [x_{k+1}, x_k]) + ([x_{k+1}, x_k] - [2x_{k+1} - x_k, x_k]))\|$$
$$\leq \beta\gamma^2 + (\|x_0 - x_{k+1}\| + \|x_0 - x_k\| + \|x_k - x_{k+1}\|)\alpha$$
$$\leq \beta\gamma^2 + 2(t_0 - t_{k+1})\alpha < 2\beta\gamma^2 + 2\alpha r \leq 1. \qquad (8.199)$$

It follows by the Banach lemma on invertible operators and (8.199) that A_{k+1}^{-1} exists, so that

$$\|A_{k+1}^{-1} A_0\| \leq [1 - \beta\gamma^2 - (\|x_0 - x_{k+1}\| + \|x_0 - x_k\| + \|x_k - x_{k+1}\|)\alpha]^{-1}. \qquad (8.200)$$

We can also obtain

$$\|A_0^{-1}([x_{k+1}, x_k] - A_k)\|$$
$$= \|A_0^{-1}([x_{k+1}, x_k] - [x_k, x_k]$$
$$+ [x_k, x_k] - [x_k, x_{k-1}] + [x_k, x_{k-1}] - [2x_k - x_{k-1}, x_{k-1}])\|$$
$$= \|A_0^{-1}(([x_{k+1}, x_k] - [x_k, x_k])$$
$$+ ([x_k, x_{k-1}, x_k] - [2x_k - x_{k-1}, x_{k-1}, x_k])(x_k - x_{k-1}))\|$$
$$\leq \alpha\|x_k - x_{k+1}\| + \beta\|x_{k-1} - x_k\|^2. \qquad (8.201)$$

Using (8.149), (8.200) and (8.201) we get

$$\|x_{k+2} - x_{k+1}\|$$
$$= \|A_{k+1}^{-1} F(x_{k+1})\| = \|A_{k+1}^{-1}(F(x_{k+1}) - F(x_k) - A_k(x_{k+1} - x_k))\|$$
$$\leq \|A_{k+1}^{-1} A_0\| \, \|A_0^{-1}([x_{k+1}, x_k] - A_k)\| \cdot \|x_k - x_{k+1}\|$$
$$\leq \frac{\alpha\|x_k - x_{k+1}\| + \beta\|x_{k-1} - x_k\|^2}{1 - \beta\gamma^2 - \alpha(\|x_0 - x_{k+1}\| + \|x_0 - x_k\| + \|x_k - x_{k+1}\|)} \|x_k - x_{k+1}\|$$
$$\leq \frac{[\alpha(t_k - t_{k+1}) + \beta(t_{k-1} - t_k)^2](t_k - t_{k+1})}{1 - \beta\gamma^2 - 2(t_0 - t_{k+1})\alpha} \leq t_{k+1} - t_{k+2}, \qquad (8.202)$$

which together with (8.188) completes the induction.

It follows from (8.189) that sequence $\{x_n\}$ ($n \geq -1$) is Cauchy in a Banach space X and as such it converges to some $x^* \in \overline{U}(x_0, r_0)$ (since $\overline{U}(x_0, r_0)$ is a closed set). By letting $k \to \infty$ in (8.202) we obtain $F(x^*) = 0$.

Finally to show uniqueness, define operator

$$B_0 = \int_0^1 [y^* + t(x^* - y^*), y^* + t(x^* - y^*)] dt \qquad (8.203)$$

where y^* is a solution of equation (8.148) in $\overline{U}(x_0, r_0)$. We can have

$$\|A_0^{-1}(A_0 - B_0)\|$$
$$\leq \alpha \int_0^1 [\|2x_0 - x_{-1} - y^* - t(x^* - y^*)\| + \|x_{-1} - y^* - t(x^* - y^*)\|] dt$$
$$\leq \alpha \left[\|x_0 - x_{-1}\| + \|x_0 - y^*\| + \frac{\|x_0 - y^*\| + \|x_0 - x^*\|}{2} \right.$$
$$\left. + \|x_0 - x_{-1}\| + \|x_0 - y^*\| + \frac{\|x_0 - x^*\| + \|x_0 - y^*\|}{2} \right]$$
$$\leq 2\alpha(\gamma + 2r_0) < 1. \qquad (8.204)$$

It follows from the Banach lemma on invertible operators and (8.204) that linear operator B is invertible.

We deduce from (8.203) and the identity

$$F(x^*) - F(y^*) = B_0(x^* - y^*) \qquad (8.205)$$

that

$$x^* = y^*. \qquad (8.206)$$

The proof of Theorem 8.7 is now complete. $\square$

Remark 8.10 *(a) It follows from (8.189), (8.190), (8.197) and (8.202) that the order of convergence of scalar sequence $\{t_n\}$ and iteration $\{x_n\}$ is quadratic.*

(b) The conclusions of Theorem 8.7 hold in a weaker setting. Indeed assume:

$$\|A_0^{-1}([x_0,x_0] - [x,x_0])\| \leq \alpha_0\|x - x_0\|, \qquad (8.207)$$

$$\|A_0^{-1}([x,x_0] - [x,y])\| \leq \alpha_1\|y - x_0\|, \qquad (8.208)$$

$$\|A_0^{-1}([y,x] - [2y - x,x])\| \leq \alpha_2\|y - x\|, \qquad (8.209)$$

$$\|A_0^{-1}([y,x] - [x,x])\| \leq \alpha_3\|y - x\|, \qquad (8.210)$$

$$\|A_0^{-1}([2x_0 - x_{-1}, x_0] - [x,y])\| \leq \alpha_4(\|2x_0 - x_{-1} - x\| + \|x_0 - y\|) \qquad (8.211)$$

and

$$\|A_0^{-1}([2x_0 - x_{-1}, x_{-1}, x_0] - [x_0, x_{-1}, x_0])\| \leq \beta_0\|x_0 - x_{-1}\| \qquad (8.212)$$

for all $x, y \in D_0$.

It follows from (8.178), (8.179) and (8.207) – (8.212) that

$$\alpha_i \leq 2\alpha, \quad i = 1,2,3,4 \qquad (8.213)$$

and

$$\beta_0 \leq \beta. \qquad (8.214)$$

For the derivation of: (8.200), we can use (8.207) – (8.209) and (8.212) instead of (8.178) and (8.179), respectively; (8.201), we can use (8.210) instead of (8.178; (8.204), we can use (8.212) instead of (8.178). The resulting majorizing sequence call it $\{s_n\}$ is also converging to zero and is finer than $\{t_n\}$ because of (8.213) and (8.214).

Therefore if (8.157), (8.207) – (8.212) are used in Theorem 8.7 instead of (8.178) we draw the same conclusions but with weaker conditions, and corresponding error bounds are such that:

$$\|x_{n+1} - x_n\| \leq s_n - s_{n+1} \leq t_{n+1} - t_n \qquad (8.215)$$

and

$$\|x_n - x^*\| \leq s_n \leq t_n \qquad (8.216)$$

for all $n \geq 0$.

(c) Condition (8.179) can be replaced by the stronger (not really needed in the proof) but more popular,

$$\|A_0^{-1}([v,x,y]-[u,x,y])\| \le \beta_1 \|u-v\| \qquad (8.217)$$

for all $v,u,x,y \in D$.

(d) As already noted at the end of Remark 8.9, conditions (8.157) and (8.187) can be replaced by

$$U_2 = U(x_0, R_0) \subseteq D_0 \quad \text{with} \quad R_0 = 3r_0 \qquad (8.2.40')$$

provided that $x_{-1} \in U_2$.

Indeed if $x_{n-1}, x_n \in U_0$ $(n \ge 0)$ then

$$\|2x_n - x_{n-1}\| \le 2\|x_n - x_0\| + \|x_{n-1} - x_0\| < 3r_0.$$

That is $2x_n - x_{n-1} \in U_2$ $(n \ge 0)$.

We can also provide a posteriori estimates for method (8.149):

Proposition 8.2 *Assume hypotheses of Theorem 8.150 hold. Define scalar sequences $\{p_n\}$ and $\{q_n\}$ for all $n \ge 1$ by:*

$$p_n = \alpha \|x_{n-1} - x_n\|^2 + \beta \|x_{n-1} - x_{n-2}\|^2 \|x_{n-1} - x_n\| \qquad (8.218)$$

and

$$q_n = 1 - 2\alpha \|x_n - x_0\| + \beta \gamma^2. \qquad (8.219)$$

Then the following error bounds hold for all $n \ge 1$:

$$\|x_n - x^*\| \le \varepsilon_n \le t_n, \qquad (8.220)$$

where,

$$\varepsilon_n = 2\{q_n + (q_n^2 - 4\alpha p_n)^{1/2}\}^{-1} p_n. \qquad (8.221)$$

Proof. As in (8.199) we can have in turn:

$$\begin{aligned}
\|A_0^{-1}(A_0 - [x_n, x^*])\| &= \|A_0^{-1}(A_0 - [x_0, x_0] + [x_0, x_0] - [x_n, x^*])\| \\
&\le \beta \gamma^2 + \alpha(\|x_0 - x_n\| + \|x_0 - x^*\|) \\
&\le \beta \gamma^2 + \alpha(2t_0 - t_n) \\
&< \beta \gamma^2 + 2\alpha r_0 \le 1. \qquad (8.222)
\end{aligned}$$

It follows from (8.218) and the Banach lemma on invertible operators that linear operator $[x_n, x^*]$ is invertible, and

$$\|[x_n, x^*]^{-1} A_0\| \leq (q_n - \alpha \|x_n - x^*\|)^{-1}. \tag{8.223}$$

Using (8.149) we obtain the approximation:

$$x_n - x^* = ([x_n, x^*]^{-1} A_0)(A_0^{-1} F(x_n)). \tag{8.224}$$

By (8.201), (8.223) and (8.224) we obtain the left-hand side estimate of (8.220).

Moreover we can have in turn:

$$\|x_n - x^*\|$$
$$\leq \frac{p_n}{q_n - \alpha \|x_n - x^*\|}$$
$$\leq \frac{\alpha(t_{n-1} - t_n)^2 + \beta(t_{n-1} - t_n)(t_{n-1} - t_{n-2})^2}{1 - \beta\gamma^2 - 2\alpha(t_0 - t_n) - \alpha t_n}$$
$$\leq \frac{\{[\gamma_0 - \beta(2t_{n-1} + t_n)](t_{n-1} - t_n) + \beta(t_{n-2} - t_{n-1})^2\}(t_{n-1} - t_n)}{1 - \beta\gamma^2 - 2\alpha(t_0 - t_n) - \alpha t_n - (3r_0 + \gamma)(t_0 - t_n)\beta - \beta t_n^2 - \beta\gamma r_0}$$
$$\leq \frac{-\beta t_n^3 + \alpha t_n^2 + \gamma_1}{-\beta r_n^2 + \alpha t_n + \gamma_1} = t_n. \tag{8.225}$$

That completes the proof of Proposition 8.2. □

A simple numerical example follows to show:

(a) how to choose divided difference in method (8.149);
(b) method (8.149) is faster than the Secant method

$$x_{n+1} = x_n - [x_n, x_{n-1}]^{-1} F(x_n) \quad (n \geq 0) \tag{8.226}$$

(c) method (8.149) can be as fast as Newton's method

$$x_{n+1} = x_n - F'(x_n)^{-1} F(x_n) \quad (n \geq 0). \tag{8.227}$$

Note that the analytical representation of $F'(x_n)$ may be complicated which makes the use of method (8.149) very attractive.

Example 8.9 Let $X = Y = \mathbf{R}$, and define function F on $D_0 = D = (.4, 1.5)$ by

$$F(x) = x^2 - 6x + 5. \tag{8.228}$$

Moreover define divided difference of order one appearing in method (8.149) by

$$[2y - x, x] = \frac{F(2y - x) - F(x)}{2(y - x)}. \qquad (8.229)$$

In this case method (8.149) becomes

$$x_{n+1} = \frac{x_n^2 - 5}{2(x_n - 3)}, \qquad (8.230)$$

and coincides with Newton's method (8.227) applied to F. Furthermore Secant method (8.226) becomes:

$$x_{n+1} = \frac{x_{n-1}x_n - 5}{x_{n-1} + x_n - 6}. \qquad (8.231)$$

Choose $x_{-1} = .6$ and $x_0 = .7$. Then we obtain:

n	Method 2	Secant method (8.231)
1	.980434783	.96875
2	.999905228	.997835498
3	.999999998	.99998323
4	$1 = x^*$	.999999991
5	–	1

We conclude this section with an example involving a nonlinear integral equation:

Example 8.10 Let $H(x, t, x(t))$ be a continuous function of its arguments which is sufficiently many times differentiable with respect to x. It can easily be seen that if operator F in (8.148) is given by

$$F(x(s)) = x(s) - \int_0^1 H(s, t, x(t))dt, \qquad (8.232)$$

then divided difference of order one appearing in (8.149) can be defined as

$$h_n(s, t) = \frac{H(s, t, 2x_n(t) - x_{n-1}(t)) - H(s, t, x_{n-1}(t))}{2(x_n(t) - x_{n-1}(t))}, \qquad (8.233)$$

provided that if for $t = t_m$ we get $x_n(t) = x_{n-1}(t)$, then the above function equals $H'_x(s, t_m, x_n(t_m))$. Note that this way $h_n(s, t)$ is continuous for all $t \in [0, 1]$.

We refer the reader to Chapter 2 for the concepts concerning partially ordered topological spaces (POTL-spaces).

The monotone convergence of method (8.149) is examined in the next result.

Theorem 8.8 *Let F be a nonlinear operator defined on an open subset of a regular POTL-space X with values in a POTL-space Y. Let x_0, y_0, y_{-1} be points of D such that:*

$$x_0 \leq y_0 \leq y_{-1}, \quad D_0 = \langle x_0, y_{-1} \rangle \subseteq D, \quad F(x_0) \leq 0 \leq F(y_0). \quad (8.234)$$

Moreover assume: there exists a divided difference $[\cdot, \cdot] \colon D \to L(X, Y)$ such that for all $(x, y) \in D_0^2$ with $x \leq y$

$$2y - x \in D_0, \quad (8.235)$$

and

$$F(y) - F(x) \leq [x, 2y - x](y - x). \quad (8.236)$$

Furthermore, assume that for any $(x, y) \in D_0^2$ with $x \leq y$, and $(x, 2y - x) \in D_0^2$ the linear operator $[x, 2y - x]$ has a continuous non-singular, non-negative left subinverse.

Then there exist two sequences $\{x_n\}$ $(n \geq 1)$, $\{y_n\}$ $(n \geq 1)$, and two points x^*, y^* of X such that for all $n \geq 0$:

$$F(y_n) + [y_{n-1}, 2y_n - y_{n-1}](y_{n+1} - y_n) = 0, \quad (8.237)$$
$$F(x_n) + [y_{n-1}, 2y_n - y_{n-1}](x_{n+1} - x_n) = 0, \quad (8.238)$$
$$F(x_n) \leq 0 \leq F(y_n), \quad (8.239)$$
$$x_0 \leq x_1 \leq \cdots \leq x_n \leq x_{n+1} \leq y_{n+1} \leq y_n \leq \cdots \leq y_1 \leq y_0, \quad (8.240)$$
$$\lim_{n \to \infty} x_n = x^*, \quad \lim_{n \to \infty} y_n = y^*. \quad (8.241)$$

Finally, if linear operators $A_n = [y_{n-1}, 2y_n - y_{n-1}]$ are inverse non-negative, then any solution of the equation $F(x) = 0$ from the interval D_0 belongs to the interval $\langle x^*, y^* \rangle$ (i.e., $x_0 \leq v \leq y_0$ and $F(v) = 0$ imply $x^* \leq v \leq y^*$).

Proof. Let $\overline{A}_0$ be a continuous non-singular, non-negative left subinverse of A_0. Define the operator $Q \colon \langle 0, y_0 - x_0 \rangle \to X$ by

$$Q(x) = x - \overline{A}_0[F(x_0) + A_0(x)].$$

It is easy to see that Q is isotone and continuous. We also have:

$$Q(0) = -\overline{A}_0 F(x_0) \geq 0,$$
$$Q(y_0 - x_0) = y_0 - x_0 - \overline{A}_0(F(y_0)) + \overline{A}_0(F(y_0) - F(x_0) - A_0(y_0 - x_0))$$
$$\leq y_0 - x_0 - \overline{A}_0(F(y_0)) \leq y_0 - x_0.$$

According to Kantorovich's theorem on POTL-spaces [99], [272], operator Q has a fixed point $w \in \langle 0, y_0 - x_0 \rangle$. Set $x_1 = x_0 + w$. Then we get

$$F(x_0) + A_0(x_1 - x_0) = 0, \quad x_0 \leq x_1 \leq y_0. \tag{8.242}$$

By (8.236) and (8.242) we deduce:

$$F(x_1) = F(x_1) - F(x_0) + A_0(x_0 - x_1) \leq 0.$$

Consider the operator $H \colon \langle 0, y_0 - x_1 \rangle \to X$ given by

$$H(x) = x + \overline{A}_0(F(y_0) - A_0(x)).$$

Operator H is clearly continuous, isotone and we have:

$$H(0) = \overline{A}_0 F(y_0) \geq 0,$$
$$H(y_0 - x_1) = y_0 - x_1 + \overline{A}_0 F(x_1) + \overline{A}_0[F(y_0) - F(x_1) - A_0(y_0 - x_1)]$$
$$\leq y_0 - x_1 + \overline{A}_0 F(x_1) \leq y_0 - x_1.$$

By Kantorovich's theorem there exists a point $z \in \langle 0, y_0 - x_1 \rangle$ such that $H(z) = z$. Set $y_1 = y_0 - z$ to obtain

$$F(y_0) + A_0(y_1 - y_0) = 0, \quad x_1 \leq y_1 \leq y_0. \tag{8.243}$$

Using (8.236), (8.243) we get:

$$F(y_1) = F(y_1) - F(y_0) - A_0(y_1 - y_0) \geq 0.$$

Proceeding by induction we can show that there exist two sequences $\{x_n\}$ $(n \geq 1)$, $\{y_n\}$ $(n \geq 1)$ satisfying (8.237)–(8.240) in a regular space X, and as such they converge to points $x^*, y^* \in X$ respectively. We obviously have $x^* \leq y^*$. If $x_0 \leq u \leq y_0$ and $F(u) = 0$, then we can write

$$A_0(y_1 - u) = A_0(y_0) - F(y_0) - A_0(u) = A_0(y_0 - u) - (F(y_0) - F(u)) \geq 0$$

and

$$A_0(x_1 - u) = A_0(x_0) - F(x_0) - A_0(u) = A_0(x_0 - u) - (F(x_0) - F(u)) \leq 0.$$

If the operator A_0 is inverse non-negative then it follows that $x_1 \leq u \leq y_1$. Proceeding by induction we deduce that $x_n \leq u \leq y_n$ holds for all $n \geq 0$. Hence we conclude

$$x^* \leq u \leq y^*.$$

That completes the proof of Theorem 8.8. □

In what follows we give some natural conditions under which the points x^* and y^* are solutions of equation $F(x) = 0$.

Proposition 8.3 *Under the hypotheses of Theorem 3, assume that F is continuous at x^* and y^* if one of the following conditions is satisfied:*

(a) $x^* = y^*$;
(b) X is normal, and there exists an operator $T\colon X \to Y$ $(T(0) = 0)$ which has an isotone inverse continuous at the origin and such that $A_n \leq T$ for sufficiently large n;
(c) Y is normal and there exists an operator $Q\colon X \to Y$ $(Q(0) = 0)$ continuous at the origin and such that $A_n \leq Q$ for sufficiently large n;
(d) operators A_n $(n \geq 0)$ are equicontinuous.

Then we deduce

$$F(x^*) = F(y^*) = 0. \tag{8.244}$$

Proof. (a) Using the continuity of F and (92) we get

$$F(x^*) \leq 0 \leq F(y^*).$$

Hence, we conclude

$$F(x^*) = 0.$$

(b) Using (90)–(93) we get

$$0 \geq F(x_n) = A_n(x_n - x_{n+1}) \geq T(x_n - x_{n+1}),$$
$$0 \leq F(y_n) = A_n(y_n - y_{n+1}) \leq T(y_n - y_{n+1}).$$

Therefore, it follows:

$$0 \geq T^{-1}F(x_n) \geq x_n - x_{n+1}, \quad 0 \leq T^{-1}F(y_n) \leq y_n - y_{n+1}.$$

By the normality of X, and

$$\lim_{n \to \infty}(x_n - x_{n+1}) = \lim_{n \to \infty}(y_n - y_{n+1}) = 0,$$

we get $\lim_{n\to\infty} T^{-1}F(x_n)) = \lim_{n\to\infty} T^{-1}(F(y_n)) = 0$. Using the continuity of F we obtain (8.244).

(c) As before for sufficiently large n

$$0 \geq F(x_n) \geq Q(x_n - x_{n+1}), \quad 0 \leq F(y_n) \leq Q(y_n - y_{n+1}).$$

By the normality of Y and the continuity of F and Q we obtain (8.244).

(d) It follows from the equicontinuity of operator A_n that $\lim_{n\to\infty} A_n v_n = 0$ whenever $\lim_{n\to\infty} v_n = 0$. Therefore, we get $\lim_{n\to\infty} A_n(x_n - x_{n+1}) = \lim_{n\to\infty} A_n(y_n - y_{n+1}) = 0$. By (8.237), (8.238), and the continuity of F at x^* and y^* we obtain (8.244).

That completes the proof of Proposition 8.3. □

Remark 8.11 *Hypotheses of Theorem 8.8 can be weakened along the lines of Remarks 8.9 and 8.10 above and the works in [236, pp.102-105], or Chapter 5 on the monotone convergence of Newton-like methods. However, we leave the details to the motivated reader.*

Remark 8.12 *We finally note that (8.149) is a special case of the class of methods of he form:*

$$x_{n+1} = x_n - [(1+\lambda_n)x_n - \lambda_n x_{n-1}, x_{n-1}]^{-1} F(x_n) \quad (n \geq 0) \quad (8.245)$$

where λ_n are real numbers depending on x_{n-1} and x_n, i.e.

$$\lambda_n = \lambda(x_{n-1}, x_n) \quad (n \geq 0), \quad \lambda\colon X^2 \to R, \quad (8.246)$$

and are chosen so that in practice, e.g.,

$$\text{for all } x, y \in D \Rightarrow (1+\lambda(x,y))y - \lambda(x,y)x \in D. \quad (8.247)$$

Note that setting $\lambda(x,y) = 1$ for all $x, y \in D$ in (8.245) we obtain (8.149).

Using (8.245) instead of (8.149) all the results obtained here can immediately be reproduced in this more general setting.

8.3 Exercises

8.1. Let $F : S \subseteq X \to Y$ be a three times Fréchet-differentiable operator defined on an open convex domain S of Banach space X with values in a Banach space Y. Assume $F'(x_0)^{-1}$ exists for some $x_0 \in S$, $\|F'(x_0)^{-1}\| \leq \beta$, $\|F'(x_0)^{-1}F(x_0)\| \leq \eta$, $\|F''(x)\| \leq M$, $\|F'''(x)\| \leq N$,

$\|F'''(x) - F'''(y)\| \leq L\|x - y\|$ for all $x, y \in S$, and $\bar{U}(x_0, r\eta) \subseteq S$, where

$$A = M\beta\eta, \quad B = N\beta\eta^2, \quad C = L\beta\eta^3,$$
$$a_0 = 1 = c_0, \quad b_0 = \tfrac{2A}{3}, \quad d_0 = \tfrac{A}{2}(1+A),$$
$$a_{n+1} = \tfrac{a_n}{1 - Aa_n(c_n + d_n)}, \quad b_{n+1} = \tfrac{2A}{3}a_{n+1}c_{n+1},$$
$$c_{n+1} = \tfrac{32}{2187} \cdot \frac{27\left[4 + \left(1 + \tfrac{3}{2}b_n\right)^2\right] A^3 a_n^2 + 18ABa_n + 17c}{b_n^4 \left(1 + \tfrac{3}{2}b_n\right)^4} a_{n+1} d_n^4,$$
$$d_{n+1} = \tfrac{3}{4}b_{n+1}\left(1 + \tfrac{3}{2}b_{n+1}\right)c_{n+1} \quad (n \geq 0),$$

and $r = \lim_{n\to\infty} \sum_{i=0}^{n}(c_i + d_i)$. If $A \in [0, \tfrac{1}{2}]$, $B = [0, \tfrac{1}{18A}(P(A) - 17c)]$ and $c \in [0, \tfrac{P(A)}{17}]$, where $P(A) = 27(A-1)(2A-1)(A^2 + A + 2)(A^2 + 2A + 4)$. Then show [145]: Chebysheff-Halley method given by

$$y_n = x_n - F'(x_n)^{-1}F(x_n)$$
$$H_n = F'(x_n)^{-1}\left[F'\left(x_n + \tfrac{2}{3}(y_n - x_n)\right) - F'(x_n)\right]$$
$$x_{n+1} = y_n - \tfrac{3}{4}H_n\left[I - \tfrac{3}{2}H_n\right](y_n - x_n)$$

is well defined, remains in $U(x_0, r\eta)$ and converges to a solution $x^* \in \bar{U}(x_0, r\eta)$ of equation $F(x) = 0$. Moreover, the solution x^* is unique in $U\left(x_0, \tfrac{2}{M\beta} - r\eta\right)$. Furthermore, the following error estimates hold for all $n \geq 0$

$$\|x_n - x^*\| \leq \sum_{i=n}^{\infty}(c_i + d_i)\eta \leq \tfrac{3}{2A}\left[1 + \tfrac{A}{2}(1+A)\right]\tfrac{b_1}{\gamma^{1/3}}\sum_{i=1}^{\infty}\gamma^{4^{i-1}/3},$$

where $\gamma = \tfrac{b_2}{b_1}$.

8.2. Consider the scalar equation [147]

$$f(x) = 0.$$

Using the degree of logarithmic convexity of f

$$Lf(x) = \frac{f(x)f''(x)}{f'(x)^2},$$

the convex acceleration of Newton's method is given by

$$x_{n+1} = F(x_n) = x_n - \frac{f(x_n)}{f'(x_n)}\left[1 + \frac{Lf(x_n)}{2(1 - Lf(x_n))}\right] \quad (n \geq 0)$$

for some $x_0 \in \mathbb{R}$. Let $k \geq 1754877$, the interval $[a, b]$ satisfying $a + \frac{2k-1}{2(k-1)} \frac{f(b)}{f'(b)} \leq b$ and $x_0 \in [a, b]$ with $f(x_0) > 0$, and $x_0 \geq a + \frac{2k-1}{2(k-1)} \frac{f(b)}{f'(b)}$. If $|Lf(x)| \leq \frac{1}{k}$ and $Lf'(x) \in \left[\frac{1}{k}, 2(k-1)^2 - \frac{1}{k}\right)$ in $[a, b]$, then show: Newton's method converges to a solution x^* of equation $f(x) = 0$ and $x_{2n} \geq x^*$, $x_{2n+1} \leq x^*$ for all $n \geq 0$.

8.3. Consider the midpoint method [68], [149]:

$$y_n = x_n = \Gamma_n F(x_n), \quad \Gamma_n = F'(x_n)^{-1},$$
$$z_n = x_n + \tfrac{1}{2}(y_n - x_n),$$
$$x_{n+1} = x_n - \bar{\Gamma}_n F(x_n), \quad \bar{\Gamma}_n = F'(z_n)^{-1} \quad (n \geq 0),$$

for approximating a solution x^* of equation $F(x) = 0$. Let $F : \Omega \subseteq X \to Y$ be a twice Fréchet-differentiable operator defined on an open convex subset of a Banach space X with values in a Banach space Y. Assume:

(1) $\Gamma_0 \in L(Y, X)$ for some $x_0 \in \Omega$ and $\|\Gamma_0\| \leq \beta$;
(2) $\|\Gamma_0 F(x_0)\| \leq \eta$;
(3) $\|F''(x)\| \leq M$ $(x \in \Omega)$;
(4) $\|F''(x) - F''(y)\| \leq K \|x - y\|$ $(x, y \in \Omega)$.

Denote $a_0 = M\beta\eta$, $b_0 = K\beta\eta^2$. Define sequence $a_{n+1} = a_n f(a_n)^2 g(a_n, b_n)$, $b_{n+1} = b_n f(a_n)^3 g(a_n, b_n)^2$, $f(x) = \frac{2-x}{2-3x}$, and $g(x, y) = \frac{x^2}{(2-x)^2} + \frac{7y}{24}$. If $0 < a_0 < \frac{1}{2}$, $b_0 < h(a_0)$, where

$$h(x) = \frac{96(1-x)(1-2x)}{7(2-x)^2}, \quad \bar{U}(x_0, R\eta) \subseteq \Omega,$$
$$R = \frac{2}{2-a_0} \frac{1}{1-\Delta}, \quad \Delta = f(a_0)^{-1},$$

then show: midpoint method $\{x_n\}$ $(n \geq 0)$ is well defined, remains in $\bar{U}(x_0, R\eta)$ and converges at least R-cubically to a solution x^* of equation $F(x) = 0$. The solution x^* is unique in $U(x_0, \frac{2}{M\beta} - R\eta) \cap \Omega$ and

$$\|x_{n+1} - x^*\| \leq \frac{2}{2-a_0} \gamma^{\frac{3^n-1}{2}} \frac{\Delta^n}{1-\Delta} \eta.$$

8.4. Consider the multipoint method [173]:

$$y_n = x_n - \Gamma_n F(x_n), \quad \Gamma_n = F'(x_n)^{-1},$$
$$z_n = x_n + \theta(y_n - x_n),$$
$$H_n = \tfrac{1}{\theta}\Gamma_n [F'(x_n) - F'(z_n)], \quad \theta \in (0,1],$$
$$x_{n+1} = y_n + \tfrac{1}{2}H_n(y_n - x_n) \quad (n \geq 0),$$

for approximating a solution x^* of equation $F(x) = 0$. Let F be a twice-Fréchet-differentiable operator defined on some open convex subset Ω of a Banach space X with values in a Banach space Y. Assume:

(1) $\Gamma_0 \in L(Y, X)$, for some $x_0 \in X$ and $\|\Gamma_0\| \leq \beta$;
(2) $\|\Gamma_0 F(x_0)\| \leq \eta$;
(3) $\|F''(x)\| \leq M$, $(x \in \Omega)$;
(4) $\|F''(x) - F''(y)\| \leq K\|x - y\|^p$, $(x, y) \in \Omega$, $K \geq 0$, $p \in [0, 1]$.

Denote $a_0 = M\beta\eta$, $b_0 = K\beta\eta^{1+p}$ and define sequence

$$a_{n+1} = a_n f(a_n)^2 g_\theta(a_n, b_n),$$
$$b_{n+1} = b_n f(a_n)^{2+p} g_\theta(a_n, b_n)^{1+p},$$
$$f(x) = \tfrac{2}{2 - 2x - x^2} \text{ and}$$
$$g_\theta(x, y) = \tfrac{x^3 + 4x^2}{8} + \tfrac{[2 + (p+2)\theta^p]y}{2(p+1)(p+2)}.$$

Suppose $a_0 \in (0, \tfrac{1}{2})$ and $b_0 < h_p(a_0, \theta)$, where

$$h_p(x, \theta) = \tfrac{(p+1)(p+2)}{4[2+(p+2)\theta^p]}(1 - 2x)(8 - 4x^2 - x^3).$$

Then, if $\bar{U}(x_0, R\eta) \subseteq \Omega$, where $R = \left(1 + \tfrac{a_0}{2}\right)\tfrac{1}{1-\gamma\Delta}$, $\Delta = f(a_0)^{-1}$ show: iteration $\{x_n\}$ $(n \geq 0)$ is well defined, remains in $\bar{U}(x_0, R\eta)$ for all $n \geq 0$ and converges with R-order at least $2 + p$ to a solution x^* of equation $F(x) = 0$. The solution x^* is unique in $U(x_0, \tfrac{2}{M\beta} - R\eta) \cap \Omega$. Moreover, the following error bounds hold for all $n \geq 0$

$$\|x_n - x^*\| \leq \left[1 + \tfrac{a_0}{2}\gamma^{\left(\tfrac{(2+p)^n - 1}{1+p}\right)}\right]\gamma^{\left(\tfrac{(2+p)^n - 1}{1+p}\right)}\tfrac{\Delta^n}{1 - \gamma^{(2+p)^n}\Delta}\eta,$$

where $\gamma = \tfrac{a_1}{a_0}$.

8.5. Consider the multipoint iteration [176]:

$$y_n = x_n - \Gamma_n F(x_n), \quad \Gamma_n = F'(x_n)^{-1},$$
$$z_n = x_n - \tfrac{2}{3}\Gamma_n F(x_n),$$
$$H_n = \Gamma_n [F'(z_n) - F'(x_n)],$$
$$x_{n+1} = y_n - \tfrac{3}{4}\left[I + \tfrac{3}{2}H_n\right]^{-1} H_n (y_n - x_n) \quad (n \geq 0),$$

for approximation equation $F(x) = 0$. Let $F : \Omega \subseteq X \to Y$ be a three times Fréchet-differentiable operator defined on some convex subset Ω of a Banach space X with values in a Banach space Y. Assume $F'(x_0)^{-1} \in L(Y, X)$ $(x_0 \in \Omega)$, $\|\Gamma_0\| \leq \alpha$, $\|\Gamma_0 F(x_0)\| \leq \beta$, $\|F''(x)\| \leq M$, $\|F'''(x)\| \leq N$, and $\|F''''(x) - F''''(y)\| \leq k\|x - y\|$ for all $x, y \in \Omega$. Denote $\theta = M\alpha\beta$, $w = N\alpha\beta^2$ and $\delta = K\alpha\beta^3$. Define sequences

$$a_0 = c_0 = 1, \quad b_0 = \tfrac{2}{3}\theta, \quad d_0 = \tfrac{2-\theta}{2(1-\theta)},$$
$$a_{n+1} = \tfrac{a_n}{1-\theta a_n d_n}, \quad b_{n+1} = \tfrac{2}{3}\theta a_{n+1} c_{n+1},$$
$$c_{n+1} = \tfrac{8(2-3b_n)^4}{(4-3b_n)^4}\left[\tfrac{a_n^2}{(2-3b_n)^2}\theta^3 + \tfrac{17}{108}\delta + \tfrac{wa_n}{3(2-3b_n)}\theta\right] a_{n+1} d_n^4$$

and

$$d_{n+1} = \tfrac{4-3b_{n+1}}{4-6b_{n+1}} c_{n+1} \quad (n \geq 0).$$

Moreover, assume: $\bar{U}(x_0, R\beta) \subseteq \Omega$, where

$$R = \lim_{n \to \infty} \sum_{i=0}^{n} d_i, \quad \theta \in (0, \tfrac{1}{2}),$$
$$0 \leq \delta < \tfrac{27(2\theta-1)(\theta^3 - 8\theta^2 + 16\theta - 8)}{17(1-\theta)^2},$$
$$0 \leq \omega < \tfrac{3(2\theta-1)(\theta^3 - 8\theta^2 + 16\theta - 8)}{4\theta(1-\theta)^2} - \tfrac{17\delta}{36\theta}.$$

Then show: iteration $\{x_n\}$ $(n \geq 0)$ is well defined, remains in $U(x_0, R\beta)$ for all $n \geq 0$ and converges to a solution x^* of equation $F(x) = 0$. Furthermore, the solution x^* is unique in $U\left(x_0, \tfrac{2}{\alpha M} - R\beta\right)$ and for all $n \geq 0$

$$\|x_n - x^*\| \leq \sum_{i \geq n} d_i \beta \leq \beta \tfrac{(2-\theta)}{2(1-\theta)} \tfrac{1}{\gamma^{1/3}} \sum_{j \geq n} \gamma^{4^{i+1}/3},$$

where $\gamma = \tfrac{b_1}{b_0}$.

8.6. Consider the multipoint iteration [148]:

$$y_n = x_n - F'(x_n)^{-1} F(x_n)$$
$$G_n = [F'(x_n + p(y_n - x_n)) - F'(x_n)](y_n - x_n), \quad p \in (0, 1],$$
$$x_{n+1} = y_n = \frac{1}{2p} F'(y_n)^{-1} G_n \quad (n \geq 0)$$

for approximating a solution x^* of equation $F(x) = 0$. Let $F : \Omega \subseteq X \to Y$ be a continuously Fréchet-differentiable operator in an open convex domain Ω which is a subset of a Banach space X with values in a Banach space Y. Let $x_0 \in \Omega$ such that $\Gamma_0 = F'(x_0)^{-1} \in L(Y, X)$; $\|\Gamma_0\| \leq \beta$, $\|y_0 - x_0\| \leq \eta$, $p = \frac{2}{3}$, and $\|F'(x) - F'(y)\| \leq K \|x - y\|$ for all $x, y \in \Omega$. For $b_0 = K\beta\eta$, define $b_n = b_{n-1} f(b_{n-1})^2 g(b_{n-1})$, where

$$f(x) = \frac{2(1-x)}{x^2 - 4x + 2} \quad \text{and} \quad g(x) = \frac{x(x^2 - 8x + 8)}{8(1-x)^2}.$$

If $b_0 < r = .2922...$, where r is the smallest positive root of the polynomial $q(x) = 2x^4 - 17x^3 + 48x^2 - 40x + 8$, and $\bar{U}(x_0, \frac{1}{K\beta}) \subseteq \Omega$, then show: iteration $\{x_n\}$ $(n \geq 0)$ is well defined, remains in $\bar{U}(x_0, \frac{1}{K\beta})$ and converges to a solution x^* of equation $F(x) = 0$, which is unique in $U(x_0, \frac{1}{K\beta})$.

8.7. Consider the biparametric family of multipoint iterations [150]:

$$y_n = x_n - \Gamma_n F(x_n), \quad z_n = x_n + p(y_n - x_n), \quad p \in [0, 1],$$
$$H_n = \frac{1}{p} \Gamma_n [F'(z_n) - F'(x_n)],$$
$$x_{n+1} = y_n - \frac{1}{2} H_n (I + \alpha H_n)(y_n - x_n), \quad (n \geq 0)$$

where $\Gamma_n = F'(x_n)^{-1}$ $(n \geq 0)$ and $\alpha = -2\beta \in \mathbb{R}$. Assume $\Gamma_0 = F'(x_0)^{-1} \in L(Y, X)$ exists at some $x_0 \in \Omega_0 \subseteq X$, $F : \Omega_0 \subseteq X \to Y$ twice Fréchet-differentiable, X, Y Banach spaces, $\|\Gamma_0\| \leq \beta$, $\|\Gamma_0 F(x_0)\| \leq \eta$, $\|F''(x)\| \leq M$, $x \in \Omega_0$ and $\|F''(x) - F''(y)\| \leq K \|x - y\|$ for all $x, y \in \Omega_0$. Denote $a_0 = M\beta\eta$, $b_0 = k\beta\eta^2$. Define sequences

$$a_{n+1} = a_n f(a_n)^2 g(a_n, b_n), \quad b_{n+1} = b_n f(a_n)^3 g(a_n, b_n)^2,$$

where

$$f(x) = 2\left[2 - 2x - x^2 - |\alpha| x^3\right]^{-1},$$

and

$$g(x, y) = \frac{|\alpha|^2}{8} x^5 + \frac{|\alpha|}{4} x^4 + \frac{1 + 4|\alpha|}{8} x^3 + \frac{|1 + \alpha|}{2} x^2 + \frac{1-p}{4} xy + \frac{2 + 3p}{12} y$$

for some real parameters α and p. Assume:

$$a_0 \in \left(0, \tfrac{1}{2}\right), \quad b_0 < P = \frac{3(8 - 16a_0 - 4a_0^2 + 7a_0^3 + 2a_0^4)}{3a_0 + 2},$$

$|\alpha| < \min\{6, r\}$, $p \in (0, 1]$ and $p < h(|\alpha|)$, where r is a positive root of

$$h(x) = \frac{1}{6b_0(1-a_0)} \Big[(24 - 48a_0 - 12a_0^2 + 21a_0^3 + 6a_0^4 - 2b_0(3a_0 + 2))$$
$$+ 6a_0^2 (2a_0^3 + 3a_0^2 - 6a_0 - 2) x + 3a_0^5 (2a_0 - 1) x^2 \Big],$$

$\bar{U}(x_0, R\eta) \subseteq \Omega_0$, $R = \left[1 + \tfrac{a_0}{2}(1 + |\alpha| a_0)\right] \tfrac{1}{1 - \gamma \Delta}$, $\gamma = \tfrac{a_1}{a_0}$, $\Delta = f(a_0)^{-1}$.

Then show: iteration $\{x_n\}$ $(n \geq 0)$ is well defined, remains in $\bar{U}(x_0, R\eta)$ for all $n \geq 0$ and converges to a unique solution x^* of equation $F(x) = 0$ in $U(x_0, \tfrac{2}{M\beta} - R\eta) \cap \Omega_0$. The following estimates hold for all $n \geq 0$:

$$\|x_n - x^*\| \leq \left[1 + \tfrac{1}{2}\gamma^{\frac{3^n-1}{2}} a_0 \left(1 + |\alpha|\gamma^{\frac{3^n-1}{2}} a_0\right)\right] \gamma^{\frac{3^n-1}{2}} \frac{\Delta^n}{1 - \gamma^{3^n}\Delta}\eta.$$

8.8. Let f be a real function, x^* a simple root of f and G a function satisfying $G(0) = 1$, $G'(0) = \tfrac{1}{2}$ and $|G''(0)| < +\infty$. Then show [158]: iteration

$$x_{n+1} = x_n - G(Lf(x_n))\frac{f(x_n)}{f'(x_n)}, \quad (n \geq 0)$$

where

$$Lf(x) = \frac{f(x) f''(x)}{f'(x)^2}$$

is of third order for an appropriate choice of x_0. This result is due to Gander. Note that function G can be chosen

$G(x) = 1 + \tfrac{x}{2}$ (Chebyshev method);
$G(x) = 1 + \tfrac{x}{2-x}$ (Halley method);
$G(x) = 1 + \tfrac{x}{2(1-x)}$ (Super-Halley method).

8.9. Consider the super-Halley method [68] for all $n \geq 0$ in the form:

$$F(x_n) + F'(x_n)(y_n - x_n) = 0,$$
$$3F(x_n) + 3F'(x_n)\left[x_n + \tfrac{2}{3}(y_n - x_n)\right](y_n - x_n)$$
$$+ 4F'(y_n)(x_{n+1} - y_n) = 0,$$

for approximating a solution x^* of equation $F(x) = 0$. Let $F : \Omega \subseteq X \to Y$ be a three-times Fréchet-differentiable operator defined on an open convex subset Ω of a Banach space X with values in a Banach space Y. Assume:

(1) $\Gamma_0 = F'(x_0)^{-1} \in L(Y, X)$ for some $x_0 \in \Omega$ with $\|\Gamma_0\| \leq \beta$;
(2) $\|\Gamma_0 F(x_0)\| \leq \eta$;
(3) $\|F''(x)\| \leq M \ (x \in \Omega)$;
(4) $\|F'''(x) - F'''(y)\| \leq L \|x - y\| \ (x, y \in \Omega), \ (L \geq 0)$.

Denote by $a_0 = M\beta\eta$, $c_0 = L\beta\eta^3$, and define sequences

$$a_{n+1} = a_n f(a_n)^2 g(a_n, c_n),$$
$$c_{n+1} = c_n f(a_n)^4 g(a_n, c_n)^3,$$

where

$$f(x) = \frac{(1-x)}{x^2 - 4x + 2} \text{ and } g(x, y) = \frac{1}{8}\left[\frac{x^3}{(1-x)^2} + \frac{17}{27}y\right].$$

Suppose: $a_0 \in \left(0, \frac{1}{2}\right)$, $c_0 < h(a_0)$, where

$$h(x) = \frac{27(2x-1)(x-1)(x-3+\sqrt{5})(x-3-\sqrt{5})}{17(1-x)^2},$$

$$\bar{U}(x_0, R\eta) \subseteq \Omega, \quad R = \left[1 + \frac{a_0}{2(1-a_0)}\right]\frac{1}{1-\Delta},$$

and $\Delta = f(a_0)^{-1}$. then show: iteration $\{x_n\}$ $(n \geq 0)$ is well defined, remains in $\bar{U}(x_0, R\eta)$ for all $n \geq 0$ and converges to a solution x^* of equation $F(x) = 0$. the solution x^* is unique in $U(x_0, \frac{2}{M\beta} - R\eta) \cap \Omega$ and

$$\|x_n - x^*\| \leq \left[1 + \frac{a_0 \gamma^{\frac{4^n-1}{3}}}{2(1-a_0)}\right] \gamma^{\frac{4^n-1}{3}} \frac{\Delta^n}{1 - \gamma^{4^n}\Delta}\eta \ (n \geq 0),$$

where $\gamma = \frac{a_1}{a_0}$.

8.10. Consider the multipoint iteration method [68], [151]:

$$y_n = x_n - F'(x_n)^{-1} F(x_n)$$
$$G_n = F'(x_n)^{-1}\left[F'\left(x_n + \tfrac{2}{3}(y_n - x_n)\right) - F'(x_n)\right],$$
$$x_{n+1} = y_n - \tfrac{3}{4}G_n\left[I - \tfrac{3}{2}G_n\right](y_n - x_n) \ (n \geq 0),$$

for approximating a solution x^* of equation $F(x) = 0$. Let $F : \Omega \subseteq X \to Y$ be a three times Fréchet-differentiable operator defined on an open convex subset Ω of a Banach space X with values in a Banach space Y. Assume:

(1) $\Gamma_0 = F'(x_0)^{-1} \in L(Y, X)$ exists for some $x_0 \in \Omega$ and $\|\Gamma_0\| \leq \beta$;
(2) $\|\Gamma_0 F(x_0)\| \leq \eta$;
(3) $\|F''(x)\| \leq M$ $(x \in \Omega)$;
(4) $\|F'''(x)\| \leq N$ $(x \in \Omega)$;
(5) $\|F'''(x) - F'''(y)\| \leq L \|x - y\|$ $(x, y \in \Omega)$, $(L \geq 0)$.

Denote by $a_0 = M\beta\eta$, $b_0 = N\beta\eta^2$ and $c_0 = L\beta\eta^3$. Define the sequence

$$a_{n+1} = a_n f(a_n)^2 g(a_n, b_n, c_n),$$
$$b_{n+1} = b_n f(a_n)^3 g(a_n, b_n, c_n)^2,$$
$$c_{n+1} = c_n f(a_n)^4 g(a_n, b_n, c_n)^3,$$

where

$$f(x) = \frac{2}{2 - 2x - x^2 - x^3},$$

and

$$g(x, y, z) = \frac{1}{216} \left[27 x^3 \left(x^2 + 2x + 5 \right) + 18 xy + 17 z \right].$$

If $a_0 \in \left(0, \frac{1}{2}\right)$, $17 c_0 + 18 a_0 b_0 < p(a_0)$, where

$$p(x) = 27 (1-x)(1-2x)(x^2 + x + 2)(x^2 + 2x + 4),$$
$$\bar{U}(x_0, R\eta) \subseteq \Omega, \quad R = \left[1 + \frac{a_0}{2}(1 + a_0)\right] \frac{1}{1 - \gamma\Delta},$$
$$\gamma = \frac{a_1}{a_0}, \quad \Delta = f(a_0)^{-1},$$

then show: iteration $\{x_n\}$ $(n \geq 0)$ is well defined, remains in $\bar{U}(x_0, R\eta)$ for all $n \geq 0$ and converges to a solution x^* of equation $F(x) = 0$, which is unique in $U(x_0, \frac{2}{M\beta} - R\eta) \cap \Omega$. Moreover the following error bounds hold for all $n \geq 0$:

$$\|x_n - x^*\| \leq \left[1 + \frac{a_0}{2} \gamma^{\frac{4^n - 1}{3}} \left(1 + a_0 \gamma^{\frac{4^n - 1}{3}}\right)\right] \gamma^{\frac{4^n - 1}{3}} \frac{\Delta^n}{1 - \gamma^{4^n} \Delta} \eta.$$

8.11. Consider the Halley method [68], [126]

$$x_{n+1} = x_n - [I - L_F(x_n)]^{-1} F'(x_n)^{-1} F(x_n) \quad (n \geq 0)$$

where

$$L_F(x) = F'(x)^{-1} F''(x) F'(x)^{-1} F(x),$$

for approximating a solution x^* of equation $F(x) = 0$. Let $F : \Omega \subseteq X \to$ be a twice Fréchet-differentiable operator defined on an open convex subset of a Banach space X with values in a Banach space Y. Assume:

(1) $F'(x_0)^{-1} \in L(Y, X)$ exists for some $x_0 \in \Omega$;
(2) $\|F'(x_0)^{-1} F(x_0)\| \leq \beta$;
(3) $\|F'(x_0)^{-1} F''(x_0)\| \leq \gamma$;
(4) $\|F'(x_0)^{-1} [F''(x) - F''(y)]\| \leq M \|x - y\|$ $(x, y \in \Omega)$.

If

$$\beta \leq \frac{2[2\sqrt{\gamma^2+2M}+\gamma]}{3[\sqrt{\gamma^2+2M}+\gamma]^2}, \quad \bar{U}(x_0, r_1) \subseteq \Omega,$$

$(r_1 \leq r_2)$ where r_1, r_2 are the positive roots of $h(t) = \beta - t + \frac{\gamma}{2}t^2 + \frac{M}{6}t^3$, then show: iteration $\{x_n\}$ $(n \geq 0)$ is well defined, remains in $\bar{U}(x_0, r_1)$ for all $n \geq 0$ and converges to a unique solution x^* of equation $F(x) = 0$ in $\bar{U}(x_0, r_1)$. Moreover, the following error bounds hold for all $n \geq 0$:

$$\|x^* - x_{n+1}\| \leq (r_1 - t_{n+1}) \left(\frac{\|x^* - x_n\|}{r_1 - t_n}\right)^3,$$

$$\frac{(\lambda_2 \theta)^{3^n}}{\lambda_2 - (\lambda_2 \theta)^{3^n}} (r_2 - r_1) \leq r_1 - t_n \leq \frac{\lambda_1 \theta}{\lambda_1 - (\lambda_1 \theta)^{3^n}} (r_2 - r_1),$$

$$\theta = \frac{r_1}{r_2}, \quad \lambda_1 = \sqrt{\frac{(r_0 - r_2)^2 + r_0 r_2}{(r_0 - r_1)^2 + r_0 r_1}} \leq 1, \quad \lambda_2 = \frac{\sqrt{3}}{2},$$

$-r_0$ is the negative root of h, and $t_{n+1} = H(t_n)$, where

$$H(t) = t - \frac{h(t)/h'(t)}{1 - \frac{1}{2} L_h(t)}, \quad L_h(t) = \frac{h(t)/h''(t)}{h'(t)^2}.$$

8.12. Consider the iteration [68], [149]

$$x_{n+1} = x_n - [I + T(x_n)] \Gamma_n F(x_n) \quad (n \geq 0),$$

where $\Gamma_n = F'(x_n)^{-1}$ and $T(x_n) = \frac{1}{2} \Gamma_n A \Gamma_n F(x_n)$ $(n \geq 0)$, for approximating a solution x^* of equation $F(x) = 0$. Here $A : X \times X \to Y$ is a bilinear operator with $\|A\| = \alpha$, and $F : \Omega \subseteq X \to Y$ is a Fréchet-differentiable operator defined on an open convex subset Ω of a Banach space X with values in a Banach space Y. Assume:

(1) $F'(x_0)^{-1} = \Gamma_0 \in L(Y, X)$ exists for some $x_0 \in \Omega$ with $\|\Gamma_0\| \leq \beta$;
(2) $\|\Gamma_0 F(x_0)\| \leq \eta$;
(3) $\|F'(x) - F'(y)\| \leq k \|x - y\|$ $(x, y \in \Omega)$.

Let a, b be real numbers satisfying $a \in [0, \frac{1}{2})$, $b \in (0, \sigma)$, where

$$\sigma = \frac{2\left[2a^2 - 3a - 1 + \sqrt{1 + 8a - 4a^2}\right]}{a(1 - 2a)}.$$

Set $a_0 = 1$, $c_0 = 1$, $b_0 = \frac{b}{2}$ and $d_0 = 1 + \frac{b}{2}$. Define sequence

$$a_{n+1} = \frac{a_n}{1 - aa_n d_n}, \qquad c_{n+1} = \frac{a_{n+1}}{2}\left[a + \frac{b}{(1 + b_n)^2}\right] d_n^2,$$

$$b_{n+1} = \frac{b}{2} a_{n+1} c_{n+1}, \qquad d_{n+1} = (1 + b_{n+1}) c_{n+1}$$

and $r_{n+1} = \sum_{k=0}^{n+1} d_k$ $(n \geq 0)$. If $a = \kappa\beta\eta \in [0, \frac{1}{2})$, $\bar{U}(x_0, r\eta) \subseteq \Omega$, $r = \lim_{n \to \infty} r_n$, $\alpha \in [0, \frac{\sigma}{\beta\eta})$, then show: iteration $\{x_n\}$ $(n \geq 0)$ is well defined, remains in $\bar{U}(x_0, r\eta)$ and converges to a solution x^* of equation $F(x) = 0$, which is unique in $U(x_0, \frac{2}{\kappa\beta} - r\eta)$. Moreover the following error bounds hold for all $n \geq 0$

$$\|x_{n+1} - x_n\| \leq d_n \eta$$

and

$$\|x^* - x_{n+1}\| \leq (r - r_n)\eta = \sum_{k=n+1}^{\infty} d_k \eta.$$

8.13. Consider the Halley-method [68], [164] in the form:

$$y_n = x_n - \Gamma_n F(x_n), \qquad \Gamma_n = F'(x_n)^{-1},$$
$$x_{n+1} = y_n + \tfrac{1}{2} L_F(x_n) H_n (y_n - x_n) \qquad (n \geq 0),$$
$$L_F(x_n) = \Gamma_n F''(x_n) \Gamma_n F(x_n), \qquad H_n = [I - L_F(x_n)]^{-1} \qquad (n \geq 0),$$

for approximating a solution x^* of equation $F(x) = 0$. Let $F : \Omega \subseteq X \to Y$ be a twice Fréchet-differentiable operator defined on an open convex subset Ω of a Banach space X with values in a Banach space Y. Assume:

(1) $\Gamma_0 \in L(Y, X)$ exists for some $x_0 \in \Omega$ with $\|\Gamma_0\| \leq \beta$;
(2) $\|F''(x)\| \leq M$ $(x \in \Omega)$;
(3) $\|F''(x) - F''(y)\| \leq N\|x - y\|$ $(x, y \in \Omega)$;
(4) $\|\Gamma_0 F(x_0)\| \leq \eta$;
(5) the polynomial $p(t) = \frac{k}{2}t^2 - \frac{1}{\beta}t + \frac{\eta}{\beta}$, where $M^2 + \frac{N}{2\beta} \leq k^2$ has two positive roots r_1 and r_2 with $(r_1 \leq r_2)$.

Let sequences $\{s_n\}$, $\{t_n\}$, $(n \geq 0)$ be defined by

$$s_n = t_n - \frac{p(t_n)}{p'(t_n)}, \qquad t_{n+1} = s_n + \frac{1}{2}\frac{L_p(t_n)}{1 - L_p(t_n)}(s_n - t_n) \quad (n \geq 0).$$

If, $\bar{U}(x_0, r_1) \subseteq \Omega$, then show: iteration $\{x_n\}$ $(n \geq 0)$ is well defined, remains in $U(x_0, r_1)$ for all $n \geq 0$ and converges to a solution x^* of equation $F(x) = 0$. Moreover if $r_1 < r_2$ the solution x^* is unique in $\bar{U}(x_0, r_2)$. Furthermore, the following error bounds hold for all $(n \geq 0)$

$$\|x^* - x_n\| \leq r_1 - t_n = \frac{(r_2 - r_1)\theta^{4^n}}{1 - \theta^{4^n}}, \qquad \theta = \frac{r_1}{r_2}.$$

8.14. Consider the two-point method [68], [144]:

$$y_n = x_n - F'(x_n)^{-1} F(x_n),$$
$$H_n = \tfrac{1}{p} F'(x_n)^{-1} [F'(x_n + p(y_n - x_n)) - F'(x_n)], \qquad p \in (0, 1],$$
$$x_{n+1} = y_n - \tfrac{1}{2} H_n [I + H_n]^{-1} (y_n - x_n) \qquad (n \geq 0),$$

for approximating a solution x^* of equation $F(x) = 0$. Let $F: \Omega \subseteq X \to Y$ be a twice-Fréchet-differentiable operator defined on an open convex subset Ω of a Banach space X with values in a Banach space Y. Assume:

(1) $\Gamma_0 = F'(x_0)^{-1} \in L(Y, X)$ exists for some $x_0 \in \Omega$ with $\|\Gamma_0\| \leq \beta$;
(2) $\|\Gamma_0 F(x_0)\| \leq \eta$;
(3) $\|F''(x)\| \leq M$ $(x \in \Omega)$;
(4) $\|F''(x) - F''(y)\| \leq K \|x - y\|$, $\quad x, y \in \Omega$.

Denote by $a_0 = M\beta\eta$, $b_0 = K\beta\eta^2$. Define sequences

$$a_{n+1} = a_n f(a_n)^2 g_p(a_n, b_n),$$
$$b_{n+1} = b_n f(a_n)^3 g_p(a_n, b_n)^2,$$

where $f(x) = \frac{2(1-x)}{x^2 - 4x + 2}$ and $g_p(x, y) = \frac{3x^3 + 2y(1-x)[(1-6p)x + (2+3p)]}{24(1-x)^2}$. If $a_0 \in (0, \tfrac{1}{2})$, $b_0 < h_p(h_0)$, where

$$h_p(x) = \frac{3(2x-1)(x-2)(x - 3 + \sqrt{5})(x - 3 - \sqrt{5})}{2(1-x)[(1-6p)x + 2 + 3p]}, \qquad \bar{U}(x_0, \tfrac{\eta}{a_0}) \subseteq \Omega$$

then show: iteration $\{x_n\}$ $(n \geq 0)$ is well defined, remains in $\bar{U}(x_0, \tfrac{\eta}{a_0})$ for all $n \geq 0$ and converges to a solution x^* of equation $F(x) = 0$, which is unique in $U(x_0, \frac{\eta}{a_0})$. Moreover, the following error bounds hold for all $n \geq 0$:

$$\|x^* - x_n\| \leq \left[1 + \frac{a_0 \gamma^{\frac{3^n-1}{2}}}{2(1-a_0)}\right] \gamma^{\frac{3^n-1}{2}} \frac{\Delta^n}{1-\Delta} \eta,$$

where $\gamma = \frac{a_1}{a_0}$ and $\Delta = f(a_0)^{-1}$.

8.15. Consider the two-step method:

$$y_n = x_n - F'(x_n)^{-1} F(x_n)$$
$$x_{n+1} = y_n - F'(x_n)^{-1} F(y_n) \quad (n \geq 0);$$

for approximating a solution x^* of equation $F(x) = 0$. Let $F : \Omega \subseteq X \subseteq Y$ be a Fréchet-differentiable operator defined on an open convex subset Ω of a Banach space X with values in a Banach space Y. Assume:

(1) $\Gamma_0 = f'(X_0)^{-1} \in L(Y, X)$ for some $x_0 \in \Omega$, $\|\Gamma_0\| \leq \beta$;
(2) $\|\Gamma_0 F(x_0)\| \leq \eta$;
(3) $\|F'(x) - F'(y)\| \leq K \|x - y\|$ $(x, y \in \Omega)$.

Denote $a_0 = k\beta\eta$ and define the sequence $a_{n+1} = f(a_n)^2 g(a_n) a_n$ $(n \geq 0)$, where $f(x) = \frac{2}{2-2x-x^2}$ and $g(x) = x^2(x+4)/8$. If $a_0 \in (0, \frac{1}{2})$, $\bar{U}(x_0, R\eta) \subseteq \Omega$, $R = \frac{1+\frac{a_0}{2}}{1-\gamma\Delta}$, $\gamma = \frac{a_1}{a_0}$ and $\Delta = f(a_0)^{-1}$, then show: iteration $\{x_n\}$ $(n \geq 0)$ is well defined, remains in $\bar{U}(x_0, R\eta)$ for all $n \geq 0$ and converges to a solution x^* of equation $F(x) = 0$, which is unique in $U(x_0, \frac{2}{K\beta} - R\eta) \cap \Omega$. Moreover, the following error bounds hold for all $n \geq 0$

$$\|x_n - x^*\| \leq \left[1 + \frac{a_0}{2}\gamma^{\frac{3^n-1}{2}}\right] \gamma^{\frac{3^n-1}{2}} \frac{\Delta^n}{1-\gamma^{3^n}\Delta} \eta \, (n \geq 0).$$

Chapter 9

Variational Inequalities

We study the convergence of Newton's method to solve variational inequalities.

9.1 Generalized Equations Using Newton's Method

In this section we are concerned with the problem of approximating a locally unique solution x^* of the generalized equation

$$F(x) + G(x) \ni 0, \qquad (9.1)$$

where, $F\colon D_0 \subseteq D \subseteq H \to H$ is a continuous operator which is Fréchet-differentiable at each point of the interior D_0 of a closed convex subset D of a Hilbert space H with values in H, and G is a multivalued maximal monotone operator from H into H (to be precised later) [180].

The generalized Newton iteration

$$F'(x_n)(x_{n+1}) + G(x_{n+1}) \ni F'(x_n)(x_n) - F(x_n) \quad (n \geq 0) \qquad (9.2)$$

has already been used to generate a sequence approximating x^*. In particular Uko [267] has provided local and semilocal convergence results for method (9.2) as well as a procedure for the computation of the inner-iterative procedures for the computation of the generalized iterates x_n ($n \geq 0$). This way he extended the classical Newton–Kantorovich results to hold for nonsmooth generalized equations. His results extend earlier works on nonsmooth equations [180], [249], [267]. As in the classical cases Uko used Lipschitz differentiability conditions on F' and the maximality properties of G.

Here using a combination of center-Lipschitz and Lipschitz conditions we provide local and semilocal convergence results for method (9.2) with the following advantages over earlier works and in particular [267]:

(a) our results hold whenever the corresponding ones in [267] hold but not vice versa;
(b) in the semilocal case our Newton–Kantorovich hypotheses sufficient for the convergence of (9.2) is weaker than the corresponding one in [267];
(c) our error bounds on the distances $\|x_{n+1} - x_n\|$, $\|x_n - x^*\|$ are finer and the information on the location of the solution x^* more precise;
(d) in the local case and under weaker hypotheses our convergence radius can be larger. This observation is very important in computational mathematics (see also Remark 9.3).

Problems that are special cases of equation (9.1) have been in the literature for a long time. For example if $H = \mathbf{R}^j$ and $G(x_1,...,x_j) = G_1(x_1) \times \cdots \times G_j(x_j)$, where G_i, are suitable functions $i = 1, 2, ..., j$ then (9.1) is called separable [249]. Moreover set

$$F(x_1, x_2, ..., x_j) = (F_1(x_1,...,x_j),...,F_j(x_1,...,x_j)),$$

in which case (9.1) reduces to

$$F_i(x_1,...,x_j) + G_i(x_i) \ni 0, \quad i = 1,...,j.$$

Moreover as in [267] let

$$G_i = \{0\} \times (-\infty, 0] \cup (0, \infty) \times \{0\} \quad (i \geq 0)$$

to obtain the complementarity problem

$$F_i(x_1,...,x_j) \geq 0, \quad x_i \geq 0, \quad i = 1,...,j, \quad \sum_{i=1}^{j} x_i F_i(x_1,...,x_j) = 0.$$

These type of special cases of (9.1) have been studied extensively [249]. Furthermore if $\phi : H \to (-\infty, \infty]$ is a proper lower semicontinuous convex operator and

$$G(x) = \partial\varphi(x) = \{y \in H : \varphi(v) - \varphi(w) \leq \langle y, v - w \rangle, \text{ for all } w \in H\},$$

then (9.1) becomes the variational inequality

$$F(x) + \partial\varphi(x) \ni 0.$$

Other examples of special cases of (9.1) can be found in the references above.

Throughout this section we assume:
$$\|F'(x) - F'(y)\| \leq q\|x - y\| \tag{9.3}$$
$$\|F'(x) - F'(x_0)\| \leq q_0\|x - x_0\| \tag{9.4}$$

for all $x, y \in D_0$ and some fixed $x_0 \in D_0$. G is a nonempty subset of $H \times H$ so that there exists $a \geq 0$ such that

$$[x, y] \in G \text{ and } [v, w] \in G \Rightarrow \langle w - y, v - x \rangle \geq a\|x - v\|^2, \tag{9.5}$$

and which is not contained in any larger subset of $H \times H$.

We will use Lemma 2.2. from [267, pp. 256]:

Lemma 9.1 *Let G be a maximal monotone operator satisfying (9.5), and let M be a bounded linear operator from H into H. If there exists $c \in \mathbf{R}$ such that $c > -a$, and*

$$\langle M(x), x \rangle \geq c\|x\|^2 \quad \text{for all } x \in H, \tag{9.6}$$

then there exists a unique $z \in H$ for any $b \in H$ such that

$$M(z) + G(z) \ni b. \tag{9.7}$$

We provide the following result on majorizing sequences (see also Lemma 8.1):

Lemma 9.2 *Assume: there exist parameters $L \geq 0$, $L_0 \geq 0$ with $L_0 \leq L$, $\eta \geq 0$, and $\delta \in [0, 1]$ such that:*

$$h_\delta = (\delta L_0 + L)\eta \leq \delta. \tag{9.8}$$

Then, iteration $\{t_n\}$ $(n \geq 0)$ given by

$$t_0 = 0, \quad t_1 = \eta, \quad t_{n+2} = t_{n+1} + \frac{L(t_{n+1} - t_n)^2}{2(1 - L_0 t_{n+1})} \quad (n \geq 0) \tag{9.9}$$

is non-decreasing, bounded above by $t^{**} = \frac{2\eta}{2-\delta}$ and converges to some t^* such that

$$0 \leq t^* \leq t^{**}. \tag{9.10}$$

Moreover, the following error bounds hold for all $n \geq 0$

$$0 \leq t_{n+2} - t_{n+1} \leq \frac{\delta}{2}(t_{n+1} - t_n) \leq \left(\frac{\delta}{2}\right)^{n+1} \eta. \tag{9.11}$$

Proof. The result clearly holds if $\delta = 0$, or $L = 0$ or $\eta = 0$. Let us assume $\delta \neq 0$, $L \neq 0$ and $\eta \neq 0$. We must show for all $n \geq 0$

$$L(t_{n+1} - t_n) + \delta L_0 t_{n+1} \leq \delta, \quad t_{n+1} - t_n \geq 0 \quad \text{and} \quad 1 - L_0 t_{n+1} > 0. \quad (9.12)$$

Estimate (9.11) can then follow immediately from (9.9) and (9.12). Using induction on the integer n we have for $n = 0$

$$L(t_1 - t_0) + \delta L_0 t_1 = L\eta + \delta L_0 \eta \leq \delta, t_1 \geq t_0, \quad \text{and} \quad 1 - L_0\eta > 0 \quad \text{(by (8))}.$$

But then (9.9) gives

$$0 \leq t_2 - t_1 \leq \frac{\delta}{2}(t_1 - t_0).$$

Let us assume (9.11) and (9.12) holds for all $n \leq k + 1$.

We can have in turn

$$L(t_{k+2} - t_{k+1}) + \delta L_0 t_{k+2}$$
$$\leq L\eta \left(\frac{\delta}{2}\right)^{k+1} + \delta L_0 \left[t_1 + \frac{\delta}{2}(t_1 - t_0) + \left(\frac{\delta}{2}\right)^2 (t_1 - t_0)\right.$$
$$\left. + \cdots + \left(\frac{\delta}{2}\right)^{k+1} (t_1 - t_0)\right]$$
$$\leq L\eta \left(\frac{\delta}{2}\right)^{k+1} + \delta L_0 \eta \frac{1 - \left(\frac{\delta}{2}\right)^{k+2}}{1 - \frac{\delta}{2}}$$
$$= L\eta \left(\frac{\delta}{2}\right)^{k+1} + \frac{2\delta L_0 \eta}{2-\delta}\left[1 - \left(\frac{\delta}{2}\right)^{k+2}\right]$$
$$= \left\{L\left(\frac{\delta}{2}\right)^{k+1} + \frac{2L_0\delta}{2-\delta}\left[1 - \left(\frac{\delta}{2}\right)^{k+2}\right]\right\}\eta. \quad (9.13)$$

By (9.8) and (9.13) it suffices to show

$$L\left(\frac{\delta}{2}\right)^{k+1} + \frac{2L_0\delta}{2-\delta}\left[1 - \left(\frac{\delta}{2}\right)^{k+2}\right] \leq L + \delta L_0$$

or

$$\delta L_0\left\{\frac{2}{2-\delta}\left(1 - \left(\frac{\delta}{2}\right)^{k+2}\right) - 1\right\} \leq L\left[1 - \left(\frac{\delta}{2}\right)^{k+1}\right]$$

or

$$\left[\frac{L_0\delta^2}{2-\delta} - L\right]\left[1 - \left(\frac{\delta}{2}\right)^{k+1}\right] \leq 0,$$

or

$$\frac{L_0\delta^2}{2-\delta} \leq L, \quad (9.14)$$

Variational Inequalities

which is true by the choice of δ. Hence, the first estimate in (9.12) holds for all $n \geq 0$. We must also show:

$$t_k \leq t^{**}.$$

For $k = 0, 1, 2$ we have

$$t_0 = 0 \leq t^{**}, \quad t_1 = \eta \leq t^{**} \quad \text{and} \quad t_2 \leq \eta + \tfrac{\delta}{2}\eta = \tfrac{2+\delta}{2}\eta \leq t^{**}.$$

It follows from (9.11) that for all $k \geq 0$,

$$\begin{aligned}
t_{k+2} &\leq t_{k+1} + \tfrac{\delta}{2}(t_{k+1} - t_k) \leq t_k + \tfrac{\delta}{2}(t_k - t_{k-1}) + \tfrac{\delta}{2}(t_{k+1} - t_k) \\
&\leq \cdots \leq t_1 + \tfrac{\delta}{2}(t_1 - t_0) + \cdots + \left(\tfrac{\delta}{2}\right)(t_k - t_{k-1}) + \tfrac{\delta}{2}(t_{k+1} - t_k) \\
&\leq \eta + \tfrac{\delta}{2}\eta + \left(\tfrac{\delta}{2}\right)^2 \eta + \cdots + \left(\tfrac{\delta}{2}\right)^{k+1} \eta \\
&\leq \left[1 + \tfrac{\delta}{2} + \left(\tfrac{\delta}{2}\right)^2 + \cdots + \left(\tfrac{\delta}{2}\right)^{k+1}\right] \eta \\
&\leq \frac{1 - \left(\tfrac{\delta}{2}\right)^{k+2}}{1 - \tfrac{\delta}{2}} \eta < \tfrac{2}{2-\delta}\eta = t^{**}.
\end{aligned}$$

Moreover, we have

$$L_0 t_{k+2} < \tfrac{2L_0 \eta}{2-\delta} \leq 1 \quad \text{(by (8))}. \tag{9.15}$$

Hence, sequence $\{t_n\}$ ($n \geq 0$) is bounded above by t^{**}. It also follows from (9.9) that $\{t_n\}$ ($n \geq 0$) is non-decreasing and as such it converges to some t^* satisfying (9.10).

That completes the proof of Lemma 9.2. □

Remark 9.1 *It follows immediately from the proof of Lemma 9.2 that condition (9.8) can be replaced by the weaker*

$$h_\delta \leq \delta, \quad \frac{L_0 \delta^2}{2-\delta} \leq L, \quad \frac{2L_0 \eta}{2-\delta} \leq 1 \quad \text{and} \quad \delta \in [0, 2). \tag{9.16}$$

We present the main semilocal convergence theorem for method (9.2) using Lipschitz (9.3) and center-Lipschitz conditions (9.4).

Theorem 9.1 *Let F and G satisfy (9.3), (9.4) and (9.5), (9.6) respectively, for $M = F'(x_0)$. For $x_0 \in D_0$ assume there exists $y_0 \in H$ such that $G(x_0) \ni y_0$ and $\|F(x_0) + y_0\| \leq b_0$ for $b_0 > 0$. Moreover suppose (9.8) holds for*

$$L_0 = \frac{q_0}{c_0 + a}, \quad c_0 = c, \quad L = \frac{q}{c_0 + a}, \quad \eta = \frac{b_0}{c_0 + a}, \tag{9.17}$$

and
$$\overline{U}(x_0, t^*) \subseteq D. \tag{9.18}$$

Then sequence $\{x_n\}$ $(n \geq 0)$ generated by generalized Newton's method (9.2) is well defined, remains in $\overline{U}(x_0, t^*)$ for all $n \geq 0$, and converges to a unique solution x^* of equation (9.1) in $\overline{U}(x_0, t^*)$. Moreover the following error bounds hold for all $n \geq 0$

$$\|x_{n+1} - x_n\| \leq t_{n+1} - t_n \tag{9.19}$$

and

$$\|x_n - x^*\| \leq t^* - t_n, \tag{9.20}$$

where $\{t_n\}$ is given by (9.9).

Proof. We use induction on $k = 0, 1, 2, \ldots$ to show:

$$x_k \in \overline{U}(x_0, t^*), \tag{9.21}$$
$$\|x_{k+1} - x_k\| \leq t_{k+1} - t_k, \tag{9.22}$$
$$\overline{U}(x_{k+1}, t^* - t_{k+1}) \subseteq \overline{U}(x_k, t^* - t_k), \tag{9.23}$$
$$\exists y_k \in H \text{ such that } y_k \in G(x_k), \tag{9.24}$$
$$\exists b_k > 0 \text{ such that } \|F(x_k) + y_k\| \leq b_k, \tag{9.25}$$
$$\exists c_k > -a \text{ such that } \langle F'(x_k)(x), x \rangle \geq c_k \|x\|^2 \text{ for all } x \in H. \tag{9.26}$$

The induction is true if $k = 0$ for (9.21), (9.24)–(9.26) by the hypotheses of the theorem. It then follows from (9.26) and Lemma 9.1 that there exists a unique $x_1 \in H$ satisfying (9.2). By (9.5), (9.6), (9.9), (9.17) and (9.2) we obtain in turn

$$a\|x_1 - x_0\|^2 + \langle y_0 + F(x_0) - F'(x_0)(x_0 - x_1), x_1 - x_0 \rangle \leq 0,$$

or

$$a\|x_1 - x_0\|^2 + \langle F'(x_0)(x_1 - x_0), x_1 - x_0 \rangle \leq \langle -F(x_0) - y_0, x_1 - x_0 \rangle \tag{9.27}$$

or

$$\|x_1 - x_0\| \leq a_0 = \frac{b_0}{c_0 + a} = t_1 - t_0. \tag{9.28}$$

For every $z \in \overline{U}(x_1, t^* - t_1)$,

$$\|z - x_0\| \leq \|z - x_1\| + \|x_1 - x_0\| \leq t^* - t_1 + t_1 = t^* - t_0, \tag{9.29}$$

implies $z \in \overline{U}(x_0, t^* - t_0)$. It follows from (9.28) and (9.29) that (9.22) and (9.23) hold for $k = 0$. Given they hold for $n = 0, \ldots, k$ and again using (9.26) and Lemma 9.1 we conclude that there exists a unique $x_{k+1} \in H$ satisfying (9.2),

$$\|x_{k+1} - x_0\| \leq \sum_{i=1}^{k+1} \|x_i - x_{i-1}\| \leq \sum_{i=1}^{k+1}(t_i - t_{i-1}) = t_{k+1} - t_0 = t_{k+1} \leq t^*,$$
(9.30)

$$\|x_k + \theta(x_{k+1} - x_k) - x_0\| \leq t_k + \theta(t_{k+1} - t_k) < t^* \quad \theta \in [0, 1].$$
(9.31)

Hence (9.21) holds if k is replaced by $k + 1$. As in (9.27) we obtain in turn

$$a\|x_{k+1} - x_k\|^2 + \langle y_k + F(x_k) - F'(x_k)(x_k - x_{k+1}), x_{k+1} - x_k \rangle \leq 0$$

or

$$a\|x_{k+1} - x_k\|^2 + \langle F'(x_k)(x_{k+1} - x_k), x_{k+1} - x_k \rangle \leq \langle -F(x_k) - y_k, x_{k+1} - x_k \rangle$$
(9.32)

or

$$\|x_{k+1} - x_k\| \leq t_{k+1} - t_k.$$
(9.33)

That is (9.22) and (9.23) hold for k replaced by $k + 1$.

By (9.4) and (9.30) we get

$$\|F'(x_{k+1}) - F'(x_0)\| \leq q_0 \|x_{k+1} - x_0\| \leq q_0 t_{k+1}.$$
(9.34)

Set

$$c_{k+1} = c_0 - q_0 t_k.$$
(9.35)

Then by hypothesis (9.8) we get

$$c_{k+1} > -a.$$
(9.36)

Therefore

$$\langle F'(x_0)(x) - F'(x_{k+1})(x), x \rangle \leq \|F'(x_0) - F'(x_{k+1})\| \, \|x\|^2 \leq q_0 t_k \|x\|^2,$$
(9.37)

for all $x \in H$. Hence (9.26) holds for k replaced by $k + 1$.

Define

$$y_{k+1} = -F(x_k) - F'(x_k)(x_{k+1} - x_k).$$
(9.38)

Then (9.24) holds by (9.7) and

$$\|F(x_{k+1}) + y_{k+1}\| \le \|F(x_{k+1}) - F(x_k) - F'(x_k)(x_{k+1} - x_k)\|$$
$$= \left\| \int_0^1 [F'(x_k + \theta(x_{k+1} - x_k)) - F'(x_k)](x_{k+1} - x_k) dt \right\|$$
$$\le \frac{q}{2}\|x_{k+1} - x_k\|^2 = b_{k+1}, \qquad (9.39)$$

where

$$a_k = \frac{b_k}{c_k + a} \quad (k \ge 0). \qquad (9.40)$$

Thus for every $z \in \overline{U}(x_{k+1}, t^* - t_{k+1})$, we have

$$\|z - x_k\| \le \|z - x_{k+1}\| + \|x_{k+1} - x_k\| \le t^* - t_{k+1} + t_{k+1} - t_k = t^* - t_k. \quad (9.41)$$

That is

$$z \in \overline{U}(x_k, t^* - t_k). \qquad (9.42)$$

The induction for (9.21)–(9.26) is now completed.

Lemma 9.2 implies that $\{t_n\}$ $(n \ge 0)$ is a Cauchy sequence. By (9.9) and (9.33) it follows that $\{x_n\}$ $(n \ge 0)$ is a Cauchy sequence too, and as such it converges to some $x^* \in \overline{U}(x_0, t^*)$ (since $\overline{U}(x_0, t^*)$ is a closed set). By letting $m \to \infty$ in

$$\|x_{k+m} - x_k\| \le \sum_{i=k}^{k+m-1} \|x_{i+1} - x_i\| \le \sum_{i=k}^{k+m-1} (t_{i+1} - t_i) = t_{k+m} - t_k \quad (9.43)$$

we obtain (9.20). Moreover, since $\lim_{k \to \infty} x_{k+1} = x^*$,

$$\lim_{k \to \infty} [F'(x_k)(x_k - x_{k+1}) - F(x_k)] = -F(x^*), \quad \text{and}$$

$$G(x_{k+1}) \ni F'(x_k)(x_{k+1} - x_k) - F(x_k)$$

it follows that $G(x^*) \ni -F(x^*)$. Hence x^* is a solution of (9.1).

Finally to show uniqueness in $\overline{U}(x_0, t^*)$, let us assume there exists a solution $y^* \in \overline{U}(x_0, t^*)$. Then we obtain in turn

$$a\|x_{k+1} - y^*\|^2 + \langle F'(x_k)(x_{k+1} - y^*), x_{k+1} - y^* \rangle$$
$$\le \langle F(y^*) - F(x_k) - F'(x_k)(y^* - x_k), x_{k+1} - y^* \rangle$$

or (as in (9.32))

$$\|x_{k+1} - y^*\| \leq \frac{q}{2(c_k + a)}\|x_k - y^*\|^2 < \|x_k - y^*\| \qquad (9.44)$$

(since $\frac{q}{2(c_k+a)}\|x_k - y^*\| < 1$ by (9.8). Hence we get $x^* = \lim_{k \to \infty} x_k = y^*$.
That completes the proof of Theorem 9.1. □

Remark 9.2 Note that t^* can be replaced by $\frac{2\eta}{2-\delta}$ in condition (9.18).

Remark 9.3 In order for us to compare our Theorem 9.1 with earlier ones, and in particular to Theorem 2.11 in [267] we define the scalar function p by

$$p(s) = \frac{L}{2}s^2 - s + a_0, \qquad (9.45)$$

where L is given by (9.17). Uko's Newton-Kantorovich hypothesis (see (2.14) in [267]) becomes

$$h = 2La_0 \leq 1, \qquad (9.46)$$

whereas ours for $\delta = 1$ reduces to

$$h_1 = (L + L_0)a_0 \leq 1. \qquad (9.47)$$

But

$$L_0 \leq L \qquad (9.48)$$

holds in general. Hence (9.46) always implies (9.47) but not vice versa unless if $L = L_0$. If strict inequality holds in (9.48) then (9.47) may hold but not (9.46). Moreover define sequence $\{s_n\}$ by

$$s_{n+1} = s_n + \frac{\frac{L}{2}s_n^2 - s_n + a_0}{1 - Ls_n}, \quad s_0 = 0 \quad (n \geq 0), \qquad (9.49)$$

and

$$s^* = \lim_{n \to \infty} s_n. \qquad (9.50)$$

Then it is known (see Chapter 4) that

$$s^* = \frac{1 - \sqrt{1 - 2L_0 a_0}}{L}, \qquad (9.51)$$

$$s_{n+1} - s_n = -\frac{p(s_n)}{p'(s_n)} = \frac{\frac{L}{2}(s_n - s_{n-1})^2}{1 - Ls_n} \quad (n \geq 1), \qquad (9.52)$$

and

$$s^* - s_{n+1} = \frac{\frac{L}{2}(s^* - s_n)^2}{1 - Ls_n} \leq \frac{1}{L2^{n+1}} h^{2^{n+1}} \quad (n \geq 0). \qquad (9.53)$$

Uko essentially showed error bounds (9.19) and (9.20) with sequence $\{s_n\}$, and point s^* replacing $\{t_n\}$, and point t^* respectively.

That is for all $n \geq 0$:

$$\|x_{n+1} - x_n\| \leq s_{n+1} - s_n \qquad (18)'$$

and

$$\|x_n - x^*\| \leq s^* - s_n. \qquad (19)'$$

We show that our error bounds are finer and the location of the solution x^* more precise:

Proposition 9.1 Under hypotheses of Theorem 9.1 (for $\ell_0 < \ell$) and (9.46) the following error bounds hold:

$$t_{n+1} < s_{n+1} \quad (n \geq 1), \qquad (9.54)$$
$$t_{n+1} - t_n < s_{n+1} - s_n \quad (n \geq 1), \qquad (9.55)$$
$$t^* - t_n \leq s^* - s_n \quad (n \geq 0), \qquad (9.56)$$
$$t^* \leq s^*, \qquad (9.57)$$
$$0 \leq t_{n+1} - t_n \leq \alpha^{2^{n-1}}(s_{n+1} - s_n) \ (n \geq 1), \quad \alpha = \frac{1 - \ell\eta}{1 - \ell_0\eta} \in [0, 1) \qquad (9.58)$$

and

$$0 \leq t^* - t_n \leq \alpha^{2^{n-1}}(s^* - s_n) \quad (n \geq 1). \qquad (9.59)$$

Moreover we have: $t_n = s_n$ $(n \geq 0)$ if $\ell = \ell_0$.

Proof. We use induction on the integer n to show (9.54) and (9.55) first. For $n = 0$ in (9.9) we obtain

$$t_2 - \eta = \frac{\ell\eta^2}{2(1 - \ell_0\eta)} \leq \frac{\ell\eta^2}{2(1 - \ell\eta)} = s_2 - s_1$$

and

$$t_2 < s_2.$$

Assume:
$$t_{k+1} < s_{k+1}, \quad t_{k+1} - t_k < s_{k+1} - s_k \quad (k \leq n+1).$$

Using (9.9), and (9.49) we get
$$t_{k+2} - t_{k+1} = \frac{\frac{\ell}{2}(t_{k+1} - t_k)^2}{1 - \ell_0 t_{k+1}} < \frac{\frac{\ell}{2}(s_{k+1} - s_k)^2}{1 - \ell s_{k+1}} = s_{k+2} - s_{k+1}$$

and
$$t_{k+2} - t_{k+1} < s_{k+2} - s_{k+1}.$$

Let $m \geq 0$, we can obtain
$$\begin{aligned} t_{k+m} - t_k &< (t_{k+m} - t_{k+m-1}) + (t_{k+m-1} - t_{k+m-2}) + \cdots + (t_{k+1} - t_k) \\ &< (s_{k+m} - s_{k+m-1}) + (s_{k+m-1} - s_{k+m-2}) + \cdots + (s_{k+1} - s_k) \\ &< s_{k+m} - s_k. \end{aligned} \quad (9.60)$$

By letting $m \to \infty$ in (9.60) we obtain (9.56). For $n = 1$ in (9.56) we get (9.57).

Finally, (9.58) and (9.59) follow easily from (9.9) and (9.49). Note also that (9.58) holds as a strict inequality if $n \geq 2$.

That completes the proof of Proposition 9.1. $\square$

Remark 9.4 We now use two numerical examples when $G = 0$ on D. In the first one we show that hypothesis (9.46) fails whereas (9.47) holds. In the second example we show estimates (9.19), (9.20) compare favorably with (18)', (19)', respectively.

Example 9.1 Let $H = \mathbf{R}$, $D = [\sqrt{2} - 1, \sqrt{2} + 1]$, $x_0 = \sqrt{2}$ and define function F on D by
$$F(x) = \frac{1}{6}x^3 - \left(\frac{2^{3/2}}{6} + .23\right). \quad (9.61)$$

Using (9.3), (9.4), (9.5) and (9.6) we obtain
$$a = 0, \quad c = 2, \quad a_0 = .23, \quad L = 2.4142136, \quad L_0 = 1.914213562,$$
$$h = 2La_0 = 1.1105383 > 1, \quad (9.62)$$

and (9.8) for $\delta = 1$
$$(L + L_0)a_0 = .995538247 < 1. \quad (9.63)$$

That is, there is no guarantee that Newton's method $\{x_n\}$ $(n \geq 0)$ starting at x_0 converges to a solution x^* of equation $F(x) = 0$, since (9.46) is violated. However since (9.63) holds, Theorem 1 guarantees the convergence of Newton's method to $x^* = 1.614507018$.

Example 9.2 Let $H = \mathbf{R}$, $x_0 = 1.3$, $D = [x_0 - 2\eta, x_0 + 2\eta]$ and define function F on D by

$$F(x) = \frac{1}{3}(x^3 - 1). \tag{9.64}$$

As in Example 9.1 we obtain

$a_0 = .236094674, \quad L = 2.097265501, \quad L_0 = 1.817863519$

$h = 2L\eta = .990306428 < 1, \quad h_1 = (L + L_0)\eta = .92434111 < 1, \quad \text{(for } \delta = 1\text{)}$

$t^* = .369677842 \quad \text{and} \quad s^* = .429866445.$

That is, we provide a better information on the location of the solution x^* since

$$\overline{U}(x_0, t^*) \subset \overline{U}(x_0, s^*). \tag{9.65}$$

Moreover using (9.2) and (9.64) we can tabulate the following results:

COMPARISON TABLE

x_n	Estimates (9.19)	Estimates (9.20)	Estimates (18)'	Estimates (19)'
$x_1 = 1.0639053254$	.236094674	.133583172	.236094674	.193771771
$x_2 = 1.0037617275$	.102400629	.031182539	.115780708	.0779910691
$x_3 = 1.0000140800$	.028585756	.002596783	.053649732	.024342893
$x_4 = 1.0000000002$	.002575575	.000021208	.020186667	.004156226
$n = 5$	.000021207	.000000001	.003987206	.00016902
$n = 6$	.000000001	0	.000166761	.000002259

Throughout the rest of the section we assume:

$$\|F'(x) - F'(x^*)\| \leq \ell\|x - x^*\| \quad \text{for all } x \in D_0. \tag{9.66}$$

We can show the following local result for method (9.2):

Theorem 9.2 *Let G be a maximal monotone operator satisfying (9.5). Suppose (9.6) holds for $M = F'(x^*)$ and the generalized equation (9.1) has a solution x^* in D_0 such that*

$$\overline{U}(x^*, r^*) \subseteq D_0, \tag{9.67}$$

where,
$$r^* = \frac{2}{3\ell}(a+c). \tag{9.68}$$

Then sequence $\{x_n\}$ $(n \geq 0)$ generated by generalized Newton's method (9.2) is well defined, remains in $U(x^*, r^*)$ for all $n \geq 0$, and converges to x^* provided that $x_0 \in U(x^*, r^*)$.

Moreover the following error bounds hold for all $n \geq 0$:

$$\|x_1 - x^*\| \leq \frac{\ell}{2(a+c)} \|x_0 - x^*\|^2, \tag{9.69}$$

$$\|x_{n+1} - x^*\| \leq \frac{\ell}{2(a+c-\ell\|x_0 - x^*\|)} \|x_n - x^*\|^2 \leq d_0 d^{2^n}, \quad (n \geq 1) \tag{9.70}$$

where,
$$d_0 = \frac{2(a+c-\ell\|x_0 - x^*\|)}{\ell}, \tag{9.71}$$

and
$$d = \|x_0 - x^*\| d_0^{-1}. \tag{9.72}$$

Proof. We first establish the existence of solution x_1. Using (9.6) and (9.66) we obtain in turn for all $x \in H$

$$\langle [F'(x^*) - F'(x_0)](x), x \rangle \leq \|F'(x^*) - F'(x_0)\| \|x\|^2$$
$$\leq \ell \|x_0 - x^*\| \|x\|^2 \tag{9.73}$$

or
$$c\|x\|^2 - \ell\|x_0 - x^*\| \|x\|^2 \leq \langle F'(x_0)(x), x \rangle$$

or
$$(c - \ell\|x_0 - x^*\|)\|x\|^2 \leq \langle F'(x_0)(x), x \rangle. \tag{9.74}$$

It follows by the choice of x_0 that
$$-\ell\|x_0 - x^*\| > -a. \tag{9.75}$$

Hence by Lemma 9.1 x_1 exists, and solves (9.1). By (9.5) we obtain
$$a\|x_1 - x^*\|^2 \leq \langle F(x^*) - F(x_0) - F'(x_0)(x_1 - x_0), x_1 - x^* \rangle$$

or

$$a\|x_1 - x^*\|^2 + \langle F'(x_0)(x_1 - x^*), x_1 - x^* \rangle$$
$$\leq \langle F(x^*) - F(x_0) - F'(x_0)(x^* - x_0), x_1 - x^* \rangle \qquad (9.76)$$

or

$$(a+c)\|x_1 - x^*\| \leq \frac{\ell}{2}\|x_0 - x^*\|^2,$$

which shows (9.69), $x_1 \in U(x^*, r^*)$, and in particular

$$\|x_1 - x^*\| < \|x_0 - x^*\|. \qquad (9.77)$$

Assume $x_k \in U(x^*, r^*)$, x_k solves (9.1) and

$$\|x_k - x^*\| < \|x_0 - x^*\| \quad (k \geq 1). \qquad (9.78)$$

As in (9.75) we get in turn

$$\langle [F'(x^*) - F'(x_k)](x), x \rangle \leq \|F'(x^*) - F'(x_k)\| \, \|x\|^2 \leq \ell\|x^* - x_k\| \, \|x\|^2$$

or

$$\langle F'(x^*)(x), x \rangle - \langle F'(x_k)(x), x \rangle \leq \ell\|x_0 - x^*\| \, \|x\|^2$$

or

$$(c - \ell\|x_0 - x^*\|)\|x\|^2 \leq \langle F'(x_k)(x), x \rangle$$

which establishes the existence of x_{k+1}. Moreover by (9.5) we get

$$a\|x_{k+1} - x^*\|^2 \leq \langle F(x^*) - F(x_k) - F'(x_k)(x_{k+1} - x_k), x_{k+1} - x^* \rangle$$

or

$$a\|x_{k+1} - x^*\|^2 + \langle F'(x_k)(x_{k+1} - x^*), x_{k+1} - x^* \rangle$$
$$\leq \langle F(x^*) - F(x_k) - F'(x_k)(x^* - x_k), x_{k+1} - x^* \rangle$$

or

$$(a + c - \ell\|x_0 - x^*\|)\|x_{k+1} - x^*\| \leq \frac{\ell}{2}\|x_k - x^*\|^2,$$

which shows (9.70), $x_{k+1} \in U(x^*, r^*)$, and $\lim_{k \to \infty} x_k = x^*$.
That completes the proof of Theorem 9.2. $\qquad \square$

Remark 9.5 *A local result similar to Theorem 9.2 is given in [268, Thm. 2.5] where the stronger and more difficult to verify conditions are used:*

$$\|F'(x) - F'(y)\| \leq \gamma \|x - y\| \quad \text{for all } x \in D_0 \tag{9.79}$$

there exists $c_1 > -a$ such that $\langle F'(z)(x), x \rangle \geq c_1 \|x\|^2$ for all $x \in H, z \in D_0$. (9.80)
The coercivity condition (9.80) which implies $F'(x)^{-1}$ exists for all $x \in D_0$ is rather strong, and may not hold in many problems occuring in applications. Note also that it is possible to obtain a larger convergence radius despite the fact that we use weaker conditions.

9.2 Exercises

9.1. Consider the problem of approximating a locally unique solution of the variational inequality

$$F(x) + \partial \varphi(x) \ni 0, \tag{1}$$

where F is a Gâteaux differentiable operator defined on a Hilbert space H with values in H; $\varphi \colon H \to (-\infty, \infty]$ is a lower semicontinuous convex function.
We approximate solutions x^* of (1) using the generalized Newton method in the form

$$F'(x_n)(x_{n+1}) + \partial \varphi(x_{n+1}) \ni F'(x_n)(x_n) - F(x_n) \tag{2}$$

to generate a sequence $\{x_n\}$ $(n \geq 0)$ converging to x^*.
Define: the set

$$D(\varphi) = \{x \in H \colon \varphi(x) < \infty\} \quad \text{and assume } D(\varphi) \neq \phi;$$

the subgradient

$$\partial \varphi(x) = \{z \in H \colon \varphi(x) - \varphi(y) \leq \langle z, x - y \rangle, y \in D(\varphi)\};$$

and the set

$$D(\partial \varphi) = \{x \in D(\varphi) \colon \partial \varphi(x) \neq 0\}.$$

Function $\partial\varphi$ is multivalued and for any $\lambda > 0$, $(1+\lambda\partial\varphi)^{-1}$ exists (as a single valued function) and satisfies

$$\|(1+\lambda\partial\varphi)^{-1}(x) - (I+\lambda\partial\varphi)^{-1}(y)\| \le \|x-y\| \quad (x,y \in H).$$

Moreover $\partial\varphi$ is monotone:

$$f_1 \in \partial\varphi(x_1), f_2 \in \partial\varphi(x_2) \Rightarrow \langle f_1 - f_2, x_1 - x_2 \rangle \ge 0.$$

Furthermore, we want $D(\overline{\varphi}) = D(\overline{\partial}\varphi)$, so that $D(\partial\varphi)$ is sufficient for our purposes.

We present the following local result for variational inequalities and twice Gâteaux differentiable operators:

(a) Let $F\colon H \to H$ be a twice Gâteaux differentiable function. Assume:

(1) variational inequality (1) has a solution x^*;
(2) there exist parameters $a \ge 0$, $b > 0$, $c > 0$ such that

$$\|F''(x) - F''(y)\| \le a\|x-y\|,$$
$$\|F''(x^*)\| \le b,$$

and

$$c\|y-z\|^2 \le \langle F'(x)(y-z), y-z \rangle$$

for all $x, y, z \in H$;
(3) $x_0 \in D(\varphi)$ and $x_0 \in U(x^*, r) = \{x \in H \mid \|x - x^*\| < r\}$, where

$$r = 4c\left[b + \sqrt{b^2 + \tfrac{16ac}{3}}\right]^{-1}.$$

Then show: generalized Newton method (2) is well defined, remains in $U(x^*, r)$ and converges to x^* with

$$\|x_n - x^*\| \le p \cdot d^{2^n}, \quad (n \ge 0)$$

where,

$$p^{-1} = \tfrac{1}{c}\left[\tfrac{1}{3}a\|x^* - x_0\| + \tfrac{b}{2}\right] \quad \text{and} \quad d = p^{-1}\|x^* - x_0\|.$$

(b) We will approximate x^* using the generalized Newton method in the form

$$f''(x_n)(x_{n+1}) + \partial\varphi(x_{n+1}) \ni f''(x_n)(x_n) - \nabla f(x_n). \qquad (3)$$

We present the following semilocal convergence result for variational inequalities involving twice Gâteaux differentiable operators. Let $f\colon H \to R$ be twice Gâteaux differentiable Assume:

(1) for $x_0 \in D(\varphi)$ there exist parameters $\alpha \geq 0$, $\beta > 0$, $c > 0$ such that

$$\langle (f'''(x) - f'''(x_0))(y,y), z \rangle \leq \alpha \|x - x_0\| \|y\|^2 \|z\|,$$

$$\|f'''(x_0)\| \leq \beta$$

and

$$c\|y - z\|^2 \leq \langle f''(x)(y - z), y - z \rangle$$

for all $x, y, z \in H$;

(2) the first two terms of (3) x_0 and x_1, are such that for

$$\eta \geq \|x_1 - x_0\|$$

$$\eta \leq \begin{cases} c[\beta + 2\sqrt{\alpha c}]^{-1}, & \beta^2 - 4\alpha c \neq 0 \\ c(2\beta)^{-1}, & \beta^2 - 4\alpha c = 0. \end{cases}$$

Then show: generalized Newton's method (3) is well defined, remains in $U(x_0, r_0)$ for all $\eta \geq 0$, where c_0 is the small zero of function δ,

$$\delta(r) = \alpha\eta r^2 - (c - \beta\eta)r + c\eta,$$

and converges to a unique solution x^* of inclusion $\nabla f(x) + \partial\varphi(x) \ni 0$. In particular $x^* \in \bar{U}(x_0, r_0)$. Moreover the following error bounds hold for all $n \geq 0$

$$\|x_n - x^*\| \leq \gamma d^{2^n},$$

where,

$$\gamma^{-1} = \tfrac{\alpha r_0 + \beta}{c} \quad \text{and} \quad d = \eta\gamma^{-1}.$$

9.2. Let M, $\langle \cdot, \cdot \rangle$, $\|\cdot\|$ denote the dual, inner product and norm of a Hilbert space H, respectively. Let C be a closed convex set in H. Consider an operator $a\colon H \times H \to [0, +\infty)$. If a is continuous bilinear and satisfies

$$a(x, y) \geq c_0 \|y\|^2, \quad y \in H, \tag{1}$$

and
$$a(x,y) \leq c_1\|x\| \cdot \|y\|, \quad x,y \in H, \tag{2}$$

for some constants $c_0 > 0$, $c_1 > 0$ then a is called a coercive operator. Given $z \in M$, there exists a unique solution $x \in C$ such that:

$$a(x, x-y) \geq \langle z, x-y \rangle, \quad y \in C. \tag{3}$$

Inequality (3) is called variational. It is well known that x^* can be obtained by the iterative procedure

$$x_{n+1} = P_C(x_n - \rho F(G(x_n) - z)), \tag{4}$$

where P_C is a projection of H into C, $\rho > 0$ is a constant, F is a canonical isomorphism from M onto H, defined by

$$\langle z, y \rangle = \langle F(z), y \rangle, \quad y \in H, \; z \in M, \tag{5}$$

and

$$a(x,y) = \langle G(x), y \rangle, \quad y \in H. \tag{6}$$

Given a point-to-set operator C from H into M we define the quasi-variational inequality problem to be: find $x \in C(x)$ such that:

$$a(x, y-x) \geq \langle z, y-x \rangle \quad y \in C(x). \tag{7}$$

Here, we consider $C(x)$ to be of the form

$$C(x) = f(x) + C, \tag{8}$$

where f is a point-to-point operator satisfying

$$\|f(x^*) - f(y)\| \leq c_2 \|x^* - y\|^\lambda \tag{9}$$

for some constants $c_2 \geq 0$, $\lambda \geq 1$, all $y \in H$ and x^* a solution of (7). We will extend (4) to compute the approximate solution to (7).

(a) Show: For fixed $z \in H$, $x \in C$ satisfies

$$\langle x - z, y - x \rangle \geq 0 \quad y \in C \tag{10}$$

$$\Leftrightarrow x = P_C(z), \tag{11}$$

where P_C is the projection of H into C.

(b) P_C given by (11) is non-expansive, that is

$$\|P_C(x) - P_C(y)\| \leq \|x - y\|, \quad x, y \in H. \tag{12}$$

(c) For C given by (8), $x \in C(x)$ satisfies (7) $\Leftrightarrow$

$$x = f(x) + P_C(x - \rho F(G(x) - z)). \tag{13}$$

Result (c) suggests the iterative procedure

$$x_{n+1} = f(x_n) + P_C(x_n - \rho F(G_n) - z) - f(x_n)) \tag{14}$$

for approximating solutions of (7).
Let us define the expression

$$\theta = \theta(\lambda, \rho) = 2c_2\|x_0 - x\|^{\lambda-1} + \sqrt{1 + \rho^2 c_1^2 - 2c_0\rho}.$$

It is simple algebra to show that $\theta \in [0, 1)$ in the following cases:

(1) $\lambda = 1$, $c_2 \leq \frac{1}{2}$, $c_0 \geq 2c_1\sqrt{c_2(1-c_2)}$, $0 < \rho < \frac{c_0 + \sqrt{c_0^2 - 4c_1^2 c_2(1-c_2)}}{c_1^2}$;

(2) $\lambda > 1$, $c_0 \leq c_1$, $\|x_0 - x\| < \left[\frac{1}{2c_2}\left(1 + \sqrt{1 - \left(\frac{c_0}{c_1}\right)^2}\right)\right]^{1/\lambda-1} = c_3$,

$$0 < \rho < \frac{c_0 + \sqrt{c_0^2 + qc_1^2}}{c_1^2}, \quad q = \left(1 - 2c_2\|x_0 - x\|^{\lambda-1}\right)^2 - 1;$$

(3) $\lambda > 1$, $c_0 > c_1$, $\|x_0 - x\| \leq \left(\frac{1}{2c_2}\right)^{1/\lambda-1} = c_4$, $0 < \rho < \frac{c_0 + \sqrt{c_0^2 + qc_1^2}}{c_1^2}$.

Denote by H_0, H_1 the sets

$$H_0 = \{y \in H \mid \|y - x^*\| \leq c_3\} \quad \text{and} \quad H_1 = \{y \in H \mid \|y - x^*\| \leq c_4\}.$$

(d) Let operator f satisfy (9) and C be a non-empty closed convex subset of H. If $a(x, y)$ is a coercive, continuous bilinear operator on H, x^* and x_{n+1} are solutions of (7) and (14) respectively, then x_{n+1} converges to x^* strongly in H if (1) or (2) or (3) above hold.

It follows from (d) that a solution x^* of (7) can be approximated by the iterative procedure

(1) $x^* \in C(x^*)$ is given,

(2) $x_{n+1} = f(x_n) + P_C(x_n - \rho F(G(x_n) - z) - f(x_n))$,
where ρ, x_0 are as in (1) or (2) or (3).

If $\lambda = 1$ our result (d) reduces to Theorem 3.2 in [223] (provided that (9) is replaced by $\|f(x) - f(y)\| \leq c_2^1 \|x - y\|$ for all $x, y \in H$). Note also that since $c_2^1 \geq c_2$ in general our error bounds on the distances $\|x_n - x^*\|$ $(n \geq 0)$ are smaller. Moreover, if $C(x)$ is independent of x, then $f = 0$ and $c_2 = 0$, in which case (c) and (d) reduce to the ones in [222].

9.3. Let $x_0 \in D$ and $R > 0$ be such that $D \equiv U(x_0, R)$. Suppose that f is m-times Fréchet-differentiable on D, and its mth derivative $f^{(m)}$ is in a certain sense uniformly continuous:

$$\|f^{(m)}(x) - f^{(m)}(x_0)\| \leq w(\|x - x_0\|), \quad \text{for all } x \in D, \quad (1)$$

for some monotonically increasing positive function w satisfying

$$\lim_{t \to \infty} w(r) = 0, \quad (2)$$

or, even more generally, that

$$\|f^{(m)}(x) - f^{(m)}(x_0)\| \leq w(r, \|x - x_0\|), \quad \text{for all } x \in \bar{D}, \, r \in (0, R), \quad (3)$$

for some monotonically increasing in both variables positive function w satisfying

$$\lim_{t \to 0} w(r, t) = 0, \quad r \in [0, R]. \quad (4)$$

Let us define function θ on $[0, R]$ by

$$\theta(r) = \frac{1}{c+\alpha} \left[\eta + \frac{\alpha_m}{m!} r^m + \cdots + \frac{\alpha_2}{2!} r^2 \right. $$
$$\left. + \int_0^{v_{m-2}} \cdots \int_0^{v_1} w(v_{m-1})(r - v_1) dv_1 \cdots dv_{m-1} \right] - r \quad (5)$$

for some constants α, c, η, α_i, $i = 2, \ldots, m$; the equation,

$$\theta(r) = 0; \quad (6)$$

and the scalar iteration $\{r_n\}$ $(n \geq 0)$ by

$$r_0 = 0, \quad r_{n+1} = r_n - \frac{\theta(r_n)}{\theta'(r_n)}. \quad (7)$$

Let g be a maximal monotone operator satisfying $L(z) + g(z) \ni y$, and suppose: (1) holds, there exist α_i ($i = 2, \ldots, m$) such that

$$\|F^{(i)}(x_0)\| \leq \alpha_i, \tag{8}$$

and equation (6) has a unique $r^* \in [0, R]$ and $\theta(R) \leq 0$.
Then show: the generalized Newton's method $\{x_n\}$ ($n \geq 0$) generated by

$$f'(x_n)x_{n+1} + g(x_{n+1}) \ni f'(x_n)(x_n) - f(x_n) \quad (n \geq 0), \ (x_0 \in D)$$

is well defined, remains in $V(x_0, r^*)$ for all $n \geq 0$, and converges to a solution x^* of

$$f(x) + g(x) \ni x.$$

Moreover, the following error bounds hold for all $n \geq 0$:

$$\|x_{n+1} - x_n\| \leq r_{n+1} - r_n, \tag{9}$$

and

$$\|x_n - x^*\| \leq r^* - r_n, \quad r^* = \lim_{n \to \infty} r_n. \tag{10}$$

9.4. Let $x_0 \in D$ and $R > 0$ be such that $D \equiv U(x_0, R)$. Suppose that f is Fréchet-differentiable on D, and its derivative f' is in a certain sense uniformly continuous as an operator from D into $L(H, H)$; the space of linear operators from H into H. In particular we assume:

$$\|f'(x) - f'(y)\| \leq w(\|x - y\|), \quad x, y \in D, \tag{1}$$

for some monotonically increasing positive function w satisfying

$$\lim_{t \to \infty} w(r) = 0,$$

or, even more generally, that

$$\|f'(x) - f'(y)\| \leq w(r, \|x - y\|), \quad x, y \in \bar{D}, \ r \in (0, R), \tag{2}$$

for some monotonically increasing in both variables positive function w satisfying

$$\lim_{t \to 0} w(r, t) = 0, \quad r \in [0, R].$$

Conditions of this type have been studied in the special cases $w(t) = dt^\lambda$, $w(r, t) = d(r)t^\lambda$, $\lambda \in [0, 1]$ for regular equations; and for $w(t) = dt$

for generalized equations of the form $f(x) + g(x) \ni x$. The advantages of using (1) or (2) have been explained in great detail in the excellent paper [6]. It is useful to pass from the function w to

$$\bar{w}(r) = \sup\{w(t) + w(s) : t + s = r\}.$$

The function may be calculated explicitly in some cases. For example if $w(r) = dr^\lambda$ $(0 < \lambda \leq 1)$, then $\bar{w}(r) = 2^{1-\lambda} dr^\lambda$. More generally, if w is a concave function on $[0, R]$, then $\bar{w}(r) = 2w\left(\frac{r}{2}\right)$, and $\bar{w}$ is increasing, convex and $\bar{w}(r) \geq w(r)$, $r \in [0, R]$.

Let us define the functions $\theta, \bar{\theta}$ on $[0, R]$ by

$$\theta(r) = \tfrac{1}{c+\alpha}\left[\eta + \int_0^r w(t)dt\right] - r, \quad \text{for some } \alpha > 0, \eta > 0, c > 0$$

$$\bar{\theta}(r) = \tfrac{1}{c+\alpha}\left[\eta + \int_0^r \bar{w}(t)dt\right] - r$$

and the equations

$$\theta(r) = 0,$$
$$\bar{\theta}(r) = 0.$$

Let g be a maximal monotone operator satisfying

there exists $c > -\alpha$ such that $\langle f'(z)(x), x \rangle \geq c\|x\|^2$,

for all $x \in H, z \in D,$

and suppose: (1) holds and equation $\bar{\theta}(r) = 0$ has a unique solution $r^* \in [0, R]$.

Then, show: the generalized Newton method $\{x_n\}$ $(n \geq 0)$ generated by

$$f'(x_n)x_{n+1} + g(x_{n+1}) \ni f'(x_n)(x_n) - f(x_n) \quad (n \geq 0), \ (x_0 \in D)$$

is well defined, remains in $U(x_0, r^*)$ for all $n \geq 0$, and converges to a solution x^* of $f(x) + g(x) \ni x$. Moreover, the following error bounds hold for all $n \geq 0$:

$$\|x_{n+1} - x_n\| \leq r_{n+1} - r_n$$

and

$$\|x_n - x^*\| \leq r^* - r_n,$$

where,
$$r_0 = 0, \ r_{n+1} = r_n - \frac{\bar{\theta}(r_n)}{\theta'(r_n)},$$
and
$$\lim_{n\to\infty} r_n = r^*.$$

Chapter 10

Special Topics

We discuss the convergence of certain iterative methods to solutions of equations involving outer or generalized inverses.

10.1 Methods Involving Outer or Generalized Inverses

In this chapter we are concerned with the problem of approximating a solution x^* of the equation

$$F'(x_0)^\# F(x) = 0, \tag{10.1}$$

where F is an m-times Fréchet-differentiable operator ($m \geq 2$ an integer) defined on an open convex subset of a Banach space X with values in a Banach space Y, and $x_0 \in D$. Operator $F'(x)^\#$ ($x \in D$) denotes an outer inverse of $F'(x)$ ($x \in D$). Many authors have provided local and semilocal results for the convergence of Newton's method to x^* using hypotheses on the Fréchet-derivative [68], [117].

Here we provide local convergence theorems for Newton's method using outer or generalized inverses given by

$$x_{n+1} = x_n - F'(x_n)^\# F(x_n) \quad (n \geq 0) \quad (x_0 \in D). \tag{10.2}$$

Our Newton-Kantorovich type convergence hypothesis is different from the corresponding famous condition used in the above-mentioned works (see Remark 10.1 (b)). Hence, our results have theoretical and practical value. In fact we show using a simple numerical example that our convergence ball contains earlier ones. This way, we have a wider choice of initial guesses than before. Our results can be used to solve undetermined systems, nonlinear least squares problems and ill-posed nonlinear operator equations [68], [117].

In this section we restate some of the definitions and lemmas given in the elegant paper [117].

Let $A \in L(X,Y)$. A linear operator $B\colon Y \to X$ is called an inner inverse of A if $ABA = A$. A linear operator B is an outer inverse of A if $BAB = B$. If B is both an inner and an outer inverse of A, then B is called a generalized inverse of A. There exists a unique generalized inverse $B = A^\dagger_{P,Q}$ satisfying $ABA = A$, $BAB = B$, $BA = I - P$, and $AB = Q$, where P is a given projector on X onto $N(A)$ (the null set of A) and Q is a given projector of Y onto $R(A)$ (the range of A). In particular, if X and Y are Hilbert spaces, and P, Q are orthogonal projectors, then $A^\dagger_{P,Q}$ is called the Moore-Penrose inverse of A.

We will need five lemmas of Banach-type and perturbation bounds for outer inverses and for generalized inverses in Banach spaces. The Lemmas 10.1–10.5 stated here correspond to Lemmas 2.2–2.6 in [117] respectively. See also [104] for a comprehensive study of inner, outer and generalized inverses.

Lemma 10.1 *Let $A \in L(X,Y)$ and $A^\# \in L(Y,X)$ be an outer inverse of A. Let $B \in L(X,Y)$ be such that $\|A^\#(B - A)\| < 1$. Then $B^\# = (I + A^\#(B - A))^{-1} A^\#$ is a bounded outer inverse of B with $N(B^\#) = N(A^\#)$ and $R(B^\#) = R(A^\#)$. Moreover, the following perturbation bounds hold:*

$$\|B^\# - A^\#\| \leq \frac{\|A^\#(B-A)A^\#\|}{1 - \|A^\#(B-A)\|} \leq \frac{\|A^\#(B-A)\|\,\|A^\#\|}{1 - \|A^\#(B-A)\|}$$

and

$$\|B^\# A\| \leq (1 - \|A^\#(B-A)\|)^{-1}.$$

Lemma 10.2 *Let $A, B \in L(X,Y)$ and $A^\#, B^\# \in L(Y,X)$ be outer inverses of A and B, respectively. Then $B^\#(I - AA^\#) = 0$ if and only if $N(A^\#) \subseteq N(B^\#)$.*

Lemma 10.3 *Let $A \in L(X,Y)$ and suppose X and Y admit the topological decompositions $X = N(A) \oplus M$, $Y = R(A) \oplus S$. Let $A^\dagger \ (= A^\dagger_{M,S})$ denote the generalized inverse of A relative to these decompositions. Let $B \in L(X,Y)$ satisfy*

$$\|A^\dagger(B - A)\| \leq 1$$

and

$$(I + (B-A)A^\dagger)^{-1}B \quad \text{maps } N(A) \text{ into } R(A).$$

Then $B^\dagger = B^\dagger_{R(A^\dagger),N(A^\dagger)}$ exists and is equal to

$$B^\dagger = A^\dagger(I + TA^\dagger)^{-1} = (I + A^\dagger T)^{-1}A^\dagger,$$

where $T = B - A$. Moreover, $R(B^\dagger) = R(A^\dagger)$, $N(B^\dagger) = N(A^\dagger)$ and $\|B^\dagger A\| \leq (1 - \|A^\dagger(B-A)\|)^{-1}$.

Lemma 10.4 *Let $A \in L(X,Y)$ and $A^\dagger$ be the generalized inverse of Lemma 10.3. Let $B \in L(X,Y)$ satisfy the conditions $\|A^\dagger(B-A)\| < 1$ and $R(B) \subseteq R(A)$. Then the conclusion of Lemma 10.3 holds and $R(B) = R(A)$.*

Lemma 10.5 *Let $A \in L(X,Y)$ and $A^\dagger$ be a bounded generalized inverse of A. Let $B \in L(X,Y)$ satisfy the condition $\|A^\dagger(B-A)\| < 1$. Define $B^\# = (I + A^\dagger(B-A))^{-1}A^\dagger$. Then $B^\#$ is a generalized inverse of B if and only if $\dim N(B) = \dim N(A)$ and $\operatorname{codim} R(B) = \operatorname{codim} R(A)$.*

Let $A \in L(X,Y)$ be fixed. Then, we will denote the set on nonzero outer inverses of A by

$$\Delta(A) = \{B \in L(Y,X) : BAB = B, B \neq 0\}.$$

In [68], we showed the following semilocal convergence theorem for Newton's method (10.2) using outer inverses for m-Fréchet-differentiable operators ($m \geq 2$ an integer).

Theorem 10.1 *Let $F: D \subseteq X \to Y$ be an m-times Fréchet-differentiable operator ($m \geq 2$ an integer). Assume:*

(a) *there exist an open convex subset D_0 of D, $x_0 \in D_0$, a bounded outer inverse $F'(x_0)^\#$ of $F'(x_0)$, and constants α_i, $\eta \geq 0$ such that for all $x, y \in D_0$ the following conditions hold:*

$$\|F'(x_0)^\#(F^{(m)}(x) - F^{(m)}(y))\| \leq q, \ q > 0, \quad \forall x \in U(x_0, \delta_0), \delta_0 > 0, \tag{10.3}$$

$$\|F'(x_0)^\# F(x_0)\| \leq \eta, \tag{10.4}$$

$$\|F'(x_0)^\# F^{(i)}(x_0)\| \leq \alpha_i, \quad i = 2, 3, \ldots, m; \tag{10.5}$$

the positive zero s of $p'(s) = 0$ is such that:

$$p(s) \leq 0, \tag{10.6}$$

where
$$f(t) = \eta - t + \frac{\alpha_2}{2!}t^2 + \cdots + \frac{\alpha_m + q}{m!}t^m. \qquad (10.7)$$

Then polynomial p has only two positive zeros denoted by t^*, t^{**} ($t^* \leq t^{**}$).

(b)
$$\bar{U}(x_0, t^*) = \{x \in X : \|x - x_0\| \leq \delta\} \subseteq D_0, \quad \delta = \max\{\delta_0, t^*, t^{**}\}. \qquad (10.8)$$

(c) $\delta_0 \in [t^*, t^{**}]$ or $\delta_0 > t^{**}$.

Then

(i) Newton's method $\{x_n\}$ $(n \geq 0)$ generated by (10.2) with
$$F'(x_n)^\# = [I + F'(x_0)^\# (F'(x_n) - F'(x_0))]^{-1} F'(x_0)^\# \quad (n \geq 0)$$
is well defined, remains in $U(x_0, t^*)$ and converges to a solution $x^* \in \bar{U}(x_0, t^*)$ of equation $F'(x_0)^\# F(x) = 0$;

(ii) the following error bounds hold for all $n \geq 0$
$$\|x_{n+1} - x_n\| \leq t_{n+1} - t_n \qquad (10.9)$$
and
$$\|x_n - x^*\| \leq t^* - t_n, \qquad (10.10)$$
where $\{t_n\}$ $(n \geq 0)$ is a monotonically increasing sequence generated by
$$t_0 = 0, \quad t_{n+1} = t_n - \frac{f(t_n)}{f'(t_n)}; \qquad (10.11)$$

(iii) equation $F'(x_0)^\#$ has a unique solution in $\tilde{U} \cap \{R(F'(x_0)^\#) + x_0\}$, where
$$\tilde{U} = \begin{cases} \bar{U}(x_0, t^*) \cap D_0 & \text{if } \delta_0 \in [t^*, t^{**}], \\ U(x_0, t^{**}) \cap D_0 & \text{if } \delta_0 > t^{**} \end{cases} \qquad (10.12)$$
and
$$R(F'(x_0)^\#) + x_0 := \{x + x_0 : x \in R(F'(x_0)^\#)\}.$$

We provide a local convergence theorem for Newton's method $\{x_n\}$ $(n \geq 0)$ generated by (10.2) for m-Fréchet-differentiable operators.

Theorem 10.2 *Let $F: D \subseteq X \to Y$ be an m-times Fréchet-differentiable operator ($m \geq 2$ an integer). Assume:*

(a) $F^{(i)}(x)$, $i = 2, 3, \ldots, m$ satisfies

$$\begin{aligned} \|F^{(m)}(x) - F^{(m)}(y)\| &\leq q_0, \\ \|F^{(i)}(x) - F^{(i)}(y)\| &\leq b^i \|x - y\|, \quad \text{for all } x, y \in D; \end{aligned} \quad (10.13)$$

(b) there exists $x^* \in D$ such that $F(x^*) = 0$ and

$$\|F^{(i)}(x^*)\| \leq b_i, \quad i = 2, 3, \ldots, m; \quad (10.14)$$

(c) let r_0 be the positive zero of equation $g'(t) = 0$, where

$$g(t) = p\left[\tfrac{b_m + q_0}{m!} t^m + \cdots + \tfrac{b_2}{2!} t^2\right] - t + b_0, \quad \text{for any } b_0, p > 0 \quad (10.15)$$

and such that $U(x^*, r_0) \subseteq D$;

(d) there exists an $F'(x^*)^\# \in \Delta(F'(x^*))$ such that

$$\|F'(x^*)^\#\| \leq p, \quad (10.16)$$

and for any $x \in U(x^*, r_1)$, for given $\varepsilon_0 > 1$, r_1 is the positive zero of equation $g_1(t) = 0$, where

$$g_1(t) = p\varepsilon_0 \left[\tfrac{b_m + q_0}{(m-1)!} t^m + \cdots + b_2 t\right] + (1 - \varepsilon_0), \quad (10.17)$$

the set $\Delta(F'(x))$ contains an element of minimal mean.

Then, there exists $U(x^*, r) \subseteq D$ with $r \in (0, r_1)$ such that for any $x_0 \in U(x^*, r)$, Newton's method $\{x_n\}$ $(n \geq 0)$ generated by (10.2) for

$$F'(x_0)^\# \in \arg\min\{\|B\| : B \in \Delta(F'(x_0))\}$$

with $F'(x_n)^\# = [I + F'(x_0)^\#(F'(x_n) - F'(x_0))]^{-1} F'(x_0)^\#$, converges to $y \in U(x_0, r_0) \cap \{R(F'(x_0)^\#) + x_0\}$ such that $F'(x_0)^\# F(y) = 0$. Here, we denote

$$R(F'(x_0)^\#) + x_0 = \{x + x_0 : x \in R(F'(x_0)^\#)\}.$$

is an outer inverse $F'(x)$, and

$$\|F'(x)^\#\| \leq \frac{\|F'(x^*)^\#\|}{1 - p\left[\tfrac{b_m + q_0}{(m-1)!} r_1^{m-1} + \cdots + b_2 r_1\right]} \leq p\varepsilon_0, \quad (10.18)$$

by the choice of r_1 and ε_0. That is, for any $x_0 \in U(x^*, r)$, the outer inverse

$$F'(x_0)^\# \in \arg\min\{\|B\| : B \in \Delta(F'(x_0))\} \quad \text{and} \quad \|F'(x_0)^\#\| \leq p\varepsilon_0.$$

We can then obtain for all $x, y \in D$

$$\|F'(x_0)^\#(F^{(m)}(x) - F^{(m)}(y))\| \le p\varepsilon_0 \|F^{(m)}(x) - F^{(m)}(y)\| \le p\varepsilon_0 q_0 = q,$$
$$\|F'(x_0)^\# F^{(m)}(x_0)\| \le p\varepsilon_0 \|F^{(m)}(x_0)\| \le p\varepsilon_0 [b_m + b_{m+1} r_1] = \alpha_m$$
(by (10.13) and (10.14)),

and

$$\|F'(x_0)^\# F^{(i)}(x_0)\| \le p\varepsilon_0(b_i + b^i r_1) = \alpha_i, \quad i = 2, 3, \ldots, m-1,$$
$$\eta \le \|F'(x_0)^\# F(x_0)\| \le p\varepsilon_0 \le s - \tfrac{\alpha_2}{2!} s^2 - \cdots - \tfrac{\alpha_{m+q}}{m!} s^m, \tag{10.19}$$

by the choice of ε and ε_1. Hence, there exists a minimum positive zero $t^* < r_1$ of polynomial f given by (6.7). It also follows from (6.15), (6.17) and the choice of ε_2 that $f(r_0 - r_1) \le 0$. That is,

$$r_1 + t^* \le r_0. \tag{10.20}$$

Hence, for any $x \in U(x_0, t^*)$ we have

$$\|x^* - x\| \le \|x_0 - x^*\| + \|x_0 - x\| \le r_1 + t^* \le r_0 \quad \text{(by (6.20))}. \tag{10.21}$$

It follows from (6.21) that $U(x_0, t^*) \subseteq U(x^*, r_0) \subseteq D$. Consequently Newton's method $\{x_n\}$ ($n \ge 0$) stays in $U(x_0, t^*)$ for all $n \ge 0$ and converges to a solution y of equation $F'(x_0)^\# F(x) = 0$.

In the next theorem we examine the order of convergence of Newton method $\{x_n\}$ ($n \ge 0$).

Theorem 10.3 *Under the hypotheses of Theorem 10.2, if $F'(x_0)^\# F(y) = 0$, then*

$$\|y - x_{n+1}\| \le \frac{\tfrac{\alpha_{m+q}}{m!}\|x_n - y\|^{m-2} + \cdots + \tfrac{\alpha_2}{2!}}{1 - \alpha_2 \|x_n - y\| - \cdots - \tfrac{\alpha_{m+q}}{(m-1)!}\|x_n - y\|^m} \|y - x_n\|^2,$$
for all $n \ge 0$, \hfill (10.22)

and, if $y \in U(x_0, r_2)$, where r_2 is the positive zero of equation $g_2(t) = 0$,

$$g_2(t) = \tfrac{(\alpha_m + q)(m+1)}{m!} t^{m-1} + \cdots + \tfrac{3\alpha_2}{2!} t - 1, \tag{10.23}$$

then, sequence $\{x_n\}$ ($n \ge 0$) converges to y quadratically.

Proof. We first note that $r_2 < r_0$. By Lemma 10.1 we get $R(F'(x_0)^\#) = R(F'(x_n)^\#)$ $(n \geq 0)$. We have

$$x_{n+1} - x_n = F'(x_n)^\# F(x_n) \in R(F'(x_n)^\#) \quad (n \geq 0),$$

from which it follows

$$x_{n+1} \in R(F'(x_n)^\#) + x_n = R(F'(x_{n-1})^\#) + x_n = R(F'(x_0)^\#) + x_0,$$

and $y \in R(F'(x_n)^\#) + x_{n+1}$ $(n \geq 0)$. That is, we conclude that

$$y \in R(F'(x_0)^\#) + x_0 = R(F'(x_n)^\#) + x_0,$$

and

$$F'(x_n)^\# F'(x_n)(y - x_{n+1})$$
$$= F'(x_n)^\# F'(x_n)(y - x_0) - F'(x_n)^\# F'(x_n)(x_{n+1} - x_0)$$
$$= y - x_{n+1}.$$

We also have by Lemma 10.2 $F'(x_n)^\# = F'(x_n)^\# F'(x_0) F'(x_0)^\#$. By $F'(x_0)^\# F(y) = 0$ and $N(F'(x_0)^\#) = N(F'(x_n)^\#)$, we get $F'(x_n)^\# F(y) = 0$. Using the estimate

$$\|y - x_{n+1}\|$$
$$= \|F'(x_n)^\# F'(x_n)(y - x_{n+1})\|$$
$$= \|F'(x_n)^\# F'(x_n)[y - x_n + F'(x_n)^\#(F(x_n) - F(y))]\|$$
$$\leq \|F'(x_n)^\# F'(x_0)\| \cdot$$
$$\cdot \left\| F'(x_0)^\# \left\{ \int_0^1 [F''[x_n + t(y - x_n)] - F''(x^*)](1-t)dt(y - x_n)^2 \right.\right.$$
$$\left.\left. + \tfrac{1}{2} F''(x^*)(y - x_n)^2 \right\} \right\|$$
$$\leq \frac{\frac{\alpha_{m+q}}{m!}\|x_n - y\|^{m-2} + \cdots + \frac{\alpha_2}{2!}}{1 - \alpha_2\|x_n - y\| - \cdots - \frac{\alpha_{m+1}}{m!}\|x_n - y\|^m}\|y - x_n\|^2 \quad (n \geq 0),$$

which shows (10.22) for all $n \geq 0$. By the choice of r_2 and (10.22) there exists $\alpha \in [0, 1)$ such that $\|y - x_{n+1}\| \leq \alpha \|y - x_n\|$ $(n \geq 0)$, which together with (10.22) show that $x_n \to y$ as $n \to \infty$ quadratically. $\square$

We provide a result corresponding to Theorem 10.2 but involving generalized instead of outer inverses.

Theorem 10.4 *Let F satisfy the hypotheses of Theorems 10.2 and 10.3 except (d) which is replaced by*

(d)′ the generalized inverse $F'(x^*)$ exists, $\|F'(x^*)^\dagger\| \leq p$,

$$\dim N(F'(x)) = \dim N(F'(x^*)) \qquad (10.24)$$

and

$$\operatorname{codim} R(F'(x)) = \operatorname{codim} R(F'(x^*)) \qquad (10.25)$$

for all $x \in U(x^*, r_1)$.

Then, the conclusions of Theorems 10.2 and 10.3 hold with

$$F'(x_0)^\# \in \{B : B \in \Delta(F'(x_0)), \|B\| \leq \|F'(x_0)^\dagger\|\}. \qquad (10.26)$$

Proof. In Theorem 10.2 we showed that the outer inverse $F'(x)^\# \in \operatorname{argmin}\{\|B\| : B \in \Delta(F'(x))\}$ for all $x \in U(x^*, r)$, $r \in (0, r_1)$ and $\|F'(x)^\#\| \leq p\varepsilon_0$. We must show that under (d)′ the outer inverse

$$F'(x)^\# \in \{B : B \in \Delta(F'(x)), \|B\| \leq \|F'(x)^\dagger\|\}$$

satisfies $\|F'(x)^\#\| \leq p\varepsilon_0$. As in (10.21), we get

$$\|F'(x^*)^\dagger(F'(x) - F'(x^*))\| \leq p\left[\tfrac{b_m + g_0}{(m-1)!}r_0^{m-1} + \cdots + b_2 r_0\right] < 1.$$

Moreover, by Lemma 10.5

$$F'(x)^\dagger = [I + F'(x^*)^\dagger (F'(x) - F'(x^*))]^{-1} F'(x^*)^\dagger \qquad (10.27)$$

is the generalized inverse of $F'(x)$. Furthermore, by Lemma 10.1 as in (10.18) $\|F'(x)^\dagger\| \leq p\varepsilon_0$. That is, the outer inverse

$$F'(x_0)^\# \in \{B : B \in \Delta(F'(x_0)), \|B\| \leq \|F'(x_0)^\dagger\|\}$$

satisfies $\|F'(x_0)^\#\| \leq p\varepsilon_0$, provided that $x_0 \in U(x^*, r)$.
The rest follows exactly as in Theorems 10.2 and 10.3. □

Remark 10.1

(a) We note that Theorem 6.4 was proved in [68] with the weaker condition

$$\|F'(x_0)^\#(F^{(m)}(x) - F^{(m)}(x_0))\| \leq \bar{\alpha}_{m+1}\|x - x_0\|$$

replacing (6.3).

(b) Our conditions $(6.3) - (6.7)$ differ from the corresponding ones in [110] (see, for example, Theorem 3.1) unless if $\alpha_i = 0$, $i = 2, 3, \ldots, m$, $q = 0$, in which case our condition (6.6) becomes the Newton-Kantorovich hypothesis (3.3) in [118, p. 450]:

$$K\eta \leq \tfrac{1}{2}, \qquad (10.28)$$

where K is such that

$$\|F'(x_0)^\#(F'(x) - F'(y))\| \leq K\|x - y\| \qquad (10.29)$$

for all $x, y \in D$. Similarly (if $\alpha_i = 0$, $i = 2, 3, \ldots, m$), our r_0 equals the radius of convergence in Theorem 3.2 [118, p. 450].

(c) In Theorem 3.2 [118] the condition

$$\|F'(x) - F'(y)\| \leq c_0\|x - y\| \quad \text{for all} \quad x, y \in D \qquad (10.30)$$

was used instead of (6.29). The ball used there is $U(x^*, r^*)$, (corresponding to $U(x^*, r_0)$) where

$$r^* = \tfrac{1}{c_0 p}. \qquad (10.31)$$

Finally, for convergence $x_0 \in U(x^*, r_1^*)$, where

$$r_1^* = \tfrac{2}{3} r^*. \qquad (10.32)$$

Below we consider such a case. For simplicity we have taken $F'(x)^\# = F'(x)^{-1}$ ($x \in D$) and $m = 2$.

Remark 10.2 *Methods/routines of how to construct the appropriate actions of the required outer generalized inverses of the derivative can be found at a great variety in the elegant paper [4].*

Example 10.1 Let us consider the system of equations

$$F(x, y) = 0,$$

where $F : \mathbb{R}^2 \to \mathbb{R}^2$,

$$F(x, y) = (xy - 1, xy + x - 2y).$$

Then, we get

$$F'(x, y) = \begin{bmatrix} y & x \\ y+1 & x-2 \end{bmatrix},$$

and

$$F'(x, y)^{-1} = \tfrac{1}{x+2y} \begin{bmatrix} 2-x & x \\ y+1 & -y \end{bmatrix},$$

provided that (x, y) does not belong on the straight line $x + 2y = 0$. The second derivative is a bilinear operator on $\mathbb{R}^2$ given by the following matrix

$$F''(x, y) = \begin{bmatrix} 0 & 1 \\ 1 & 0 \\ - & - \\ 0 & 1 \\ 1 & 0 \end{bmatrix}.$$

We consider the max-norm in $\mathbb{R}^2$. Moreover in $L(\mathbb{R}^2, \mathbb{R}^2)$ we use for

$$A = \begin{bmatrix} a_{11} & a_{12} \\ a_{21} & a_{22} \end{bmatrix}$$

the norm,

$$\|A\| = \max\{|a_{11}| + |a_{12}|, |a_{21}| + |a_{22}|\}.$$

As in [6], we define the norm of a bilinear operator B on $\mathbb{R}^2$ by

$$\|B\| = \sup_{\|z\|=1} \max_i \sum_{j=1}^{2} \left| \sum_{k=1}^{2} b_i^{jk} z_k \right|,$$

where,

$$z = (z_1, z_2) \quad \text{and} \quad B = \begin{bmatrix} b_1^{11} & b_1^{12} \\ b_1^{21} & b_1^{22} \\ - & - \\ b_2^{11} & b_2^{12} \\ b_2^{21} & b_2^{22} \end{bmatrix}.$$

For $m = 2$ and $(x^*, y^*) = (1, 1)$, we get $c_0 = \tfrac{4}{3}$, $r_1^* = .5$, $\alpha_2 = 1$. We can set $q = .001$ to obtain $r_2 = .666444519$. Note that $r_2 > r_1^*$.

10.2 Exercises

10.1. (a) Assume there exist non-negative parameters K, M, L, ℓ, μ, η, $\delta \in [0, 1]$ such that:

$$L \leq K, \tag{1}$$
$$\ell + 2\mu < 1, \tag{2}$$

and

$$h_\delta \equiv \left(K + L\delta + \tfrac{4M}{2-\delta}\right)\eta + \delta\ell + 2\mu \leq \delta. \tag{3}$$

Show: iteration $\{t_n\}$ $(n \geq 0)$ given by

$$t_0 = 0, \quad t_1 = \eta,$$
$$t_{n+2} = t_{n+1} + \frac{K(t_{n+1} - t_n) + 2(Mt_n + \mu)}{2(1 - \ell - Lt_{n+1})}(t_{n+1} - t_n) \quad (n \geq 0)$$

is nondecreasing, bounded above by t^{**} and converges to some t^* such that

$$0 \leq t^* \leq \tfrac{2\eta}{2-\delta} \equiv t^{**}.$$

Moreover, the following error bounds hold for all $n \geq 0$

$$t_{n+2} - t_{n+1} \leq \tfrac{\delta}{2}(t_{n+1} - t_n) \leq \left(\tfrac{\delta}{2}\right)^{n+1} \eta.$$

(b) Let $F \colon D \subseteq X \to Y$ be a Fréchet-differentiable operator. Assume: there exist an approximation $A(x) \in L(X, Y)$ of $F'(x)$, an open convex subset D_0 of D, $x_0 \in D_0$, a bounded outer inverse $A^{\#}$ of $A(x_0)$, and parameters $\eta > 0$, $K > 0$, $M \geq 0$, $L \geq 0$, $\mu \geq 0$, $\ell \geq 0$ such that (1)–(3) hold

$$\|A^{\#} F(x_0)\| \leq \eta,$$
$$\|A^{\#}[F'(x) - F'(y)]\| \leq K\|x - y\|,$$
$$\|A^{\#}[F'(x) - A(x)]\| \leq M\|x - x_0\| + \mu,$$

and

$$\|A^{\#}[A(x) - A(x_0)]\| \leq L\|x - x_0\| + \ell$$

for all $x, y \in D_0$, and

$$\bar{U}(x_0, t^*) \subseteq D_0.$$

Show: sequence $\{x_n\}$ ($n \geq 0$) generated by Newton-like method with

$$A(x_n)^\# = [I + A^\#(A(x_n) - A(x_0))]^{-1} A^\#$$

is well defined, remains in $U(x_0, s^*)$ for all $n \geq 0$ and converges to a unique solution x^* of equation $A^\# F(x) = 0$, $\bar{U}(x_0, t^*) \cap D_0$ Moreover, the following error bound hold for all $n \geq 0$

$$\|x_{n+1} - x_n\| \leq t_{n+1} - t_n,$$

and

$$\|x_n - x^*\| \leq t^* - t_n.$$

(c) Assume:
- there exist an approximation $A(x) \in L(X, Y)$ of $F'(x)$, a simple solution $x^* \in D$ of equation (1), a bounded outer inverse $A_1^\#$ of $A(x^*)$ and non-negative parameters $\bar{K}, \bar{L}, \bar{M}, \bar{\mu}, \bar{\ell}$, such that:

$$\|A_1^\# [F'(x) - F'(y)]\| \leq \bar{K} \|x - y\|,$$
$$\|A_1^\# [F'(x) - A(x)]\| \leq \bar{M} \|x - x^*\| + \bar{\mu},$$

and

$$\|A_1^\# [A(x) - A(x^*)]\| \leq \bar{L} \|x - x^*\| + \bar{\ell}$$

for all $x, y \in D$;
- equation

$$\left(\frac{\bar{K}}{2} + \bar{M} + \bar{L}\right) r + \bar{\mu} + \bar{\ell} - 1 = 0$$

has a minimal non-negative zero r^* satisfying

$$\bar{L} r + \bar{\ell} < 1,$$

and

$$U(x^*, r^*) \subseteq D.$$

Show: sequence $\{x_n\}$ ($n \geq 0$) generated by Newton-like method is well defined, remains in $U(x^*, r^*)$ for all $n \geq 0$ and converges to x^* provided that $x_0 \in U(x^*, r^*)$. Moreover, the following error bounds hold for all $n \geq 0$:

$$\|x^* - x_{n+1}\|$$
$$\leq \frac{1}{1 - \bar{L}\|x^* - x_n\| - \bar{\ell}} \left[\frac{\bar{K}}{2}\|x^* - x_n\| + (\bar{M}\|x^* - x_n\| + \bar{\mu})\right] \|x^* - x_n\|$$
$$< \frac{(\frac{\bar{K}}{2} + \bar{M})r^* + \bar{\mu}}{1 - \bar{L}r^* - \bar{\ell}} \|x^* - x_n\|.$$

10.2. (a) Let $F: D \subseteq X \to Y$ be an m-times Fréchet-differentiable operator ($m \geq 2$ integer).
Assume:

(a_1) there exist an open convex subset D_0 of D, $x_0 \in D_0$, a bounded outer inverse $F'(x_0)^\#$ of $F'(x_0)$, and constants $\eta > 0$, $\alpha_i \geq 0$, $i = 2, ..., m+1$ such that for all $x, y \in D_0$ the following conditions hold:

$$\|F'(x_0)^\# (F^{(m)}(x) - F^{(m)}(x_0))\| \leq \varepsilon, \quad \varepsilon > 0, \tag{1}$$

for all $x \in U(x_0, \delta_0)$ and some $\delta_0 > 0$.

$$\|F'(x_0)^\# F(x_0)\| \leq \eta,$$
$$\|F'(x_0)^\# F^{(i)}(x_0)\| \leq \alpha_i$$

the positive zeros s of p' is such that

$$p(s) \leq 0,$$

where,

$$p(t) = \eta - t + \frac{\alpha_2 t^2}{2!} + \cdots + \frac{\alpha_m + \varepsilon}{m!} t^m.$$

Show: polynomial p has only two positive zeros denoted by t^*, t^{**} ($t^* \leq t^{**}$).

(a_2)
$$\bar{U}(x_0, t^*) = \{x \in X \mid \|x - x_0\| \leq \delta\} \subseteq D_0, \quad \delta = \max\{\delta_0, t^*, t^{**}\}.$$

(a_3) $\delta_0 \in [t^*, t^{**}]$ or $\delta_0 > t^{**}$.

Moreover show: sequence $\{x_n\}$ ($n \geq 0$) generated by Newton's method with $F'(x_n)^\# = [I + F'(x_0)^\# (F'(x_n) - F'(x_0))]^{-1} F'(x_0)^\#$ ($n \geq 0$) is well defined, remains in $U(x_0, t^*)$ and converges to a solution $x^* \in \bar{U}(x_0, t^*)$ of equation $F'(x_0)^\# F(x) = 0$;

- the following error bounds hold for all $n \geq 0$

$$\|x_{n+1} - x_n\| \leq t_{n+1} - t_n$$

and

$$\|x_n - x^*\| \leq t^* - t_n,$$

where $\{t_n\}$ $(n \geq 0)$ is a monotonically increasing sequence converging to t^* and generated by

$$t_0 = 0, \quad t_{n+1} = t_n - \frac{p(t_n)}{p'(t_n)}.$$

(b) Let $F \colon D \subseteq X \to Y$ be an m-times Fréchet-differentiable operator ($m \geq 2$ an integer). Assume:

(b$_1$) condition (1) holds;

(b$_2$) there exists an open convex subset D_0 of D, $x_0 \in D_0$, and constants $\alpha, \beta, \eta \geq 0$ such that for any $x \in D_0$ there exists an outer inverse $F'(x)^{\#}$ of $F'(x)$ satisfying $N(F'(x)^{\#}) = N(F'(x_0)^{\#})$ and

$$\|F'(x_0)^{\#} F(x_0)\| \leq \eta,$$

$$\left\| F'(y)^{\#} \int_0^1 F''[x + t(y-x)](1-t)dt(y-x)^2 \right\| \leq$$

$$\leq \left[\tfrac{\alpha_m + \varepsilon}{m!} \|y - x\|^{m-2} + \cdots + \tfrac{\alpha_2}{2!}\right] \|y - x\|^2,$$

for all $x, y \in D_0$,

$$\left[\tfrac{\alpha_m + \varepsilon}{m!} \eta^{m-2} + \cdots + \tfrac{\alpha_2}{2!}\right] \eta < 1,$$

and

$$\bar{U}(x_0, r) \subseteq D_0 \quad \text{with} \quad r = \min\left\{\tfrac{\eta}{1-r_0}, \delta_0\right\},$$

where,

$$r_0 = \left[\tfrac{\alpha_m + \varepsilon}{m!} \eta^{m-2} + \cdots + \tfrac{\alpha_2}{2!}\right] \eta.$$

Show: sequence $\{x_n\}$ $(n \geq 0)$ generated by Newton's method is well defined, remains in $\bar{U}(x_0, r)$ for all $n \geq 0$ and converges to a solution x^* of $F'(x_0)^{\#} F(x) = 0$ with the iterates satisfying $N(F'(x_n)^{\#}) =$

$N(F'(x_0)^\#)$ $(n \geq 0)$. Moreover, the following error bounds hold for all $n \geq 0$

$$\|x_{n+1} - x_n\| \leq r_0^n \|x_1 - x_0\|,$$
$$\|x^* - x_n\| \leq \frac{r_0^n}{1-r_0} \|x_1 - x_0\|,$$

and

$$\|x_n - x_0\| \leq \frac{1-r_0^n}{1-r_0} \|x_1 - x_0\| \leq \frac{1-r_0^n}{1-r_0} \eta \leq r.$$

10.3. Let X and Y be Banach spaces, and let L be a bounded linear operator on X into Y. a linear operator $M : Y \to X$ is said to be an inner inverse of L if $LMA = L$. A linear operator $M : Y \to X$ is an outer inverse of L if $MLM = M$. Let L be an $m \times n$ matrix, with $m > n$. Any outer inverse M of L will be an $n \times m$ matrix. Show:

(a) If $\text{rank}(L) = n$, then L can be written as

$$L = A \begin{bmatrix} I \\ 0 \end{bmatrix},$$

where I is the $n \times n$ identity matrix, and A is an $m \times m$ invertible matrix. The $n \times m$ matrix

$$M = \begin{bmatrix} I & B \end{bmatrix} A^{-1}$$

is an outer inverse of L for any $n \times (m-n)$ matrix B.

(b) If $\text{rank}(L) = r < n$, then L can be written as

$$L = A \begin{bmatrix} I & 0 \\ 0 & 0 \end{bmatrix} C,$$

where A is an $m \times m$ invertible matrix, I is the $r \times r$ identity matrix, and C is an $n \times n$ invertible matrix. If E is an outer (inner) inverse of the matrix $\begin{bmatrix} I & 0 \\ 0 & 0 \end{bmatrix}$, then the $n \times m$ matrix

$$M = C^{-1} E A^{-1}$$

is an outer (inner) inverse of L.

(c) E is both an inner and an outer inverse of $\begin{bmatrix} I & 0 \\ 0 & 0 \end{bmatrix}$ if and only if E can be written in the form

$$E = \begin{bmatrix} I & M \\ C & CM \end{bmatrix} = \begin{bmatrix} I \\ C \end{bmatrix} \begin{bmatrix} I & M \end{bmatrix}.$$

(d) For any $(n-r) \times r$ matrix T, the matrix $E = \begin{bmatrix} I & 0 \\ T & 0 \end{bmatrix}$ is an outer inverse of $\begin{bmatrix} I & 0 \\ 0 & 0 \end{bmatrix}$.

10.4. Let $F : D \subseteq X \to Y$ be a Fréchet-differentiable operator between two Banach space X and Y, $A(x) \in L(X,Y)$ ($x \in D$) be an approximation to $F'(x)$. Assume that there exist an open convex subset D_0 of D, $x_0 \in D_0$, a bounded outer inverse $A^{\#}$ of $A (= A(x_0))$ and constants $\eta, k > 0, M, L, \mu, l \geq 0$ such that for all $x, y \in D_0$ the following conditions hold:

$$\|A^{\#} F(x_0)\| \leq \eta, \quad \|A^{\#}(F'(x) - F'(y))\| \leq k\|x - y\|,$$
$$\|A^{\#}(F'(x) - A(x))\| \leq M\|x - x_0\| + \mu,$$
$$\|A^{\#}(A(x) - A)\| \leq L\|x - x_0\| + l, \quad b := \mu + l < 1.$$

Assume $h = \sigma\eta \leq \frac{1}{2}(1-b)^2$, $\sigma := \max(k, M+L)$, and $\bar{U} = \bar{U}(x_0, t^*) \subseteq D_0, t^* = \frac{1-b-\sqrt{(1-b)^2 - 2h}}{\sigma}$. Then show

- (i) sequence
  $\{x_n\}$ ($n \geq 0$) generated by $x_{n+1} = x_n - A(x_n)^{\#} F(x_n)$ ($n \geq 0$) with $A(x_n)^{\#} = [I + A^{\#}(A(x_n) - A)]^{-1} A^{\#}$ remains in U and converges to a solution $x^* \in \bar{U}$ of equation $A^{\#} F(x) = 0$.
- (ii) equation $A^{\#} F(x) = 0$ has a unique solution in $\bar{U} \cap \{R(A^{\#}) + x_0\}$, where

$$\tilde{U} = \begin{cases} (x_0, t^*) \cap D_0 & \text{if } h = \frac{1}{2}(1-b)^2 \\ U(x_0, t^{**}) \cap D_0 & \text{if } h < \frac{1}{2}(1-b)^2, \end{cases}$$

$$R(A^{\#}) + x_0 := \{x + x_0 : x \in R(A^{\#})\},$$

and

$$t^* = \frac{1 - b + \sqrt{(1-b)^2 - 2h}}{\sigma}$$

- (iii) $\|x_{n+1} - x_n\| \leq t_{n+1} - t_n$, $\|x^* - x_n\| \leq t^* - t_n$, where $t_0 = 0$, $t_{n+1} = t_n + \frac{f(t_n)}{g(t_n)}$, $f(t) = \frac{\sigma}{2}t^2 - (1-b)t + \eta$, and $g(t) = 1 - Lt - \ell$.

10.5. Let $F : D \subseteq X \to Y$ be a Fréchet-differentiable operator between two Banach spaces X and Y and let $A(x) \in L(X,Y)$ be an approximation of $F'(x)$. Assume that there exist an open convex subset D_0 of D, a point $x_0 \in D_0$ and constants $\eta, k > 0$ such that for any $x \in D_0$ there exists an outer inverse $A(x)^{\#}$ of $A(x)$ satisfying $N(A(x)^{\#}) = N(A^{\#})$, where $A = A(x_0)$ and $A^{\#}$ is a bounded outer inverse of A, and for this outer inverse the following conditions hold:

$$\|A^{\#} F(x_0)\| \leq \eta,$$

$$\|A(y)^{\#} (F'(x + t(y-x)) - F'(x))\| \leq kt \|x - y\|$$

for all $x, y \in D_0$ and $t \in [0, 1]$, $h = \frac{1}{2} k \eta < 1$ and $\bar{U}(x_0, r) \subseteq D_0$ with $r = \frac{\eta}{1-h}$. Then show sequence $\{x_n\}$ $(n \geq 0)$ generated by $x_{n+1} = x_n - A(x_n)^{\#} F(x_n)$ $(n \geq 0)$ with $A(x_n)^{\#}$ satisfying $N(A(x_n)^{\#}) = N(A^{\#})$ remains in $\bar{U}(x_0, r)$ and converges to a solution x^* of equation $A^{\#} F(x) = 0$.

10.6. Show that Newton's method with outer inverses $x_{n+1} = x_n - F'(x_n)^{\#}$ $(n \geq 0)$ converges quadratically to a solution $x^* \in \widetilde{U} \cap \{R(F'(x_0)^{\#}) + x_0\}$ of equation $F'(x_0)^{\#} F(x) = 0$ under the conditions of Exercise 10.4 with $A(x) = F'(x)$ $(x \in D_0)$.

10.7. Let $F : D \subseteq X \to Y$ be Fréchet-differentiable and assume that $F'(x)$ satisfies a Lipschitz condition

$$\|F'(x) - F'(y)\| \leq L \|x - y\|, \quad x, y \in D.$$

Assume $x^* \in D$ exists with $F(x^*) = 0$. Let $a > 0$ such that $U\left(x^*, \frac{1}{a}\right) \subseteq D$. Suppose there is an

$$F'(x^*)^{\#} \in \Omega(F'(x^*)) = \{B \in L(Y, X) : BF'(x^*)B = B, \ B \neq 0\}$$

such that $\|F'(x^*)^{\#}\| \leq a$ and for any $x \in U\left(x^*, \frac{1}{3La}\right)$, the set $\Omega(F'(x))$ contains an element of minimum norm. Then show there exists a ball $U(x^*, r) \subseteq D$ with $cr < \frac{1}{3aL}$ such that for any $x_0 \in U(x^*, r)$ the sequence $\{x_n\}$ $(n \geq 0)$ $x_{n+1} = x_n - F'(x_n)^{\#} F(x_n)$ $(n \geq 0)$ with

$$F'(x_0)^{\#} \in \mathrm{argmin}\,\{\|B\| \,|\, B \in \Omega(F'(x_0))\}$$

and with $F'(x_n)^{\#} = (I + F'(x_0)^{\#} (F'(x_n) - F'(x_0)))^{-1} F'(x_0)^{\#}$ converges quadratically to $\bar{x}^* \in U\left(x_0, \frac{1}{La}\right) \cap \{R(F'(x_0)^{\#}) + x_0\}$, which is a solution of equation $F'(x_0)^{\#} F(x) = 0$. Here, $R(F'(x_0)^{\#}) + x_0 = \{x + x_0 : x \in R(F'(x_0)^{\#})\}$.

Chapter 11

Operator Equations and Their Discretizations

11.1 The Mesh Independence Principle Under Hölder continuity

Consider the equation

$$F(x) = 0 \qquad (11.1)$$

where F is a nonlinear operator defined between two Banach spaces E_1, E. The Newton's method

$$x_{n+1} = x_n - F'(x_n)^{-1} F(x_n), \quad n = 0, 1, 2, ... \qquad (11.2)$$

has been used extensively to approximate a solution x^* of (11.1). The iterates $\{x_n\}$, $n = 0, 1, ...$ can rarely be computed in infinite dimensional spaces.

That is why we replace (11.1) by a family of discretized equations

$$F_h(z) = 0, \quad h > 0 \qquad (11.3)$$

where F_h is a nonlinear operator between two finite dimensional spaces E_h^1 and E_h. The discretization on E_1 is defined by the linear operators $L_h : E_1 \to E_h^1$.

The Newton's iteration for (11.3) is given by

$$z_0^h = L_h(x_0), \quad z_{n+1}^h = z_n^h - F_h'(z_n^h), \quad n = 0, 1, ... \qquad (11.4)$$

In the excellent paper in reference [2], it is shown that under certain assumptions the solution z_h^* and the iterates z_n^h satisfy the relations

$$z_h^* = L_n(x^*) + 0(h^q),$$
$$z_h^h - z_h^* = L_n(x_n - x^*) + 0(h^q),$$
$$z_{n+1}^h - z_h^h = L_h(x_{n+1} - x_n) + 0(h^q), \quad q > 0,$$

and for any $\epsilon > 0$

$$\left| \min\{n \geq 0, \|x_n - x^*\| < \epsilon\} - \min\{n \geq 0, \|z_n^h - z_h^*\| < \epsilon\} \right| \leq 1$$

for h sufficiently small and x_0 in a ball centered at x^* and of some specific radius $r > 0$.

One of the basic assumptions in [2] is that the Fréchet derivative of F is Lipschitz continuous on a subset $E_2 \subset E_1$.

Here we show that the above results can be extended to include the case when the Fréchet-derivative of F is only (γ, λ)–Hölder continuous (to be precised later) for some $\gamma > 0$ and $\lambda \in [0,1]$. Our results reduce to the ones in [2] for $\lambda = 1$.

An example is also provided for $\lambda = \frac{1}{2}$ for a scalar, second order, two-point boundary value problem, where our results apply where the ones in [2] do not.

To make the section as self-contained as possible we will use some of the techniques developed in the proofs of the results in [2], [68], [99].

The norms in all spaces will be denoted by the same symbol $\|\ \|$.

Definition 11.1 We say that the Fréchet-derivative $F'(x)$ of F is (γ, λ)–Hölder continuous on $E_2 \subset E_1$ if for some $\gamma > 0$, $\lambda \in [0,1]$

$$\|F'(x) - F'(y)\| \leq \gamma \|x - y\|^\lambda \quad \text{for all} \quad x, y \in E_2. \tag{11.5}$$

We then say that $F'(\cdot) \in H_{E_2}(\gamma, \lambda)$.

It is well-known that if E_2 is convex then

$$\|F(x) - F(y) - F'(x)(x - y)\| \leq \frac{\gamma}{1 + \lambda} \|x - y\|^{1+\lambda} \quad \text{for all} \quad x, y \in E_2. \tag{11.6}$$

We assume that (11.1) has a solution $x^* \in E_2$ which is simple in the sense that $F'(x^*)$ has a bounded inverse with norm $\gamma = \|F'(x^*)^{-1}\|$.

Theorem 11.1 *Let $F : E_1 \to E$. Assume $F'(\cdot) \in H_{E_2}(\gamma, \lambda)$ on a convex set $E_2 \subset E_1$. If $x^* \in E_2$ is a solution of*

$$F(x) = 0$$

for which $F'(x^*)$ is nonsingular, set

$$U^* = \overline{U}(x^*, r^*),$$

with

$$0 < r^* < \left[\frac{1+\lambda}{(2+\lambda)\beta\gamma}\right]^{1/\lambda}, \lambda \in (0, 1] \tag{11.7}$$

such that $U^* \subset E_2$.

Then, for any $x_0 \in U^*$, Newton's iteration (11.2) converges to x^* and the iterates satisfy

$$\|x_{n+1} - x^*\| \le \frac{\beta\gamma}{1+\lambda} \frac{\|x_n - x^*\|^{1+\lambda}}{1 - \beta\gamma\|x_n - x^*\|^\lambda}, \quad n = 0, 1, \ldots. \tag{11.8}$$

Proof. By the Banach Lemma on invertible operators it follows that $F'(x)$ is nonsingular in U^* and

$$\left\|F'(x)^{-1}\right\| \le \frac{\beta}{1 - \beta\gamma\|x - x^*\|^\lambda}, \quad \text{for all} \quad x \in U^*. \tag{11.9}$$

Hence Newton's iteration function

$$P(x) = x - F'(x)^{-1} F(x), \quad x \in U^*$$

is well defined on U^* and from

$$\|P(x) - x^*\| \le \left\|F'(x)^{-1}\right\| \|F(x^*) - F(x) - F'(x)(x^* - x)\|$$

$$\le \frac{\beta}{1 - \beta\gamma\|x - x^*\|^\lambda} \gamma \|x - x^*\|^{1+\lambda}$$

$$\le a(r) \|x - x^*\|, \quad \text{for all} \quad x \in U^*$$

and

$$a(r) = \frac{\beta\gamma(r^*)^\lambda}{(\lambda+1)\left(1 - \beta\gamma(r^*)^\lambda\right)} < 1$$

we obtain the results. □

We now state a theorem that can be found in [187, pp. 145], whose proof follows exactly as the proof of the Newton-Kantorovich theorem for $\lambda = 1$ [187, pp. 143].

Theorem 11.2 Let $F : E_1 \to E$. Assume:

(a) the linear operator $F'(\cdot) \in H_{E_3^*}(\gamma, \lambda)$, where $E_3^* = U(x_0, R) \subset E_1$ for some $x_0 \in E_1$ and $R > 0$;

(b) the linear operator $F'(x_0)^{-1}$ exists and satisfies

$$\left\| F'(x_0)^{-1} \right\| \leq b_0, \quad \left\| F'(x_0)^{-1} F(x_0) \right\| \leq \eta_0, \quad \ell_0 = b_0 \gamma \eta_0^\lambda \leq s \quad (11.10)$$

where s is the minimum positive root of the equation

$$\left(\frac{s}{1+\lambda} \right)^\lambda = (1-s)^{1+\lambda} \quad in \quad \left(0, \frac{1}{2} \right) \quad with \quad 0 < \lambda < 1. \quad (11.11)$$

If

$$R \geq r_0 = \frac{\eta_0}{1 - p_0}, \quad where \quad P_0 = \frac{\ell_0}{(1+\lambda)(1-\ell_0)} \quad (11.12)$$

then Newton's iteration (11.2) converges to a unique solution x^* of the equation

$$F(x) = 0$$

in $\overline{U}(x_0, r_0)$.

As in [2] consider a subset $W^* \subset E_1$ such that

$$x^* \in W^*, \quad x_n \in W^*, \quad x_n - x^* \in W^*, \quad x_{n+1} - x_n \in W^*, \quad n = 0, 1, 2, \dots. \quad (11.13)$$

Consider the discretization method given by the family

$$\{F_h, L_h, \overline{L}_n\}, \quad h > 0 \quad (11.14)$$

where

$$F_h : D_h \subset E_h^1 \to E_h, \quad h > 0$$

are nonlinear operators and

$$L_h : E_1 \to E_h^1, \quad \overline{L}_h : E \to E_h, \quad h > 0,$$

are bounded linear discretization operators such that

$$L_h (W^* \cap U^*) \subset D_h, \quad h > 0. \quad (11.15)$$

The discretization (11.14) is called λ–Hölder uniform if there exist constants $w > 0$, $\ell > 0$ such that

$$\overline{U}(L_h(x^*), w) \subset D_h, \quad h > 0 \quad (11.16)$$

and

$$\|F'_h(w_1) - F'_h(w_2)\| \le \ell \|w_1 - w_2\|^\lambda, \quad \lambda \in [0,1), \quad h > 0, \quad (11.17)$$
$$w_1, w_2 \in \overline{U}(L_h(x^*), w).$$

Moreover, the discretization (11.14) is called: <u>bounded</u> if there is a constant $b > 0$ such that

$$\|L_h(u)\| \le b \|u\|, \quad u \in W^*, \quad h > 0, \quad (11.18)$$

<u>stable</u> if there is a constant d.0 such that

$$\left\|F'_h(L_h(u))^{-1}\right\| \le d, \quad u \in W^* \cap U^*, \quad h > 0, \quad (11.19)$$

<u>consistent of order $q > 0$</u> if there are two constants $c_0 > 0$, $c_1 > 0$ such that

$$\|\overline{L}_h(F(x)) - F_h(L_h(x))\| \le c_0 h^q, \quad x \in W^* \cap U^*, \quad h > 0 \quad (11.20)$$

$$\|\overline{L}_h(F'(u))(v) - F'_h(L_h(u))L_h(v)\| \le c_1 h^q, \quad \in W^* \cap U^*,$$
$$v \in W^*, \quad h > 0. \quad (11.21)$$

We can now prove the main result:

Theorem 11.3 *Let $F : E_2 \subset E_1 \to E$ be an operator satisfying the hypotheses of Theorem 11.1 and consider a uniform discretization (11.14) which is bounded, stable and consistent of order q. Then (11.3) has a locally unique solution*

$$z_h^* = L_h(x^*) + 0(h^q) \quad (11.22)$$

for all $h > 0$ satisfying

$$0 < h \le h_0 = \min\left[\left(\frac{e}{d^2 \ell c_0}\right)^{1/q}, \left(\frac{w}{mdc_0}\right)^{1/q}\right] \quad (11.23)$$

with $e = \frac{1}{2}$ and $m = \frac{(1+\lambda)(1-e)}{(1+\lambda)(1-e)-e}$.
Moreover, if the following condition are satisfied:

$$T \equiv A\left(\frac{B-C}{2A}\right)^{\lambda+1} + C\left(\frac{B-C}{2A}\right)^\lambda - B\left(\frac{B-C}{2A}\right) < 0, \quad C < B \quad (11.24)$$

with

$$A = (\lambda + 2) d\ell$$
$$B = \lambda + 1, \quad \lambda \in (0,1)$$
$$C = 2r^* (\lambda + 1) \ell b$$
$$D = 2(\lambda + 1) c, \quad c = \max(c_0, c_1).$$

Then there exist constants $h_1 \in (0, h_0]$, $r_1 \in (0, r^*]$ such that Newton's iteration (11.4) converges to z_h^* and that

$$z_n^h = L_h(x_n) + 0(h^q), \quad n = 0, 1, 2, \ldots \tag{11.25}$$
$$z_n^h - z_h^* = L_h(x_n - x^*) + 0(h^q), \quad n = 0, 1, 2, \ldots \tag{11.26}$$

for all $h \in (0, h_1]$, and all starting points $z_0 \in \overline{U}(z^*, r_1)$.

Proof. For simplicity, we will prove the theorem for $\lambda \in (0, 1)$. By Theorem 11.2, when

$$\ell_0 = \ell_0(h) = d\ell \, \|F_h'(L_h(x^*))\| \le s = s(h) < e \tag{11.27}$$
$$r_0 = r_0(h) = \frac{\eta_0(h)}{1 - p_0(h)} \le w,$$

with

$$p_0(h) = \frac{\ell_0}{(1 + \lambda)(1 - \ell_0)} \tag{11.28}$$

then (11.3) has a unique root $z_h^* \in \overline{U}(L_h(x^*), r_0)$.

By (11.20), (11.21) and (11.23) we get

$$\ell_0 \le d^2 \ell \, \|F_h(L_h(x^*)) - \overline{L}_h(F(x^*))\| \le d^2 \ell c_0 h^q < e$$

and

$$r_0 \le m d c_0 h^q \le w, \tag{11.29}$$

which shows that (11.27) and (11.28) hold for all h satisfying (11.23).

Thus (11.22) follows from

$$\|z_h^* - L_h(x^*)\| \le r_0 \le m d c_0 h^q. \tag{11.30}$$

By applying Theorem 11.3 to (11.3) we see that the Newton sequence (11.4) converges to z_h^* if

$$\|L_h(x_0) - z_h^*\| < \left(\frac{e}{\ell \left\|F_h'(z_h^*)^{-1}\right\|}\right)^{1/\lambda} \tag{11.31}$$

$$\overline{U}(z_h^*, \|L_h(x_0) - z_h^*\|) \subset \overline{U}(L_h(x^*), w). \tag{11.32}$$

But (11.32) holds if

$$\|z_h^* - L_h(x^*)\| + \|L_h(x_0) - z_h^*\| \le w, \tag{11.33}$$

and by (11.18) and (11.30) we have

$$\|L_h(x_0) - z_h^*\| \le \|L_h(x_0) - L_h(x^*)\| + \|L_h(x^*) - z_h^*\|$$
$$\le b\|x_0 - x^*\| + mdc_0 h^q. \tag{11.34}$$

Hence (11.33) is satisfied if

$$b\|x_0 - x^*\| + 2mdc_0 h^q \le w. \tag{11.35}$$

Since,

$$F_h'(z_h^*) = F_h'(L_h(z^*))\left[I - F_h'(L_h(z^*))^{-1}(F_h'(L_h(x^*)) - F_h'(z_h^*))\right]$$

using (11.17), (11.14) and (11.30) we get

$$\left\|F_h'(z_h^*)^{-1}\right\| \le \frac{\left\|F_h'(L_h(z^*))^{-1}\right\|}{1 - \ell \left\|F_h(L_h(z^*))^{-1}\right\| \|L_h(x^*) - z_h^*\|^\lambda}$$
$$\le \frac{d}{1 - \ell d (mdc_0 h^q)^\lambda} \tag{11.36}$$

Thus, (11.31) holds when

$$b\|x_0 - x^*\| + 2mdc_0 h^q < \left[\frac{e}{\ell}\left(\frac{1 - \ell d (mdc_0 h^q)^\lambda}{d}\right)\right]^{1/\lambda}. \tag{11.37}$$

By setting,

$$h_2 = \min\left[\left(\frac{w}{4mdc}\right)^{1/q}, \left(\frac{1}{4mdc}\right)^{1/q}\left(\frac{e(1 - \ell d w^\lambda)}{\ell d}\right)^{1/\lambda q}\right], \tag{11.38}$$

and

$$r_2 = \min\left[\frac{w}{2b}, \frac{1}{2b}\left(\frac{e(1-\ell dw^\lambda)}{\ell d}\right)^{1/\lambda}\right], \quad (11.39)$$

it can easily be verified that (11.34) and (11.37) hold for all $h \in (0, h_2]$ and $x_0 \in U(x^*, r_2)$.

That is, for these h and x_0, the sequence (11.4) converges to z_h^*. Let us now define the function v by

$$v = v(h) = c_2 h^q, \quad c_2 > 0.$$

We now prove that for $h \in (0, h_1)$ and $x_0 \in \overline{U}(x^*, r_1)$ and all $n = 0, 1, \ldots$ the estimate

$$\|z_n^h - L_h(x_n)\| \le v \quad (11.40)$$

holds, where

$$h_1 = \min\left[h_0, h_2, \left(\frac{(C-B)^2}{4AD}\right)^{1/q}, \left(\frac{-T}{D}\right)^{1/2}, \left(\frac{1}{\ell d c_2^\lambda}\right)^{1/q\lambda}\right] \quad (11.41)$$

and

$$r_1 = \min(r_2, r^*). \quad (11.42)$$

we use induction. for $n = 0$ (11.40) is trivially true.
Consider the identity,

$$z_{i+1}^h - L_h(x_{i+1})$$
$$= F_h'(z_i^h)^{-1}\left\{[F_h'(z_i^h)(z_i^h - L_h(x_i)) - F_h(z_i^h) + F_h(L_h(x_i))]\right.$$
$$+ \left[(F_h'(z_i^h) - F_h'(L_h(x_i)))L_h\left(F'(x_i)^{-1}F(x_i)\right)\right]$$
$$+ \left[F_h'\left(L_h(x_i)L_h\left(F'(x_i)^{-1}F(x_i)\right) - \overline{L}_h(F(x_i))\right)\right]$$
$$+ \left.[\overline{L}_h(F(x_i)) - F_h(L_h(x_i))]\right\}. \quad (11.43)$$

As in (11.36) we can obtain

$$\|F_h'(z_i^h)^{-1}\| \le \frac{d}{1-\ell dv^\lambda}. \quad (11.44)$$

Using a standard argument we have that

$$\|F'_h(z_i^h)(z_i^h - L_h(x_i)) - F'_h(z_i^h) + F_h(L_h(x_i))\|$$
$$\leq \tfrac{1}{\lambda+1}\ell \|z_i^h - L_h(x_i)\|^{\lambda+1}$$
$$\leq \tfrac{1}{\lambda+1}\ell v^{\lambda+1}. \qquad (11.45)$$

Also,

$$\left\|\left(F'_h(z_i^h) - F'_h(L_h(x_i))\right)\left(L_h\left(F'(x_i)^{-1} F(x_i)\right)\right)\right\|$$
$$\leq \ell b \|z_i^h - L_h(x_i)\|^{\lambda} \|x_i - x_{i+1}\|$$
$$\leq 2\ell b v^{\lambda} \|x_0 - x^*\|$$
$$\leq 2\ell b v^{\lambda} r_1 \qquad (11.46)$$

(since by Theorem 11.1 $\|x_{i+1} - x^*\| \leq \|x_i - x^*\|$).
Finally, from (11.20) and (11.21) we obtain

$$\left\|F'_h(L_h(x_i)) L_h\left(F'(x_i)^{-1}(F(x_i)) - \overline{L}_h F(x_i)\right)\right\| \leq c_1 h^q \leq c h^q \qquad (11.47)$$

and

$$\|\overline{L}_h F(x_i) - F_h(L_h(x_i))\| \leq c_0 h^q \leq c h^q. \qquad (11.48)$$

Using the above estimates in (11.43) we obtain that

$$\|z_{i+1}^h - L_h(x_{i+1})\| \leq \frac{d}{1 - \ell d v^{\lambda}} \left[\frac{1}{\lambda+1}\ell v^{\lambda+1} + 2\ell b v^{\lambda} r^* + 2 c h^q\right]. \qquad (11.49)$$

Define the real functions f and g by

$$f(v) = A v^{\lambda+1} - B v + C v^{\lambda} + D h^q \qquad (11.50)$$

and

$$g(v) = A v^2 + (C - B) v + D h^q. \qquad (11.51)$$

By the choice of r_1 and h_1

$$C < B, \qquad (11.52)$$
$$(C - B)^2 - 4 A D h^q > 0, \qquad (11.53)$$

and f has two positive solutions.
Therefore, the function g has a minimum at

$$v_m = \frac{B - C}{2A} \qquad (11.54)$$

and
$$f(v_m) = T + Dh^q \qquad (11.55)$$

which according to (11.24) and the choice of h is negative. Since $f(v)$ is continuous, $f(0) = 0$ and $f(v) > 0$ for v sufficiently large we are assured that $f(v)$ has two positive solutions. Denote by v_1 the smallest positive root. then the right hand side of (11.49) is equal to v_1.

Moreover,
$$v_1 \left[Av_1^\lambda - B + Cv_1^{\lambda-1}\right] = -Dh^q \qquad (11.56)$$

or
$$v_1 = \frac{D}{B - Av_1^\lambda - Cv_1^{\lambda-1}} h^q \qquad (11.57)$$

with
$$B - Av_1^\lambda - Cv_1^{\lambda-1} > 0. \qquad (11.58)$$

By (11.58), there exist v_2, v_3 sufficiently close to v_1 with $v_2 \leq v_1 \leq v_3$ such that
$$B - Av_3^\lambda - Cv_2^{\lambda-1} > 0. \qquad (11.59)$$

Therefore, by (11.57), we obtain that
$$v_1 \leq \frac{D}{B - Av_3^\lambda - Cv_2^{\lambda-1}} h^q = c_2 h^q \qquad (11.60)$$

by setting
$$c_2 = \frac{D}{B - Av_3^\lambda - Cv_2^{\lambda-1}}.$$

This proves (11.25) since, we have
$$\left\|z_n^h - L_h(x_n)\right\| \leq v_1 \leq c_2 h^q. \qquad (11.61)$$

Finally, by (11.30), (11.40), and (11.61), we get
$$\left\|(z_n^h - z_h^*) - L_h(x_n - x^*)\right\| \leq \left\|z_n^h - L_h(x_n)\right\| + \left\|z_h^* - L_h(x^*)\right\|$$
$$\leq mdc_0 h^q + c_2 h^q = c_3 h^q \qquad (11.62)$$

by setting $c_3 = mdc_0 + c_2$, which shows (11.26) and that completes the proof of the theorem. $\square$

We can now prove the following to justify the claims made in the introduction.

Theorem 11.4 *Assume:*

(a) *the hypotheses of Theorem 11.3 are true;*
(b) *there exists a $\delta > 0$ such that*

$$\liminf_{h>0} \|L_h(u)\| \geq \delta \|u\| \quad \text{for each} \quad u \in W^*. \tag{11.63}$$

Then for some $\bar{r} \in (0, r_1)$, and for any fixed $\epsilon > 0$ and $x_0 \in U(x^*, \bar{r})$ there exists a constant $\bar{h}$ depending on ϵ and z_0 with $\bar{h} \in (0, h_1]$ such that

$$\left|\min\{n \geq 0, \|x_n - x^*\| < \epsilon\} - \min\{n \geq 0, \|z_n^h - z_h^*\| < \epsilon\}\right| \leq 1 \tag{11.64}$$

for all $h \in (0, \bar{h}]$.

Proof. Let k be the unique integer defined by

$$\|x_{k+1} - x^*\| < \epsilon \leq \|x_i - x^*\| \tag{11.65}$$

and $h_3 > 0$ such that

$$\|L_h(x_i - x^*)\| \geq \delta \|x_i - x^*\|, \quad \text{with} \quad 0 < h < h_3. \tag{11.66}$$

Set,

$$\bar{r} = \min\left(r_1, \frac{1}{2}\left(\frac{\bar{b}}{1+d\ell\bar{b}}\right)^{1/\lambda}\right), \tag{11.67}$$

$$a = \frac{d\ell}{1+\lambda},$$

$$\bar{b} = \min\left(b, \frac{1}{2b}, \frac{\delta}{2}\right), \tag{11.68}$$

and

$$\bar{h} = \min\left[h_1, h_3, \left(\frac{\bar{b}}{1+d\ell\bar{b}}\right)^{1/\lambda q}\left(\frac{1}{2mdc_0}\right)^{1/q}, \left(\frac{\bar{b}\epsilon}{c_3}\right)^{1/q}\right]. \tag{11.69}$$

We will prove the theorem for the above choices of $\bar{r}$ and $\bar{h}$.

By (11.62) and (11.69) we obtain that

$$\|z_{i+1}^h - z_h^*\| \leq \|L_h(x_{i+1} - x^*)\| + c_3 h^q \leq b\epsilon + c_3 h^q \leq 2b\epsilon, \tag{11.70}$$

and from (11.34), (11.67), (11.69) and (11.8)

$$\begin{aligned}\left\|z_{i+2}^h - z_h^*\right\| &\leq \frac{d\ell \left\|z_{i+1} - z_h^*\right\|^{1+\lambda}}{(1+\lambda)\left[1 - d\ell \left\|z_{i+1} - z_h^*\right\|^{\lambda}\right]} \\ &\leq \left(\frac{d\ell}{1+\lambda}\right) \frac{\left\|z_0^h - z_h^*\right\|^{\lambda}}{\left(1 - d\ell \left\|z_0^h - z_h^*\right\|^{\lambda}\right)} \left\|z_{i+1} - z_h^*\right\| \\ &\leq 2\bar{b}b\epsilon < \epsilon.\end{aligned} \qquad (11.71)$$

By (11.66) and (11.62) we get

$$\epsilon \leq \|x_i - z^*\| \leq \frac{1}{\delta}\|L_h(x_i - x^*)\| \leq \frac{1}{\delta}\left(\left\|z_i^h - z_h^*\right\| + c_3 h^q\right) \qquad (11.72)$$

or

$$\left\|z_i^h - z_h^*\right\| \geq \delta\epsilon - c_3 h^q \geq \delta\epsilon - \frac{\delta\epsilon}{2} = \frac{\delta\epsilon}{2}. \qquad (11.73)$$

If $\left\|z_{i-1}^h - z_h^*\right\| < \epsilon$, then as in (11.71) we get

$$\left\|z_i^h - z_h^*\right\| < \frac{\bar{b}\epsilon}{2} \leq \frac{\delta\epsilon}{2} \qquad (11.74)$$

which contradicts (11.74). That is,

$$\left\|z_{i-1}^h - z_h^*\right\| \geq \epsilon. \qquad (11.75)$$

The result now follows from (11.65), (11.71), and (11.75). $\square$

Remark 11.1 *(a) The condition (11.63) follows from*

$$\lim_{h \to 0} \|L_h(u)\| = \|u\|, \quad u \in W^* \qquad (11.76)$$

(b) For some discretizations we have

$$\lim_{h \to 0} \|L_h(u)\| = \|u\| \quad \text{uniformly for} \quad u \in W^*. \qquad (11.77)$$

Both conditions above hold in many discretization studies [2], [68], [99], [227], [247].

The following result is now immediate:

Corollary 11.1 *Assume:*

(a) the hypotheses of Theorem 11.3 are satisfied;
(b) the condition (11.77) holds uniformly for $u \in W^*$.

Then there exists $\bar{r}_1 \in (0, r_1]$ and, for any fixed $\epsilon > 0$, some $\bar{h}_1 = \bar{h}_1(\epsilon) \in (0, h_1]$ such that

$$\left|\min\{n \geq 0, \|x_n - x^*\| < \epsilon\} - \min\{n \geq 0, \|z_n^h - z_h^*\| < \epsilon\}\right| \leq 1$$

holds for all $h \in (0, \bar{h}_1]$ and all $z_0 \in \bar{U}(x^*, \bar{r}_1)$.

Example 11.1 Consider the differential equation

$$y'' + y^{1+\lambda} = 0, \quad \text{for} \quad \lambda \in (0, 1)$$
$$y(0) = y(1) = 0.$$

Define the operator

$$F : M \subset C^2[0, 1] \to C[0, 1] \times \mathbb{R}^2,$$
$$F(y) = \{y'' + y^{1+\lambda}; 0 \leq x \leq 1, y(0), y(1)\}.$$

Assume that M is such that the equation

$$F(y) = 0$$

has a unique solution $x^* \in M$ and set

$$U(x^*, w) = \{(x_1^*, x_2^*, x_3^*) \in \mathbb{R}^3;$$
$$0 \leq x_1^* \leq 1, \ |x_2^* - x^*(x_1^*)| \leq w, \ |x_3 - x^*(x_1^*)| \leq w\}.$$

It can easily be seen that $x^* \in C^3[0, 1]$.

The Fréchet derivative of F is given by

$$F'(y)u = \left\{u'' + (1+\lambda)y(\bar{t}_n)^\lambda u, \ 0 \leq x, \ \bar{t}_n \leq 1, \ u(0), \ u(1)\right\}$$

and hence Newton's iteration becomes

$$x_{n+1}'' = -x_n^{1+\lambda} + (1+\lambda)x_n^\lambda(\bar{t}_n)(x_n - x_{n+1})$$

with

$$x_{n+1}(0) = x_{n+1}(1) = 0.$$

Define the norm on $C^m[0, 1]$, $m \geq 0$ with

$$\|u\| = \{(\max |u^i(x)|, \ 0 \leq x \leq 1, \ i = 0, 1, ..., m)\}.$$

Choose $x_0 \in C^2[0,1]$ then $x_{n+1} \in C^3[0,1]$, $n = 0,1,2,...$. We will assume also that $x_0 \in C^3[0,1]$. By the convergence of x_n to x^* in the norm of $C^2[0,1]$, there exists $K > 0$ such that

$$x_n \in W_k = \left\{x \in C^3[0,1]; \sup_t \left|x^{(i)}(t)\right| \leq K, \ i = 0,1,2,3 \right\}, \ n = 0,1,2,...$$

By choosing sufficiently large K we assume

$$x^* \in W_K, \quad x_n - x^* \in W_k \quad \text{and} \quad x_n - x_{n+1} \in W_K, \quad n = 0,1,... \ .$$

We now divided the interval $[0,1]$ into n subintervals and set $h = \frac{1}{n}$. we denote the points of subdivision by

$$p_0 = 0 < p_1 < \cdots < p_n = 1$$

with the corresponding values of the function $y_i = y(p_i)$, $i = 0,1,2,...,n$.

A simple approximation for the derivative at these points is

$$y_i'' \simeq \frac{y_{i-1} - 2y_i + y_{i+1}}{h^2}, \quad i = 1,2,...,n-1.$$

Since, $y_0 = y_n = 0$ this leads to the following system of nonlinear equations

$$h^2 y_1^{1+\lambda} - 2y_1 + y_2 = 0,$$
$$y_{i-1} + h^2 y^{1+\lambda} - 2y_i + y_{i+1} = 0, \quad i = 2,3,...,n-1,$$
$$y_{n-2} + h^2 y_{n-1}^{1+\lambda} - 2y_{n-1} = 0.$$

We therefore have an operator $H : \mathbb{R}^{n-1} \to \mathbb{R}^{n-1}$ whose Fréchet-differential may be written as

$$H'(y) = \begin{bmatrix} (1+\lambda)h^2 y_1^\lambda - 2 & 1 & 0 \cdots & & 0 \\ & & & & \vdots \\ 1 & (1+\lambda)h^2 y^\lambda - 2 & 1 & & \\ 0 & & & & 0 \\ \vdots & & & & 1 \\ 0 & \cdots & 0 & 1 & (1+\lambda)h^2 y_{n-1}^\lambda - 2 \end{bmatrix}$$

Choose $\lambda = \frac{1}{2}$ for simplicity and let $x \in \mathbb{R}^{n-1}$ with norm given by

$$\|x\| = \max_{1 \leq j \leq n-1} |x_j|.$$

The corresponding norm on $Q \in \mathbb{R}^{n-1} \times \mathbb{R}^{n-1}$ is

$$\|Q\| = \max_{1 \leq j \leq n-1} \sum_{k=1}^{n-1} |Q_{jk}|.$$

Then for all $y, z \in \mathbb{R}$ with $|y_i| > 0$, $|z_i| > 0$, $i = 1, 2, ..., n-1$

$$\|H'(y) - H'(z)\| = \left\| diag \left\{ \frac{3}{2}h^2 \left(y_j^{1/2} - z_j^{1/2} \right) \right\} \right\|$$

$$= \frac{3}{2}h^2 \max_{1 \leq j \leq n-1} \left| y_j^{1/2} - z_j^{1/2} \right|$$

$$\leq \frac{3}{2}h^2 \left[\max_{1 \leq j \leq n-1} |y_j - z_j| \right]^{1/2}$$

$$= \frac{3}{2}h^2 \|y - z\|^{1/2}.$$

Here $\ell = \frac{3}{2}h^2$ and $\lambda = \frac{1}{2}$, therefore the results in [2], [227], [247] cannot be applied here. As in [2] the discretization method $\{T_h, L_h, \overline{L}_h\}$ is defined as follows:

$$G_h = \{p_i = ih, \ i = 0, 1, ..., n\}, \quad G_h^0 = G_h \setminus \{0, 1\},$$
$$E_h^1 = \{\eta : G_h \to \mathbb{R}\}, \quad \eta_i = \eta(p_i), \ i = 0, 1, ..., n,$$
$$E_h = \{(\eta, a, b) ; \eta \in G_h^0 \to \mathbb{R}, \ a, b \in \mathbb{R}\},$$
$$L_h(y) = y/G_h, \quad \overline{L}_h(y, a, b) = (y/G_h^0, a, b),$$
$$F_h(\eta) = \left\{ \frac{\eta_{i+1} - 2\eta_i + \eta_{i-1}}{h^2} + \eta_i^{3/2}; \ i = 1, 2, ..., n-1, \eta_0, \eta_n \right\}.$$

The following norms are used in the corresponding spaces

$$\|y\| = \max\{|y^i(x)|, \ 0 \leq x \leq 1, \ i = 0, 1, 2\}, \ y \in C^2[0, 1]$$
$$\|\gamma\| = \max\{|u(x)|, \ a, b; \ 0 \leq x \leq 1\}, \ \gamma = (u, a, b) \in C[0, 1] \times \mathbb{R}^2$$
$$\|\eta\| = \max\left\{|\eta_i|, \left|\frac{\eta_{i+1} - 2\eta_i + \eta_{i-1}}{h^2}\right|, \ i = 1, 2, ..., n-1\right\}, \ \eta \in E_h^1$$
$$\|\xi\| = \max\{|a|, |b|, |\eta_i|, \ i = 1, 2, ..., n-1\}, \ \xi = (\eta, a, b) \in E_h.$$

It can now easily be seen that (11.8) is satisfied for $b = 1$ and (11.20), (11.21) are satisfied with $q = 3/2$.

Moreover, we can easily see with the above norms that

$$\|L_h(u)\| \le \|u\| \le \|L_h(u)\| + K\left(\frac{1}{6}h+1\right)h \quad \text{for} \quad u \in W_K,$$

that is, (11.77) is satisfied.

Therefore, Theorem 11.3 and the corollary may now apply.

Further examples on differential equations can be found in the references.

11.2 Exercises

11.1. Consider an equation

$$F(z) = 0 \tag{11.78}$$

where F is a nonlinear operator between the Banach spaces E, $\widehat{E}$. Unde certain conditions, Newton's method

$$z_{n+1} = z_N - F'(z_N)^{-1} F(z_n), \quad n = 0, 1, ..., \tag{11.79}$$

produces a sequence which converges quadratically to a solution z^* of (11.78). since the formal procedure (11.79) can rarely be executed in infinite-dimensional spaces (11.78) is replaced in practice by a family od discretized equations

$$\phi_h(\zeta) = 0 \tag{11.80}$$

—indexed by some real numbers $h > 0$—where now ϕ_h is a nonlinear operator between finite-dimensional spaces E_h, $\widehat{E}_h$. Let the discretization on E be defined by the bounded linear operators $\Delta_h : E \to E_h$. Then, under appropiate assumptions, the equations (11.80) have solutions

$$\zeta_h^* = \Delta_h z^* + O(h^p)$$

which are the limit of the Newton sequence applied to (11.80) and started at $\Delta_h z_0$; that is,

$$\zeta_0^h = \Delta_h z_0, \ \zeta_{n+1}^h = \zeta_n^h - \phi_h'(\zeta_n^h), \ n = 0, 1, ... \tag{11.81}$$

In many applications it turns out that the solution z^* of (11.78) as well as the Newton interates $\{z_n\}$ have "better smoothness" properties than the elements of E. This is a motivation for considering a subset

$W^* \subset E$ such that

$$z^* \in W^*, \ z_n \in W^*, \ z_n - z^* \in W^*, \ z_{n+1} - z_n \in W^*, n = 0, 1, \dots . \tag{11.82}$$

The discretization methods are described by a family of triplets.

$$\left\{\phi_h, \Delta_h, \hat{\Delta}_h\right\}, \ h > 0 \tag{11.83}$$

where

$$\phi_h : D_h \subset E_h \to \hat{E}_h, \ h > 0$$

are nonlinear operators and

$$\Delta_h : E \to E_h, \ \hat{\Delta}_h : \hat{E} \to \hat{E}_h, \ h > 0,$$

are bounded linear (discretization) operators such that

$$\Delta_h (W^* \cap B^*) \subset D_h, \ h > 0. \tag{11.84}$$

The discretization (11.83) is called *Lipschitz uniform* if there exist scalars $\rho > 0$, $L > 0$ such that

$$\overline{B}(\Delta_h, z^*, \rho) \subset D_h, \ h > 0, \tag{11.85}$$

and

$$\|\phi'_h(\eta) - \phi'_h(\xi)\| \leq L \|\eta - \xi\|, \ h > 0, \ \eta, \xi \in \overline{U}(\Delta_h z^*, \rho). \tag{11.86}$$

Moreover, the discretization family (11.83) is called: *bounded* if there is a constant $q > 0$ such that

$$\|\Delta_h u\| \leq q \|u\|, \ u \in W^*, \ h > 0, \tag{11.87}$$

stable if there is a constant $\sigma > 0$ such that

$$\left\|\phi'_h(\Delta_h u)^{-1}\right\| \leq \sigma, \ u \in W^* \cap B^*, \ h > 0, \tag{11.88}$$

consistent of order p if there are two constants $c_0 > 0$, $c_1 > 0$ such that

$$\left\|\hat{\Delta}_h F(z) - \phi_h(\Delta_h z)\right\| \leq c_0 h^p, \ z \in W^* \cap B^*, \ h > 0, \tag{11.89}$$

$$\left\|\hat{\Delta}_h (F'(u)v - \phi'_h(\Delta_h u) \Delta_h v)\right\| \leq c_1 h^p, \ u \in W^* \cap B^*, \ v \in W^*, \tag{11.90}$$

$h > 0$.

Let $F : D \subset E \to \hat{E}$ be a nonlinear operator such that F' is γ Lipschitz continuous on $U(z^*, r^*) \subseteq D$ with z^* such that $F(z^*) = 0$, $\left\| F'(z^*)^{-1} \right\| = \beta$ and $r^* = \frac{2}{3\beta\gamma}$, and consider a Lipschitz uniform discretization (11.83) which is bounded, stable, and consistent of order p. Then

Show:

(a) (11.80) has a locally unique solution

$$\zeta_h^* = \Delta_h z^* + O(h^p) \qquad (11.91)$$

for all $h > 0$ satisfying

$$0 < h \leq h_0 = \left[\frac{1}{2\sigma c_0} \min\left(\rho, (\sigma L)^{-1} \right) \right] \qquad (11.92)$$

(b) there exist constants $h_1 \in (0, h_0]$, $r_1 \in (0, r^*]$ such that the discrete process (11.81) converges to ζ_h^*, and that

$$\zeta_n^h = \Delta_h z_n + O(h^p), \quad n = 0, 1, ..., \qquad (11.93)$$

$$\phi_h(\zeta_n^h) = \hat{\Delta}_h F(z_n) + O(h^p), \quad n = 0, 1, ..., \qquad (11.94)$$

$$\zeta_n^h - \zeta_h^* = \Delta_h(z_n - z^*) + O(h^p), \quad n = 0, 1, ..., \qquad (11.95)$$

for all $h \in (0, h_1]$, and all starting points $z_0 \in B(z^*, r_1)$.

11.2. Suppose that the hypotheses of Exercise 11.1 hold and that there is a constant $\delta > 0$ for such

$$\lim_{h > 0} \| \Delta_h u \| \geq 2\delta \|u\| \quad \text{for each} \quad u \in W^*. \qquad (11.96)$$

Then show that for some $\bar{r} \in (0, r_1]$, and for any fixed $\varepsilon > 0$ and $z_0 \in U(z^*, \bar{r})$ there exists a constant $\bar{h} = \bar{h}(\varepsilon, z_0) \in (0, h_1]$ such that

$$\left| \min\{ n \geq 0, \|z_n - z^*\| < \varepsilon \} - \min\{ n \geq 0, \|\zeta_n^h - \zeta_h^*\| < \varepsilon \} \right| \leq 1 \qquad (11.97)$$

for all $h \in (0, \bar{h}]$.

11.3. Suppose that the hypothesis of Exercise 11.1 is satisfied and that $\lim_{h \to 0} \|\Delta_h u\| = \|u\|$ holds uniformly for $u \in W^*$. Then show there exists a constant $\bar{r}_1 \in (0, r_1]$ and, for any fixed $\varepsilon > 0$, some $\bar{h}_1 = \bar{h}(\varepsilon) \in (0, h_1]$ such that (11.97) holds for all $h \in (0, \bar{h}_1]$ and all starting points $z^0 \in U(z^*, \bar{r}_1)$.

11.4. Consider the operator

$$F: D \subset C^2[0,1] \to C[0,1] \times \mathbb{R}^2,$$
$$F(y) = \{y'' - f(x, y, y'); 0 \le x \le 1, y(0) - \alpha, y(1) - \beta\},$$

where D and f are assumed to be such that (11.78) has a unique solution $z^* \in D$, and

$$f \in C^3(U(z^*, \rho)),$$
$$U(z^*, \rho) = \{(x_1, x_2, x_3) \in \mathbb{R}^3; 0 \le x_1 \le 1,$$
$$|x_2 - z^*(x_1)| \le \rho, |x_3 - z^{*\prime}(x_1)| \le \rho\}.$$

Under these assumptions it follows that $z^* \in C^5[0,1]$. Indeed from $z^{*\prime\prime} = f(x, z^*, z^{*\prime})$ we deduce that $z^{*\prime\prime\prime}$ exists and

$$z^{*\prime\prime\prime} = f^{(1,0,0)}(x, z^*, z^{*\prime}) + f^{(0,1,0)}(x, z^*, z^{*\prime})$$
$$+ f^{(0,0,1)}(x, z^*, z^{*\prime}) z^{*\prime\prime}$$

which, in turn, gives the existence of $z^{*(iv)}$, etc. Here $f^{(1,0,0)}$ etc. denote the partial derivatives of f.

As usual we equip $C^k[0,1]$, $k \ge 0$, with the norm $\|u\| = \{(\max |u^i(x)|, 0 \le x \le 1), i = 0, ..., k\}$. The Fréchet derivative of F is

$$F'(y)u = \{u'' - f^{(0,1,0)}(x, y, y')u - f^{(0,0,1)}(x, y, y')u',$$
$$0 \le x \le 1, u(0), u(1)\}$$

and hence, for given $z_n \in D$, Newton's method specifies z_{n+1} as the solution of the linear equation

$$z''_{n+1} = f(x, z_n, z'_n) - f^{(0,1,0)}(x, z_n, z'_n)(z_n - z_{n+1})$$
$$- f^{(0,0,1)}(x, z_n, z'_n)(z'_n - z'_{n+1}) \qquad (11.98)$$

subject to the boundary conditions $z_{n+1}(0) = \alpha$, $z_{n+1}(1) = \beta$.

From (11.98) it follows easily that if $z_0 \in C^3[0,1]$ then $z_{n+1} \in C^4[0,1]$, $n = 0, 1, 2, ...$ .We shall assume also that $z_0 \in C^4[0,1]$. Moreover, (11.98) and the fact that z_n converges to z^* in the norm of $C^2[0,1]$ imply that there exists a constant $K > 0$ such that

$$z_n \in W_K = \left\{ z \in C^4[0,1]; \sup_x \left| z^{(i)}(x) \right| \le K, i = 0, 1, 2, 3, 4 \right\},$$

$n = 0, 1, \ldots$. By choosing, if necessary, a larger K it is not restrictive to assume that $z^* \in W_K$, $z_n - z^* \in W_K$ and $z_n - z_{n+1} \in W_K$, $n = 0, 1, \ldots$ which is (11.82).

The discretization method $\{\phi_h, \Delta_h, \overline{\Delta}_h\}$ is specified as follows

$$h = 1/n, \quad n = 1, 2, \ldots,$$
$$G_h = \{x_i = ih, \ i = 0, 1, \ldots, n\}, \quad \mathring{G}_h = G_h \setminus \{0, 1\},$$
$$E_h = \{\eta : G_h \to \mathbb{R}\}, \quad \eta_i = \eta(x_i), \quad i = 0, 1, \ldots, n,$$
$$\hat{E}_h = \left\{(\eta, a, b); \eta : \mathring{G}_h \to \mathbb{R}, \ a, b \in \mathbb{R}\right\},$$
$$\Delta_h y = y|_{G_h}, \quad \hat{\Delta}_h(y, a, b) = \left(y|_{\mathring{G}_h}, a, b\right),$$
$$\phi_h(\eta) = \left\{\left[\frac{\eta_{i+1} - 2\eta_i + \eta_{i-1}}{h^2} - f\left(x_i, \eta_i, \frac{\eta_{i+1} - \eta_{i-1}}{2h}\right)\right];\right.$$
$$\left. i = 1, 2, \ldots, n-1; (\eta_0 - \alpha), (\eta_n - \beta)\right\}.$$

We use the following norms

$$\|y\| = \max\left\{\left|y^{(i)}(x)\right|, 0 \le x \le 1, i = 0, 1, 2\right\}, \quad y \in C^2[0, 1],$$
$$\|v\| = \max\{|u(x)|, a, b; 0 \le x \le 1\}, \quad v = (u, a, b) \in C[0, 1] \times \mathbb{R}^2,$$
$$\|\eta\| = \left\{|\eta_0|, |\eta_n|, |\eta_i|, \left|\frac{\eta_{i+1} - \eta_{i-1}}{2h}\right|, \left|\frac{\eta_{i+1} - 2\eta_i + \eta_{i-1}}{h^2}\right|, i = 1, \ldots, n-1\right\},$$

$\eta \in E_h$. It is easily seen that for $y \in W_K$ we have

$$\left|\frac{y_{i+1} - y_{i-1}}{2h}\right| \le \tfrac{1}{6}Kh^2, \quad \left|\frac{y_{i+1} - 2y_i + y_{i-1}}{h^2} - y_i''\right| \le \tfrac{1}{12}Kg^2,$$

where $y_i = y(x_i)$, $y_i' = y'(x_i)$, $y_i'' = y''(x_i)$, $i = 1, 2, \ldots, n-1$. It is now difficult to prove that, with the above norms, (11.87) holds with $q = 1$ and (11.89), (11.90) are satisfied with $p = 2$. It is also easily seen that

$$\|\Delta_h u\| \le \|u\| \le \|\Delta_h u\| + K\left(\frac{1}{6}(h+1)\right)h$$

for $u \in W_K$ and hence that $\lim_{h \to 0} \|D_h u\| = \|u\|$.

Thus the conclusions of Exercises 11.1 and 11.2.

11.5.

(a) Let F be a Fréchet-differentiable operator defined on a convex subset D of a Banach space X with values in a Banach space Y. Assume that the equations $F(x) = 0$ has a simple zero $x^* \in D$ in the sense that $F'(x^*)$ has an inverse $F'(x^*)^{-1} \in L(Y, X)$. Moreover assume

$$\|F'(x^*)^{-1}[F'(x) - F'(x^*)]\| \le \ell_1 \|x - x^*\| \quad \text{for all } x \in D. \quad (11.99)$$

Then, show: sequence $\{x_n\}$ ($n \geq 0$) generated by Newton's method is well defined, remains in $U(x^*, r^*)$ for all $n \geq 0$ and converges to x^* with

$$\|x_{n+1} - x^*\| \leq \frac{3\ell_1}{2[1 - \ell_1\|x_n - x^*\|]}\|x_n - x^*\|^2, \quad r^* = \frac{2}{5\ell_1} \quad (n \geq 0) \tag{11.100}$$

provided that $x_0 \in U(x^*, r^*)$ and $U = U(x^*, r^*) \subseteq D$.

(b) Let F be as in (a). Assume:
(1) there exists $x_0 \in D$ such that $F'(x_0)^{-1} \in L(Y, X)$;
(2)

$$\|F'(x_0)^{-1}[F'(x) - F'(x_0)]\| \leq \ell_0 \|x - x_0\| \quad \text{for all } x \in D; \tag{11.101}$$

(3)

$$\|F'(x_0)^{-1}F(x_0)\| \leq \eta \quad \text{some } \eta \geq 0, \tag{11.102}$$

$$(5 + 2\sqrt{6})\ell_0\eta \leq 1, \tag{11.103}$$

(4) $U(x_0, r_2) \subseteq D$, where, r_1, r_2 are the real zeros ($r_1 \leq r_2$) of equations

$$f(t) = 3\ell_0 r^2 - (1 + \ell_0\eta)r + \eta = 0. \tag{11.104}$$

Then, show: sequence $\{x_n\}$ ($n \geq 0$) generated by Newton's method is well defined, remains in $U(x_0, r_1)$ and converges to a unique solution x^* of equation $F(x) = 0$ in $\overline{U}(x_0, r_1)$. Moreover the following error bounds hold for all $n \geq 0$. The solution x^* is unique in $U(x_0, r_2)$.

$$\|x_{n+2} - x_{n+1}\| \leq \frac{2\ell_0 r_1}{1 - \ell_0 r_1}\|x_{n+1} - x_n\| \tag{11.105}$$

and

$$\|x_{n+1} - x^*\| \leq \frac{c^n}{1-c}\|x_n - x^*\|, \tag{11.106}$$

where,

$$c = \frac{2\ell_0 r_1}{1 - \ell_0 r_1}. \tag{11.107}$$

(c) Let $F \colon D \subseteq X \to Y$ be a nonlinear operator satisfying the hypotheses of (a), and consider a Lipschitz uniform discretization which is bounded, stable and consistent of order p. Then equation $T_h(v) = 0$ has a locally unique solution

$$y_h^* = L_h(x^*) + O(h^p)$$

for all $h > 0$ satisfying

$$0 < h \leq h_0 = \left[\frac{1}{c_0\sigma} \min\left\{\frac{\rho}{2}, \frac{5-2\sqrt{6}}{l\sigma}\right\}\right]^{1/p}.$$

Moreover, there exist constants $h_1 \in (0, h_0]$ and $r_3 \in (0, r^*]$ such that the discrete process converges to y_h^* for all $h \in (0, h_1]$ and all starting points $x_0 \in U(x^*, r_1)$.

11.6.

(a) Let F be a Fréchet-differentiable operator defined on a convex subset D of a Banach space X with values in a Banach space Y. Assume that the equations $F(x) = 0$ has a simply zero $x^* \in D$ in the sense that $F'(x^*)$ has an inverse $F'(x^*)^{-1} \in L(Y,X)$. Then

(1) for all $\varepsilon_1 > 0$ there exists $\ell_1 > 0$ such that

$$\|F'(x^*)^{-1}[F'(x) - F'(y)]\| < \varepsilon_1$$

for all $x, y \in U(x^*, \ell_1) \subseteq D$.

(2) If $\varepsilon_1 \in \left[0, \frac{1}{2}\right)$ and $x_0 \in U(x^*, \ell_1)$ then sequence $\{x_n\}$ $(n \geq 0)$ generated by Newton's method is well defined, remains in $U(x^*, \ell_1)$ for all $n \geq 0$ and converges to x^* with

$$\|x_{n+1} - x^*\| \leq \frac{\varepsilon_1}{1 - \varepsilon_1}\|x_n - x^*\| \quad (n \geq 0).$$

(b) Let F be as in (a). Assume:

(1) there exist $\eta \geq 0$, $x_0 \in D$ such that $F'(x_0)^{-1} \in L(Y,X)$,

$$\|F'(x_0)^{-1}F(x_0)\| \leq \eta;$$

Then, for all $\varepsilon_0 > 0$ there exists $\ell_0 > 0$ such that

$$\|F'(x_0)^{-1}[F'(x) - F'(y)]\| < \varepsilon_0$$

for all $x, y \in U(x_0, \ell_0) \subseteq D$;

(2)

$$\frac{\eta(1 - \varepsilon_0)}{1 - 2\varepsilon_0} \leq \ell_0$$

and
$$\varepsilon_0 < \frac{1}{2}.$$

Then, show: sequence $\{x_n\}$ ($n \geq 0$) generated by Newton's method is well defined, remains in $U(x_0, \ell_0)$ and converges to a unique solution x^* of equation $F(x) = 0$ in $\overline{U}(x_0, \ell_0)$. Moreover the following error bounds hold for all $n \geq 0$

$$\|x_{n+2} - x_{n+1}\| \leq \frac{\varepsilon_0}{1 - \varepsilon_0} \|x_{n+1} - x_n\|$$

and

$$\|x_{n+1} - x^*\| \leq \frac{c^n}{1 - c} \|x_n - x^*\|,$$

where

$$c = \frac{\varepsilon_0}{1 - \varepsilon_0}.$$

(c) Let $F: D \subseteq X \to Y$ be a nonlinear operator satisfying a Fréchet uniform discretization $\{T_h, L_h, \hat{L}_h\}$, $h > 0$ which is bounded, stable and consistent of order p. Then show equation $T_h(v) = 0$ has a locally unique solution

$$y_h^* = L_h(x^*) + O(h^p)$$

for all $h > 0$ satisfying

$$0 < h \leq h_0 = \left[\frac{\rho(1 - 2\ell\sigma)}{(1 - \ell\sigma)\sigma c_0}\right]^{1/p}.$$

Moreover, there exist constants $h_1 \in (0, h_0]$ and $r_1 \in (0, r^*]$ such that the discrete process (4) converges to y_h^* for all $h \in (0, h_1]$ and all starting points $x_0 \in U(x^*, r_1)$, where $r^* = \min \rho\{\frac{1}{2q}, 1\}$.

11.7.

(a) Let $F: D \subseteq X \to Y$ be continuously Fréchet differentiable with $F'(x)$ invertible for all $x \in D$, D open and convex. Moreover, assume:

$$\left\|\int_0^1 F'(y)^{-1}[F'(x + t(y - x)) - F'(x)](y - x)dt\right\| \leq \varepsilon \|y - x\|$$

for all $x, y \in U(x_0, \delta)$; \hfill (11.108)

$$\varepsilon \in [0, 1);$$
$$\frac{\eta}{1-\varepsilon} \leq \delta,$$

where,
$$\|F'(x_0)^{-1}F(x_0)\| \leq \eta;$$

and
$$\overline{U}(x_0, \delta) \subseteq D.$$

Show: sequence $\{x_n\}$ $(n \geq 0)$ generated by Newton's method is well defined, remains in $U(x_0, \delta)$ for all $n \geq 0$ and converges to a solution $x^* \in \overline{U}(x_0, \delta)$ of equation $F(x) = 0$. Moreover the following error bounds hold:

$$\|x_{n+1} - x_n\| \leq \varepsilon \|x_n - x_{n-1}\| \quad (n \geq 1)$$

and

$$\|x_n - x^*\| \leq \frac{\varepsilon^n}{1-\varepsilon} \|x_0 - x^*\| \quad (n \geq 0).$$

Furthermore, if the linear operator

$$L = \int_0^1 F'(x + t(y-x))dt, \quad x, y \in \overline{U}(x_0, \delta)$$

is invertible, then x^* is the unique solution of equation $F(x) = 0$ in $\overline{U}(x_0, \delta)$.

(b) Show: condition (11.108) can be replaced by the stronger

$$\|F'(y)^{-1}[F'(x + t(y-x)) - F'(x)](y-x)\| \leq 2t\varepsilon \|y - x\| \quad (11.109)$$

or

$$\|F'(y)^{-1}[F'(x + t(y-x)) - F'(x)](y-x)\| \leq \varepsilon \|y - x\| \quad (11.110)$$

for all $x, y \in U(x_0, \delta) \subseteq D;\ t \in [0, 1]$.

11.8. Let F be a twice continuously Fréchet-differentiable operator defined on an open convex subset D of a Banach space X with values in a Banach space Y. Assume:

$$F'(x)^{-1} \in L(Y, X) \quad (x \in D);$$

there exist constants $a \geq 0$, $b \geq 0$, $\eta > 0$, $x_0 \in D$ such that

$$\|F'(z)^{-1}[F''(u) - F''(x)](u-x)^2\| \leq a\|u-x\|^2, \quad u, x, z \in D \text{ collinear}$$

$$\|F'(y)^{-1}F''(x)\| \leq b, \quad x, y \in D,$$
$$\|F'(x_0)^{-1}F(x_0)\| \leq \eta;$$

$$h = \frac{c}{2}\eta < 1, \quad c = a + b,$$
$$\overline{U}(x_0, r) \subset D,$$

where,

$$r = \eta \sum_{i=0}^{\infty} h^{2^i - 1} \leq \frac{\eta}{1-h}.$$

Show, sequence $\{x_n\}$ ($n \geq 0$) generated by Newton's method is well defined, remains in $U(x_0, r)$ and converges to a solution $x^* \in \overline{U}(x_0, r)$ of equation $F(x) = 0$, which is unique in S, where S is given by $S = \bigcup_{i=0}^{\infty} U\left(x_k, \frac{2}{c}\right) \cap D$. Moreover the following error bounds hold for all $n \geq 0$:

$$\|x_{n+2} - x_{n+1}\| \leq \frac{c}{2}\|x_{n+1} - x_n\|^2$$

and

$$\|x_{n+1} - x^*\| \leq \frac{c}{2}\|x_n - x^*\|^2.$$

Chapter 12

Convergence on Generalized Spaces

The local and semilocal convergence of iterative methods in generalized spaces with a convergence structure under weak conditions is examined in this Chapter.

12.1 Iterative Methods on Banach Spaces with a Convergence Structure

In this section, we are concerned with approximating a solution x^* of the nonlinear operator equation

$$F(x) + Q(x) = 0, \tag{12.1}$$

where F is a Fréchet-differentiable operator defined on a convex subset D of a Banach space X with values in X, and Q is a non-differentiable nonlinear operator with the same domain and range as F.

We generate a sequence $\{x_n\}$ $(n \geq 0)$ using the perturbed Newton-like method scheme given by

$$x_{n+1} = x_n + \delta_n \quad (n \geq 0) \tag{12.2}$$

where the correction δ_n satisfies

$$A(x_n)\delta_n = -(F(x_n) + Q(x_n)) + r_n \quad (n \geq 0) \tag{12.3}$$

for a suitable $r_n \in X$ $(n \geq 0)$ called residual. The importance of studying perturbed Newton-like methods comes from the fact that variants of Newton's method can be considered as procedures of this type. In fact, the approximations (12.2) and (12.3) characterize any iterative process in which the corrections are taken as approximate solutions of the Newton equations.

This happens when approximation (12.2) is solved by any iterative method or when there in the derivative is replaced by a suitable approximation. In [99] we provided sufficient conditions for the convergence of iteration (12.2) to a solution x^* of equation (12.1), by assuming that X is a Banach space with a convergence structure (to be precised later).

Here we derive sufficient conditions for controlling the residuals r_n in such a way that the convergence of the sequence $\{x_n\}$ $n \geq 0$ to a solution of equation (7.164) is ensured.

We also refer the reader to [99], [199], [200] and the references there for relevant work, which however is valid on a Banach space X without a convergence structure. The advantages of working on a Banach space with a convergence structure have been explained in some detail [99], [199], [200].

Preliminaries

We will need the definitions:

Definition 12.1 The triple (X, V, E) is a Banach space with a convergence structure if

(C_1) $X, \|\cdot\|$ is a real Banach space;

(C_2) $(V, C, \|\cdot\|_v)$ is a real Banach space which is partially ordered by the closed convex cone C; the norm $\|\cdot\|_v$ is assumed to be monotone on C;

(C_3) E is a closed convex cone in $X \times V$ satisfying $\{0\} \times C \subseteq E \subseteq X \times C$;

(C_4) the operator $|\cdot| : D \to C$ is well defined:

$$|x| = \inf \{q \in C |\, (x, q) \in E\}$$

for

$$x \in S = \{x \in X |\, \exists q \in C : (x, q) \in E\};$$

and

(C_5) for all $x \in S$, $\|x\| \leq \||x|\|_v$.

The set

$$U(a) = \{x \in X \,|(x, a) \in E\}$$

defines a sort of generalized neighborhood of zero.

Definition 12.2 An operator $L \in C^1(V_1 \to V)$ defined on an open subset V_1 of an ordered Banach space V is order convex on $[a, b] \subseteq V_1$ if

$$c, d \in [a, b], \ c \leq d \Longrightarrow L'(d) - L'(c) \in L_+(V),$$

where for $n \geq 0$

$$L_+(V^n) = \{L \in L(V^n) \mid 0 \leq x_i \Longrightarrow 0 \leq L(x_1, x_2, ..., x_n)\}$$

and $L(V^n)$ denotes the space of multilinear, symmetric, bounded operators on V.

Definition 12.3 The set of bounds for an operator $A \in L(X^n)$ is defined to be

$$B(A) = \{L \in L_+(V^n) \mid (x_i, q_i) \in E \Longrightarrow [A(x_1, ..., x_n), L(q_1, ..., q_n)] \in E\}.$$

Definition 12.4 A partially ordered topological space V is called regular if every order bounded increasing sequence has a limit.

We will need the following proposition due to Kantorovich [183].

Proposition 12.1 *Let V be a regular partially ordered topological space (POTL-space) and let x, y be two points of V such that $x \leq y$. If $H : [x, y] \to V$ is a continuous isotone operator having the property that $x \leq Hx$ and $y \geq Hy$, then there exists a point $z \in [x, y]$ such that $z = Hz$.*

Convergence Analysis

We will need the following basic result:

Lemma 12.1 *Let V be a regular partially ordered topological space, an operator $L \in C^1(V_1 \to V)$ with $[0, a] \subseteq V_1 \subseteq V$ for some $a \in V$, an operator $M \in C(V_1 \to V)$, operator $R < I$, $B > I$, T, $K \in L_+(V)$, a point $c \in V$ with $c > 0$ and a point $p \in [0, a]$.*
Assume:

(a) The equation

$$\begin{aligned} g(q) = BT[&L(p+q) - L(p) - L'(p)q + M(p+q) \\ &- M(p) + K(p+q) - K(p)] \\ &- (I-r)q + c = 0 \end{aligned} \quad (12.4)$$

has solutions in the interval $[0, a]$ and denote by q^* the least of them.

(b) Let $G \in L_+(V)$ be given and $R_+ \in L_+(V)$, $c_+, p_+ \in V$ be such that the following conditions are satisfied:

$$R_+ < \min\{(G-2I)BTL'(p) + BTGL'(p_+) + G(R-I) + I, I\} = \alpha \tag{12.5}$$

$$0 < c_+ \leq BTG(L(p_+) + M(p_+) + K(p_+)) + GBT(L(p) + M(p)) \\ + K(p)) - 2BT(L(p) + M(p) + K(p)) \\ + (R_+ + G - I - BTGL'(p_+))c = \beta \tag{12.6}$$

and

$$0 \leq p_+ \leq p + c, \tag{12.7}$$

where α and β are functions of the operators and points involved.

(c) The following estimate is true

$$M(p) \leq M(p+q) \tag{12.8}$$

for all $p, q \in [0, a]$.

Then the equation

$$g_+(q) = BGT\Big[L(p_+ + q) - L(p_+) - L'(p_+)q + M(p_+ + q) - M(p_+) \\ + K(p_+ + q) - K(p_+)\Big] - (I - R_+)q + c_+ = 0 \tag{12.9}$$

has nonnegative solutions and the least of them, denoted by q_+^* lies in the interval $[c_+, q^* - c]$.

Proof. Using the hypotheses $g(q^*) = 0$ and $R < I$ we deduce from (12.4) that $c \leq q^*$. We will show that

$$g_+(q^* - c) \leq 0. \tag{12.10}$$

From equation (12.9), and using (12.6), we obtain in turn

$$g_+(q^* - c) \leq \Big[BGTL(p + c + q^* - c) - L(p_+) - L'(p_+)(q^* - c) \\ + M(p + c + q^* - c) - M(p_+) \\ + K(p + c + q^* - c) - K(p_+)\Big] - (I - R_+)(q^* - c) + c_+ \\ = g(q^*) - BT\Big[L(p + q^*) - L(p) - L'(p)q^* + M(p + q^*) \\ - M(p) + K(p + q^*) - K(p)\Big]$$

$$+ BTG\Big[L(p+q^*) - L(p_+) - L'(p_+)(q^* - c) + M(p+q^*) - M(p_+)$$
$$+ K(p+q^*) - K(p_+)\Big] + (I-R)q^* - c - (I-R_+)(q^* - c) + c_+$$
$$= [BTL'(p) - BTGL'(p_+) + (I-R) - (I-R_+)]q^*$$
$$+ (G-I)\Big\{(I-R)q^* - c - BT[L(p) + M(p) + K(p) + L'(p)q^*]\Big\}$$
$$+ BT\Big[L(p) + M(p) + K(p) - GL(p_+) - GM(p_+)\Big]$$
$$+ BTGL'(p_+)c + (I-R_+)c + c_+ - c$$
$$= \Big[(2I-G)BTL'(p) - BTGL'(p_+) + R_+ + G(I-R) - I\Big]q^*$$
$$+ 2BT(l(p) + M(p) + K(p)) - BTG(L(p_+) + M(p_+) + K(p_+))$$
$$- GBT(L(p) + M(p) + K(p)) + (BTGL'(p_+) - G + I - R_+)c + c_+$$
$$\leq 0,$$

since (12.5), and (12.6) are satisfied.

Moreover from (12.9) for $q = c_+$, we obtain

$$g_+(c_+) \geq 0. \tag{12.11}$$

By inequalities (12.10), (12.11), the fact that g is continuous and isotone on $[c_+, q^* - c]$, and since V is a regular partially ordered topological space, from the proposition, we deduce that there exists a point q_+^* with

$$g_+(q_+^*) = 0 \tag{12.12}$$

and

$$c_+ \leq q_+^* \leq q^* - c. \tag{12.13}$$

We can assume that q_+^* denotes the least of the solutions of equation (12.9). That completes the proof of the lemma. □

The following result is a consequence of the above lemma.

Theorem 12.1 *Let $\{c_n\} \in V$, $\{T_n\}$, $\{R_n\}$, $\{G_n\} \in L_+(V)$ $(n \geq 0)$ be sequences and V as in the above lemma. Assume:*

(a) *There exists a sequence $\{p_n\} \in [0, a] \subseteq V_1 \subseteq V$ for some $a \in V$ with $p_0 = 0$, and*

$$p_{n+1} \leq \sum_{j=0,1,\ldots,n} c_j \text{ for } n \geq 0. \tag{12.14}$$

(b) $R_0 < I$ and the function

$$g_0(q) = BT_0\left[L(p_0+q) - L(p_0) - L'(p_0)q + M(p_0+q)\right. \quad (12.15)$$
$$\left. - M(p_0) + K(p_0+q) - K(p_0)\right] - (I-R)q + c_0 = 0$$

has root on $[0, a]$, where B, L, M, K are as in the above lemma. Denote by q_0^* the least of them.

(c) The following conditions are satisfied for all $n \geq 0$

$$R_{n+1} < \alpha_{n+1}, \quad (12.16)$$
$$0 < c_{n+1} \leq \beta_{n+1}, \quad (12.17)$$

and

$$0 \leq p_{n+1} \leq p_n + c_n. \quad (12.18)$$

(d) The linear operators T_n are boundedly invertible for all $n \geq 0$, and set $G_n = T_{n+1}T_n^{-1}$ $(n \geq 0)$.

(e) Condition (12.8) is satisfied.

Then, the equation

$$g_n(q) = BG_nT_n\left[L(p_n+q) - L(p_n) - L'(p_n)q + M(p_n+q)\right. \quad (12.19)$$
$$\left. - M(p_n) + K(p_n+q) - K(p_n)\right] - (I-R_n)q + c_n = 0$$

has solution in $[0, a]$ for every $n \geq 0$ and denoting by q_n^* the least of them, we have

$$\sum_{j=n,\dots,\infty} c_j \leq q_n^* \quad (n \geq 0). \quad (12.20)$$

Proof. Let us assume that for some nonnegative integer n, $I - R_n > 0$, $g_n(q)$ has roots on $[0, a]$ and denote by q_n^* the least of them. We use introduction on n. We also observe that this is true by hypothesis (b) for $n = 0$. Using (12.14), (12.16), (12.17), (12.18), the lemma, and setting $c = c_n$, $c_+ = c_{n+1}$, $R = R_n$, $R_+ = R_{n+1}$, $G = G_n$ we deduce that q_{n+1}^* exists, and

$$c_{n+1} \leq q_{n+1}^* \leq q_n^* - c_n. \quad (12.21)$$

The induction is now complete and (12.20) follows immediately from (12.21).

That completes the proof of the theorem. From now on we assume that X is a Banach space with a convergence structure in the sense of [200]. □

The following result is an immediate consequence of Theorem 12.1.

Theorem 12.2 *Assume:*

(a) *the hypotheses of Theorem 12.1 are satisfied;*
(b) *there exists a sequence $\{x_n\}$ $(n \geq 0)$ in a Banach space X with a convergence structure such that $|x_{n+1} - x_n| \leq c_n$.*

Then,

(i) *the sequence $\{x_n\}$ $(n \geq 0)$ converges to some point x^*;*
(ii) *moreover the following error bounds are true*

$$|x^* - x_n| \leq q_n^*, \qquad (12.22)$$

and

$$|x^* - x_{n+1}| \leq q_n^* - c_n, \quad \text{for all } n \geq 0. \qquad (12.23)$$

We can introduce the main result:

Theorem 12.3 *Let X be a Banach space with convergence structure (X, V, E) with $V = (V, C, \|\cdot\|_v)$, an operator $F \in C^1(D \to X)$ with $D \subseteq X$, an operator $Q \in C(D \to X)$, an operator $A(x) \in C(X \to D)$, an operator $L \in C^1(V_1 \to V)$ with $V_1 \subseteq V$, an operator $M \in C(V_1 \to V)$, an operator $K \in L_+(V)$, and a point $a \in C$ such that the following conditions are satisfied:*

(a) *the inclusions $U(a) \subseteq D$, and $[0, a] \subseteq V_1$ are true;*
(b) *L is order-convex on $[0, a]$, and satisfies*

$$K + L'|x| + |y| - L'(|x|) \in B(A(x) - F'(x + y)) \qquad (12.24)$$

for all $x, y \in U(a)$ with $|x| + |y| \leq a$;
(c) *M satisfies the condition*

$$M(|x| + |y|) - M(|x|) \in B(Q(x) - Q(x + y)), \quad M(0) = 0 \qquad (12.25)$$

for all $x, y \in U(a)$ with $|x| + |y| \leq a$;
(d) *for the sequences $\{c_n\}, \{T_n\}, \{R_n\}, \{G_n\}, \{p_n\}$ $(n \geq 0)$ the hypotheses (12.14), (b), (12.16), (12.18) and (d) of Theorem 12.1 are satisfied;*

(e) the following conditions are also satisfied

$$|\delta_n| \leq c_n \leq T_n |-(F(x_n)+Q(x_n))| \leq \gamma_n \leq \beta_n \quad (n \geq 1), \quad (12.26)$$
$$|-r_n| \leq T_n^{-1} R_n c_n, \quad (12.27)$$

where,

$$\gamma_n = T_n \Big[L(p_n + c_n) - L(p_n) - L'(p_n) c_n + M(p_n + c_n) - M(p_n)$$
$$+ K(p_n + c_n) - K(p_n) \Big] + R_n c_n \quad (n \geq 1). \quad (12.28)$$

Then,

(i) the sequence $\{x_n\}$ $(n \geq 0)$ generated by

$$x_{n+1} = x_n + \delta_n, \quad \text{with } x_0 = 0$$

remains in $\bar{U}(x_0, t_0^*)$ and converges to a solution x^* of equation $F(x) = 0$;

(ii) moreover, the error estimates (12.22) and (12.23) are true where q_n^* is the least root in $[0, \alpha]$ of the function $g_n(q)$ defined in (12.19), with $p_n = \|x_n - x_0\|$ $(n \geq 0)$.

Proof. Let us assume that $x_n, x_{n+1} \in U(x_0, q_0^*)$, where the existence of q_0^* is guaranteed from hypotheses (d). We note that $|\delta_0| \leq c_0$. Using the approximation

$$-(F(x_{n+1}) + Q(x_{n+1})) =$$
$$= (F(x_n) - F(x_{n+1}) + A(x_n)(x_{n+1} - x_n)) + (Q(x_n) - Q(x_{n+1})) - r_n, \quad (12.29)$$

(12.24), (12.25), (12.27) and setting $p_n = |x_n - x_0|$, we obtain in turn

$$|-(F(x_{n+1}) + Q(x_{n+1}))|$$
$$\leq |F(x_n) - F(x_{n+1}) + A(x_n)(x_{n+1} - x_n)| + |Q(x_n) - Q(x_{n+1})| + |-r_n|$$
$$\leq \int_0^1 [L'(p_n + t|x_{n+1} - x_n|) - L'(p_n) + K] |x_{n+1} - x_n| \, dt$$
$$+ M(p_n + |x_{n+1} - x_n|) - M(p_n) + |-r_n|$$
$$\leq L(p_n + c_n) - L(p_n) - L'(p_n) c_n + Kc_n + M(p_n + c_n) - M(p_n)$$
$$+ T_n^{-1} R_n c_n.$$

hence, by (12.26) we get

$$c_{n+1} \leq T_{n+1} \left|-(F(x_{n+1}) + Q(x_{n+1}))\right| \leq \gamma_n \leq \beta_n \quad (n \geq 1),$$

which shows (12.17).

It can easily be seen that by using induction on n, the hypotheses of Theorem 12.1 are satisfied. Hence, by (12.24) and (12.27) the iteration $\{x_n\}$ $(n \geq 0)$ remains in $U(x_0, q_0^*)$ and converges to x^* so that (12.22), and (12.23) satisfied. Moreover, from the estimate

$$\left|-(F(x_n) + Q(x_n))\right| \leq |A(x_n) - F'(x_n)| c_n + |F'(x_n)| c_n + |-r_n|,$$

(12.24), (12.27), the continuity of F, F', A, T_n, R_n, and $c_n \to 0$ as $n \to \infty$, we deduce that

$$F(x^*) + Q(x^*) = 0.$$

That completes the proof of the theorem. □

We will complete this study with an example that shows how to choose L, M, in practical applications. For simplicity, we assume that $A(x) = F'(x)$ for all $x \in D$.

Example 12.1 We discuss the case of a real Banach space with norm $\|\cdot\|$. Assume that $F'(0) = I$ and there exists a monotone operator

$$f : [0, a] \to \mathbb{R}$$

such that

$$\|F''(x)\| \leq f(\|x\|), \quad \text{for all } x \in U(a) \tag{12.30}$$

and a continuous nondecreasing function ℓ on $[0, r]$, $r \leq a$ such that

$$\|Q(x) - Q(y)\| \leq \ell(r) \|x - y\| \tag{12.31}$$

for all $x, y \in U\left(\frac{r}{2}\right)$.

We showed in [99], (see also [220]) that (12.18) implies that

$$\|Q(x+h) - Q(x)\| \leq h(r + \|h\|) - h(r), \quad x \in U(a), \quad \|h\| \leq a - r, \tag{12.32}$$

where

$$h(r) = \int_0^r \ell(t)\, dt. \tag{12.33}$$

Conversely, it is not hard to see that we may assume, without loss of generality, that the function h and all functions $h(r+t) - h(r)$ are monotone in r. Hence, we may assume that $h(r)$ is convex and hence differentiable from the right. Then, as in [99], we show that (12.32) implies (12.31) and $\ell(r) = h'(r+0)$. Hence, we can set

$$L_1(q) = L(q) - \|F(0) + Q(0)\| + \int_0^q ds \int_0^s f(t)\, dt, \qquad (12.34)$$

and

$$M(q) = \int_0^q \ell(t)\, dt. \qquad (12.35)$$

Define the functions f_1, f_2, f_3 on $[0, a]$ by

$$f_1(q) = L(q) - q, \quad f_2(q) = f_1(q) + h(q) \text{ and } f_3(q) = \int_0^q f(t)\, dt.$$

Choose $B = T_0 = 1$, $R_0 = K = 0$ and $p_0 = 0$. Then by (12.15) we get

$$g_0(q) = f_2(q).$$

It can easily be seen that with the above choices of L and M conditions (12.24) and (12.25) are satisfied.

Suppose that the function g_0 has a unique zero q_0^* in $[0, a]$ and $g_0(a) \leq 0$. It is then known [99], [200] that equation (12.1) admits a solution x^* in $U(q_0^*)$, this solution is unique in $U(a)$, and the iteration $\{x_n\}$ $(n \geq 0)$ given by (12.2) is well defined, remains in $U(q_0^*)$ for all $n \geq 0$ and converges to x^*. By applying the Banach lemma on invertible operators we can show that $A(x_n)$ is invertible and $\|A(x_n)^{-1}\| \leq -f_1'(\|x_n\|)^{-1} = T_n$ $(n \geq 0)$.

Moreover, suppose that (12.30) is satisfied. furthermore, assume that instead of conditions (12.16) and (12.17), the weaker condition (12.10) is satisfied. Using (12.30) and the approximation

$$r_n = [(F'(x_n) - F'(x_0)) + F'(x_0)]\delta_n + F(x_n) + Q(x_n),$$

12.2 Exercises

12.1. Let L, M, M_1 be operators such that $L \in C^1(V_1 \to V)$, $M_1 \in L_+(V)$, $M \in C(V_1 \to V)$, and x_n be points in D. It is convenient for us to define the sequences c_n, d_n, a_n, b_n $(n \geq 0)$ by

$$d_{n+1} = (L + M + M_1)(d_n) + L'(|x_n|) c_n, \quad d_0 = 0,$$
$$c_n = |x_{n+1} - x_n|,$$
$$a_n = (L + M + M_1)^n (a) \quad \text{for some} \quad a \in C,$$
$$b_n = (L + M + M_1)^n (0),$$

and the point b by

$$b = (L + M + M_1)^\infty (0).$$

Prove the result:
Let X be a Banach space with convergence structure (X, V, E) with $V = (V, C, \|\cdot\|_v)$, an operator $F \in C^1(D \to X)$ with $D \subseteq X$, an operator $Q \in C(D \to X)$, an operator $L \in C^1(V_1 \to V)$ with $V_1 \subseteq V$, an operator $M \in C(V_1 \to V)$, an operator $M_1 \in L_+(V)$, a point $a \in C$, and a null sequence $\{z_n\} \in D$ such that the following conditions are satisfied:

(C_6) the inclusions $U(a) \subseteq D$ and $[0, a] \subseteq V1$ are true;
(C_7) L is order-convex on $[0, a]$, and satisfies

$$L'(|x| + |y|) - L'(|x|) \in B(F'(x) - F'(x+y))$$

for all $x, y \in U(a)$ with $|x| + |y| \leq a$;
(C_8) M satisfies the conditions

$$0 \leq (Q(x) - Q(x+y)), \quad M(|x| + |y|) - M(|x|) \in E$$

for all $x, y \in U(a)$ with $|x| + |y| \leq a$, and

$$M(w_1) - M(w_2) \leq M(w_3) - M(w_4) \quad \text{and} \quad M(w) \geq 0$$

for all $w, w_1, w_2, w_3, w_4 \in [0, a]$ with $w_1 \leq w_3$, $w_2 \leq w_4$, $w_2 \leq w_1$, $w_4 \leq w_3$;
(C_9) M_1, x_n, z_n satisfy the inequality

$$0 \leq (F'(x_n)(z_n) - F'(x_{n-1})(z_{n-1}), M_1(d_n - d_{n-1})) \in E$$

for all $n \geq 1$;
(C_{10}) $L'(0) \in B(I - F'(0))$, and
$(-(F(0) + Q(0) + f'(0)(z_0)), L(0) + M(0) + M_1(0)) \in E$;
(C_{11}) $(L + M + M_1)(a) \leq a$ with $0 \leq L + M + M_1$; and
(C_{12}) $(M + M_1 + L'(a))^n a \to 0$ as $n \to \infty$.

Then,

(i) The sequence $(x_n, d_n) \in (X \times V)^N$ is well defined, remains in E^N, is monotone, and satisfies for all $n \geq 0$

$$d_n \leq b$$

where b is the smallest fixed point of $L + M + M_1$ in $[0, a]$.

(ii) Moreover, the iteration $\{x_n\}$ $(n \geq 0)$ generated by

$$x_{n+1} = x_n + F'(x_n)^* [-(F(x_n) + Q(x_n))] - z_n, \quad z_0 = 0$$

converges to a solution $x^* \in U(b)$ of the equation $F(x) + Q(x) = 0$, which is unique in $U(a)$.

(iii) Furthermore, the following estimates are true for all $n \geq 0$:

$$b_n \leq d_n \leq b,$$
$$b_n \leq a_n,$$
$$|x_{n+1} - x_n| \leq d_{n+1} - d_n,$$
$$|x_n - x^*| \leq b - d_n,$$

and

$$|x_n - x^*| \leq a_n - b_n, \quad \text{for } M_1 = 0, \text{ and } z_n = 0 \ (n \geq 0).$$

12.2. We will now introduce results on a posteriori estimates for the iteration introduced in Exercise 12.1. It is convenient to define the operator

$$R_n(q) = (I - L'(|x_n|))^* S_n(q) + c_n$$

where,

$$S_n(q) = (L + M + M_1)(|x_n| + q) - (L + M + M_1)(|x_n|) - L'(|x_n|)(q),$$

and the interval

$$I_n = [0, a - |x_n|].$$

Show:

(a) operators S_n are monotone on I_n;

(b) operators $R_n : [0, a - d_n] \to [0, a - d_n]$ are well defined, and monotone. *Hint*: Verify the scheme:

$$d_n + c_n \leq d_{n+1} \implies R(a) - d_{n+1}$$
$$\leq a - d_n - c_n \implies S_n(a - d_n) + L'(|x_n|)(a - d_n - c_n)$$
$$\leq a - d_n - c_n \quad (n \geq 0);$$

(c) if $q \in I_n$ satisfy $R_n(q) \leq q$, then

$$c_n \leq R_n(q) = p \leq q,$$

and

$$R_{n+1}(p - c_n) \leq p - c_n \quad \text{for all } n \geq 0;$$

(d) under the hypotheses of Exercise 12.1, let $q_n \in I_n$ be a solution of $R_n(q) \leq q$, then

$$|x^* - x_m| \leq a_m \quad (m \geq n),$$

where

$$a_n = q_n \text{ and } a_{m+1} = R_m(a_m) - c_m;$$

and

(e) under the hypotheses of Exercise 12.1, any solution $q \in I_n$ of $R_n(q) \leq q$ is such that

$$|x^* - x_n| s < R_n^\infty(0) \leq q.$$

12.3. Let $A \in L(X \to X)$ be a given operator. Define the operators $P, T(D \to X)$ by

$$P(x) = AT(x + u),$$
$$T(x) = G(x) + R(x), \quad P(x) = F(x) + Q(x),$$

and

$$F(x) = AG(x + u), \quad Q(x) = AR(x + u),$$

where $A \in L(X \to X) G, R$ are as F, Q respectively. We deduce immediately that under the hypotheses of Exercise 12.1 the zero x^* of P is also a zero of AT, if $u = 0$.

We will now provide a monotonicity result to find a zero x^* of AT. the space X is assumed to be partially ordered and satisfies the conditions for V given in (C_1)–(C_5). Moreover, we set $X = V$, $D = C^2$ so that $|\cdot|$ turns out to be I.

Prove the result:

Let V be a partially ordered Banach space satisfying conditions (C_1)–(C_5), Y be a Banach space, $G \in C^1(D \to Y)$, $R \in C(D \to Y)$ with $D \subseteq V$, $A \in L(X \to V)$, $M \in C(D \to V)$, $M_1 \in L_+(V)$ and $u, v \in V$ such that:

(C_{13}) $[u, v] \subseteq D$;
(C_{14}) $I - AG'(u) + M + M_1 \in L_+(V)$;
(C_{15}) for all $w_1, w_2 \in [u, v] : w_1 \leq w_2 \Longrightarrow AG'(w_1) \geq AG'(w_2)$;
(C_{16}) $AT(u) + AG'(u)(z_0) \leq 0$, $AT(v) + AG'(v)(z_0) \geq 0$ and $AT(v) - M_1(v - u) \geq 0$;
(C_{17}) condition (C_8) is satisfied, and $M(v - u) \leq -Q(v - u)$;
(C_{18}) condition (C_9) is satisfied;
(C_{19}) the following initial condition is satisfied

$$-(Q(0) + AG'(u)(z_0)) \leq (M + M_1)(0);$$

and

(C_{20}) $(I - AG'(v) + m + M_1)^n (v - u) \to 0$ as $n \to \infty$.

Then the Newton sequence

$$y_0 = u, \quad y_{n+1} = y_n + (AG'(y_n))^* [-AT(y_n)] - z_n \quad (n \geq 0)$$

is well defined for all $n \geq 0$, monotone and converges to a unique zero x^* of AT in $[u, v]$.

12.4. Let X be a Banach space with convergence structure (X, V, E) with $V = (V, C, \|\cdot\|_v)$, an operator $F \in C^1(X_F \to X)$ with $X_F \subseteq X$, an operator $L \in C^1(V_L \to V)$ with $V_L \subseteq V$ and a point of C such that

(a) $U(a) \subseteq X_F$, $[0, a] \subseteq V_L$;
(b) L is order convex on $[0, a]$ and satisfies for $x, y \in U(a)$ with $|x| + |y| \leq a$;

$$L'(|x| + |y|) - L'(|x|) \in B(F'(x) - F'(x + y));$$

(c) $L'(0) \in B(I - F'(0))$, $(-F(0), L(0)) \in E$;
(d) $L(a) \leq a$;
(e) $L'(a)^n \to 0$ as $n \to \infty$.

Then show Newton's sequence $x_0 := 0$, $x_{n+1} = x_n + F'(x_n)^*(-F(x_n))$ is well defined, and converges to the unique zero z of F in $U(a)$.

12.5. Under the hypotheses of Exercise 12.4 consider the case of a Banach space with a real norm $\|\cdot\|$. Let $F'(0) = I$ and define a monotone operator

$$k : [0, a] \to \mathbb{R} \ \forall x \in U(a) : \|F''(x)\| \le k(\|x\|)$$

and

$$L(t) = \|F(0)\| + \int_0^t ds \int_0^s d\theta k(\theta).$$

Show (d) above is equivalent to $\|F(0)\| + .5k(a)a^2 \le a$. Under what conditions is this inequality true.

If conditions (a)–(c) or Exercise 7.4 are satisfied, and

$$\exists t \in (0, 1) : L(a) \le ta,$$

then show there exists $a' \in [0, ta]$ satisfying (a)–(e). The zero $z \in U(a')$ is unique in $U(a)$.

12.6. Let $L \in L_+(V)$ and $a, e \in C$ be given such that: Let $a \le e$ and $L^n e \to 0$ as $n \to \infty$. then show: operator

$$(I - L)^* : [0, a] \to [0, a],$$

is well defined and continuous.

12.7. Let $A \in L(X)$, $L \in B(A)$, $y \in D$, and $e \in C$ such that

$$Le + |y| \le e \text{ and } L^n e \to 0 \text{ as } n \to \infty.$$

Then show $x := (I - A)^* y$ is well defined, $x \in D$, and $|x| \le (I - L)^* |y| \le e$.

12.8. Let V be a partially ordered Banach space, Y a Banach space, $G \in C^1(V_G \to Y)$, $A \in L(X \to Y)$ and $u, v \in V$ such that

(a) $[u, v] \subseteq V_G$;
(b) $I - AG'(u) \in L_+(V)$;
(c) $\forall w_1, w_2 \in [u, v] : w_1 \le w_2 \Longrightarrow AG'(w_1) \ge AG'(w_2)$;
(d) $AG(u) \le 0$ and $AG'(v) \ge 0$;
(e) $(I - AG'(v))^n (v - u) \to 0$ as $n \to \infty$.

Then show: the Newton sequence

$$u_0 = u, \quad u_{n+1} := u_+ [AG'(u_n)]^* [-AG(u_n)] \quad (n \geq 0)$$

is well defined, monotone, and converges to the unique zero z of AG in $[u, v]$.

12.9. Consider the two boundary value problem

$$-x''(s) = 4\sin(x(s)) + f(s), \quad x(0) = x(1) = 0$$

as a possible application of Exercises 12.1–12.8 in the space $X = C[0,1]$ and $V = X$ with natural partial ordering; the operator $/\cdot/$ is defined by taking absolute values. Let $G \in L_+(C[0,1])$ be given by

$$Gx(s) = \frac{1}{2\sin(2)} \left\{ \int_0^s \sin(2t) \sin(2-2s) x(t) \, dt \right.$$

$$\left. + \int_0^s \sin(2-2t) \sin(2s) x(t) \, dt \right\}$$

satisfying $Gx = y \iff -y'' - 4y = x, \; y(0) = y(1) = 0$. Define the operator

$$F : C[0,1] \to C[0,1], \quad F(x) := x - G(4\sin(x) - 4x + f).$$

Let

$$L : C[0,1] \to C[0,1], \quad L(e) = 4G(e - \sin(e)) + |Gf|.$$

For $x, y, w \in C[0,1]$

$$|x|, |x| + |y| \leq .5\pi \Rightarrow$$
$$|[F'(x) - F'(x+y)]w| \leq [L'(|x|+|y|) - L'L'(|x|)] |w|.$$

Further we have $L(0) = |Gf|$ and $L'(0) = 0$. We have to determine $a \in C_+[0,1]$ with $|a| \leq .5\pi$ and $s \in (0,1)$ such that $L(a) = 4G(a - \sin(a)) + |Gf| \leq sa$. We seek a constant function as a solution. For $e_0(s) = 1$, we compute

$$p : \|Ge_0\|_\infty = .25 \left(\frac{1}{\cos(e)} - 1 \right).$$

Show that $a = te_0$ will be a suitable solution if

$$4p(t - \sin(t)) + \|Gf\|_\infty < t.$$

12.10. (a) Assume: given a Banach space X with a convergence structure (X, V, E) with $V = (V, C, \|\cdot\|_V)$, an operator $F \in C^1(X_0 \subseteq X \to X)$, operators M, L_0, $L \in L^1(V_0 \subseteq V \to V)$, and a point $p \in C$ such that the following conditions hold:

$$U(p) \subseteq X_0, \quad [0, p] \subseteq V_0;$$

M, L_0, L are order convex on $[0, p]$, and such that for $x, y, z \in U(p)$ with $|x| \leq p$, $|y| + |x| \leq p$

$$L_0'(|x|) - L_0'(0) \in B(F'(0) - F'(x)),$$
$$L'(|y| + |x|) - L'(|y|) \in B(F'(y) - F'(y + x)),$$
$$M(p) \leq p$$
$$L_0(p_0) \leq M(p_0) \quad \text{for all} \quad p_0 \in [0, p],$$
$$L_0'(p_0) \leq M'(p_0) \quad \text{for all} \quad p_0 \in [0, p],$$
$$L_0' \in B(I - F'(0)), \quad (-F(0), L_0(0)) \in E,$$
$$M'(p)^n(p) \to 0 \text{ as } n \to \infty,$$
$$M'(d_n)(b - d_n) + L(d_n) \leq M(b) \quad \text{for all } n \geq 0,$$

where

$$d_0 = 0, \quad d_{n+1} = L(d_n) + L_0'(|x_n|)(c_n), \quad c_n = |x_{n+1} - x_n| \quad (n \geq 0)$$

and

$$b = M^\infty(0).$$

Show: sequence $\{x_n\}$ $(n \geq 0)$ generated by Newton's method is well defined, remains in $E^{\mathbb{N}}$, is monotone, and converges to a unique zero x^* in $U(b)$, where b is the smallest fixed point of M in $[0, p]$. Moreover the following bounds hold for all $n \geq 0$

$$d_n \leq b,$$
$$|x^* - x_n| \leq b - d_n,$$
$$|x^* - x_n| \leq M^n(p) - M^n(0),$$
$$c_n + d_n \leq d_{n+1},$$

and

$$c_n = |x_{n+1} - x_n|.$$

(b) If $r \in [0, p - |x_n|]$ satisfies $R_n(r) \le r$ then show: the following holds for all $n \ge 0$:

$$c_n \le R_n(r) = p \le r, \qquad (1)$$

and

$$R_{\{n+1\}}(p - c_n) \le p - c_n$$

(c) Assume hypotheses of (a) hold and let $r_n \in [0, p - |x_n|]$ be a solution (1). Then show the following a posteriori estimates hold for all $n \ge 0$

$$|x^* - x_m| < q_m,$$

where

$$q_n = r_n, \quad q_{m+1} = R_m(q_m) - c_m \quad (m \ge n).$$

(d) Under hypotheses of (a) show any solution $r \in [0, p - |x_n|]$ of $R_n(r) \le r$ yields the a posteriori estimate

$$|x^* - x_n| \le R_n^\infty(0) \le r \quad (n \ge 0).$$

Let X be partially ordered and set

$$X = V, \quad D = C^2 \text{ and } |\cdot| = I.$$

(e) Let X be partially ordered, Y a Banach space, $G \in C^1(X_0 \subseteq X \to Y)$, $A \in L(Y \to X)$ and $x, y \in X$ such that:

1. $[x, y] \in X_0$.
2. $I - AG'(x) \in L_+(X)$,
3. $z_1 \le z_2 \implies AG'(z_1) \le AG'(z_2)$ for all $z_1, z_2 \in [x, y]$,
4. $AG(x) \le 0$, $AG(y) \ge 0$,
5. $(I - AG'(y))^n (y - x) \to 0$ as $n \to \infty$.

Show: sequence $\{y_n\}$ $(n \ge 0)$ generated by

$$y_0 = x, \quad y_{n+1} = y_n + AG'(y_n)^* [-AG(y_n)]$$

is well defined for all $n \ge 0$, monotone and converges to a unique zero y^* of AG in $[x, y]$.

12.11. Let there be given a Banach space X with convergence structure (X, V, E) where $V = (V, C, \|\cdot\|_V)$, and operator $F \in C''(X_0 \to X)$ with $X_0 \subseteq X$, and operator $M \in C'(V_0 \to V)$ with $V_0 \subseteq V$, and a point $p \in C$ satisfying:

$$U(p) \subseteq X_0, \quad [0, p] \subseteq V_0;$$

M is order-convex on $[0, p]$ and for all $x, y \in U(p)$ with $|x| + |y| \leq p$

$$M'(|x+y|) - M'(x) \in B([F''(x) - F''(x+y)](y))$$
$$M'(0) \in B(I - F'(0)), \quad (-F(0), M(0)) \in E;$$
$$M(p) \leq p;$$

and

$$M'(p)^n p \to 0 \quad \text{as} \quad n \to \infty.$$

Then, show sequence $(x_n, t_n) \in (X \times V)^{\mathbb{N}}$, where $\{x_n\}$ is generated by Newton's method and $\{t_n\}$ ($n \geq 0$) is given by

$$t_0 = 0, \quad t_{n+1} = M(t_n) + M'(|x_n|)(a_n), \quad a_n = |x_{n+1} - x_n|$$

is well defined for all $n \geq 0$, belongs in $E^{\mathbb{N}}$, and is monotone. Moreover the following hold for all $n \geq 0$

$$t_n \geq b,$$

where,

$$b = M^\infty(0),$$

is the smallest fixed point of M in $[0, p]$.

b) Show corresponding results as in Exercises 12.10 (b)–(c).

12.12. Assume hypotheses of Exercise 12.10 hold for $M = L$.

Show: (a) conclusions (a)–(e) hold (under the revised hypothesis) and the last hypothesis in (a) can be dropped;

(b) error bounds $|x^* - x_n|$ ($n \geq 0$) obtained in this setting are finer and the infromation on the location of the solution more precise than the corresponding ones in Exercise 12.4 provided that $L_0(p_0) < L(p_0)$ or $L'_0(p_0) < L'(p_0)$ for all $p \in [0, p]$.

As in Exercise 12.5 assume there exists a monotone operator $k_0 : [0, p] \to R$ such that

$$\|F'(x) - F'(0)\| \leq k_0(\|x\|) \|x\|, \quad \text{for all } x \in U(p),$$

and define operator L_0 by

$$L_0(t) = \|F(0)\| + \int_0^t ds \int_0^s d\theta k_0(\theta).$$

Sequence $\{d_n\}$ given by $d_0 = 0$, $d_{n+1} = L(d_n) + L'_0(|x_n|) c_n$ converges to some $p^* \in [0, p]$ provided that

$$\left(k_0(p) + \tfrac{k(p)}{2}\right) \|F(0)\| \leq 1.$$

Conclude that the above semilocal convergence condition is weaker than (d) in Exercise 12.4 or equivalently (for $p = a$)

$$2k(p) \|F(0)\| \leq 1.$$

Finally, conclude that in the setting of Exercise 12.12 and under the same computational cost we always obtain and under weaker conditions: finer error bounds on the distances $|x_n - x^*|$ $(n \geq 0)$ and a better information on the location of the solution x^* than in Exercise 12.5 (i.e., [199], see also [99], [200]).

Chapter 13

Dynamic Processes

In this Chapter we examine the point-to-set convergence.

13.1 On Time Dependent Multistep Dynamic Processes

In the context of nonlinear programming Zangwill [297] presented a general theory on convergence of iteration processes based on point-to-set mappings. He investigated only one-step stationary iterations, and he proved that the process either terminates after a finite number of steps or the limit of any convergent subsequence is a solution. Special but practically useful criteria were derived for example by Brock and Scheinkman [106], Fujimoto [155], Szidarovszky and Okuguchi [259] based on special selections of the Liapunov function.

In this section the convergence theorem of Zangwill is generalized and extended to nonstationary multistep iteration processes in partially ordered topological spaces. In addition, monotone convergence and the speed of the convergence of the processes are examined.

Let $S \subset X$ be a set such that $u^* \in \overline{S}$, and for $k \geq 0$ the point-to-set mappings $f(k;\cdot)$ are defined on

$$S^\ell = S \times S \times S \cdots \times S,$$

and for all $t^{(1)}, \cdots, t^{(\ell)} \in S$ and $k \geq \ell - 1$, $f\left(k; t^{(1)}, \cdots, t^{(\ell)}\right)$ is nonempty in S. Define the iteration sequence

$$x_{k+1} \in f\left(k_{k-\ell+1}, x_{k-\ell+2}, \cdots, x_k\right) \tag{13.1}$$

where $k \geq \ell - 1, x_0, x_1, ..., x_{\ell-1} \in S$, and an arbitrary element from the set can be selected as the successor of x_k.

Definition 13.1 A function $v : S^\ell \to R_+$ is called the *Liapunov function* of process (13.1), if for arbitrary $t^{(i)} \in S$ ($i = 1, 2, ..., \ell, t^{(\ell)} \neq u^*$) and $y \in f\left(k; t^{(1)}, ..., t^{(\ell)}\right)$ $(k \geq \ell - 1)$,

$$V\left(t^{(2)}, ..., t^{(\ell)}, y\right) < V\left(t^{(1)}, t^{(2)}, ..., t^{(\ell)}\right).$$

Definition 13.2 The Liapunov-function V is called *closed*, if it is defined on $\overline{S} = \overline{S} \times \overline{S} \times \cdots \times \overline{S}$, furthermore, if $k_i \to \infty$, $t_i^{(j)} \to t^{(j)^*}$

$$\left(t_i^{(j)} \in S \text{ for } i \geq 0 \text{ and } j = 1, 2, ..., \ell, t^{(\ell)^*} \neq u^*\right),$$

$y_i \in f\left(k_i; t_i^{(1)}, ..., t_i^{(\ell)}\right)$ $(i \geq 0)$ and $y_i \to y^*$, then

$$v\left(t^{(2)^*}, ..., t^{(\ell)^*}, y^*\right) < v\left(t^{(1)^*}, ..., t^{(\ell)^*}\right).$$

Remark 13.1 Assume that $f(k; \cdot) \equiv f(\cdot)$ for all k, $S = \overline{S}$, and mapping $f(\cdot)$ is closed (for the definition of closed mappings see e.g. Zangwill, [297, pp. 88], then any Liapunov-function is also closed.

Our main convergence result can be formulated as follows:

Theorem 13.1 *Assume that X is a topological space and*

(A) *For all $k \geq \ell - 1$, $f\left(k; t^{(1)}, ..., t^{(\ell)}, u^*\right) = \{u^*\}$ with arbitrary $t^{(1)}, ..., t^{(\ell-1)} \in S$, if $u^* \in S \subseteq X$;*

(B) *The iteration process (13.1) has a continuous, closed Liapunov function;*

(C) *There exists a compact set C in X and that for all $k \geq \ell - 1$, $x_k \in C$. Then $x_k \to u^*$ as $k \to \infty$.*

Proof. Condition (A) implies that if for $k \geq \ell - 1$, $x_k = u^*$, then all successors of x_k are also equal to u^*. Hence we assume that $x_k \neq u^*$ ($k \geq \ell - 1$). assume that $x_k \not\to u^*$, then since the sequence is in a compact set there is a subsequence x_{k_i} which tends to $x^* \neq u^*$. The construction of the iteration sequence and the definition of the Liapunov function imply that for all $i \geq 0$,

$$V\left(x_{k_{i+1}-\ell+1}, ..., x_{k_{i+1}}\right) \leq V\left(x_{k_i-\ell+2}, ..., x_{k_i}, x_{k_i+1}\right)$$
$$\leq V\left(x_{k_i-\ell+1}, ..., x_{k_i}\right). \qquad (13.2)$$

Without loss of generality assume that all sequences $\{x_{k_i-\ell+1}\}$, $\{x_{k_i-\ell+2}\}, ..., \{x_{k_i}\}$, and $\{x_{k_i+1}\}$ are also convergent, otherwise take further subsequences of $\{x_{k_i}\}$. Let $x_{\ell-1}^*, ..., x_1^*$ and y^* denote the limits of the above subsequences, then the continuity of the Liapunov-function and relation (13.2) imply that

$$V\left(x_{\ell-2}^*, ..., x_1^*, x^*, y^*\right) = V\left(x_{\ell-1}^*, ..., x_1^*, x^*\right).$$

Since the Liapunov-function is closed, strict inequality must hold in the above relation. This contradiction completes the proof. □

Remark 13.2 *Assumption $u^* \in \overline{S}$ is need in order to obtain u^* as the limit of sequences from S. Assumption (A) guarantees that if at any iteration step the solution u^* is obtained, then the process remains at the solution. We may also show that the existence of a Liapunov function is not a too strong assumption. Consider the special case when X is a normed space and f is point-to-point from S to S, and assume that starting from arbitrary initial solution $x_0 \in S$ the process converges to the solution u^* of equation $x = f(x)$. Let $V : S \to R_+$ be constructed as follows. with selecting $x_0 = x$ consider sequence $x_{k+1} = f(x_k)$ $(k \geq 0)$, and define*

$$V(x) = \begin{cases} 0 & \text{if } x = u^* \\ \max_k \|x_k - u^*\| & \text{otherwise.} \end{cases}$$

Obviously $V(f(x)) \leq V(x)$ for all $x \in S$. The continuity-type assumptions in (B) are also natural, since without certain continuity conditions no convergence can be established. Assumption (C) is necessarily satisfied, for example, if $x = R^n$, and either S is bounded or if for every $B > 0$ there exists a $Q > 0$ such that $t^{(1)}, ..., t^{(\ell)} \in S$ and $\|t^{(j)}\| > Q$ (for at least one index j) imply relation

$$V\left(t^{(1)}, ..., g^{(\ell)}\right) > B.$$

In the case of one-step processes (that is, if $\ell = 1$) this last condition can be reformulated as

$$\lim_{\substack{\|x\| \to \infty \\ x \in S}} V(x) = \infty.$$

Remark 13.3 *Iteration processes in this general form have real practical importance. Note first that one of the most popular solvers of nonlinear equations is the secant method, which is actually a two-step process. Many* dynamic economic processes are based on the selection of optimal strategies by the participants at each time period. If the optimal solution is not unique, then the strategy for the next period can be selected from the set of optimal solutions. hence the iteration is based on a set-valued mapping. In addition, if the participants' decisions are based on extrapolative expectations on the other's behavior, then the process becomes multistep. Time dependency of the process follows from price changes, technological development, etc. For the description of such models in the oliopoly theory see Okuguchi Szidarovszky [260].

Assume next that the iteration process is stationary, that is, in recursion (13.1) function f does not depend on k. In this case Theorem 13.1 reduces to the following.

Theorem 13.2 *Assume that X is a topological space, $S \subset X$, furthermore*

(i) $S = \overline{S}$;
(ii) *For all*

$$t^{(1)}, ..., t^{(\ell-1)} \in S,$$
$$f\left(t^{(1)}, ..., t^{(\ell-1)}, u^*\right) = \{u^*\};$$

(iii) *Function f is closed on S;*
(iv) *The iteration process has a continuous Liapunov function;*
(v) *There exists a compact set $C \subset X$ such that for all $k \geq \ell - 1$, $x_k \in C$.*

Then $x_k \to u^*$ as $k \to \infty$.

Remark 13.4 *This result in the special case of $\ell = 1$ can be considered as the discrete-time-scale counterpart of the famous stability result of Uzawa [271].*

Sufficient conditions will be now given for the monotone convergence of the iteration scheme (13.1). Assume now that X is a partially ordered topological space, and $S \subset X$.

Definition 13.3 *The sequence of point-to-set mappings $f(k; \cdot)$ from S to S is called increasingly isotone on S if for arbitrary $k \geq \ell - 1$, $t^{(i)} \in S$ ($i = 1, 2, ..., \ell + 1$) such that $t^{(\ell+1)} \geq t^{(\ell)} \cdots \geq t^{(2)} \geq t^{(1)}$ and for any $y_1 \in f\left(k; t^{(1)}, ..., t^{(\ell)}\right)$ and $y_2 \in f\left(k+1; t^{(2)}, ..., t^{(\ell+1)}\right)$, $y_1 \leq y_2$.*

Definition 13.4 *Point-to-set mapping* $f(k; \cdot) : S^\ell \to S$, *for a fixed* k $(k \geq \ell - 1)$, *is called increasingly isotone if* $t^{(i)} \in S$ $(i = 1, 2, ..., \ell + 1)$ *such that* $t^{(\ell+1)} \geq t^{(\ell)} ... \geq t^{(2)} \geq t^{(1)}$ *and* $y_1 \in f(k; t^{(1)}, ..., t^{(\ell)})$ *and* $y_2 \in f(k; t^{(2)}, ..., t^{(\ell+1)})$ *imply that* $y_1 \leq y_2$.

Remark 13.5 *Note that if* $f(k; \cdot)$ *does not depend on* k, *then Definition 13.3 and 13.4 are equivalent.*

Remark 13.6 *In the literature a point-to-set mapping* $f(k; \cdot)$ *is called isotone if for all* $t^{(i)} \in S$, $s^{(i)} \in S$ *such that* $t^{(i)} \leq s^{(i)}$ $(i = 1, 2, ..., \ell)$, $y_1 \leq y_2$ *for all* $y_1 \in f(k; t^{(1)}, ..., t^{(\ell)})$, $y_2 \in f(k; s^{(1)}, ..., s^{(\ell)})$. *It is obvious that an isotone mapping is increasingly isotone, but the reverse is not necessarily true, as the example of set* $S = [0, 1] \subset R^1$ *and function*

$$g\left(t^{(1)}t^{(2)}\right) = \begin{cases} t^{(2)} & \text{if } t^{(1)} \geq 2t^{(2)} - 1 \\ t^{(1)} - t^{(2)} + 1, & \text{if } t^{(1)} < 2t^{(2)} - 1 \end{cases}$$

illustrates. Let partial order $\leq$ *be defined as* $\left(s^{(1)}, s^{(2)}\right) \leq \left(t^{(1)}, t^{(2)}\right)$ *if and only if* $s^{(1)} \leq t^{(1)}$ *and* $s^{(2)} \leq t^{(2)}$. *First we show that* g *is increasingly monotone. Select* $t^{(1)} \leq t^{(2)} \leq t^{(3)}$. *Note first that* $g\left(t^{(1)}, t^{(2)}\right) \leq t^{(2)}$. *If* $t^{(1)} \geq 2t^{(2)} - 1$, *then* $g\left(t^{(1)}, t^{(2)}\right) = t^{(2)}$; *and if* $t^{(1)} < 2t^{(2)} - 1$, *then*

$$g\left(t^{(1)}, t^{(2)}\right) = t^{(1)} - t^{(2)} + 1 < 2t^{(2)} - 1 - t^{(2)} + 1 = t^{(2)}.$$

Note next that $g\left(t^{(2)}, t^{(3)}\right) \geq t^{(2)}$. *If* $t^{(2)} \geq 2t^{(3)} - 1$, *then* $g\left(t^{(2)}, t^{(3)}\right) = t^{(3)} \geq t^{(2)}$; *and if* $t^{(2)} < 2t^{(3)} - 1$, *then*

$$g\left(t^{(2)}, t^{(3)}\right) = t^{(2)} - t^{(3)} + 1 \geq t^{(2)}.$$

Hence

$$g\left(t^{(1)}, t^{(2)}\right) \leq t^{(2)} \leq g\left(t^{(2)}, t^{(3)}\right).$$

We can also verify that mapping q *is not isotone on* S. *consider points* $(t, 1)$ *and* $(t, 1 - \epsilon)$ $(t, \epsilon > 0;\ t + 2\epsilon < 1)$. *Then* $g(t, 1) = t$ *and* $g(t, 1 - \epsilon) = t - (1 - \epsilon) + 1 = t + \epsilon > g(t, 1)$. *Hence* g *is not isotone.*

Theorem 13.3 *Assume that in iteration (13.1) the sequence of mappings* $f(k; \cdot)$ *is increasingly isotone, furthermore* $x_i \in S$ $(0 \leq i < \ell - 1)$ *and* $x_0 \leq x_1 \cdots, \leq x_{\ell-1} \leq x_\ell$. *Then for all* $k \geq 0$, $x_{k+1} \geq x_k$.

Proof. By induction, assume that for i $(i < k)$, $x_{i+1} \geq x_i$. Then relations $x_k \in f(k-1, x_{k-\ell}, ..., x_{k-1})$, $x_{k+1} \in f(x_{k-\ell+1}, ..., x_k)$ and the definition of increasingly monotone family of mappings imply that $x_k \leq x_{k+1}$. since this inequality holds for $k = 0, 1, ..., \ell - 1$, the proof is completed. □

Consider next the modified iteration scheme

$$y_{k+1} \in f(k; y_k, y_{k-1}, ..., y_{k-\ell+1}). \qquad (13.3)$$

Using finite induction, similarly to Theorem 13.3, we may prove the following:

Theorem 13.4 *Assume that the sequence of mappings $f(k; \cdot)$ is increasingly isotone, furthermore $y_i \in S$ $(0 \leq i \leq \ell - 1)$ and*

$$y_0 \geq y_1 \geq \cdots, \geq y_{\ell-1} \geq y_\ell.$$

Then for all $k \geq 0$, $y_{k+1} \geq y_k$.

Corollary 13.1 *Assume that $X = R^n$ and for $k \to \infty$, the sequence $\{x_k\}$ and $\{y_k\}$ have the same limit u^*, and $\leq$ is the usual partial order of vectors. (That is, $a = (a(i)) \leq b = b(i)$ if and only if $(a(i) \leq b(i)$ for all i). Under the conditions of Theorem 13.3 and 13.4, for all $k \geq 0$,*

$$x_k \leq u^* \leq y_k.$$

This relation is very useful in the error analysis of the iteration methods (13.1) and (13.3), since for all coordinates $x_k(i)$, $y_k(i)$ and $u^*(i)$ of vectors x_k, y_k, u^* respectively,

$$0 \leq u^*(i) - x_k(i) \leq y_k(i) - x_k(i)$$

and

$$0 \leq y_k(i) - u^*(i) \leq y_k(i) - x_k(i).$$

Furthermore, we can show:

Theorem 13.5 *Assume that X is a regular POB-space, $S \subset X$ and*

(A) *The sequence of mappings $f(k; \cdot)$ is increasingly isotone in iteration (13.1) with $x_i \in S$ $(0 \leq i \leq \ell - 1)$ and*

$$x_0 \leq x_1 \leq \cdots \leq x_{\ell-1} \leq x_\ell;$$

(B) *There exists a set H_1 defined by $H_1 = \{x \in S; \ x \leq x_0\}$ with the property that if for any points $t^{(1)}, t^{(2)}, ..., t^{(\ell)}$ in H_1 with*

$$t^{(1)} \leq t^{(2)} \leq \cdots \leq t^{(\ell)} \leq x_0,$$

then
$$x_{k+1} \leq x_0 \quad \text{for any} \quad x_{k+1} \in f\left(k; t^{(1)}, t^{(2)}, ..., t^{(\ell)}\right), \quad k \geq \ell - 1.$$

Then the sequence $\{x_n\}$, $n \geq 0$ generated by the iteration (13.1) process (13.1) is monotonically increasing, remains in H_1 and converges to some $u^* \in H_1$.

Proof. From (A) and Theorem 13.3 it follows that the sequence $\{x_n\}$, $n \geq 0$ is monotonically increasing, whereas from (B) we get that the sequence is bounded above by x_0. since X is a regular POB-space the sequence $\{x_n\}$, $n \geq 0$ converges to some u^* with $u^* \leq x_0$. hence $u^* \in H_1$.
That completes the proof of the theorem. $\square$

These monotonic properties of the iteration processes are very useful, but in cases where the convergence is very slow the above methods have only very limited practical importance. In the next section of this paper the convergence speed of the above iteration schemes is estimated and practical error estimates are derived.

We can now formulate estimates on the speed of convergence using following theorem.

Theorem 13.6 *Assume that X is a normal POB-space, $S \subset X$ and*

(A) *the sequence of mappings $f(k; \cdot)$ is increasing isotone in iteration (13.1) with $x_i \in S$ ($0 \leq i \leq \ell - 1$) and $x_0 \leq x_1 \leq \cdots \leq x_{\ell-1} \leq x_\ell$.*
(B) *There exists a constant b with $0 \leqq b < 1$ such that*

$$x_{n+2} - x_{n+1} \leq b(x_{n+1} - x_n), \quad \text{for all} \quad n \geq 0. \qquad (13.4)$$

Then the sequence $\{x_n\}$, $n \geq 0$ generated by the iteration process (13.1) is monotonically increasing and converges to some u^* with

$$\|x_n - u^*\| \leq \frac{\alpha \|x_1 - x_0\|}{1 - b} b^n, \quad n \geq 0. \qquad (13.5)$$

Proof. From (A) and Theorem 13.3 it follows that the sequence $\{x_n\}$ is monotonically increasing and inequality (13.4) can be rewritten as

$$0 \leq x_{n+2} - x_{n+1} \leq b(x_{n+1} - x_n), \quad n \geq 0.$$

Using the above inequality we get

$$0 \leq x_{n+p} - x_n = \sum_{i=0}^{p-1}(x_{n+i+1} - x_{n+1}) \leq \frac{x_1 - x_0}{1 - b} b^n, \quad p \geq 0.$$

Since X is normal we deduce

$$\|x_{n+p} - x_n\| \leq \alpha \frac{\|x_1 - x_0\|}{1-b} b^n.$$

It now follows that the sequence $\{x_n\}$, $n \geq 0$ is a Cauchy sequence in a Banach space and as such it converges to some u^*. By letting $p \to \infty$ we obtain (13.5).

That completes the proof of the theorem. $\square$

Note that an identical theorem can be proven if the assumptions (A) and (B) in the above theorem are replaced with the condition

$$0 \leq x_{n+2} - x_{n+1} \leq b(x_{n+1} - x_n),$$

for all $n \geq 0$ and some b, $0 \leq b < 1$. Let us define the set H_2 by $H_2 = \{x \in S;\ x_0 \leq x,\ \|x - x_0\| \leq h\}$ for some $h > 0$.

Then we can show the following theorem.

Theorem 13.7 *Let X be a normal POB-space, $S \subset X$ and assume that the following conditions are satisfied*

$$\alpha \frac{\|x_1 - x_0\|}{1 - c_2} \leq h, \tag{13.6}$$

$$c_1(x_n - x_{n-1}) \leq x_{n+1} - x_n \leq c_2(x_n - x_{n-1}), \quad n \geq 1, \quad 0 \leq c_1 \leq c_2 < 1. \tag{13.7}$$

$$x_0 \leq x_n \tag{13.8}$$

$$\|x_n - x_0\| \leq h \tag{13.9}$$

and

$$0 \leq x_{n+1} - x_n \leq c_2^n(x_1 - x_0) \quad \text{for all} \quad n = 0, 1, 2, ..., \ell - 1. \tag{13.10}$$

Then the sequence $\{x_n\}$, $n \geq 0$ generated by the iteration process (13.1) is monotonically increasing and converges to some u^* with

$$\|x_n - u^*\| \leq \alpha \frac{\|x_1 - x_0\|}{1 - c_2} c_2^n \quad \text{for all} \quad n \geq 0. \tag{13.11}$$

Proof. We will show that the estimates (13.8), (13.9) and (13.10) are true for all $n \geq 0$. For $n = 0, 1, 2, ..., \ell - 1$, they hold by hypothesis. Let us suppose that they are true for $n = 0, 1, 2, ..., k$ with $k \geq \ell - 1$. From (13.8) and (13.10) for $n = k$ it follows that

$$x_0 \leq x_k \leq x_{k+1}$$

and thus (13.8) is true for $n = k+1$.

Using (13.10), the above inequality, and the properties of the partial order $\leq$, we have successively:

$$0 \leq x_{k+1} - x_0 = \sum_{i=0}^{k}(x_{i+1} - x_i) \leq (x_1 - x_0)\sum_{i=1}^{k} c_2^i \leq \frac{x_1 - x_0}{1 - c_2}$$

where from (13.6) we deduce that (13.9) is true for $n = k+1$.

From (13.7), (13.10) and the induction hypothesis we get

$$0 \leq x_{k+2} - x_{k+1} \leq c_2 c_2^k (x_1 - x_0) = c_2^{k+1}(x_1 - x_0).$$

It now follows that (13.10) is true for $n = k+1$. Moreover for $p \geq 0$ we get

$$0 \leq x_{n+p} - x_n = \sum_{i=0}^{p-1}(x_{n+i+1} - x_{n+i}) \leq \frac{x_1 - x_0}{1 - c_2} c_2^n$$

where from we obtain

$$\|x_{n+p} - x_n\| \leq \alpha \frac{\|x_1 - x_0\|}{1 - c_2} c_2^n \leq h c_2^n. \tag{13.12}$$

The above inequality shows that the sequence $\{x_n\}$, $n \geq 0$ is Cauchy in a POB-space and as such it converges to some u^*. By letting $p \to \infty$ in (13.12) we obtain (13.11).

That completes the proof of the theorem. □

Remark 13.7 *Note that a similar theorem can be proven if the condition (13.7) is replaced by the relation*

$$0 \leq x_{n+1} - x_n \leq c_2(x_n - x_{n-1}), \quad n \geq 1, \quad 0 \leq c_2 < 1.$$

Remark 13.8 *Assume that there exists a sequence $c_{2n}^{(n)}$, $n \geq 0$, such that more generally*

$$0 \leq x_{n+1} - x_n \leq c_2^{(n)}(x_n - x_{n-1}), \quad n \geq 1, \quad 0 \leq c_2^{(n)} \leq q < 1.$$

Then similarly to (13.12) we have that

$$x_{n+p} - x_n = \sum_{i=0}^{p-1}(x_{n+i+1} - x_{n+i})$$
$$\leq (x_1 - x_0)(v_n + v_{n+1} + \cdots + v_{n+p-1}),$$

where

$$v_n = c_2^{(1)} c_2^{(2)} \cdots c_2^{(n)}.$$

Hence

$$x_{n+p} - x_n \leq (x_1 - x_0) v_n \left(1 + q + q^2 + \cdots\right)$$
$$= \frac{x_1 - x_0}{1 - q} v_n,$$

and therefore (13.11) is modified as

$$\|x_n - u^*\| \leq \alpha \frac{\|x_1 - x_0\|}{1 - q} v_n.$$

In the special case when $c_2^{(n)}$ is a decreasing sequence we may select $q = c_2^{(1)}$.

13.2 The Monotone Convergence of General Newton-Like Methods

This section examines conditions for the monotone convergence of Newton-like methods. Using the famous Kantorovich lemma on monotone mappings, (see Chapter 2) we derive several convergence results. The speed of convergence of these processes is also examined.

In particular, let us consider the Newton-Like iterates

$$z_{k+1} = G_k(z_k) \quad (k \geq 0), \tag{13.13}$$

where

$$G_k(z_k) = z_k - A_k(z_k, z_{k-1})^{-1} f_k(x_k) \ (k \geq 0). \tag{13.14}$$

Here $f_k, G_k : D \subseteq B \to B_1$ $(k \geq 0)$ are nonlinear mappings acting between two partially ordered linear topological spaces (POL-spaces), whereas $A_k(u,v)(\cdot) : D \to B_1$ $(k \geq 0)$ are invertible linear mappings. We provide sufficient conditions for the convergence of iteration (13.13) to O. we may have this assumption without loosing generality, since any solution x^* can be transformed into O by introducing the transformed mapping $q_k(x) = f_k(x + x^*) - x^*$ $(k \geq 0)$. Iterations of the above type are extremely important in solving optimization problems, as well as linear and nonlinear equations. A very important field of such applications can also be found in solving equilibrium problems, in economy and in solving nonlinear input-output systems (see e.g. Fujimoto [155], Krasnoselskii [184],

Okuguchi [226], Szidarovszky [259], [260], Tishydhigama, et al [262], Tarski [265]. Our results can be reduced to the ones obtained earlier by Argyros [68], Baluev [101], Dennis [132], Kantorovich & Akilov [183], Krasnoselskii [184], Potra [237], Slugin [256], Tishydhigama et al., [262] when $f_k = f\,(k \geq 0)$.

We assume that the following conditions hold:

(A) Consider mappings $f_k : D \subset B \to B_1$ where B is a regular POTL-space and B_1 is a POTL-space. Let x_0, y_0, y_{-1} be three points of D such that

$$x_0 \leq y_0 \leq y_{-1},\ \langle x_0, y_{-1}\rangle \subset D, f_0\,(x_0) \leq O \leq f_0\,(y_0), \qquad (13.15)$$

and denote

$$S_1 = \{(x,y) \in B^2 / x_0 \leq x \leq y \leq y_0\},$$
$$S_2 = \{(u, y_{-1}) \in B^2 / x_0 \leq u \leq y_0\},$$
$$s_3 = S_1 \cup S_2.$$

assume mappings $A_k\,(.,.) : S_3 \to LB\,(B, B_1)$ such that

$$f_k\,(y) - f_k\,(x) \leq A_k\,(w, z)\,(y - x) \qquad (13.16)$$

for all $k \geq 0$, (x,y), $(y, w) \in S_1$, $(w, z) \in S_3$.
Suppose for any $(u, v) \in S_3$ the linear mappings $A_k\,(u,v)\,(k \geq 0)$ have a continuous nonsingular nonnegative subinverse. Assume furthermore that

$$f_k\,(x) \leq f_{k-1}\,(x) \quad \text{for all} \quad x \in \langle x_0, y_0\rangle\,(k \geq 1),\ f_{k-1}\,(x) \leq O, \tag{13.17}$$

$$f_k\,(y) \geq f_{k-1}\,(y) \quad \text{for all} \quad y \in \langle x_0, y_0\rangle\,(k \geq 1),\ f_{k-1}\,(y) \geq O. \tag{13.18}$$

We can now formulate then main result.

Theorem 13.8 *Assume condition (A) is satisfied. Then there exist two sequences $\{x_k\}$, $\{y_k\}\,(k \geq 0)$ and points x^*, y^*,*

x_1^*, y_1^* such that for all $k \geq 0$;

$$f_k(y_k) + A_k(y_k, y_{k-1})(y_{k+1} - y_k) = O, \qquad (13.19)$$

$$f_k(x_k) + A_k(y_k, y_{k-1})(x_{k+1} - x_k) = O, \qquad (13.20)$$

$$f_k(x_k) \leq f_{k-1}(x_{k-1}) \leq O \leq f_{k-1}(y_{k-1}) \leq f_k(y_k) \ (k \geq 1) \qquad (13.21)$$

$$x_0 \leq x_1 \leq \ldots \leq x_k \leq x_{k+1} \leq y_{k+1} \leq y_k \leq \ldots \leq y_1 \leq y_0 \qquad (13.22)$$

$$x_k \to x^*, \quad y_k \to y^* \quad as \quad k \to \infty, \ x^* \leq y^* \qquad (13.23)$$

and

$$f_k(x_k) \to x_1^*, \quad f_k(y_k) \to y_1^* \quad as \quad k \to \infty, \ with \ x_1^* \leq O \leq y_1^*. \qquad (13.24)$$

Proof. Let L_0 be a continuous nonsingular nonnegative left subinverse of $A_0(y_0, y_{-1}) \equiv A_0$ and consider mapping $P : \langle 0, y_0 - x_0 \rangle \to B$ defined by

$$P(x) = x - L_0(f_0(x_0) + A_0(x)),$$

where $A_0(x)$ denotes the image of x with respect to mapping $A_0 = A_0(y_0, y_{-1})$. It is easy to see that P is isotone and conditions. We also have

$$p(O) = -L_0(f_0(x_0)) \geq O,$$
$$P(y_0 - x_0) = y_0 - x_0 - L_0(f_0(y_0)) + L_0(f_0(y_0) - f_0(x_0)) - A_0(y_0 - x_0)$$
$$\leq y_0 - x_0 - L_0(f_0(y_0)) \leq y_0 - x_0.$$

According to Kantorovich theorem of fixed points (see Chapter 2) mapping P has a fixed point $w \in \langle O, y_0 - x_0 \rangle$. Taking $x_1 = x_0 + w$, we have

$$f_0(x_0) + A_0(x_1 - x_0) = 0, \quad x_0 \leq x_1 \leq y_0.$$

Using (13.16), (13.17) and the above relation we get

$$f_1(x_1) \leq f_0(x_1) = f_0(x_1) - f_0(x_0) + A_0(x_0 = x_1) \leq 0.$$

Consider now mapping $Q : \langle O, y_0 - x_1 \rangle \to B$ given by

$$Q(x) = x + L_0(f_0(y_0) - A_0(x)).$$

Q is clearly continuous, isotone, and

$$Q(O) = L_0 f_0(y_0) \geq O,$$
$$Q(y_0 - x_1) = y_0 - x_1 + L_0 f_0(x_1) + L_0(f_0(y_0) - f_0(x_1) - A_0(y_0 - x_1))$$
$$\leq y_0 - x_1 + L_0 f_0(x_1) \leq y_0 - x_1.$$

Applying the Kantorovich lemma again, we deduce existence of a point $z \in \langle O, y_0 - x_1 \rangle$ such that $Q(z) = z$. Taking $y_1 = y_0 - z$,

$$f_0(y_0) + A_0(y_1 - y_0) = O, \quad x_1 \le y_1 \le y_0.$$

Using the above relations and conditions (13.16), (13.18) we obtain

$$f_1(y_1) \le f_0(y_1) = f_0(y_1) - f_0(y_0) + A_0(y_0 - y_1) \ge O.$$

By induction it is easy to show there exist four sequences $\{x_k\}$, $\{y_k\}$, $\{f_k(x_k)\}$, $\{f_k(y_k)\}$ $(k \ge 0)$, satisfying (13.19)-(13.22). Since space B is regular, from (13.21) and (13.22) we know that there exist x^*, y^*, x_1^*, $y_1^* \in B$ satisfying (13.23)-(13.24), which completes the proof. □

In the next part we give some natural conditions which guarantee that points x^*, y^* are common solutions of equations $f_k(x) = O$ $(k \ge 0)$.

Theorem 13.9 *Under the hypotheses of Theorem 13.8, assume furthermore that*

(i) *There exists $u \in B$ such that $x_0 \le u \le y_0$ and $f_k(u) = O$ $(k \ge 0)$;*
(ii) *Linear mappings $A_k(w, z)$ $(k \ge 0)$, $(w, z) \in S_3$ are inverse nonnegative.*

Then

$$x_k \le u \le y_k \quad (k \ge 0)$$

and

$$x^* \le u \le y^*.$$

Moreover if $x^* = y^*$, then $x^* = u = y^*$.

Proof. Using (i)

$$\begin{aligned} A_0(y_1 - u) &= A_0(y_0) - f_0(y_0) - A_0(u) \\ &= A_0(y_0 - u) - (f_0(y_0) - f_0(u)) \ge O \end{aligned}$$

and

$$\begin{aligned} A_0(x_1 - u) &= A_0(x_0) - f_0(x_0) - A_0(u) \\ &= A_0(x_0 - u) - (f_0(x_0) - f_0(u)) \le 0. \end{aligned}$$

By (ii) it follows $x_1 \leq u \leq y_1$. By induction it is easy to show that $x_k \leq u \leq y_k$ for all $k \geq 0$. Hence, $x^* \leq u \leq y^*$. Moreover if $x^* = y^*$, then $x^* = u = y^*$, which completes the proof. □

Moreover we can show:

Theorem 13.10 *Under hypotheses of Theorem 13.9, assume that either*

(i) B *is normal and there exists mapping* $L : B \to B_1 \, (L\,(O)\,O)$ *which has an isotone inverse continuous at the origin and*

$$A_k\,(y_k, y_{k-1}) \leq L \text{ for all sufficiently large } k \geq 0;$$

or

(ii) B_1 *is normal and there exists mapping* $T : B \to B_1 \, (T\,(O) = O)$ *continuous at the origin and* $A_k\,(y_k, y_{k-1}) \leq T$ *for sufficiently large* $k \geq 0$;

or

Mappings $A_k\,(y_k, y_{k-1})\,(k \geq 0)$ *are equicontinuous.*
Then $f_k\,(x_k) \to O$, $f_k\,(y_k) \to O$ *as* $k \to \infty$.

Proof. (i) Using relations (13.18)-(13.22) we get

$$O \geq f_k\,(x_k) = A_k\,(y_k, y_{k-1})\,(x_k - x_{k+1}) \geq L\,(x_k - x_{k+1}),$$
$$O \leq f_k\,(y_k) = A_k\,(y_k, y_{k-1})\,(y_k - y_{k+1}) \leq L\,(y_k - y_{k+1}).$$

Hence,

$$O \geq L^{-1} f_k\,(x_k) \geq x_k - x_{k+1}, \quad O \leq L^{-1} f_k\,(x_k) \leq y_k - y_{k+1}.$$

Since B is normal and both $x_k - x_{k+1}$ and $y_k - y_{k+1}$ converge to zero, $L^{-1} f_k\,(x_k) \to O$, $f_k\,(y_k) \to O$ as $k \to \infty$, from which the result follows.

(ii) Using relations (13.19)-(13.22) we have

$$O \geq f_k\,(x_k) = A_k\,(y_k, y_{k-1})\,(x_k - x_{k+1}) \geq T\,(x_k - x_{k+1}),$$
$$O \leq f_k\,(y_k) = A_k\,(y_k, y_{k-1})\,(y_k - y_{k+1}) \leq T\,(y_k - y_{k+1}).$$

By letting $k \to \infty$ we obtain the result.

(iii) From equicontinuity of mappings $A_k\,(y_k, y_{k-1})$ it follows $A_k\,(y_k, y_{k-1})\,(z_k) \to O$ as whenever $z_k \to O$ as. In particular, we have

$$A_k\,(y_k, y_{k-1})\,(x_k - x_{k+1}) \to O, \quad A_k\,(y_k, y_{k-1})\,(y_k - y_{k+1}) \to O \text{ as } k \to \infty.$$

By (13.19) and (13.20) and above estimate the result follows. □

The uniqueness of a common solution of equations $f_k(x) = O$ $(k \geq 0)$ in $\langle x_0, y_0 \rangle$ can be proven assuming a condition which is complementary to (13.16). More precisely we can prove the following:

Theorem 13.11 *Let B and B_1 be two POL-spaces. Let $f_k(\cdot) : D \subset B \to B_1$ be nonlinear mappings and suppose there exist two points $x_0, y_0 \in D$ such that $x_0 \leq y_0$ and $\langle x_0, y_0 \rangle \subset D$. Denote by $S_1 = \{(x,y) \in B^2; x_0 \leq x \leq y \leq y_0\}$ and assume there exist mappings $L_k(.,.) : S_1 \to L(B, B_1)$ such that $L_k(x,y)$ has a nonnegative left superinverse for each $(x,y) \in S_1$ and*

$$f_k(y) - f_k(x) \geq L_k(x,y)(y-x) \quad \text{for all} \quad (x,y) \in S_1.$$

Under these assumptions if $(x^*, y^*) \in S_1$ and $f_k(x^*) = f_k(y^*)$, then $x^* = y^*$.

Proof. Let by $T_k(x^*, y^*)$ denote a nonnegative left superinverse of $L_k(x^*, y^*)$ for all $k \geq 0$. We have

$$O \leq y^* - x^* \leq T_k(x^*, y^*) L_k(x^*, y^*)(y^* - x^*)$$
$$\leq T_k(x^*, y^*)(f_k(y^*) - f_k(x^*)) = O.$$

Hence, $x^* = y^*$, which completes the proof. $\square$

Remark 13.9 *The conclusions of Theorem 13.8 hold if iteration (13.19) − (13.20) is modified as*

$$f_k(y_k) + A_k(y_k, y_{k+1})(y_{k+1} - y_k) = O,$$
$$f_k(x_k) + A_k(y_{k+1}, y_k)(x_{k+1} - x_k) = O \quad (k \geq 0),$$

This modification seems to be advantageous (see e.g. Slugin [256]) in many applications.

Remark 13.10 *Conditions (13.17) and (13.18) of Theorem 13.8 are very natural and they hold in many interesting problems in numerical analysis. See for example, Krasnoselskii [184]. Let us consider equations $f_k(x) = (k+1)(k+2)^{-1} x$, $k \geq 0$ on $[-1, 1] = [x_0, y_0] = \mathbb{R}$, where $\mathbb{R}$ is ordered with the usual ordering of real numbers. then for any x, y with $x \in [x_0, O]$ and $y \in [O, y_0]$, conditions (13.17) and (13.18) are satisfied. We note that when $f_k = f$ $(k \geq 0)$, the same conditions are satisfied as equalities.*

Remark 13.11 *The regularity of space B which is assumed in Theorem 13.8, is a rather restrictive condition. This condition was essentially used in proving that the iterative procedure (13.19) − (13.20) is well defined* (i.e. there exist sequences $\{x_k\}$, $\{y_k\}$, $\{f_k(x_k)\}$, $\{f_k(y_k)\}$ $\{k \geq 0\}$ satisfying (13.19) – (13.22) and they are convergent. Next, we present a method to avoid this regularity assumption. Consider now the following explicit method:

$$y_{k+1} = y_k - A_k^1(y_k, y_{k-1}) f_k(y_k) \ (k \geq 0) \qquad (13.25)$$
$$x_{k+1} = x_k - A_k^2(y_k, y_{k-1}) f_k(x_k) \ (k \geq 0). \qquad (13.26)$$

where $A_k^1(y_k, y_{k-1})$ and $A_k^2(y_k, y_{k-1})$ are nonnegative subinverses of $A_k(y_k, y_{k-1})$ $(k \geq 1)$. Without the regularity it is impossible to prove that sequences $\{x_k\}$, $\{y_k\}$, $\{f_k(x_k)\}$, $\{f_k(y_k)\}$ $(k \geq 0)$ produced by (13.25) – (13.26) are convergent. However, we can verify that for any common solution $u \in \langle x_0, y_0 \rangle$ of the equations $f_k(x) = O$ $(k \geq 0)$,

$$x_k \leq x_{k+1} \leq u \leq y_{k+1} \leq y_k \ (k \geq 0).$$

This result becomes important when the existence of the solution is proven by other methods, but it has to be enclosed monotonically (see the next section).

Theorem 13.12 *Consider mappings $f_k : D \subset B \to B_1$ $(k \geq 0)$, where B and B_1 are two POL-spaces and let x_0, y_0, y_{-1} be three points of B for which condition (13.15) holds. Define S_1, S_2, S_3 as in Theorem 13.8 and assume that there exist mappings $A_k(\cdot) : S_3 \to L(B, B_1)$ $(k \geq 0)$, satisfying conditions (13.16) – (13.18) and such that $A_k(u, v)$ has a nonnegative subinverse for any $(u, v) \in S_3$.*

Then, iteration (13.25) – (13.26) defines four sequences $\{x_k\}$, $\{y_k\}$, $\{f_k(x_k)\}$, $\{f_k(y_k)\}$ $(k \geq 0)$ and they satisfy properties (13.21) – (13.22).

Moreover for any common solution $u \in \langle x_0, y_0 \rangle$ of equations $f_k(x) = O$ $(k \geq 0)$,

$$x_k \leq u \leq y_k \ (k \geq 0).$$

Proof. For $k = 0$, by denoting $A_0^1(0; y_0, y_{-1}) = A_0^1$ and $A_0^2(O; y_0, y_{-1}) = A_0^2$ we have

$$x_0 \leq y_0, \ f_0(x_0) \leq O \leq f_0(y_0), \ A_0^1 \geq O, \ A_0^2 \geq O, \ I \geq A_0 A_0^2,$$
$$I \geq A_0^1 A_0, \ I \geq A_0 A_0^2 \ \text{and} \ I \geq A_0^2 A_0. \qquad (13.27)$$

Therefore

$$y_0 - y_1 = A_0^1 f_0(y_0) \geq O,$$
$$y_1 - x_0 = y_0 - x_0 - A_0^1 f_0(y_0) \geq y_0 - x_0 - A_0^1(f_0(y_0) - f_0(x_0))$$
$$\geq A_0^1(A_0(y_0 - x_0) - (f_0(y_0) - f_0(x_0))) \geq O \qquad (13.28)$$

$$x_1 - x_0 = -A_0^2 f_0(x_0) \geq O,$$
$$y_0 - x_1 = y_0 - x_0 + A_0^2 f_0(x_0) \geq y_0 - x_0 - A_0^2(f_0(y_0) - f_0(x_0))$$
$$\geq A_0^2(A_0(y_0 - x_0) - (f_0(y_0) - f_0(x_0))) \geq O. \qquad (13.29)$$

Hence both x_1 and y_1 belong to interval $\langle x_0, y_0 \rangle$.

From (13.16)-(13.18), (13.25), (13.26) and (13.27) we get

$$f_1(y_1) \geq f_0(y_1) = f_0(y_1) + A_0(y_0 - y_{-1} - A_0^1 f_0(y_0))$$
$$= f_0(y_1) - A_0 A_0^1 f_0(y_0) + A_0(y_0 - y_1)$$
$$\geq f_0(y_1) - f_0(y_0) + A_0(y_0 - y_1) \geq O,$$
$$f_1(x_1) \leq f_0(x_1) = f_0(x_1) - A_0(y_0, y_{-1})(x_1 - x_0 + A_0^2 f_0(x_0))$$
$$= f_0(x_1) - A_0 A_0^2 f_0(x_0) - A_0(x_1 - x_0)$$
$$\leq f_0(x_1) - f_0(x_0) - A_0(x_1 - x_0) \leq O,$$

and

$$y_0 - x_1 \geq y_1 - x_1 + A_0^1 f_1(x_1) = y_0 - x_1 + A_0^1(f_1(y_0) - f_1(x_1))$$
$$\geq A_0^1[A_0(y_0 - x_1) - (f_1(y_0) - f_1(x_1))] \geq O. \qquad (13.30)$$

Thus, we have proved $x_0 \leq x_1 \leq y_1 \leq y_0$ and

$$f_1(x_1) \leq f_0(x_1) \leq O \leq f_0(y_1) \leq f_1(y_1).$$

By induction we can easily obtain (13.21) and (13.22). Consider now $u \in [x_0, y_0]$ such that $f_k(u) = O$ $(k \geq 0)$. We have

$$y_1 - u = y_0 - u - A_0^1 f_0(y_0) + A_0^1 f_0(u)$$
$$\geq A_0^1[A_0(y_0 - u) - (f_0(y_0) - f_0(u))] \geq O,$$
$$u - x_1 = u - x_0 + A_0^2 f_0(x_0) - A_0^2 f_0(u)$$
$$\geq A_0^2[A_0(u - x_0) - (f_0(u) - f_0(x_0))] \geq O.$$

Hence, $x_1 \leq u \leq y_1$. By induction it follows $x_k \leq u \leq y_k$, which completes the proof. □

If the space B is regular then from (13.21) and (13.22) it follows that the sequences $\{x_k\}$, $\{y_k\}$, $\{f_k(x_k)\}$, $\{f_k(y_k)\}$ ($k \geq 0$) are convergent. In some cases the convergence of these sequences can follow from other conditions than regularity.

In the following theorem we provide some sufficient conditions for the convergence of iterations $\{f_k(x_k)\}$, $\{f_k(y_k)\}$ ($k \geq 0$).

Theorem 13.13 *Under hypotheses of Theorem 13.9, assume:*

(i) B_1 *is a POTL-space and* B *is a normal POTL-space;*
(ii) $x_k \to x^*$ *and* $y_k \to y^*$ *as* $k \to \infty$,
(iii) *There are two continuous nonsingular nonnegative mappings* A^1 *and* A^2 *such that* $A_k^1(y_k, y_{k-1}) \geq A^1$ *and* $A_k^2(y_k, y_{k-1}) \geq A^2$ *for sufficiently large* k.

Then

$$f_k(x_k) \to O, \quad \text{and} \quad f_k(y_k) \to O \quad \text{as} \quad k \to \infty.$$

Proof. Note first that

$$O \leq A^1 f_k(x_k) \leq A_k^1(y_k, y_{k-1}) f_k(x_k) = y_k - y_{k+1}.$$
$$x_k - x_{k+1} = A_k^2(y_k, y_{k-1}) f_k(x_k) \leq A_k^2(y_k, y_{k-1}) f_k(x_k) \leq O$$

for sufficiently large k. The normality of B implies that

$$A^1 f_k(x_k) \to O, \quad \text{and} \quad A^2 f_k(y_k) \to O \quad \text{as} \quad k \to \infty,$$

from which the result follows. □

Remark 13.12 *Instead of the algorithm* (13.25), (13.26), *we may consider, more generally, an iteration scheme of the form*

$$y_{k+1} = y_k - A_k^1(y_k, y_{k-1}) z_k^1 \quad (k \geq 0),$$
$$x_{k+1} = x_k - A_k^2(y_k, y_{k-1}) z_k^2 \quad (k \geq 0),$$

where z_k^1, z_k^2 *are arbitrarily elements satisfying the inequality*

$$f_k(x_k) \leq z_k^2 \leq O \leq z_k^1 \leq f_k(y_k) \quad (k \geq 0).$$

similar to the previous results it can be shown that under hypotheses of Theorem 13.9 this iteration is well defined and the resulting sequences satisfy (13.21) *and* (13.22). *This shows, roughly speaking, that if* $f_k(x_k)$ *is approximate from "below" and* $f_k(x_k)$ *is approximated from "above" then monotone convergence is preserved.*

This observation is important in many practical computations.

Remark 13.13 *In Theorem 13.9, we assumed that $A_k(u,v)$ $(k \geq 0)$, have nonnegative subinverses for $(u,v) \in S_3$. If we make the stronger assumption that $A_k(u;v)$ is inverse nonnegative for $(u,v) \in S_3$ then in iteration (13.25) – (13.26) $A_k^1(y_k, y_{k-1})$ and $A_k^2(y_k, y_{k-1})$ can be taken as any nonnegative right subinverses of $A_k(y_k, y_{k-1})$ $(k \geq 0)$. Note that the property that it is a left subinverse was used only in proving inequalities (13.28) – (13.30). Observing that*

$$A_0(y_1 - x_0) = A_0\left(y_0 - x_0 - A_0^1 f_0(y_0)\right)$$
$$\geq A_0(y_0 - x_0) - f_0(y_0)$$
$$\geq A_0(y_0 - x_0) - (f_0(x_0)) \geq O$$

and using the inverse nonnegativity of A_0 we deduce that $x_0 \leq y_1$.

The inequalities $x_1 \leq y_0$ and $x_1 \leq y_1$ can be proved analogously.

Remark 13.14 *Note that replacing condition (13.16) by the milder condition*

$$f_k(y) - f_k(x) \leq A_k(y,z)(y-x), k \geq 0, (x,y) \in S_1, (y,z) \in S_3 \quad (13.31)$$

we can still prove that iteration (13.19) is well defined and that iteration sequence satisfies $y_k \downarrow y^* \geq x_0$ whereas $f_k(y_k) \to O$ as $k \to \infty$. However, assumption (13.31) does not imply these properties. However by replacing (13.16) by (13.31), we can only prove that sequence (13.25) satisfies

$$x_0 \leq y_{k+1} \leq y_k \leq y_0 \quad (k \geq 0).$$

As the conclusion of this section we will now give some examples which satisfy conditions (13.16) and indicate how the general results of this section can be applied to obtain monotone convergence theorems for Newton's and secant methods.

Let us consider mappings $f_k : D \subset B \to B_1$ $(k \geq 0)$, where B and B_1 are POTL-spaces. We recall that f_k is called order-convex on an interval $\langle x_0, y_0 \rangle \subset D$ if

$$f_k(\lambda x + (1-\lambda x)y) \leq \lambda f_k(x) + (1-\lambda) f_k(y) \quad (k \geq 0) \quad (13.32)$$

for all comparable $x, y \in \langle x_0, y_0 \rangle$ and $\lambda \in [0,1]$. If B and B_1 are POTL-spaces and if f_k $(k \geq 0)$ has a linear G-derivative $f_k'(x)$ at each point $x \in$

$\langle x_0, y_0 \rangle$ then (13.32) holds if and only if

$$f_k(x)(y-x) \leq f_k(y) - f_k(x) \leq f'_k(y)(y-x) \quad (k \geq 0) \quad \text{for} \quad x_0 \leq x \leq y \leq y_0.$$

Thus, for order-convex G-differentiable mappings (13.31) is satisfied with $A_k(u, v) = f'_k(u)$. In unidimensional case (13.32) is equivalent with isotony of mapping $x \to f'_k(x)$ but in general the latter property is stronger. Assuming isotony of mapping $x \to f'_k(x)$, we have

$$f_k(y) - f_k(x) \leq f'_k(w)(y-x) \quad (k \geq 0) \quad \text{for} \quad x_0 \leq x \leq y \leq w \leq y_0$$

so, in this case condition (13.16) is satisfied for $A_k(w, z) = f'_k(w) \; (k \geq 0)$.

The above observations show that our results can be applied for the Newton method. Iteration (13.19)-(13.20) becomes

$$f_k(y_k) + f'_k(y_k)(y_{k+1} - y_k) = O, \tag{13.33}$$

$$f_k(x_k) + f'_k(y_k)(x_{k+1} - x_k) = O, \tag{13.34}$$

whereas iteration (13.25)-(13.26) becomes

$$y_{k+1} = y_k - f'_k(y_k)^{-1} f_k(y_k), \tag{13.35}$$

$$x_{k+1} = x_k - f'_k(y_k)^{-1} f_k(x_k). \tag{13.36}$$

Moreover, if in addition

$$\left\| f'_k(z)^{-1} \left(f'_k(x) - f'_k(y) \right) \right\| \leq \gamma \|x - y\|, \quad \text{for} \quad x, y, z \in \langle x_0, y_0 \rangle \tag{13.37}$$

then

$$\|y_{k+1} - x_{k+1}\| \leq .5\gamma \|y_k - x_k\|^2 \quad (k \geq 0),$$
$$\|y_{k+1} - y^*\| \leq .5\gamma \|y_k - y^*\| \quad (k \geq 0)$$

and

$$\|x_{k+1} - x^*\| \leq .5\gamma \|x_k - x^*\|^2 \quad (k \geq 0).$$

These results follow immediately by using (13.35)-(13.37), since

$$\|y_{k+1} - x_{k+1}\| = \left\| y_k - x_k - f'_k(y_k)^{-1} (f_k(y_k) - f_k(x_k)) \right\|$$

$$\left\| f'_k(y_k)^{-1} [f'_k(y_k)(y_k - x_k) - (f_k(y_k) - f_k(x_k))] \right\|$$

$$\leq .5\gamma \|y_k - x_k\|^2 \quad (k \geq 0).$$

we note that iteration (13.19)-(13.20) with $f_k = f \; (k \geq 0)$ and $A_k(u, v) = f'(u)$ is exactly the same algorithm which was proposed by Fourier in 1918, (see e.g. [183]) in the unidimensional case and was extended by Baluev [101]) in the general case.

If $f_k : [a,b] \to \mathbb{R}$ is a real mapping of a real variable then f_k $(k \geq 0)$, is order convex if and only if

$$(f_k(x) - f_k(y))(x-y)^{-1} \leq (f_k(u) - f_k(v))(u-v)^{-1}$$

for all $x, y, y, v \in [a,b]$ such that $x \leq u$ and $y \leq v$. This fact motivates the notion of convexity with respect to a divided difference discussed earlier for the case $f_k = f$ $(k \geq 0)$.

Let $f_k : D \subset B \to B_1$ be nonlinear mappings between two linear spaces B and B_1. A mapping $\delta f_k(.,.) : D \times D \to L(B, B_1)$ is called a divided difference of f_k $(k \geq 0)$ on D if

$$\delta f_k(u,v)(u-v) = f_k(u) - f_k(v) \, (k \geq 0), \quad u, v \in D.$$

If B and B_1 are topological linear spaces then linear mapping $\delta f_k(u,v)$ is supposed continuous (i.e. $\delta f_k(u,v) \in LB(B, B_1)$) Now suppose B, B_1 are two POL-spaces and assume nonlinear mapping $f_k(\cdot) : D \subset B \to B_1$ $(k \geq 0)$ has a divided difference δf_k on D $(k \geq 0)$. Then f_k $(k \geq 0)$ is called convex with respect to divided difference $\delta f_k(\cdot)$ on D if

$$\delta f_k(x,y) \leq \delta f_k(u,v) \, (k \geq 0), \quad \text{for all} \quad x, y, u.v \in D, \tag{13.38}$$

with $x \leq y$ and $y \leq v$. (13.38). Moreover mapping $\delta f_k(.,.) : D \times D \to L(B, B_1)$ $(k \geq 0)$ satisfying

$$\delta f_k(u,v)(u-v) \geq f_k(u) - f_k(v) \, (k \geq 0) \quad \text{for all comparable } u, v \in D \tag{13.39}$$

is called generalized divided difference of f_k $(k \geq 0)$ on D. If both conditions (13.38) and (13.39) are satisfied, then we say f_k $(k \geq 0)$, is convex with respect to the generalized divided difference δf_k $(k \geq 0)$. It is easily seen, that if (13.38) and (13.39) are satisfied on $D = \langle x_0, y_{-1} \rangle$ then condition (13.16) is satisfied with $A_k(u,v) = \delta f_k(u,v)$ $(k \geq 0)$. Indeed, for $x_0 \leq x \leq y \leq w \leq z \leq y_{-1}$, we have

$$\delta f_k(x,y)(y-x) \leq f_k(y) - f_k(x) \leq \delta f_k(y,x)(y-x)$$
$$\leq \delta f_k(w,z)(y-x).$$

That is, our results can be applied also for secant method.

In what follows we reformulate list two fixed point theorems which hold in arbitrary complete lattices. These theorems are due to Tarski [261].

The first theorem provides sufficient conditions for the existence of a fixed point of mapping $f : S \to S$ where S is a nonempty set. The second theorem provides sufficient conditions for the existence of a common fixed point x^* of a sequence $f_k : S \to S$ ($k \geq 0$) of mappings.

We will need some definitions:

Definition 13.5 By a *lattice* we mean a system $Q = \{S, \leq\}$ formed by a nonempty set S and a binary relation $\leq$; it is assumed that $\leq$ establishes a partial order in S and that for any two elements $a, b \in S$ there is a least upper bound (join) $a \cup b$ and a greatest lower bound (meet) $a \cap b$. The relations $\geq$, $<$ and $>$ are defined in the usual way in terms of $\leq$.

Definition 13.6 The lattice $Q = \{S, \leq\}$ is called *complete*, if every subset S_1 of S has a least upper bound $\cup S_1$ and a greatest lower bound $\cap S_1$. such a Lattice has in particular two elements 0 and 1 defined by the formulas

$$0 = \cap S \quad \text{and} \quad 1 = \cup S.$$

Given any two elements $a, b \in S$ with $a \leq b$, we denote by $[a, b]$ the interval with the end points a and b, that is the set of all elements $x \in S$ for which $a \leq x \leq b$; in symbols $[a, b] = E_x [x \in S \text{ and } a \leq x \leq b]$. system $\{[a, b], \leq\}$ is clearly a lattice; it is complete if Q is complete.

We consider functions f on S to S and, more generally on a subset S_1 of S to another subset S_2 of S. Such a function f is called increasing if, for any elements $x, y \in S_1$, $x \leq y$ implies $f(x) \leq f(y)$. Note that this assumptions is the same as isotony.

We can now present the following theorem whose proof can be found for example in Tarski, [261].

Assume that

(B$_1$) (i) $Q = \{S, \leq\}$ is a complete lattice;
(B$_2$) f is an increasing function on S to S;
(B$_3$) P is the set of all fixed points of f.

Theorem 13.14 *Assume conditions (B$_1$)-(B$_3$) are satisfied*

Then set P is not empty and system $\{P, \leq\}$ is a complete lattice. In particular,

$$\cup P = \cup E_x [f(x) \geq x] \in P$$

and
$$\cup P = \cap E_x\left[f(x) \leq x\right] \in P.$$

By the above theorem, the existence of a fixed point for every increasing function is a necessary condition for the completeness of a lattice. The question arises whether this conditions is also sufficient. It has been shown that the answer to this question is affirmative (see [261]).

A set W of functions is called commutative if

(i) All functions of W have a common domain, say, S_1 and the ranges of all functions of W are subsets of S_1;
(ii) For any $f, g \in W$,
$$f(g(x)) = g(f(x)) \quad \text{for all} \quad x \in S_1.$$

Assume that
(C_1) (i) $Q = \langle S, \leq \rangle$ be a complete Lattice;
(C_2) W is any commutative set of increasing functions on S to S;
(C_3) P is the set of all common fixed points of all functions $f \in W$.

We can provide the following.

Theorem 13.15 *Assume condition (C_1)-(C_3) are satisfied.*

Then set P is not empty and the system $\{P, \leq\}$ is a complete Lattice. In particular, we have
$$\cup P = \cup E_x\left[f(x) \geq x \quad \text{for every} \quad f \in W\right] \in P$$
and
$$\cap P = \cap E_x\left[f(x) \leq x \quad \text{for every} \quad f \in W\right] \in P.$$

The proof of this theorem is found also in Tarski [261], and it can be used in connection with the theorems of the previous section. In particular all monotone convergence methods introduced in the previous sections can be used to approximate fixed points x^* of mappings f_k $(k \geq 0)$, whose existence is guaranteed under hypotheses of the above theorems.

13.3 Convergence Methods and Point to Point Mappings

This chapter examines conditions for the convergence of special single-step methods generated by point-to-point mappings. The speed of convergence is also examined using the theory of majorants. In particular, we assume that X is a Banach space,

$$U = U(0, R) \subseteq X.$$

We will consider Newton's method

$$x_{k+1} = g_k(x_k) \quad (k \geq 0) \tag{13.40}$$

where

$$g_k(x) = x - f'_k(x)^{-1} f_k(x) \quad (k \geq 0)$$

with $f_k : U \to X$ continuously differentiable on X, and a point $x^* = 0$ is a common fixed point of functions g_k.

Note that this method is a special case of the general single step algorithm examined in Chapter 2 [68], and hence, we can directly apply the general convergence conditions. For example, note that $g'_k(0) = 0$, and therefore under continuity assumptions on the derivative we know that in a neighborhood of 0, $\|g'_k(x)\|$ is small, which according to Theorem 2.9 [68] implies local convergence. However, based on the special structure of the method more advanced convergence conditions can be derived, and better error estimates can be presented than those obtained from the general theory especially in Chapter 5 in [68]. Therefore, this advanced theory is introduced here.

Using the majorant theory, we provide sufficient conditions for the convergence of the Newton method (13.40) to 0. We also show when $f_k = f$ ($k \geq 0$), the Potra-Pták [240] estimates are simple consequences of the classical majorant method due to Kantorovich and Akilov [183]. Moreover, we provide error estimates for the speed of convergence of process (13.40).

Condition (K): We assume that for all k ($k \geq 0$), f_k is Fréchet differentiable at every point at the interior of $U(0, R)$, $R > 0$, $f'_k(0)$ is invertible, and the following condition is satisfied for each fixed $r \in [0, R]$

$$\left\| f'_k(0)^{-1} \left(f'_k(v) - f'_k(w) \right) \right\| \leq h(r) \|v - w\|, \quad \text{for all} \quad v, w \in U(0, r), \tag{13.41}$$

and

$$\left\| f'_k(0)^{-1} \left[f'_k(v) - f'_k(0) \right] \right\| \leq h_0(r) \|v\| \tag{13.42}$$

where $h, h_0 : [0, R] \to \mathbf{R}_+$ are nondecreasing functions.

Let $x_0 \in U(0, R)$. It is convenient to define the constant a by

$$\|x_0\| \leq a \qquad (13.43)$$

and the functions

$$\omega(r) = \int_0^r h(s)\,ds, \quad w_0(r) = \int_0^r h_0(s)\,ds \qquad (13.44)$$

and

$$x(r) = (w_0(r) + w(r))\,r - \int_0^r \omega(s)\,ds - r. \qquad (13.45)$$

Later developments will require the following lemma.

Lemma 13.1 *Let g be a Fréchet-differentiable function which is defined on $U(0, R)$ with values in B, such that $g'(0)^{-1}$ is invertible, and which for each fixed $r \in [0, R]$ satisfies a Lipschitz condition*

$$\left\|g'(0)^{-1}(g'(v) - g'(w))\right\| \leq d(r)\,\|v - w\|, \quad \text{for all} \quad v, w \in U(0, r) \qquad (13.46)$$

with some nondecreasing function $d : [0, R] \to \mathbf{R}_+$. Then

$$\left\|g'(0)^{-1}(g'(x + h) - g'(x))\right\| \leq G(r + \|h\|) - G(r), \quad \text{for all} \quad x \in U(z, r),$$
$$\text{and} \quad \|h\| \leq R - r \qquad (13.47)$$

where

$$G(r) = \int_0^r d(s)\,ds. \qquad (13.48)$$

Proof. Let $x \in U(0, R)$ and $\|h\| \leq R - r$. Then, by hypothesis (13.46), for any positive integer m,

$$\left\|g'^{-1}(0)(g'(x + h) - g'(x))\right\| \leq \sum_{j=1}^m \left\|g'^{-1}(0)(g'(x + jm^{-1}h) - g'(x + m^{-1}(j-1)h))\right\|$$
$$\leq \sum_{j=1}^m d(r + m^{-1}j\|h\|)\,m^{-1}\|h\|.$$

Letting m tend to infinity, by the monotonicity of d and the definition of the Riemann integral (13.48), we get estimate (13.47), which completes the proof. □

We can now formulate the main result.

Theorem 13.16 *Under condition* (K), *assume that* $x_0 \in B$ *and* $R > 0$ *exist such that* 0 *is the unique zero of function* $x(r)$ *given by* (13.45) *in* $[0, R]$. *Moreover, assume that* $a \leq R$ *and* $x(r) < 0$. *Iterates generated by relation* (13.40) *are well defined for all* k, *belong to* $U(0, R)$, *and converge to* 0 *with*

$$\|x_k\| \leq \rho_k \quad (k \geq 0), \tag{13.49}$$

where sequence ρ_k, *which is monotonically decreasing and converges to* 0, *is defined by the recursive formula*

$$\rho_{k+1} = \rho_k + x(\rho_k)(1 - w_0(\rho_k))^{-1} \quad (k \geq 0), \quad \rho_0 = R. \tag{13.50}$$

Proof. The sequence generated by relation (13.50) is monotonically decreasing and converges to 0. since 0 is the unique zero of function $x(r)$ in $[0, R)$ and $x(R) < 0$, it follows that

$$x(r) < 0 \quad \text{for all} \quad r \in (0, R]. \tag{13.51}$$

By relations (13.45) and (13.51) we obtain that for all $r \in (0, R]$,

$$0 \leq r\omega(r) - \int_0^r \omega(s)\,ds < r(1 - w_0(r)).$$

From the above inequality it follows

$$1 - w_0(r) > 0 \quad \text{for all} \quad r \in (0, R]. \tag{13.52}$$

Relations (13.51) and (13.52) imply that sequence ρ_k is monotonically decreasing. Iteration (13.50) can also be written as

$$\rho_{k+1} = \left[\omega(\rho_k)\rho_k - \int_0^{\rho_k} \omega(s)\,ds\right](1 - w_0(\rho_k))^{-1} \geq 0 \quad (k \geq 0). \tag{13.53}$$

hence, we showed that

$$0 \leq \rho_{k+1} \leq \rho_k \quad (k \geq 0),$$

from which it follows that a $\rho^* \in [0, R]$ exists such that $\rho_k \to \rho^*$ as $k \to \infty$. However, from iteration (13.50) and the uniqueness of 0 as a zero of $x(r)$ in $[0, R)$ we obtain $\rho^* = 0$.

By induction on k we will show estimate (13.49). For $k = 0$, estimate (13.49) becomes $\|x_0\| \leq \rho_0 = R$ which is true, since $a \leq R$ by hypothesis. suppose estimate (13.49) holds for $i \leq k$. Since

$$f'_k(x_k) = f'_k(x_k) - f'_k(0) + f'_k(0)$$
$$= f'_k(0)\left[I + f'_k(0)^{-1}(f'_k(x_k) - f'_k(0))\right],$$

from relation (13.42) and (13.47) we obtain

$$\left\|f'_k(0)^{-1}(f'_k(x_k) - f'_k(0))\right\| \leq \omega(\rho_k) < 1,$$

where relation (13.52) is used in the last step.

By using the Banach lemma on invertible operators we conclude that $f'_k(x_k)$ is invertible and

$$\left\|f'_k(x_k)^{-1}f'_k(0)\right\| \leq (1 - \omega(r_k))^{-1} \quad (k \geq 0). \tag{13.54}$$

Using relations (13.40)-(13.42), (13.47), (13.53) and (13.54) we find

$$\|x_{k+1}\| = \left\|\left(f'_k(x_k)^{-1}f'_k(0)\right)f'_k(0)^{-1}\int_0^1(f'_k(tx_k) - f'_k(x_k))x_k dt\right\|$$
$$\leq \int_0^1(\omega(\rho_k) - \gg w(t\rho_k))\rho_k dt(1 - \omega(\rho_k))^{-1} = \rho_{k+1}. \tag{13.55}$$

Thus, estimate (13.49) holds for $k+1$. By estimate (13.55) it follows that $x_{k+1} \in U(0, R)$. Finally, by letting $k \to \infty$ in estimate (13.55) we get $x_k \to 0$ as $k \to \infty$, which completes the proof. □

Remark 13.15 (a) The above theorem can be illustrated by the special case $h(r) = h_0(r) = h$ for all $r \in [0, R)$. Then using relations (13.44), (13.45) and (13.50) we obtain

$$w(r) = hr,$$
$$x(r) = 0.5(3hr - 2)r$$

and

$$\rho_{k+1} = 0.5h\rho_k^2(1 - h\rho_k)^{-1} \quad (k \geq 0).$$

By finite induction it is easy to show that

$$\rho_{k+1} = p(\rho_0)^{2^k}\left[1 - p(\rho_0)^{2^k-1}\right]^{-1} \quad (k \geq 0),$$

where
$$p(r) = \left[(h^2 + 8h(1-hr))^{0.5} - h\right][4(1-hr)]^{-1}.$$

Moreover, the hypotheses of Theorem 13.16 are satisfied if

$$3ah < 2, \quad \text{and} \quad a \le R < 2(3h)^{-1} \tag{13.56}$$

(b) The following estimate holds in general

$$h_0(r) \le h(r) \quad \text{for all} \quad r \in [0, R]. \tag{13.57}$$

Moreover we have shown that $\frac{h(r)}{h_0(r)}$ can be arbitrarily large. Assume again $h(r) = h$ and $h_0(r) = h_0$ with $h_0 < h$. Then conditions (13.56) are weakened since they become

$$(2h_0 + h)a < 2 \quad \text{and} \quad a \le R < \frac{2}{2h_0 + h}, \tag{13.58}$$

respectively. Furthermore, scalar sequence ρ_k converges faster to zero.

The following is a simple consequence of Theorem 13.16.

Theorem 13.17 *Using condition* (K), *assume that functions* $h(r) = h$, $h_0(r) = h_0$ *and conditions (13.56) or (13.58) are satisfied. Then the iterates generated by (13.40) are well defined for all* $k \ge 0$, *belong to* $U(0, R)$, *and converge to* 0 *with*

$$\|x_k\| \le \rho_k \quad (k \ge 0),$$

where the sequence ρ_k, *which is monotonically decreasing and converges quadratically to* 0.

As a simple application for Theorem 13.17 we consider the following example.

Example 13.1 Let $X = \mathbf{R}$ and consider iterates (13.40) where

$$f_k(x) = (k+1)(k+2)^{-1}x^2 + x \quad (k \ge 0).$$

Obviously $x^* = 0$ is a common fixed point of functions (13.40). It is easy to see that

$$\left\|f'_k(0)^{-1}(f'_k(x) - f'_k(y))\right\| = 2(k+1)(k+2)^{-1}\|x-y\|$$
$$\le 2\|x-y\| \quad (k \ge 0).$$

By setting $h(r) = h_0(r) = h = 2$, conditions (13.41) and (13.42) are satisfied, whereas conditions (13.56) become, respectively,

$$a < 3^{-1}$$

and

$$a \leq R < 3^{-1}.$$

Choose $x_0 = 0.25 = R$ and $a = \|x_0\| = 0.25$. Hypotheses of Theorem 13.17 are now satisfied with the above values. Moreover, iterates (13.40) become

$$x_{k+1} = x_k - \left(2(k+1)(k+2)^{-1} x_k + 1\right)^{-1} \left((k+1)(k+2)^{-1} x_k^2 + x_k\right),$$

$$x_0 = 0.25 \quad (k \geq 0),$$

and are dominated (in absolute value) by iterates ρ_k for $\rho_0 = R$ which converge quadratically to 0 as $k \to \infty$.

In the following part of this section we give some natural conditions under which sequence $f_k(x_k)$ ($k \geq 0$) converges to 0 as $k \to \infty$.

Theorem 13.18 *Under hypotheses of Theorem 13.16, assume that the norms of the derivatives $f_k'(x_k)$ are uniformly bounded above by some $b > 0$. Sequence $f_k(x_k)$ then converges to 0 as $k \to \infty$ and*

$$\|f_k(x_k)\| \leq b(\|x_{k+1}\| + \|x_k\|) \leq b(\rho_{k+1} + \rho_k) \quad (k \geq 0).$$

Proof. Using relations (13.40), (13.49), and the hypothesis we obtain

$$\|f_k(x_k)\| = \|f_k'(x_k)(x_k - x_{k+1})\| \leq b \|x_k - x_{k+1}\| \leq b(\|x_k\| + \|x_{k+1}\|)$$
$$\leq b(\rho_k + \rho_{k+1}),$$

which implies that $f_k(x_k) \to 0$ as $k \to \infty$. □

Remark 13.16 *The main advantage of our approach is the fact that the convergence problem of approximating a fixed point x^* of equation (13.40) under very general conditions can be reduced to the solution of a simple scalar equation $x(r) = 0$ on interval $[0, R)$, which can be carried out by completely elementary methods of classical analysis.*

We now provide further sufficient conditions for the convergence of iterations of the form (13.40). These results are analogous to those discussed earlier in the chapter, but are more applicable in certain practical cases.

Let $x_0 \in B$. We assume that (L): linear mappings $f'_k(x_0)$ are invertible on $U(x_0, R)$ for some $R > 0$ and that the following conditions are satisfied for all $k \geq 0$, $x, y \in U(x, r) \subseteq U(x, R)$

$$\left\| f'_k(x_0)^{-1} (f_k(x) - f_{k-1}(x)) \right\| \leq v_1(\|x - x_0\|) \tag{13.59}$$

and

$$\left\| f'_k(x_0)^{-1} (f'_{k-1}(x) - f'_{k-1}(y)) \right\| \leq v_2(r) \|x - y\|. \tag{13.60}$$

Here, $v_1(r)$ and $v_2(r)$ are nondecreasing functions on interval $[0, R]$.

It is convenient to define function

$$a(r) = \int_0^r v_2(s)\, ds \tag{13.61}$$

and iteration

$$\rho_{k+1} = \rho_k + [v_1(\rho_k) + a(\rho_k) - a(\rho_{k-1})$$
$$- v_2(\rho_{k-1})(\rho_k - \rho_{k-1})](1 - a(\rho_k))^{-1},$$
$$k \geq 1, \quad \rho_0 = 0, \quad 0 \leq \rho_1 < R. \tag{13.62}$$

As in Theorem 13.16, we can find several alternative sufficient conditions for the convergence of sequence ρ_k to some $\rho^* \in [0, R]$. For example, if the sequence is bounded above by R and $1 - a(R) > 0$, then it is monotonically increasing. hence, it converges to some $\rho^* \in [0, R]$.

Using the approximation

$$x_{k+1} - x_k = -f'_k(x_k)^{-1} f'_k(x_0) f'_k(x_0)^{-1} [f_k(x_k) - f_{k-1}(x_k)$$
$$+ f_{k-1}(x_k) - f_{k-1}(x_{k-1}) - f'_{k-1}(x_{k-1})(x_k - x_{k-1})],$$

and relations (13.47) and (13.59) through (13.62), by following the steps of Theorem 13.16, we can show that the following alternative convergence theorem holds.

Theorem 13.19 *Under condition (L), assume that an $x_0 \in B$ and a positive $R > 0$ exist such that*

(a) ρ_1, ρ_2 *in iteration (13.62) can be chosen such that* $\|x_1 - x_0\| \leq \rho_1 \leq \rho_2 < R$;

(b) *Sequence ρ_k given by iteration (13.62) is bounded above by R, and $1 - a(R) > 0$.*

Then

(i) Sequence ρ_k is monotonically increasing and converges to some $\rho^* \in [0, R]$;

(ii) Moreover iteration (13.40) is well defined for all $k \geq 0$, remains in $U(x_0, R)$ and converges to some $x^* \in U(x_0, R)$ such that

$$\|x_{k+1} - x_k\| \leq \rho_{k+1} - \rho_k \quad (k \geq 0)$$

and

$$\|x_k - x^*\| \leq \rho^* - \rho_k \quad (k \geq 0).$$

Remark 13.17 Consider the special case of the stationary process, and assume that f and f'^{-1} are continuous in the neighborhood of x^*. Then by letting $k \to \infty$ in recursion

$$x_{k+1} = x_k - f'(x_k)^{-1} f(x_k)$$

we get

$$f'(x^*)^{-1} f(x^*) = 0,$$

which implies that $f(x^*) = 0$.

The computation of the iterates generated by recursion (13.40) requires the evaluation of the inverses of the linear operator $f'_k(x_k)$ at each step. In many applications it is more convenient to evaluate only the inverses of the linear operators $f'_k(x_0)$ at each step. For example, if the process is stationary, $f'(x_0)^{-1}$ must be evaluated only once. We therefore consider the modified Newton iterates.

$$x_{k+1} = x_k - f'_k(x_0)^{-1} f_k(x_k) \quad (k \geq 0). \tag{13.63}$$

Our convergence theorem is based on the sequence $\{s_k\}$ defined as

$$\rho_{k+1} = \rho_k + [v_1(\rho_k) + a(\rho_k) - a(\rho_{k-1}) - v_2(\rho_{k-1})(\rho_k - \rho_{k-1})],$$
$$\rho_0 = 0, \quad 0 \leq \rho_1 < R, \quad k \geq 1. \tag{13.64}$$

By using the approximation

$$x_{k+1} - x_k = -f'_k(x_0)^{-1} [f_k(x_k) - f_{k-1}(x_k) + f_{k-1}(x_k) - f_{k-1}(x_{k-1}) \\ - f'_{k-1}(x_0)(x_k - x_{k-1})]$$

in a way similar to Theorem 13.16 we can prove the following result.

Theorem 13.20 Under condition (L), assume $x_0 \in B$ and $R > 0$ exist with:

(a) ρ_1, ρ_2 in iteration (13.64) can be chosen such that $\|x_1 - x_0\| \leq \rho_1 \leq \rho_2 < R$;
(b) Sequence ρ_k given by iteration (13.64) is bounded above by R.

Then

(i) Sequence ρ_k is monotonically increasing and converges to some $\rho^* \in [0, R]$;
(ii) Iteration (13.63) is well defined for all $k \geq 0$, remains in $U(x_0, R)$ and converges to some $x^* \in U(x_0, R)$. Furthermore,

$$\|x_{k+1} - x_k\| \leq \rho_{k+1} - \rho_k \quad (k \geq 0)$$

and

$$\|x_k - x^*\| \leq \rho^* - \rho_k \quad (k \geq 0).$$

As a conclusion of this section a new modification of Newton's method is analyzed.

Let us now define the following iterates

$$x_{k+1} = x_k - f_k'(0)^{-1} f_k(x_k) \quad (k \geq 0). \tag{13.65}$$

We can then show exactly as in Theorem 13.16 the following.

Theorem 13.21 *Assume $x_0 \in B$ and $R > 0$ exist such that 0 is the unique zero of the function $x(r) = \int_0^r w_0(s)\, ds - r$ in $[0, R]$. Moreover, assume $a \leq R$ and $x(R) < 0$. Then iterates generated by (13.65) belong to $U(0, R)$ and converge to 0 with*

$$\|x_k\| \leq q_k \quad (k \geq 0),$$

where sequence q_k which is monotonically decreasing and converges to 0 is defined by the recursive formula

$$q_{k+1} = q_k + x(q_k) \quad (k \geq 0), \quad q_0 = R.$$

Remark 13.18 *The above result can be useful, especially when the functions f_k satisfy the autonomous differential equations*

$$f_k'(x) = T_k(f_k(x)) \quad (k \geq 0),$$

$f_k'(0)$ can be evaluated without knowing the true value of the solution x^* ($x^* = 0$ was selected in the above theorem).

So f or we assumed that mappings f_k are differentiable at every point $x \in U(0, R)$. Sometimes this condition may not hold. In many such cases mapping f_k can be decomposed as

$$f_k(x) = f_{1k}(x) + f_{2k}(x),$$

where f_{1k} is differentiable, whereas the differentiability of f_{2k} is not assumed for all $k \geq 0$. Assume that for all k ($k \geq 0$) and each fixed $r \in [0, R]$, the following conditions are satisfied:

(M) $\quad \|f'_{1k}(0)^{-1}(f'_{1k}(v) - f'_{1k}(w))\| \leq h(r)\|v - w\|, \quad$ for all $\quad v, w \in U(0, r)$
$\quad \|f'_{1k}(f'_{1k}(v) - f'_{1k}(0))\| \leq h_0(r)\|v\|,$

and

$$\left\|f'_{1k}(0)^{-1}(f_{2k}(v) - f_{2k}(w))\right\| \leq h_1(r)\|v - w\|, \quad \text{for all} \quad v, w \in U(0, r)$$

where $h_0, h, h_1 : [0, R] \to \mathbf{R}_+$ are nondecreasing functions.

Define functions

$$\omega_1(r) = \int_0^r h_1(r)\,ds,$$

$$x_1(r) = (w_0(r) + w(r))r + \int_0^r \omega_1(s)\,ds - \int_0^r w(s)\,ds - r.$$

Finally, let us consider the more general process

$$x_{k+1} = x_k - f'_{1k}(x_k)^{-1}(f_{1k}(x_k) + f_{2k}(x_k)). \tag{13.66}$$

Then, following the lines of the proof of Theorem 13.16 we can show that the following theorem holds.

Theorem 13.22 *Under condition (M), assume $x_0 \in B$ and $R > 0$ exist such that 0 is the unique zero of function $x_1(r)$ in $[0, R)$. Moreover, assume $a \leq R$ and $x_1(R) < 0$. Then iterates generated by iteration (13.66) are well defined for all k, belong to $U(0, R)$, and convergence to 0 with*

$$\|x_k\| \leq \rho_k \quad (k \geq 0),$$

where sequence q_k which is monotonically decreasing and converges to 0, is defined by the recursive formula

$$\rho_{k+1} = \rho_k + x_1(\rho_k)(1 - \omega_0(\rho_k))^{-1} \quad (k \geq 0), \quad \rho_0 = R.$$

Next, the modified scheme

$$x_{k+1} = x_k - f'_{1k}(0)^{-1} (f_{1k}(x_k) + f_{2k}(x_k)) \quad (k \geq 0) \qquad (13.67)$$

is examined. Similarly to Theorem 13.21 we can show that the following result holds.

Theorem 13.23 *Under* (M), *except the* (13.64) *condition assume* $x_0 \in B$, $R > 0$ *exist such that* 0 *is the unique zero of function* $x_1(r)$ *in* $[0, R)$. *Further, assume* $a \leq R$ *and* $x_1(R) < 0$. *Then iterates generated by iteration* (13.67) *belong to* $U(0, R)$ *and converge to* 0 *with*

$$\|x_k\| \leq q_k \quad (k \geq 0),$$

where sequence q_k *which is monotonically decreasing and converges to* 0, *is defined by the recursive formula*

$$q_{k+1} = q_k + x_1(q_k) \quad (k \geq 0), \quad q_0 = R.$$

Note that in this case w is w_0 in the definition of function x_1.

In many interesting cases the inverses $f'_k(x_k)^{-1}$, $f'_k(0)^{-1}$ $(k \geq 0)$ are very difficult, expensive, or impossible to find. In such cases it is useful to consider two Newton-like iterates

$$x_{k+1} = x_k - A_k(x_k)^{-1} (f_{1k}(x_k) + f_{2k}(x_k)) \qquad (13.68)$$

for approximating a common root $x^* = 0$ of equations $f_k(x) = 0$ with

$$f_k(x) = f_{1k}(x) + f_{2k}(x) \quad (k \geq 0), \qquad (13.69)$$

where $A_k(x_k)$ denotes a linear mapping approximating the Fréchet derivative f'_{1k} of $f_{1k}(\cdot)$ at $x \in U(0, R)$. We assume that for all $k, k \geq 0$ and each fixed $r \in [0, R]$,

(N) $A_k(0)^{-1}$ exists, and for all $x, y \in U(0, r)$,

$$\left\| A_k(0)^{-1} (A_k(x) - A_k(0)) \right\| \leq w_0(\|x\|) + b, \qquad (13.70)$$

$$\left\| A_k(0)^{-1} (f'_{1k}(tx) - A_k(x)) \right\| \leq w(t\|x\|) + c, \quad t \in [0, 1] \qquad (13.71)$$

and
$$\left\|A_k(0)^{-1}(f_{2k}(x) - f_{2k}(y))\right\| \le e(r)\|x-y\|, \qquad (13.72)$$

where w_0, w, and e are nondecreasing, non-negative functions, and constants b, c are selected so that $b \ge 0$, $c \ge 0$ and $b + c < 1$. Note that the differentiability of f_{2k} is not assumed here.

The above conditions are more general than those considered in Theorems 13.16 and 13.22. Argyros (1987, 1988a, b), Zincenoko (1963), Rheinboldt (1968), Zabrejko and Zlepko (1987), Dennis (1971); and Zabrejko and Nguen (1987) have considered some special cases of the above conditions when $f_k = f$ $(k \ge 0)$. They provided sufficient conditions for the convergence of the Newton-like iterates (13.68) to 0 in this special case. We will proceed in a similar manner, but for the more general case described above.

Define now the functions
$$x^*(r) = \int_0^r w(s)\,ds + \int_0^r e(r)\,ds + (b + c - 1 + w_0(r))\,r, \qquad (13.73)$$

and
$$g(r) = 1 - b - w_0(r) \quad \text{for all} \quad r \in [0, R). \qquad (13.74)$$

Introduced the difference equation
$$\rho_{k+1} = \rho_k + x^*(\rho_k)\,g(\rho_k)^{-1} \quad (k \ge 0), \quad \rho_0 = R. \qquad (13.75)$$

In this more general case Theorem 13.16 can be modified as follows.

Theorem 13.24 *Under condition (N), assume $x_0 \in B$ and $R > 0$ exist such that 0 is the unique zero of function $x^*(r)$ given by relation (13.73) in $[0, R)$. Moreover, suppose $\|x_0\| \le a \le R$ and $x^*(R) \leqq 0$. Iterates generated by iteration (13.68) are well defined for all $k \ge 0$, belong to $U(0, R)$, and converge to 0 with*
$$\|x_k\| \le \rho_k \quad (k \ge 0), \qquad (13.76)$$

where sequence ρ_k which is monotonically decreasing and converges to 0 is given by iteration (13.75).

Proof. We will first show that the sequence generated by iteration (13.75) is monotonically decreasing and converges to 0. Because 0 is the unique zero of function $x^*(r)$ in $[0, R)$ and $x^*(R) \le 0$,
$$x^*(r) < 0 \quad \text{for all} \quad r \in [0, R). \qquad (13.77)$$

By using relation (13.73) we get

$$0 \leq \int_0^r w(s)\,ds + \int_0^r e(r)\,ds < (1 - b - c - w_0(r))r,$$

which implies that

$$g(r) > 0 \quad \text{for all} \quad r \in [0, R). \tag{13.78}$$

Using relations (13.75), (13.77), and (13.78) and finite induction we can routinely show that sequence ρ_k is monotonically decreasing. Furthermore, iteration (13.75) can also be written as

$$\rho_{k+1} = \left[\int_0^{\rho_k} w(s)\,ds + e(\rho_k)\rho_k + c\rho_k\right] g(\rho_k)^{-1} \geq 0 \quad \text{for all} \quad k \geq 0, \tag{13.79}$$

which implies that

$$0 \leq \rho_{k+} \leq \rho_k \quad (k \geq 0).$$

Hence, a $\rho^* \in [0, R)$ exists, with $\rho_k \to \rho^*$ as $k \to \infty$. Note that from iteration (13.75) and the uniqueness of 0 as a zero of $x^*(r)$ in $[0, R)$ we conclude that $\rho^* = 0$.

By induction of k we will show that estimate (13.76) holds. For $k = 0$, estimate (13.76) becomes $\|x_0\| \leq \rho_0 = R$, which is true, since $a \leq R$ by hypothesis. Assume estimate (13.76) holds for k. From relations (13.70) and (13.78) we get

$$\left\| A_k(0)^{-1}(A_k(x_k) - A_k(0)) \right\| \leq w_0(\rho_k) + b < 1.$$

By the Banach lemma on invertible mappings $A_k(x_k)$ is invertible. By using identity

$$A_k(x_k) = A_k(0)\left[I + A_k(0)^{-1}(A_k(x_k) - A_k(0))\right],$$

we see that

$$\left\|(A_k(x_k)^{-1} A_k(0))\right\| \leq g(\rho_k)^{-1} \quad (k \geq 0). \tag{13.80}$$

Using relations (13.68), (13.69) (for $x = 0$), and (13.70) through (13.75), the triangle inequality and the induction hypothesis we get

$$\|x_{k+1}\| = \left\|\left[A_k(x_k)^{-1} A_k(0)\right]\left\{\int_0^1 A_k(0)^{-1}\left[((f'_{1k}(tx_k) - A_k(x_k)), x_k dt\right.\right.\right.$$
$$\left.\left.\left. + (f_{2k}(x_k) - f_{2k}(0)))\right]\right\}\right\|$$
$$\leq \left(\int_0^1 [w(t\|x_k\|) + c + e(\|x_k\|)]\|x_k\| dt\right) g(\|x_k\|)^{-1}$$
$$\leq \left(\int_0^{\rho_k} w(s)\,ds + \int_0^{\rho_k} e(\rho_k)\,ds + c\rho_k\right) g(\rho_k)^{-1} = \rho_{k+1}. \quad (13.81)$$

Hence, estimate (13.76) holds for $k+1$. From relation (13.81) we conclude that $x_{k+1} \in U(0, R)$. Finally, by letting $k \to \infty$ in estimate (13.81) we get $x_k \to 0$, which completes the proof. $\square$

The analysis for the speed of convergence of iteration (13.76) is similar to that presented in iteration (13.56). In particular cases we can select $A_k(x_k)$ to be either $f'_{1k}(x_k)$ or $f'_{1k}(x_0)$ of $f'_{1k}(0)$ or $S_k(x_{k-1}, x_k)$ (secant mappings) or any other linear mappings satisfying relations (13.70) to (13.72).

Case studies

Case Study 1. We illustrate Theorem 13.17 by solving the following two-point boundary value problem for all $k \geq 0$

$$y''_k + \frac{k+1}{k+2} y_k^2 = 0, \quad y_k(0) = y_k(1) = 0, \quad \text{for all } k \geq 0.$$

Remark 13.19 *Note that the actual majorizing sequence is given by (13.81). A direct convergence analysis of this iteration lends to weaker convergence conditions, and finer error bounds (see earlier Chapers for point to point mappings.*

We divide the interval $[0, 1]$ into n subintervals and we let $l = (1)/(n)$. We denote the points of subdivision by $v_0 = 0 < v_1 < \ldots < v_n = 1$, with the corresponding values of the function $y_{0,k} = y_k(v_0)$, $y_{1,k} = y_k(v_1), \ldots, y_{n,k} = y_k(v_n)$. A simple approximation of the second derivative at these points is

$$y''_k = \frac{y_{i-1,k} - 2y_{i,k} + y_{i+1,k}}{l^2}, \quad i = 1, 2, \ldots, n-1.$$

Noting that $y_{0,k} = 0$ and $y_{n,k} = 0$ this leads to the following system of nonlinear equations

$$\frac{k+1}{k+2}I^2 y_{1,k}^2 - 2y_{1,k} + y_{2,k} = 0,$$

$$y_{i-1,k} + \frac{k+1}{k+2}I^2 y_{i,k}^2 - 2y_{i,k} + y_{i+1,k} = 0, \quad i = 2, 3, \ldots, n-1$$

$$y_{n-2,k} + \frac{k+1}{k+2}I^2 y_{n-1,k}^2 - 2y_{n-1,k} = 0. \tag{13.82}$$

The left-hand sides define an operator $f_k : \mathbf{R}^{n-1} \to \mathbf{R}^{n-1}$ for all $k \geq 0$ whose Fréchet derivative may be written as

$$f_k'(y) = \begin{pmatrix} 2\frac{k+1}{k+2}I^2 y_{1,k} - 2 & 1 & 0 \cdots & & 0 \\ 1 & 2\frac{k+1}{k+2}I^2 y_{2,k} - 2 & 1 \cdots & & \cdot \\ 0 & \cdot & \cdots & & \cdot \\ \cdot & \cdot & \cdots & & \cdot \\ \cdot & \cdot & \cdots & & 1 \\ 0 & \cdot & \cdots 1 & 2\frac{k+1}{k+2}I^2 y_{n-1,k} - 2 \end{pmatrix}$$

Let $x \in \mathbf{R}^{n-1}$ and select the norm

$$\|x\| = \max_{1 \leq j \leq n-1} |x_j|.$$

The corresponding matrix norm is

$$\|A\| = \max_{1 \leq j \leq n-1} \sum_{k=1}^{n-1} |a_{jk}|,$$

where a_{jk} is the (j, k) element of matrix A.

Then for all $\mathbf{y}, \mathbf{x} \in \mathbf{R}^{n-1}$ and $k \geq 0$,

$$\|f_k'(x) - f_k'(y)\| = \left\| \operatorname{diag}\left\{ 2\frac{k+1}{k+2}I^2 (x_i - y_i) \right\} \right\|$$

$$= 2\frac{k+1}{k+2}I^2 \|x - y\|.$$

Let $n = 4$, and since a solution must vanish at the end points and be positive in the interior, a reasonable choice of initial approximation seems to be $x_0 = [0, 0.1, 0.1, 0]^T$. Notice that

$$f_k'(0) = \begin{bmatrix} -2 & 1 & 0 \\ 1 & -2 & 1 \\ 0 & 1 & -2 \end{bmatrix},$$

and

$$f'_k(0)^{-1} = -\frac{1}{4}\begin{bmatrix} 3 & 2 & 1 \\ 2 & 4 & 2 \\ 1 & 2 & 3 \end{bmatrix}.$$

Since $a = 0.1$, for $h \leq 0.025$, the condition (13.57) is now satisfied. The conclusions of the theorem now apply for R satisfying condition (13.58). Note also that $x^* = (0, 0, 0, 0)$ is the common fixed point of equations $f_k(x) = 0$ where mapping f_k is defined by iteration (13.82).

Case Study 2. Select $X = \mathbf{R}$, and consider equations

$$f_k(x) = e^{(k+1)x/(k+2)} - 1 = 0 \quad (k \geq 0), \tag{13.83}$$

and the iteration equation

$$x_{k+1} = x_k - \left(e^{(k+1)x_k/(k+2)} - 1\right).$$

Notice that

$$f'_k(x) = f_k(x) + 1;$$

that is, the iteration method satisfies the conditions of the remark following Theorem 13.21. Observe also that the above iteration converges to $x^* = 0$ if x_0 is selected sufficiently small.

13.4 Exercises

13.1. Maximize $F = 240x_1 + 104x_2 + 60x_3 + 19x_4$ subject to

$$20x_1 + 9x_2 + 6x_3 + x_4 \leq 20$$
$$10x_1 + 4x_2 + 2x_3 + x_4 \leq 10$$
$$x_i \geq 0, \quad i = 1, ..., 4.$$

13.2. Minimize $F = 3x_1 + 2x_2$ subject to

$$8x_1 - x_2 \geq 8$$
$$2x_1 - x_2 \geq 6$$
$$x_1 + 3x_2 \geq 6$$
$$x_1 + 6x_2 \geq 8$$
$$x_1 \geq 0, \quad x_2 \geq 0.$$

13.3. Find an interval $[a, b]$ containing a root x^* of the equation $x = \frac{1}{2}\cos x$ such that for every $x_0 \in [a, b]$ the iteration $x_{n+1} = \frac{1}{2}\cos x_n$ will convergence to x^*. solve the equation by using Newton's or the secant method.

13.4. Solve the following nonlinear equations by the method of your choice:
  a) $\ln x = x - 4$
  b) $xe^x = 7$
  c) $e^x \ln x = 7$.

13.5. Find $(I - A)^{-1}$ (if it exists) for the matrix
$$A = \begin{pmatrix} 0.12 & 0.04 \\ 0.01 & 0.03 \end{pmatrix}.$$
Performe three steps.

13.6. Repeat Exercise 13.5 for the matrix
$$A = \begin{pmatrix} -1 & 3 \\ 2 & 4 \end{pmatrix}.$$

13.7. Solve problem
$$\dot{x} = x - y - 1 \quad x(0) = 0$$
$$\dot{y} = x + y \quad y(0) = 1.$$
Performe three steps.

13.8. solve the boundary-value problem
$$\ddot{x} = tx - \dot{x} - 1, \quad x(0) = x(1) = 1$$
by the discretization method. Select $h = 0.1$.

13.9. Solve
$$x(t) = \int_0^1 \frac{(t-s)^2}{10} x(s)\, ds - 7.$$

13.10.
$$x(t) = \int_0^t (t+s) x(s)\, ds - 4.$$

13.11. Consider a continuous map $P : \mathbf{R}^n \to \mathbf{R}^n$ such that $P \in C^1$ and $P(0) = 0$. Set $S_1 = \{x \mid \|P(x)\| < \|x\|\}$, $S_2 = \{x \mid \|P(x)\| \geq \|x\|\}$. Assume that
  a) S_1 is invariant under P, $P(S_1) \subseteq S_1$, and

b) For all $a \in S_2$, there exists a positive integer $i(a)$ such that $P^{i(a)}(a) \in S_1$.

Show that for all $x \in \mathbf{R}^n$, $P^m(x) \to 0$ as $m \to \infty$.

13.12. Assume that there exists a function $h : (0, \infty) \to \mathbf{R}$ such that $|y| \leq h(r)|x|$ for all $m \geq 0$, $r > 0$, $|x| \leq r$, $x \in X$ and $y \in F_m(x)$. Show that

$$|x_m| \leq q_m,$$

where

$$q_{m+1} = h(q_m) q_m, \quad q_0 = |x_0|.$$

Provide a convergence analysis of iteration $x_{m+1} \in F_m(x_m)$ based on the above estimate.

13.13. To find a zero for $G(x) = 0$ by iteration, where G is a real function defined on $[a, b]$ rewrite the equation as,

$$x = x + c \cdot G(x) \equiv F(x)$$

for some constant $c \neq 0$. If x^* is a root of $G(x)$ and if $G'(x^*) \neq 0$, how should c be chosen on order that the sequence $x_{m+1} = F(x_m)$ convergence to x^*?

13.14. Solve the initial value problem

$$\dot{x}(t) = 1 + \cos(x(t)), \quad x(0) = 0.$$

13.15. The predator-prey population models describe the interaction of a prey population X and a predator population Y. Assume that their interaction is modeled by the system of orderinary differential equations

$$\dot{x} = x - \frac{1}{4}x^2 - \frac{1}{10}xy + 1$$
$$\dot{y} = -\frac{1}{4}y + \frac{1}{7}xy + \frac{2}{3}.$$

(Assume $x(0) = y(0) = 0$.) solve the system.)

13.16. Assume that $F_m = F$ $(m \geq O)$, O is in the interior of X, and F is Fréchet-differentiable at O, furthermore the special radius of $F'(0)$ is less than 1. Then show that there is a neighborhood U of O such that $x_0 \in U$ implies that $x_m \to O$ as $m \to \infty$.

13.17. Let $F : \mathbf{R}^n \to \mathbf{R}^n$ be a function such that $F(0) = 0$, $F \in C^0$, and consider the difference equation $x(t+1) = F(x(t))$. If, for some norm, $\|F(x)\| \leq \|x\|$ for any $x \neq O$, then show that the origin is globally asymptotically stable equilibrium for the equation.

13.18. Assume that there exists a strictly increasing function $g : \mathbf{R} \to \mathbf{R}$ such that $g(O) = 0$, and a norm such that $g(\|F(x)\|) < g(\|x\|)$ for all $x \neq 0$. Then show that O is a globally asymptotically stable equilibrium for equation $x(t+1) = F(x(t))$, where $F(O) = 0$.

13.19. consider the following equation in $\mathbf{R}^2$:

$$x_1(t+1) = 8\sin\left(x_1(t) + \frac{\pi}{\ell}\right) + .2x_2(t)$$
$$x_2(t+1) = 8x_1(t) + .1x_2(t).$$

13.20. Solve the Fredholm-type integral equation

$$x(t) = \int_0^1 \frac{t \cdot x(s)}{10} dx + 1.$$

13.21. Solve the Volterra-type integral equation

$$x(t) = \int_0^t \frac{t \cdot x(s)}{10} ds + 1.$$

13.22. Solve equation

$$x = \frac{\sin x}{2} + 1.$$

13.23. Complete the proofs of Theorems **12.5-12.8**.

13.24. Provide the weaker conditions claimed in Remark **12.5** along the lines of the earlier chapters on single step point to point mappings.

Appendix A

Glossary of Symbols

$\mathbb{R}^n$	real n-dimensional space
$\mathbb{C}^n$	complex n-dimensional space
$X \times Y, X \times X = X^2$	Cartesian product space of X and Y
$e^1, \ldots, e^n$	the coordinate vectors of $\mathbb{R}^n$
$x = (x_1, \ldots, x_n)^T$	column vector with component x_i
x^T	the transpose of x
$\{x_n\}_{n \geq 0}$	sequence of points from X
$\|\cdot\|$	norm on X
$\|\cdot\|_p$	L_p norm
$\|\cdot\|$	absolute value symbol
$//$	norm symbol of a generalized Banach space X
$\langle x, y \rangle$	set $\{z \in X \mid z = tx + (1-t)y,\ t \in [0,1]\}$
$U(x_0, R)$	open ball $\{z \in X \mid \|x_0 - z\| < R\}$
$\bar{U}(x_0, R)$	closed ball $\{z \in X \mid \|x_0 - z\| \leq R\}$
$U(R) = U(0, R)$	ball centered at the zero element in X and of radius R
$U, \bar{U}$	open, closed balls, respectively no particular reference to X, x_0 or R
$M = \{m_{ij}\}$	matrix $1 \leq i, j \leq n$
M^{-1}	inverse of M
M^+	generalized inverse of M
$\det M$ or $|M|$	determinant of M
M^k	the kth power of M
rank M	rank of M
I	identity matrix (operator)
L	linear operator

L^{-1}	inverse
null L	null set of L
rad L	radical set of L
$F: D \subseteq X \to Y$	an operator with domain D included in X, and values in Y
$F'(x)$, $F''(x)$	first, second Fréchet-derivatives of F evaluated at x

Bibliography

Alefeld, G., Potra, F.A and Shen, Z., On the existence theorems of Kantorovich, Moore and Miranda, preprint, 2001.

Allgower, E. L., Böhmer, K., Potra, F. A., Rheinboldt, W. C., A mesh-independence principle for operator equations and their discretizations, SIAM J. Numer. Anal., **23**, (1986), no. 1, 160–169.

Amat, S., Busquier, S., and Candela, V., A class of quasi-Newton generalized Steffensen methods on Banach spaces, J. Comput. Appl. Math. **149** (2002), 397–408.

Ames, W., *Nonlinear Ordinary Differential Equations in Transport Processes*, Academic Press, New York, 1968

Anselone, P.M. and Moore, R.H., *An extension of the Newton-Kantorovich method for solving nonlinear equations with applications to elasticity*, J. Math. Anal. Appl. **13** (1996), 476-501

Appell, J., DePascale, E., Lysenko, J.V. and Zabrejko, P.P., New results on Newton–Kantorovich approximations with applications to nonlinear integral equations, Numer. Funct. Anal. and Optimiz. **18**, (1 & 2) (1997), 1–17.

Argyros, I.K., Quadratic equations and applications to Chandrasekhar's and related equations, Bull. Austral. Math. Soc. **32** (1985), 275-297

Argyros, I.K., On the cardinality of solutions of multilinear differential equations and applications, Intern. J. Math. and MAth. Sci. **9**, 4 (1986), 757-766

Argyros, I.K., On the approximation of some nonlinear equations, Aequationes Mathematicae, **32** (1987), 87-95.

Argyros, I.K., On polynomial equations in Banach space, perturbation techniques and applications, Intern. J. MAth. and Math. Sci. **10**, 1 (1987), 69-78.

Argyros, I.K., Newton-like methods under mild differentiability conditions with error analysis, Bull. Austral. Math. Soc. **37** (1987), 131-147.

Argyros, I.K., On Newton's method and nondiscrete mathematical induction, Bull. Austral. Math. Soc. **38** (1988), 131–140.

Argyros, I.K., On the Secant method and fixed points of nonlinear equations, Monatschfte für Mathematik, **106** (1988), 85-94.

Argyros, I.K., On a class of nonlinear integral equations arising in neutron transport, Aequationes Mathematicae, **36** (1988), 99-111

Argyros, I.K., On the number of solutions of some integral equations arising in radiative transfer, Intern. J. Math. and Math. Sci. **12**, 2 (1989), 297-304

Argyros, I.K., Improved error bounds for a certain class of Newton-like methods, J. Approx. Th. and Its Applic. **61** (1990), 80-98

Argyros, I.K., Error founds for the modified secant method, BIT, **20** (1990), 92-200.

Argyros, I.K., On the solution of equations with nondifferentiable operators and the Ptak error estimates, BIT, **30** (1990), 752-754.

Argyros, I.K., On some projection methods for the approximation of implicit functions, Appl. Math. Letters, **32** (1990), 5-7.

Argyros, I.K., The Newton–Kantorovich method under mild differentiability conditions and the Pták error estimates, *Monatschefte für Mathematik* **109**, 3 (1990), 110-128.

Argyros, I.K., The secant method in generalized Banach spaces, Appl. Math. and Comp. **39** (1990), 111-121.

Argyros, I, K. A mesh independence principle for operator equations and their discretizations under mild differentiability conditions, *Computing* **45**, (1990), 265-268.

Argyros, I.K., On the convergence of some projection methods with perturbation, J. Comp. and Appl. Math., **36** (1991), 255-258.

Argyros, I.K., On an application of the Zincenko method to the approximation of implicit functions, Public. Math. Debrecen, **39** (3-4) (1991), 1-7.

Argyros, I.K., On an iterative algorithm for solving nonlinear equations, Beitrage zür Numerischen Math., **10**, 1 (1991), 83-92.

Argyros, I.K., On a class of quadratic equations with perturbation, Funct. et Approx. Comm. Math. **XX** (1992), 51-63.

Argyros, I.K., Improved error bounds for the modified secant method, Intern. J. Computer Math., **43**, (1&2) (1992), 99-109.

Argyros, I.K., Some generalized projection methods for solving operator equations, J. Comp. and Appl. Math. **39**, 1 (1992), 1-6.

Argyros, I.K., On the convergence of generalized Newton-methods and implicit functions, J. Comp. and Appl. Math. **43** (1992), 335-342.

Argyros, I.K., On the convergence of inexact Newton-like methods, Publ. Math. Debrecen **42**, (1&2) (1992), 1-7.

Argyros, I.K., On the convergence of a Chebysheff-Halley-type method under Newton-Kantorovich hypothesis, Appl. Math. Letters **5**, 5 (1993), 71-74.

Argyros, I.K., Newton-like methods in partially ordered linear spaces, J. Approx. Th. and its Applic. **9**, 1 (1993), 1-10.

Argyros, I.K., On the solution of un determined systems of nonlinear equations in Euclidean spaces, Pure Math. Appl. **4**, 3 (1993), 199-209.

Argyros, I.K., A convergence theorem for Newton-like methods under generalized Chen–Yamamato-type assumptions, *Appl. Math. Comp.* **61**, 1 (1994), 25-37.

Argyros, I.K., On the discretization of Newton-like methods, *Internat. J. Computer. Math.* **52** (1994), 161-170.

Argyros, I.K., A unified approach for constructing fast two-step Newton-like methods, Mh. Math. **119** (1995), 1-22.

Argyros, I.K., Results on controlling the residuals of perturbed Newton-like methods on Banach spaces with aconvergence structure, Southwest J. Pure Appl. Math. **1** (1995), 32-38.

Argyros, I.K., On the method of tangent hyperbolas, J. Appr. Th. Appl. **12**, 1 (1996), 78-96.

Argyros, I.K., On an extension of the mesh-independence principle for operator equations in Banach space, Appl. Math. Lett. **9**, 3 (1996), 1-7.

Argyros, I.K., A generalization of Edelstein's theorem on fixed points and applications, Southwest J. Pure Appl. Math. **2** (1996), 60-64.

Argyros, I.K., Chebysheff-Halley-like methods in Banach spaces, Korean J. Comp. Appl. Math. 4, 1 (1997), 83-107.

Argyros, I.K., Concerning the convergence of inexact Newton methods, J. Comp. Appl. Math. **79** (1997), 235-247.

Argyros, I.K., General ways of constructing accelerating Newton-like iterations on partially ordered topological spaces, Southwest J. Pure Appl. Math. **2** (1997), 1-12.

Argyros, I.K., On a new Newton–Mysovskii-type theorem with applications to inexact Newton-like methods and their discretizations, *IMA J. Num. Anal.* **18** (1997), 43-47.

Argyros, I.K., On the convergence of two-step methods generated by point-to-point operators, Appl. Math. Comput. **82**, 1 (1997), 85-96.

Argyros, I.K., Improved error bounds for Newton-like iterations under Chen–Yamamoto assumptions, *Appl. Math. Letters*, **10**, 4 (1997), 97–100.

Argyros, I.K., Inexact Newton methods and nondifferentiable operator equations on Banach spaces with a convergence structure, *Approx. Th. Applic.* **13**, 3 (1997), 91–104.

Argyros, I.K., A mesh independence principle for inexact Newton-like methods and their discretizations under generalized Lipschitz conditions, Appl. Math. Comp. **87** (1997), 15-48.

Argyros, I.K., Concerning the convergence of inexact Newton methods, J. Comp. Appl. Math. **79** (1997), 235-247.

Argyros, I.K., Smoothness and perturbed Newton-like methods, Pure Math. Appl. **8**,1 (1997), 13-28.

Argyros, I.K., The asymptotic mesh independence principle for inexact Newton-Galerkin-like methods, Pure Math. Applic. **8**, (2 & 3) (1997), 169-194.

Argyros, I.K., On the convergence of a certain class of iterative procedures under relaxed conditions with applications, J. Comp. Appl. Math. **94** (1998), 13-21.

Argyros, I.K., Sufficient conditions for constructing methods faster than Newton's, Appl. Math. Comp. **93** (1998), 169-181.

Argyros, I.K., The Theory and Application of Abstract Polynomial Equations, St. Lucie/CRC/Lewis Publ. Mathematics Series, Boca Raton, Florida, 1998.

Argyros, I.K., A new convergence theorem for the Jarratt method in Banach spaces, Computers and Mathematics with Applications, **36**, 8 (1998), 13-18.

Argyros, I.K., Improving the order and rates of convergence for the Super-Halley method in Banach spaces, Comp. Appl. Math. **5**, 2 (1998), 465-474.

Argyros, I.K., Improved error bounds for a Chebysheff-Halley-type method, Acta Math. Hungarica, **84**, 3 (1999), 211-221.

Argyros, I.K., Relations between forcing sequences and inexact Newton iterates in Banach space, *Computing* **63** (1999), 131-144.

Argyros, I.K., Convergence domains for some iterative processes in Banach spaces using outer and generalized inverses, *Comput. Anal. Applic.* **1**, 1 (1999), 87-104.

Argyros, I.K., Concerning the convergence of a modified Newton-like method, Journal for Analysis and its Applications (ZAA), **18**, 3 (1999), 1-8.

Argyros, I.K., Convergence domains for some iterative processes in Banach spaces using outer and generalized inverses, *J. Comput. Anal. and Applic.* **1**, 1 (1999), 87-104.

Argyros, I.K., Concerning the radius of convergence of Newton's method and applications, Korean J. Comp. Appl. Math. **6**, 3 (1999), 451-462.

Argyros, I.K., Convergence rates for inexact Newton-like methods of singular points and applications, Appl. Math. Comp. **102** (1999), 185-201.

Argyros, I.K., Choosing the forcing sequences for inexact Newton methods in Banach space, Comput. Appl. Math. **19**, 1 (2000), 79-89.

Argyros, I.K., A survey of efficient numerical methods for solving equations and applications, Kyung Moon Publ., Seoul, Korea, 2000.

Argyros, I.K., Local convergence of inexact Newton-like iterative methods and applications, *Computers and Mathematics with Application* **39** (2000), 69-75.

Argyros, I.K., A mesh independence principle for perturbed Newton-like methods and their discretizations,Korean J. Comp. Appl. Math. **7**, 1 (2000), 139-159.

Argyros, I.K., Advances in the efficiency of computational methods and applications, World Scientific Publ. Co., River Edge, NJ, 2000.

Argyros, I.K., Newton methods on Banach spaces with a convergence structure and applications, *Computers and Math. with Appl. Intern. J. Pergamon Press* **40**, 1 (2000), 37-48.

Argyros, I.K., Forcing sequences and inexact Newton iterates in Banach space, Appl. Math. Letters. **13** (2000), 77-80.

Argyros, I.K., Choosing the forcing sequences for inexact Newton methods in Banach space, Comput. Appl. Math., **19**, 1 (2000), 79-89.

Argyros, I.K., Local convergence of inexact Newton-like iterative methods and applications, Computers and Mathematics with Application **39** (2000), 69-75.

Argyros, I.K., Semilocal convergence theorems for a certain class of iterative procedures using outer or generalized inverses, Korean J. Comp. Appl. Math. **7**, 1 (2000), 29-40.

Argyros, I.K., The effect of rounding errors on a certain class of iterative methods, Applicationes Mathematicae, **27**, 3 (2000), 369-375.

Argyros, I.K., Local convergence of Newton's method for nonlinear equations using outer or generalized inverses, Chechoslovak Math. J., **50**, (125), (2000), 603-614.

Argyros, I.K., On a class of nonlinear implicit quasivariational inequalities, Pan American Math. J. **10**, 4 (2000), 101-109.

Argyros, I.K., A new semilocal convergence theorem for Newton's method in Banach space using hypotheses on the second Fréchet-derivative, *J. Comput. Appl. Math.* **139**, (2001), 369-373.

Argyros, I.K., On the radius of convergence of Newton's method, *Intern. J. Comput. Math.* **77**, 3 (2001), 389-400.

Argyros, I.K., A Newton-Kantorovich theorem for equations involving m-Frechet differentiable operators and applications in radiative transfer, Journ. Comp. Appl. Math. **131**, 1-2 (2001), 149-159.

Argyros, I.K., Semilocal convergence theorems for Newton's method using outer inverses and hypotheses on the second Fréchet-derivative, Monatshefte fur Mathematik, **132** (2001), 183-195.

Argyros, I.K., On general auxiliary problem principle and nonlinear mixed variational inequalities, Nonlinear Functional Analysis and Applications **6**, 2 (2001), 247-256.

Argyros, I.K., On an iterative procedure for approximating solutions of quasi variational inequalities, Advances in Nonlinear Variational Inequalities, **4**, 2 (2001), 39-42.

Argyros, I.K., On generalized variational inequalities, *Advances in Nonlinear Variational Inequalities*, **4**, 2 (2001), 75-78.

Argyros, I.K., On a semilocal convergence theorem for a class of quasi variational inequalities, *Advances in Nonlinear Inequalities*, **4**, 2 (2001), 43-46.

Argyros, I.K., On the convergence of a Newton-like method based on m-Fréchet-differentiable operators and applications in radiative transfer, *J. Comput. Anal. Applic.*, **4**, 2 (2002), 141-154.

Argyros, I.K., On the convergence of Newton-like methods for analytic operators and applications, J. Appl. Math. and Computing, **10**, 1-2, (2002), 41-50.

Argyros, I.K., A unifying semilocal convergence theorem for Newton-like methods based on center Lipschitz conditions, *Comput. Appl. Math.*, **21**, 3 (2002), 789-796.

Argyros, I.K., A semilocal convergence analysis for the method of tangent hyperbolas, *Journal of Concrete and Applicable Analysis*, **1**, 2 (2002), 135-144.

Argyros, I.K., New and generalized convergence conditions for the Newton-Kantorovich method, *J. Appl. Anal.*, **9**, 2 (2003), 287-299.

Argyros, I.K., On the convergence and application of Newton's method under weak Hölder continuity assumptions, *International Journal of Computer Mathematics*, **80**, 5 (2003), 767-780.

Argyros, I.K., On a theorem of L.V. Kantorovich concerning Newton's method, *Journ. Comp. Appl. Math.*, **155**, (2003), 223-230.

Argyros, I.K., An improved error analysis for Newton-like methods under generalized conditions, *J. Comput. Appl. Math.*, **157**, 1 (2003), 169-185.

Argyros, I.K., An improved convergence analysis and applications for Newton-like methods in Banach space, *Numerical Functional analysis and Optimization*, **24**, 7 and 8, (2003), 653-672.

Argyros, I.K., and Chen, D., On the midpoint method for solving nonlinear operator equations in Banach spaces, *Appl. Math. Lett.*, **5**, 4 (1992), 7-9.

Argyros, I.K., and Chen, D., A fourth order iterative method in Banach spaces, *Appl. Math. Lett.*, **6**, 4 (1993), 97-98.

Argyros, I.K., and Chen, D., A note on the Halley method in Banach spaces, *Appl. Math. Comp.*, **58**, (1993), 215-224.

Argyros, I.K., On the Newton-Kantorovich hypothesis for solving nonlinear equations, *J. Comput. Appl. Math.*, (2004).

Argyros, I.K. and Szidarovszky, F., Convergence of general iteration schemes, *J. Math. Anal. and Applic.* **168**, (1992), 42-62.

Argyros, I.K. and Szidarovszky, F., The Theory and Application of Iteration Methods, C.R.C. Press Inc., Boca Raton, Florida, 1993.

Argyros, I.K. and Szidarovszky, F., On the convergence of modified contractions, *J. Cmput. Appl. Math.*, **55**, 2 (1994), 97-108.

Baluev, A.On the method of Chaplying, Dokl. Akad. Nauk. SSSR, **83**, (1956), 781-784.

Bartle, R.G., Newton's method in Banach spaces, Proc. Amer. Math. Soc., (1955), 827-831.

Ben-Israel, A., A Newton-Raphson method for the solution of systems of operators, J. Math. Anal. Appl., **15**, (1966), 243-252.

Ben-Israel, A. and Greville, T.N.E., Generalized Inverses: Theory and Applications, John Wiley and Sons, 1974.

Brent, R.P., Algorithms for Minimization Without Derivatives, Prentice Hall, Englewood Cliffs, New Jersey, 1973.

Brock, W.A., and Scheinkman, J.A., some results on global asymptotic stability of difference equations, J. Econ. Theory., **10**, (1975), 265-268.

Browder, F.E., and Petryshyn, W.V., The solution by iteration of linear functional equations in Banach spaces, *Bull. Amer. Math. Soc.* **72** (1996), 566-570.

Brown, P. N., A local convergence theory for combined inexact-Newton/finite-difference projection methods, *SIAM J. Numer. Anal.* **24**, (1987), 407-434.

Cătinaş, E., On some iterative methods for solving nonlinear equations, *Revue d'Analyse Numerique et de Theorie de l'Approximation*, **23**, 1 (1994), 47-53.

Cătinaş, E., A note on inexact Newton methods, *Revue d'Analyse Numérique et de Theorie de l'Approximation*, **25**, 1-2 (1996), 33-41.

Cătinaş, E., Inexact perturbed Newton methods and applications to a class of Krylov solvers, *J. Optim. Theory Appl.*, **108**, 3, (2001), 543-570.

Cătinaş, E., Affine invariant conditions for the inexact perturbed Newton methods, *Revue d'Analyse Numerique et de Theorie de l'Approximation*, **31**, 1 (2002), 17-20.

Chandrasekhar, S., Radiative Transfer, Dover Publ., New York, 1960.

Chen, D., On the convergence and optimal error estimates of King's iteration for solving nonlinear equations, *Intern. J. Computer. Math.*, **26**, (3 & 4) (1989), 229-237.

Chen, D., Kantorovich-Ostrowski convergence theorems and optimal error bounds for Jarratt's iterative method, *Intern. J. Computer. Math.* **31**, (3 & 4), (1990), 221-235.

Chen, X. and Yamamoto, T., Convergence domains of certain iterative methods for solving nonlinear equations, Numer. Funct. Anal. and Optimiz., **10** (1 & 2) (1989), 37-48.

Chen, X. and Nashed, M.Z., Convergence of Newton-like methods for singular operator equations using outer inverses, Numer. Math., **66** (1993), 235-257.

Chen, X., Nashed, Z and Qi, L., Convergence of Newton's method for singular smooth and nonsmooth equations using adaptive outer inverses, SIAM J. Optim. **7** (1997), no. 2, 445–462.

Chui, C.K. and Quak, F., Wavelets on a bounded interval. In: Numerical Methods of Approximation Theory, Vol. 9 (eds: D. Braess and Larry L. Schumaker, Intern. Ser. Num. Math., Vol. 105) Basel: Birkhäuser Verlag, pp. 53-75.

Chow, S.N. and Hale, J.K., Methods of Bifurcation Theory, Springer-Verlag, New-York, 1962.

Cianciaruso, F, De Pascale, E., Zabrejko, P. P., Some remarks on the Newton-Kantorovich approximations, Atti. Sem. Mat. Fis. Univ. Modena, **48** (2000), 207–215.

Cohen, G., Auxiliary problem principle and decomposition of optimization problems, J. Optim. Theory Appl., **32** (1980), no. 3, 277–305.

Cohen, G., Auxiliary problem principle extended to variational inequalities, J. Optim. Theory Appl., **59**, (1988), no. 2, 325–333.

Collatz, L., Functional Analysis and Numerisch Mathematik, Springer-Verlag, New York, 1964.

Danes, J., Fixed point theorems, Nemyckii and Uryson operators, and continuity of nonlinear mappings, Comment. Math. Univ. Carolinae, **11**, (1970), 481–500.

Danfu, H and Xinghua, W., The error estimates of Halley's method (submitted).

Darbo, G., Punti uniti in trasformationa codominio non compatto, Rend. Sem. Mat. Univ. Padova, **24**, (1955), 84–92.

Daubechies, I., Ten Lectures in Wavelets, (Conf. Board Math. Sci. (CBMS) Vol. 61), Society for Industrial and Applied Mathematics (SIAM), Philadelphia, PA, 1992.

Davis, H. T., Introduction to nonlinear differential and integral equations, Dover Publications, Inc., New York, 1962.

Decker, D. W., Keller, H. B., Kelley, C. T., Convergence rates of Newton's method at singular points, SIAM J. Numer. Anal., **20**, (1983), no. 2, 296–314.

Dembo, R. S., Eisenstat, S. C., Steihaug, T., Inexact Newton methods, SIAM J. Numer. Anal., **19**, (1982), no. 2, 400–408.

Dennis, J. E., Toward a unified convergence theory for Newton-like methods, In: Nonlinear Functional Anal. and Appl. (L.B. Rall, Ed.) Academic Press, New York, 1971.

Dennis, J. E., and Schnabel, R. B., Numerical methods for unconstrained optimization and nonlinear equations, Prentice-Hall, Englewood Cliffs, New Jersey, 1983.

De Pascale, E. and Zabrejko, P.P., New convergence criteria for the Newton-Kantorovich method and some applications to nonlinear integral equations, Rend. Sem. Mat. Univ. Padova, **100**, (1998), 211–230.

Deuflhard, P., A stepsize control for continuation methods and its special application to multiple shooting techniques, Numer. Math., **33**, (1979), no. 2, 115–146.

Deuflhard, P., Heindl, G., Affine invariant convergence theorems for Newton's method, and extensions to related methods, SIAM J. Numer. Anal., **16**, (1979), no. 1, 1–10.

Deuflhard, P., Potra, F. A., Asymptotic mesh independence of Newton-Galerkin methods and a refined Mysovskii theorem. SIAM J. Numer. Anal., **29**, (1992), no. 5, 1395–1412.

Diallo, O.W., On the theory of linear integro-differential equations of Barbashin type in lebesgue spaces, (Russian), VINITI, 1013, 88, Minsk, (1988).

Döring, B., Iterative lösung gewisser randwertprobleme und integralgleichungen, Apl. Mat., **24**, (1976), 1–31.

Dunford, N., Schwartz, J. T., Linear operators. Part I, Int. Publ. Leyden, (1963).

Edelstein, M., On fixed and periodic points under contractive mappings., J. London Math. Soc., **37**, (1962), 74–79.

Eisenstat, S. C., Walker, H. F., Globally convergent of inexact Newton methods, SIAM J. Optim., **4**, (1994), no. 2, 393–422.

Eisenstat, S. C., Walker, H. F., Choosing the forcing terms in an inexact Newton method, SIAM J. Sci. Comput., **17**, (1996), no. 1, 16–32.

Ezquerro, J. A., Hernández, M. A., Avoiding the computation of the second Fréchet-derivative in the convex acceleration of Newton's method., J. Comput. Appl. Math., **96**, (1998), 1–12.

Ezquerro, J.A. and Hernandez, M.A., An efficient study of convergence for a fourth order two-point iteration in Banach space (submitted).

Ezquerro, J. A., Hernández, M. A., On a convex acceleration of Newton's method, J. Optim. Theory Appl., **100**, (1999), no. 2, 311–326.

Ezquerro, J. A., Hernández, M. A., On the application of a fourth order two-point method to Chandrasekhar's integral equation, Aequationes Math., **62**, (2001), no. 1-2, 39–47.

Ezquerro, J. A., Hernández, M. A., Salanova, M. A., A discretization scheme for some conservative problems, Proceedings of the 8th International Congress on Computational and Applied Mathematics, ICCAM-98 (Leuven), J. Comput. Appl. Math., **115**, (2000), no. 1-2, 181–192.

Ezquerro, J. A., Hernández, M. A., Salanova, M. A., Recurrence relations for the midpoint method, Tamkang J. Math., **31**, (2000), no. 1, 33–41.

Ezquerro, J. A., Gutiérrez, J. M., Hernández, M. A., Salanova, M. A., A biparametric family of inverse free multipoint iterations, Comput. Appl. Math., **19**, (2000), no. 1, 109–124.

Ezquerro, J. A., Gutiérrez, J. M., Hernández, M. A. and Salanova, M. A., On the approximation of an inverse free Jarratt type approximation to nonlinear equations of Hammerstein type, (submitted).

Ferreira, O. P. and Svaiter, B.F., Kantorovich's theorem on Newton's method in Riemannian manifolds, J. Complexity, **18**, (2002), no. 1, 304–329.

Foerster, H., Frommer, A. and Mayer, G., Inexact Newton methods on a vector supercomputer, J. Comp. Appl. Math. **58**, (1995), 237-253.

Fujimoto, T., Global asymptotic stability of nonlinear difference equations I, Econ. Letters, **22**, (1987), 247-250.

Fujimoto, T., Global asymptotic stability of nonlinear difference equations II, Econ. Letters, **23**, (1987), 275-277.

Galperin, A. and Waksman, Z., Newton method under a weak smoothness assumption, J. Comp. Appl. Math., **35**, (1991), 207-215.

Galperin, A. and Waksman, Z., Regular smoothness and Newton's method, Numer Funct. Anal. Optimiz., **15**, (7 & 8), (1994), 813-858.

Gander, W., On Halley's iteration method, amer. Math. Monthly, **92**, (1985), 131-134.

Glowinski, R, Lions, J. L. and Trémolières, R., Numerical analysis of variational inequalities, North-Holland, Amsterdam, 1982.

Gragg, W. B.; Tapia, R. A., Optimal error bounds for the Newton-Kantorovich theorem, SIAM J. Numer. Anal. **11**, (1974), 10–13.

Graves, L. M., Riemann integration and Taylor's theorem in general analysis, Trans. Amer. Math. Soc. **29**, (1927), no. 1, 163–177.

Gutierez, J.M., A new semilocal convergence theorem for Newton's method, *J. Comput. Appl. Math.* **79** (1997), 131–145.

Gutiérrez, J.M., Hernández, M.A., and Salanova, M.A., Accessibility of solutions by Newton's method, *Internat. J. Comput. Math.* **57** (1995), 239–247.

Gutiérrez, J. M., Hernández, M. A., Salanova, M. A., Resolution of quadratic equations in Banach spaces, Numer. Funct. Anal. Optim., **17**, (1996), (1 & 2), 113–121.

Hackl, J., Wacker, Hj., Zulehner, W, An efficient step size control for continuation methods, BIT, **20** (1980), no. 4, 475–485.

Hartman, P., Ordinary differential equations, John Wiley & Sons, Inc., New York-London-Sydney, 1964. xiv+612 pp.

Häubler, W.M. A Kantorovich-type convergence analysis for the Gauss-Newton method, *Numer. Math.* **48** (1986), 119–125.

Hellinger, E.; Toeplitz, O. Integralgleichungen und Gleichungen mit unendlichvielen Unbekannten, (German), Chelsea Publishing Company, New York, (1953), 1335–1616.

Hernández, M. A., A note on Halley's method, Num. Math. **59**, 3 (1991), 273-276.

Hernández, M. A., Newton's Raphson's method and convexity, Zb. Rad. Prirod.- Mat. Fax. Ser. Mat. 22, 1, (1992), 159-166.

Hernández, M. A., Salanova, M. A., A family of Chebyshev-Halley type methods, Intern. J. Comp. Math., **47**, (1993), 59-63.

Hernández, M. A., Relaxing convergence conditions for Newton's method, J. Math. Anal. Appl, **249**, (2000), no. 2, 463–475.

Hernández, M. A., Chebyshev's approximation algorithms and applications, Comput. Math. Appl., **41** (2001), no. 3-4, 433–445.

Hernández, M. A., Rubio, M. J. and Ezquerro, J. A., Secant-like methods for solving nonlinear integral equations of the Hammerstein type, J. Comput. Appl. Math., **115**, (2000), 245–254.

Hernández, M. A., Rubio, M. J., Semilocal convergence of the secant method under mild convergence conditions of differentiability, Comput. Math. Appl., **44**, (2002), no. 3-4, 277–285.

Hernández, M. A. and Salanova, M. A., Sufficient conditions for semilocal convergence of a fourth order multipoint iterative method for solving equations in Banach spaces, Southwest J. Pure Appl. Math., (1999), 29–40.

Hille, E. and Philips, R.S., Functional Analysis and Smigroups, Amer. Math. Soc. Coll. Publ., New York, 1957.

Huang, Z. D., A note on the Kantorovich theorem for Newton iteration, J. Comput. Appl. Math., **47**, (1993), no. 2, 211–217.

Jarrat, P., Some efficient fourth order multipoint methods for solving equations, BIT, **9**, (1969), 119-124.

Josephy, N.H., Newton's method for generalized equations, Technical Report No. 1965, mathematics Research Center, University of Wisconsin, Madison Wisconsin, 1979.

Kanno, S., Convergence theorems for the method of tangent hyperbolas, Math. Japon., **37**, (1992), no. 4, 711–722.

Kantorovich, L.V., The method of succesive approximation for functional equations, Acta Math., **71**, (1939), 63-97.

Kantorovich, L. V., Akilov, G. P., Functional analysis in normed spaces, Pergamon Press, New York (1964).

Krasnosel'skii, M. A., Positive solutions of operator equations, Goz. Isdat. Fiz. Mat. Moscow 1962; Transl. by R. Flaherty and L. Boron, P. Noordhoff, Groningen 1964.

Krasnosel'skii, M. A., Topological Methods in the Theory of Nonlinear Integral Equations, Pergamon Press, London, 1966.

Krasnosel'skii, M. A., Approximate solution of operator equations, Walter Noordhoff Publ. Groningen, 1972.

Krasnosel'skii, M.A., Vainikko, G.M., Zabreiko, P.P., and Rutiskii, Ya.B., and Stetsenko, V.Ya., *Approximate Solution of Operator Equations*, Wolters-Noordhoff Publishing, Groningen, 1972.

Krein, S. G., Linear equations in Banach spaces, Birkhäuser Publ., Boston, MA, (1982).

Kung, H. T., The complexity of obtaining starting points for solving operator equations by Newton's method, Technical report, nr.044-422, Carnegie-Mellon Univ., Pittsburgh, Pa., october, 1975, Article in Traub, J.F., Analytic computational complexity.

Kuratowski, C., Sur les espaces complets, Fund. Math., **15**, (1930), 301-309.

Kwon, U. K. and Redheffer, R. M., Remarks on linear equations in Banach space, Arch. Rational Mech. Anal., **32**, (1969), 247–254.

Lancaster, P., Error analysis for the Newton-Raphson method, Numer. Math., **9**, (1968), 55–68.

Liusternik, L. A. and Sobolev, V. J., Elements of functional analysis, Ungar Publ., 1961.

Maiboroba, I.N., a projection iteration method of constructing two-sided approximations of solutions of operator equations, Ukrainski Mathematicheskii Zhurnal, Vol. 8, No.6, (1976), 735-744.

Marcotte, P., Wu, J. H., On the convergence of projection methods, J. Optim. Theory Appl., **85,** (1995), no. 2, 347–362.

Mayer, J., A generalized theorem of Miranda and the theorem of Newton-Kantorovich, Numer. Funct. Anal. Optim., **23,** (2002), no. 3-4, 333–357.

McCormick, S. F., A revised mesh refinement strategy for Newton's method applied to two-point boundary value problems, Lecture Notes in Mathemaics 674, Springer-Verlag, Berlin, (1978), 15–23.

Mertvecova, M.A., An analog of the process of tangent hyperbolas for general functional equations, Dokl. Akad. Nauk., SSSR, **88,** (1953), 611-614 (in Russian).

Meyer, P. W., Das modifizierte Newton-Verfahren in verallgemeinerten Banach-Räumen, Numer. Math., **43,** (1984), no. 1, 91–104.

Meyer, P. W., Newton's method in generalized Banach spaces. Numer. Funct. Anal. Optim., **9,** (1987), (3 & 4), 244–259.

Meyer, P. W., A unifying theorem on Newton's method, Numer. Funct. Anal. Optim., **13,** (1992), no. 5-6, (1992), 463–473.

Miel, G. J., Majorizing sequences and error bounds for iterative methods, Math. Comp., **34,** (1980), no. 149, 185–202.

Migovich, F.M., On the convergence of projection-iterative methods for solving nonlinear operator equations, Dopov. Akad. Nauk. Ukr. RSR, Ser. A, **1,** (1970), 20-23.

Miranda, C., Un osservatione su un teorema d, Brouwer, Ball. Unione Mat. Ital., Serr. 11, **3,** (1940), 5-7.

Mirsky, L., An Introduction to linear algebra, Clarendon Press, Oxford, England, 1955.

Moore, R. H., Approximations to nonlinear operator equations and Newton's method, Numer. Math., **12,** (1968), 23–34.

Moore, R. E., A test for existence of solutions to nonlinear systems, SIAM J. Numer. Anal., **14,** (1977), no. 4, 611–615.

Moore, R. E., Methods and applications of interval analysis, SIAM Publications, Philadelphia, Pa., 1979.

Moret, I., A note on Newton-type iterative methods, Computing **33,** (1984), 65–73.

Moret, I., On the behaviour of approximate Newton methods, Computing, **37,** (1986), no. 3, 185–193.

Moret, I., On a general iterative scheme for Newton-type methods, Numer. Funct. Anal. Optim., **9,** (1987-88), (10-12), 1115–1137.

Moret, I., A Kantorovich-type theorem for inexact Newton methods, Numer. Funct. Anal. Optim. 10 (1989), (3 & 4), 351–365.

Muroya, Y., Practical monotonous iterations for nonlinear equations, Mem. Fac. Sci. Kyushu Univ., Ser. A, **22,** (1968), 56–73.

Muroya, Y., Left subinverses of matrices and monotonous iterations for nonlinear equations, Memoirs of the Faculty of Science and Engineering, Waseda University, **34,** (1970), 157-171.

Mysovskii, I., On the convergence of Newton's method, Trudy Mat. Inst. Steklov, **28,** (1949), 145-147 (in Russian).

Natanson, I.P., The theory of functions of a Real Variable (Russian), Gostehizdat, Moscow, (1957).

Necepurenko, M.T., On Chebysheff's method for functional equations (Russian), Usephi, Mat. Nauk, **9**, (1954), 163-170.

Nerekenov, T.K., Necessary and sufficient conditions for uryson and nemytskii operators to satisfy a Lipschitz condition (Russian), VINITI 1459, **81**, Alma-Ata, (1981).

Neumaier, A. and Shen, Z., The Krawczyk operator and Kantorovich's theorem, J. Math. Anal. Appl., **149**, (1990), no. 2, 437–443.

Nguen, D.F. and Zabrejko, P.P., The majorant method in the theory of the Newton-Kantorovich approximations and the Pták error estimates, Numer. Funct. Anal. and Optimiz., **9**, (5 & 6), (1987), 671-686.

Noble, B., The numerical solution of nonlinear integral equations and related topics, University Press, Madison, WI, (1964).

Noor, K. I. and Noor, M. A., Iterative methods for a class of variational inequalities, in Numerical Analysis of singular perturbation problems, Hemker and Miller, eds., Academic Press, New-York, (1985), 441-448.

Noor, M. A., An iterative scheme for a class of quasivariational inequalities, J. Math. Anal. and Appl., **110**, (1985), no. 2, 463–468.

Noor, M.A., Generalized variational inequalities, Appl. Math. Letters, **1**, (1988), 119-122.

Ojnarov, R. and Otel'baev, M., A criterion for a Uryson operator to be a contraction (Russian), Dokl. Akad. Nauk. SSSR, **255**, (1980), 1316-1318.

Okuguchi, K., Expectations and stability in oligopoly models, Springer-Verlag, New York, (1976).

Ortega, J.M. and Rheinboldt, W.C., Iterative solution of nonlinear equations in several variables, Academic Press, New York, (1970).

Ostrowski, A.M., Solution of equations in euclidean and Banach spaces, Academic Press, New York, (1973).

Pandian, M.C., A convergence test and componentwise error estimates for Newton-type methods, SIAM J. Numer. Anal., **22**, (1985), 779-791.

Păvăloiu, I., Sur la méthode de Steffensen pour la résolution des équations opérationnelles non linéaires, Rev. Roumaine Math. Pures Appl., **13**, (1968), 6, 857–861.

Păvăloiu, I., Rezolvarea equaţiilor prin interpolare. Dacia Publ. cluj-Napoca, Romania, (1981).

Păvăloiu, I., Sur une généralisation de la méthode de Steffensen, Rev. Anal. Numér. Théor. Approx., **21**, (1992), no. 1, 59–65.

Păvăloiu, I., A converging theorem concerning the chord method, Rev. Anal. Numér. Théor. Approx., **22**, 1 (1993), 83-85.

Păvăloiu, I., Bilateral approximations for the solutions of scalar equations., Rev. Anal. Numér. Théor. Approx., **23**, (1994), no. 1, 95–100.

Potra, F.A., An error analogsis for the method, Numer. Math. **32**, (1982), 427-445.

Potra, F. A. On an iterative algorithm of order 1.839... for solving nonlinear operator equations, Numer. Funct. Anal. Optim., **7**, (1984-1985), no. 1, 75–106.

Potra, F. A., Newton-like methods with monotone convergence for solving nonlinear operator equations, Nonlinear Anal., Theory Methods and Applications, **11**, (1987), no. 6, 697–717.

Potra, F. A., On Q-order and R-order of convergence, SIAM J. Optim. Theory Appl., **63**, (1989), no. 3, 415–431.

Potra, Florian-A. and Pták, V., Sharp error bounds for Newton's method, Numer. Math., **34**, (1980), no. 1, 63–72.

Potra, F. A. and Pták, V., Nondiscrete induction and iterative processes, Pitman Publ., London, 1984.

Pták, V., The rate of convergence of Newton's process., Numer. Math., **25**, (1976), no. 3, 279–285.

Rall, L. B., Computational solution of nonlinear operator equations, Wiley, New York, (1968).

Rall, L. B., Nonlinear functional analysis and applications, Academic Press, New York, (1971).

Rall, L. B., Convergence of Stirling's method in Banach spaces, Aequationes Math., **12**, (1973), 12–20.

Rall, L. B., A comparison of the existence theorems of Kantorovich and Moore, SIAM J. Numer. Anal., **17**, (1980), no. 1, 148–161.

Reddien, G. W., On Newton's method under mild differentiability conditions with error analysis, SIAM J. Numer. Anal., **15**, (1978), no. 5, 993–996.

Rheinboldt, W. C., An adaptive continuation process for solving systems of nonlinear equations, Publish Academy of Sciences, Banach Ctr. Publ., **3**, (1977), pp.129–142.

Rheinboldt, W. C., A unified convergence theory for a class of iterative processes, SIAM J. Numer. Anal., **5**, (1968), 42-63.

Robinson, S. M., Generalized equations. In: (A. Bachem, M. Grötschel and B. Korte, eds.) Mathematical programming: the state of the art, Springer, Berlin, 1982, pp. 346–367.

Rockne, J., Newton's method under mild differentiability conditions with error analysis, Numer. Math., **18**, (1972), 401-412.

Safiev, R. A., The method of tangent hyperbolas, sov. Math. Dokl., **4**, (1963), 482-485.

Schmidt, J. W., Monotone einschliessung mit Regula-Falsi bei konvexen functionen, ZAMM, **50**, (1970), 640-643.

Schmidt, J. W. and Leonhardt, H., Eingrenzung von lösungen mit hilfe der Regula-Falsi, Computing, **6**, (1970), 318–329.

Sergeev, A.A., On the method of Chords Sibirsk, Mat. Z., **2**, (1961), 282-289.

Slugin, S. N., Approximate solution of operator equations on the basis of Caplygin method, (Russian), Dokl. Nauk SSSR, **103**, (1955), 565–568.

Slugin, S. N., Monotonic processes of bilateral approximation in a partially ordered convergence group, Soviet. Math., **3**, (1962), 1547-1551.

Smale, S., Newton's method estimates from data at one point. In the merging of disciplines in pure, applied, and computational mathematics, Springer Verlag, New York, 1986, pp.185–196,.

Stirling, J., Methodus differentialis: sive tractatus de summatione et interpolations serierum infiniterum, W. Boyer, London, 1730.

Szidarovszky, F. and Okuguchi, A note on global asymptotic stability nonlinear difference equations, Econ. Letters, **26**, (1988), 349-352.

Szidarovszky, F. and Bahill, T., Linear systems theory, CRC Press, Boca Raton, FL, (1992).

Tarskii, A.A., A lattice theoretical fixed point and its applications, Pacific J. Math., (1955), 285-309.

Tishyadhigama, S., Polak, E., Klessig, R., A comparative study of several convergence conditions for algorithms modeled by point-to-set maps, Math. Programming Stud., **10**, (1979), 172–190.

Törnig, W., Monoton konvergente Iterationsverfahren zür Lösung michtlinearer differenzen–randwertprobleme, Beiträge zür Numer. Math., **4**, (1975), 245–257.

Traub, J. F., Iterative methods for the solution of equations, Prentice-Hall Series in Automatic Computation Prentice-Hall, Inc., Englewood Cliffs, N.J. 1964, xviii+310 pp.

Traub, J. F., Analytic computational complexity, Academic Press, New York-London, 1975.

Triconi, F.G., Integral Equations, Interscience Publ., 1957.

Uko, L. U., Generalized equations and the generalized Newton method., Mathematical Programming, **73**, (1996), 251–268.

Ulm, S., Majorant principle and the secant method, I.A.N. Estonskoi S.S.R Fiz. Mat. **3**, (1964), 217-227 (in Russian).

Ulm, S., Iteration methods with divided differences of the second order, (Russian), Dokl. Akad. Nauk SSSR, **158**, (1964), 55–58.

Urabe, M., Convergence of numerical iteration in solution of equations, J. Sci. Hiroshima Univ., Ser. A, **19**, (1976), 479–489.

Uzawa, H., The stability of dynamic processes, Econometrica, **29**, (1961), 617–631.

Vandergraft, J. S., Newton's method for convex operators in partially ordered spaces, SIAM J. Numer. Anal., **4**, (1967), 406–432.

Varga, R. S., Matrix iterative analysis, Prentice-Hall, Englewood Cliffs, NJ, 1962.

Verma, R. U., Nonlinear variational and constrained hemivariational inequalities involving relaxed operators., Z. Angew. Math. Mech., **77**, (1997), no. 5, 387–391.

Verma, R. U., Approximation-solvability of nonlinear variational inequalities involving partially relaxed monotone (PRM) mappings., Adv. Nonlinear Var. Inequal., **2**, (1999), no. 2, 137–148.

Verma, R. U., A class of projection-contraction methods applied to monotone variational inequalities., Appl. Math. Lett., **13**, (2000), no. 8, 55–62.

Verma, R. U., Generalized multivalued implicit variational inequalities involving the Verma class of mappings., Math. Sci. Res. Hot-Line, **5**, (2001), no. 2, 57–64.

Walker, H.F. and Watson, L.T., Large change secant update methods for undetermined systems, SIAM J. Numer. Anal., **27**, (1990), 1227-1262.

Weiss, R., On the approximation of fixed points of nonlinear compactr operators, SIAM J. Num. Anal., **11**, (1974), 550-553.

Wilkinson, J.H., The algebraic eigenvalue problem, Clarendon Press, Oxford, 1965.

Wu, J. W and Brown, D. P., Global asymptotic stability in discrete systems, J. Math. Anal. Appl., **140**, (1989), no. 1, 224–227.

Yamamoto, T., Error bounds for computed eigenvalues and eigenvectors, Numer. Math., **39**, (1980), 189–199.

Yamamoto, T., A method for finding sharp error bounds for Newton's method under the Kantorovich assumptions, Numer. Math., **44**, (1986), 203–220.

Yamamoto, T., A convergence theorem for Newton-like methods in Banach spaces, Numer. Math., **51**, (1987), 545–557.

Yamamoto, T., A note on a posteriori error bound of Zabrejko and Nguen for Zincenkos iteration, Numer. Funct. Anal. and Optimiz., **9**, (9 and 10), (1987), 987-994.

Yamamoto, T. and Chen, Z., Convergence domains of certain iterative methods for solving nonlinear equations, Numer. Funct. Anal. Optim., **10**, (1989), 34–48.

Ypma, T.J., Numerical solution of systems of nonlinear algebraic equations, Ph. D. thesis, Oxford, 1982.

Ypma, T.J., Affine invariant convergence results for Newton's methods, BIT, **22**, (1982), 108-118.

Ypma, T.J., The effect of rounding error on Newton-like methods, IMA J. Numer. Anal., **3**, (1983), 109-118.

Ypma, T. J., Convergence of Newton-like iterative methods, Numer. Math., **45**, (1984), 241–251.

Ypma, T. J., Local convergence of inexact Newton methods, SIAM J. Numer. Anal., **21**, (1984), 583–590.

Yau, L., Ben-Israel, A., The Newton and Halley methods for complex roots, Amer. Math. Monthly, **105,** (1998), no. 9, 806–818.

Zaanen, A. C., Linear analysis, North-Holland Publ., Amsterdam, 1953.

Zabrejko, P.P. and Majorova, N.L., On the solvability of nonlinear Uryson integral equations (Russian), Kach. Pribl. Metody Issled. Oper. Uravn., **3**, (1978), 61-73.

Zabrejko, P. P.; Nguen, D. F., The majorant method in the theory of Newton-Kantorovich approximations and the Pták error estimates, Numer. Funct. Anal. Optim., **9**, (1987), no. 5-6, 671–684.

Zabrejko, P. P. and Zlepko, P.P., On majorants of Uryson integral operators (Russian), Kach. Pribl. Metody Issled. Oper. Uravn., **8** (1983), 67-76.

Zangwill, W.I., Nonlinear Programming, A unified Approach Prentice-Hall, Englewood Cliffs, N.J., (1969).

Zincenko, A.I., A class of approximate methods for solving operation equations with nondifferentiable operators, Dopovidi Akad. Nauk Ukrain. RSR (1963), 156-161.

Zlepko, P.P. and Migovich, F.M., an application of a modification of the Newton-Kantorovich method to the approximate construction of implicit functions (Ukrainian), Ukrainskii Mathematischeskii Zhürnal, **30**, 2 (1978), 222-226.

Zuhe, Shen, Wolfe, M. A., A note on the comparison of the Kantorovich and Moore theorems, Nonlinear Anal., **15**, (1990), no. 3, 229–232.

Index

additive operator, 12
analytic operator, 78, 80, 81

Banach lemma, 4, 440, 477, 486
Bilinear operator, 2, 8, 22, 24, 26, 358, 381
Bounded operator, 11, 433

Cauchy sequence, 89
Chebysheff-Halley method, 350
Cone, 432
Continuous operator, 51, 292
Contraction mapping principle, 52, 78
Convergence structure, 432, 437, 441, 444, 447, 449
Convex operator, 30

Differential equation, 482
differential equation, 74, 78, 295
Divided difference, 17, 20, 21, 24, 25, 27, 29, 30, 33, 34, 47, 216, 218, 221

Euler's identity, 296

Fibonacci sequence, 148
Fixed point, 21, 51–54, 74–76, 78–81, 147, 148, 217, 291, 442, 447, 449, 474, 478, 479, 489
Fourier, 28
Fredholm integral equation, 54
Fundamental theorem of calculus, 90

Gâteaux differentiable, 377–379
Generalized Banach space, 493
Generalized Newton's method, 379, 383
Gershgorin's theorem, 77

Halley method, 355, 357

Inner product, 12, 379
Isotone operator, 433

Laplace transform, 11
Linear functional, 2, 12–14, 32
Linear operator, 1–8, 11–14, 19–22, 27, 28, 30, 34–37, 47, 48, 76, 135, 149, 154, 292, 294, 305, 383, 420, 428, 436, 481
Lipschitz condition, 21, 22, 33, 293, 475
Lipschitz conditions, 75, 78
Logarithmic Convexity, 350

Mean value theorem, 10, 53
Midpoint method, 351
Monotone convergence, 27, 30
Mosaic, 8
Multipoint method, 352

Neumann series, 4, 296
Newton's method, 377, 378, 384, 474, 482
Newton-like method, 150, 154, 155,

294, 295, 297, 298, 302, 303, 431

Partially ordered topological space, 17, 433, 435
Picard's iteration, 78
Potra, 28, 31
Projection operator, 292

Rate of convergence, 296
Regular space, 19
Rheinboldt's theorems, 305
Riemann integral, 10, 11, 475

Secant method, 31, 149, 218–220, 222, 223
Separable Hilbert space, 293
Stirling's method, 147, 148, 291
Super-Halley method, 355

Taylor's theorem, 10, 11

Uryson operator, 15

Vector algebra, 12

Printed in the United States
By Bookmasters